A Guide to Noise in Microwave Circuits

IEEE Press
445 Hoes Lane
Piscataway, NJ 08854

A Guide to Noise in Microwave Circuits

Devices, Circuits, and Measurement

Peter Heymann

Matthias Rudolph

IEEE PRESS

WILEY

Published by John Wiley & Sons, Inc., Hoboken, New Jersey.
Published simultaneously in Canada.

Library of Congress Cataloging-in-Publication Data applied for:

ISBN: 9781119859369

Cover design by Wiley
Cover image © agsandrew/Shutterstock

Contents

Author Biographies

Peter Heymann received his Dipl.-Phys. and Dr. rer.-nat. degrees in physics from the University of Greifswald, Greifswald, Germany, in 1963 and 1969, respectively. From 1963 to 1982, he worked on different projects in the field of wave plasma interaction, which include wave propagation, RF plasma sources and heating, and microwave and far infrared plasma diagnostics. Since 1982, he has been working on GaAs microwave electronics. In 1992, he joined the Ferdinand-Braun-Institut für Höchstfrequenztechnik, Berlin, Germany, where he was responsible for measurements, characterization, and modeling of active and passive components of microwave MMICs until his retirement in 2009.

Matthias Rudolph received his Dipl.-Ing. degree from the Berlin Institute of Technology in 1996 and the Dr.-Ing. degree from Darmstadt University of Technology in 2001. In 1996, he joined the Ferdinand-Braun-Institut (FBH), Leibniz-Institut für Höchstfrequenztechnik, Berlin, Germany, where he was responsible for the modeling of III–V transistors and MMIC design. In 2009, he was appointed the Ulrich-L.-Rohde Professor at the Brandenburg University of Technology, Cottbus, Germany, and also heads the Low-Noise Components Lab at FBH.

Preface

This book is intended for engineers working in practice and for students at technical college level. Knowledge of the mode of operation of electronic components and the most important mathematical methods of electrical engineering are supposed.

There is extensive literature on the fundamentals of electronic noise and the properties of semiconductor components. These basics are also covered here. In addition to these basics, we devote ourselves in detail to questions of measurement technology in the LF, RF, and microwave range. An attempt is made to show what is hidden behind the intelligence built into measuring instruments and in the omniscient design software.

The problems cannot be understood without mathematical details, especially because of the close interconnection with the statistics of fluctuating quantities. We present them as simply and clearly as possible. In all problems, starting from simple basics, each step is presented logically in a derivation. In doing so, we always try to maintain clarity and a reference to practice and to avoid any unnecessary abstract presentation.

Knowledge and results are based on the authors' work at the Ferdinand-Braun-Institut, Leibniz-Institut für Höchstfrequenztechnik (FBH). The chapters are based on experience with measurement technology and modeling of microwave components in semiconductor technology.

These were acquired through many years of work at FBH. We thank Prof. Dr. G. Tränkle and Prof. Dr. W. Heinrich for their benevolent support and encouragement. Dipl.-Ing. R. Dœrner and Dipl.-Ing. S. Schulz have always provided valuable assistance with many technical problems.

Berlin and Cottbus
2021

Peter Heymann
Matthias Rudolph

1

Introduction

Preliminary Remarks

The spontaneous fluctuations of voltages or currents that we deal with are summarized under the term electronic noise. This is a historical term from the early days of radio technology. In those days, listeners were delighted when they had a more or less interference-free reception. In the background, or when the receiver was slightly detuned, a disturbing noise could be heard. In the age of digital data transmission, this everyday acoustic noise has largely disappeared from radio and telephony. Noise need not always be of electronic origin. We can hear it, for example, from a mountain stream, when rain falls on a roof or when an air conditioning system is in operation. A clearly perceptible acoustic impression results from the summation of a plurality of randomly occurring individual processes.

Since the publication of the fundamental work "Noise" by A. van der Ziel in 1954 [1], a number of publications on the theory and practice of electronic noise have been published. In the books [2–8] the physical, mathematical aspect is in the foreground. For the practitioner of circuit design and measurement technology, the books [9–14] are more suitable.

The RF-engineer we are addressing here knows noise as a visual impression at the screen of a spectrum analyzer or broadband oscilloscope. Even without an external signal, the display shows a statistical fluctuation, the "noise floor." With an oscilloscope, this fluctuation can be seen in the time domain. When sweeping across the screen, the spot dithers irregularly around the baseline. In the spectrum analyzer it fluctuates around an average value. At first glance, the visual impression is the same. It is clear that such an irregularity, whose time dependence is obviously unrepeatable and unpredictable, can only be treated by means of the theory of fluctuation processes. It is indispensable to work with mean values, signal statistics, probability distributions, and correlation. In most cases, time averaging is used for analyses in the time domain, while in the

A Guide to Noise in Microwave Circuits: Devices, Circuits, and Measurement, First Edition.
Peter Heymann and Matthias Rudolph.

frequency domain the ensemble average is used. Since the noise is ergodic, there is no difference between the two.

Before turning to the noise of amplifiers, receivers, and oscillators and its measurement, it is useful to understand what laws are hidden in this apparently completely chaotic process.

The reason for this is the atomistic structure of electricity. The electric current is not a continuous flow. It consists of the contributions of the individual elementary charges. Small irregular fluctuations are superimposed on the average value. This is also the case when the mean value is zero, i.e. no current flows. As a result of the thermal motion of the free electrons in the conductor, they generate a current pulse of the duration τ when flying over the distance of a free path. This current pulse corresponds to a voltage pulse at the ends of the structure. These are very short voltage pulses in short succession. In a doped semiconductor we have, e.g. $n = 10^{17}$ electrons/cm^3. The free time of flight is about $\tau = 10^{-12}$ seconds.

For a material die with a volume V of 10 μm edge length this results in z voltage impulses per second

$$z = \frac{nV}{\tau} = \frac{10^{17}10^{-9}}{10^{-12}} = 10^{20} \tag{1.1}$$

The voltage pulses generated by the individual processes are superimposed to thermal noise at the terminals of the structure. The observation of the fluctuation of the voltage at an ohmic resistance in the time domain due to the thermal motion of the electrons is therefore an obvious entry into the physics of noise processes. However, this fluctuating voltage cannot be observed without special measurement technology. Although the oscilloscope is the appropriate instrument for the time domain, it is usually not sensitive enough. The spot on the screen of an older, analog oscilloscope with, e.g. 100 MHz bandwidth, shows, even in the most sensitive range (5 mV/div), a completely smooth curve. No matter which resistor we connect to the input. A voltage fluctuating in time can only be seen on a high performance oscilloscope with extreme bandwidth. However, this noise voltage visible there is also not generated by the thermal noise of a resistor at the input, but by the amplifier chain in the device itself. Nevertheless one has a direct picture of a typical noise process in the time domain. An example is shown in Figure 1.1. Figure 1.1a is the noise voltage on the screen of a LeCroy Wave Expert SE 70 in the y-deviation 1 mV/div [15]. Figure 1.1b is the histogram of the voltage values together with the appropriate Gaussian distribution. This "elementary image" of the noise shows us some essential properties.

The mean value is zero. The amplitude distribution, the probability density function (PDF) follows the Gaussian curve. Thus one can define a standard deviation and thus an effective value of the noise voltage.

Before we deal with the origin of these $v_{RMS} = 2.8$ mV and with the statistical quantities in detail, let us look at the screen of a spectrum analyzer. Unlike the

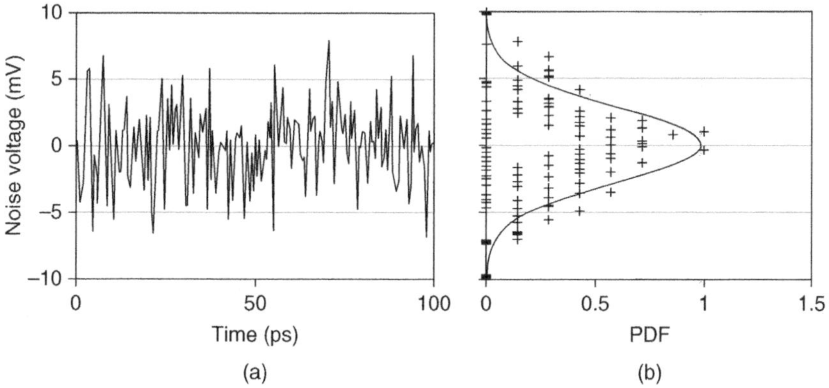

Figure 1.1 Screenshot of the noise level of a LeCroy 70 GHz oscilloscope. (a) Amplitude in the time domain. (b) Statistics of the voltage values. Gaussian distribution with $v_{RMS} = 2.8$ mV.

oscilloscope, even the simplest model shows a noise floor. What at first glance looks the same on an oscilloscope and a spectrum analyzer turns out to be quite different on closer inspection. This is generally the case with noise observations. At the output of a communication system one has a disturbing noise floor. The contributions to this come from different sources and over a wide range of channels. If one wants to minimize them, one has to understand all the elementary processes and their interaction.

Here we have a representation of the noise in the frequency domain. From the broadband noise spectrum we see a small section at the center frequency with the selected resolution bandwidth (RBW). This is not the "elementary picture" of noise as in Figure 1.1. The noise voltage in Figure 1.2 is characterized by two important networks. The IF filter with its RBW selects a narrow frequency range around the center frequency. The display is generated by a detector which displays the average value of the rectified noise voltage. The display is thus the result of data processing through a linear and a non-linear network. This also results in a change in the distribution of the amplitudes. Instead of the Gaussian distribution, we see a Rayleigh distribution on a logarithmic scale in the histogram Figure 1.2b. There is another important difference to Figure 1.1: On the spectrum analyzer [16] we see the noise from the noise. The actual noise level is the average value corresponding to the Rayleigh distribution. What we see is the remaining fluctuation that can be averaged out by reducing the video bandwidth (VBW). By reducing the VBW in relation to the RBW, the standard deviation is reduced (Figure 1.3).

Thus, by a longer averaging we can get a noise-free display of the noise. The most important use of the spectrum analyzer is to observe signals in the frequency domain. Very weak signals become visible when the noise is averaged

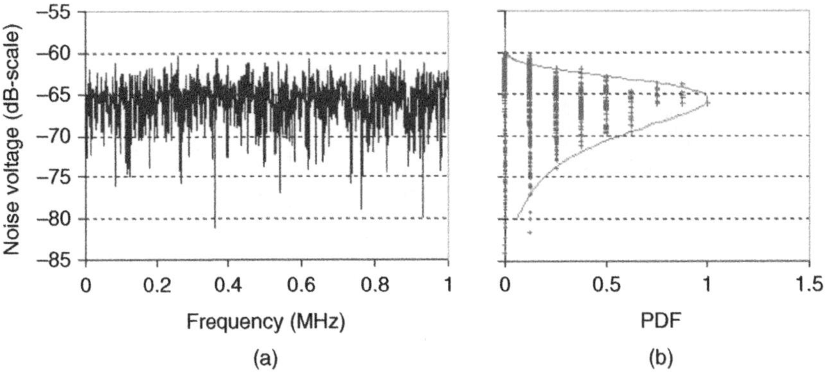

Figure 1.2 Screenshot of a spectrum analyzer (a) and histogram (b).

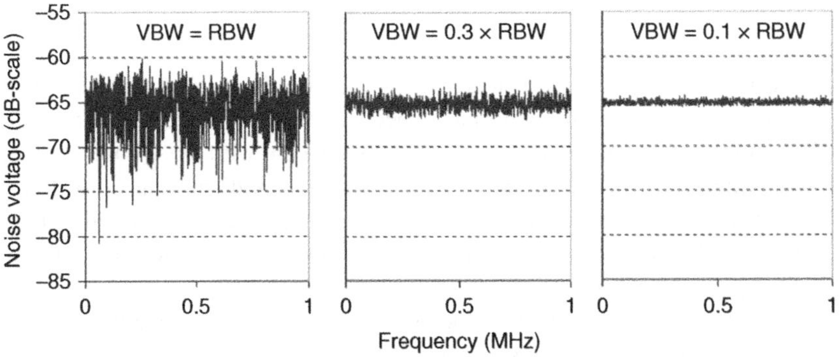

Figure 1.3 Effect of reducing "noise from noise" by reducing video bandwidth.

out. Statistical considerations make it possible to optimize bandwidth and sweep time. The level we now measure is generated by the input attenuator, the mixer, and the IF amplifier. It is easy to observe that it is directly proportional to the RBW. The larger the RBW selected, the higher the noise level.

Electronic noise is present in every component of a transmitter–receiver system, i.e. a system for transmitting information. It physically limits the transmission possibilities, since a signal that is too weak "disappears" in the noise. Even in measuring instruments, e.g. spectrum analyzers or power meters, the detection limit is given by the inherent noise. Even the mirror galvanometers previously used as sensitive current meters showed thermally induced display jitter.

A measure of the level difference between signal P_S and noise N is the signal-to-noise ratio SNR, which is usually given in dB.

$$SNR(\text{dB}) = 10 log\frac{P_S}{N} \tag{1.2}$$

Table 1.1 Values of SINR (signal to interference and noise ratio) for various applications.

Application	Radar	Mobile phone	Entertainment, video games
SNR (dB)	+15	+18	+30

Table 1.1 shows typical values for an electronic system limited only by the noise of the receiver.

A signal beyond recognition can also be caused by interfering signals that scatter into the receiving band. This can be explained using the example of mobile radio technology (Cellular Telephone Systems).

At present the most common digital technologies are UMTS (3G) in the 1.9–2.2 GHz frequency band, bandwidth $B = 5\,\text{MHz}$ with QPSK modulation and LTE (4G) in the 0.7–2.7 GHz frequency band. The bandwidth is divided into sections of 200 kHz. So in the case of $B = 5\,\text{MHz}$ in 25 so-called resource blocks. Interference signals can, for example, come from neighboring transmitting aerials, be caused by reflection or by other techniques. Formula (1.2) is expanded by the signal level of the interference P_{INT}

$$SINR(\text{dB}) = 10log\frac{P_S}{N + P_{INT}} \tag{1.3}$$

SINR is a measure of the ratio of useful signal to noise and interference. The following approximate specifications apply to the magnitude (Table 1.2).

We have already mentioned almost all the important phenomena of electronic noise. In Chapter 2, we will start from these empirical impressions and discuss the background of the description, the origin of the noise and its measurement in a form that is understandable for the practitioner. This is not possible without signal and circuit theory and the application of correlation matrices with their auto and cross spectral densities. Each derivation and calculation starts with elementary, easily understandable principles and formulas and is guided step by step to the result. In this way, an attempt is made to present this matter in a way that

Table 1.2 Values of SINR for mobile communication.

SINR (dB)	Reception quality	Comment
+40 … +20	Very good	No interference
+19 … +10	Good	Stable reception
+9 … +5	Satisfactory	Still works
+4 … 0	Poor	Detection limit

is understandable to the technician, although a high content of mathematics is necessarily inherent in it.

History

Electronic noise is a subfield of the physics of fluctuation phenomena. Fluctuations first became known with the discovery of Brownian molecular motion in 1827, when the botanist Robert Brown observed under the microscope that tiny dust particles in water are constantly dithering. Later it was discovered that the cause is the thermal motion of water molecules. It was possible to determine the average speed from the change of location Δx over a period of time Δt to $\overline{v} = {}^{\Delta x}/_{\Delta t}$, but it could not be explained by the expectation from the translational energy ${}^{3}/_{2}kT$ of the water molecules.

A. Einstein showed 1906 [17] that instead of Δx the mean square of displacement $\overline{\Delta x^2}$ must be considered. He also recognized that there must be comparable electrical phenomena. In 1906 he calculated the fluctuation around the equilibrium state of an electrical capacitor from the thermal energy.

$$\overline{\Delta E} = \frac{kT}{2} = \frac{C}{2}\overline{v^2} \tag{1.4}$$

and thus established the interrelation between thermodynamics and electrical fluctuations. However, at that time these small voltages ($v = 2$ µV at $C = 1$ nF) could not be measured. At the end of the nineteenth century, mirror galvanometers were the most sensitive electrical measuring instruments. The current to be measured flows through a small coil, on which a torque acts in a magnetic field. The coil hangs on a torsion thread, to which a small mirror is attached, which deflects a light beam. The sensitivity is limited by the dithering of the mirror. The analogy to Brownian motion did not lead to success. Then cooling or evacuation of the system should have brought an improvement. The fluctuation is obviously electrical in nature.

The connection was first established by G. Ising in 1926 [18]. He related the minimum measurable voltage of a critically damped galvanometer to the thermal energy.

$$\overline{v_{MIN}^2} = 16\overline{\Delta E}\omega_0 R \tag{1.5}$$

Here R is the internal resistance of the galvanometer coil and ω_0 is the natural frequency of the system consisting of coil mass and mirror as well as the torsion filament. The equivalent bandwidth of the critically damped galvanometer is $\Delta f \cong {}^{\omega_0}/_8$. If the thermal energy (1.4) is applied, the following results

$$\overline{v_{MIN}^2} = 16\frac{kT}{2}R \times 8\Delta f = 64kTR\Delta f \tag{1.6}$$

Except for the factor 64 instead of 4, this is already the Nyquist formula of the thermal noise of the resistor *R*.

The effect of thermal noise was finally clarified by the quantitative description of H. Nyquist 1928 [19] with the derivation of the well-known Nyquist formula. This was experimentally confirmed by J.B. Johnson 1928 [20] by measurements. Shortly afterwards E.B. Moullin [21] was able to verify the value of the Boltzmann constant k to within 1% by noise measurements. Thus the correlation between thermal energy and electronic noise was secured as an inherent effect.

W. Schottky had already dealt with the limit sensitivity of radio receivers in 1918 [22]. He was primarily interested in the noise of the electron stream in the tube amplifiers used at that time, as this was by far the strongest source of interference. Today the quantized current of the charge carriers is one of the main sources of noise in modern transistors. This shot noise is described by the Schottky formula (Chapter 3).

References

1 van der Ziel, A. (1954). *Noise*. London: Chapman and Hall.

2 van der Ziel, A. (1986). *Noise in Solid State Devices and Circuits*. New York: Wiley.

3 Bittel, H. and Storm, L. (1971). *Rauschen*. Springer Verlag.

4 Connor, F.R. (1987). *Rauschen*. Vieweg & Sohn.

5 Bendat, J.S. and Piersol, A.G. (1971). *Random Data: Analysis and Measurement*. Wiley Interscience.

6 Freman, J.J. (1958). *Principles of Noise*. Wiley.

7 Pfeifer, H. (1959). *Elektronisches Rauschen*. Leipzig: B.G. Teubner.

8 Lawson, J.L. and Uhlenbeck, G.E. (1965). *Threshold Signals*. Dover Publications Inc.

9 Müller, R. (1990). *Rauschen*. Springer Verlag.

10 Schiek, B., Rolfes, I., and Siweris, H.-J. (2006). *Noise in High-Frequency Circuits and Oscillators*. Wiley Interscience.

11 Beneking, H. (1971). *Praxis des elektronischen Rauschens*. Bibliographisches Institut Mannheim.

12 Landstorfer, F. and Graf, H. (1981). *Rauschprobleme der Nachrichtentechnik*. Oldenbourg Verlag.

13 Vasilescu, G. (2005). *Electronic Noise and Interfering Signals*. Springer Verlag.

14 Vergers, C.A. (1979). *Handbook of Electrical Noise Measurement and Technology*. TAB Books.

15 LeCroy (2015). Noise Measurement Using Your LeCroy Oscilloscope. Application Brief Lab. WM782.

16 Rauscher, C. (2007). *Grundlagen der Spektrumanalyse*. München: Rohde & Schwarz.

17 A. Einstein, 'Zur Theorie der Brownschen Bewegung,' *Ann. Phys.* Vo.19, pp. 371–381, 1906.

18 Ising, G. (1926). A natural limit for the sensibility of galvanometers. *Phil. Mag.* 1: 827–834.

19 Nyquist, H. (1928). Thermal agitation of electric charge in conductors. *Phys. Rev.* 32: 110–113.

20 Johnson, J.B. (1928). Thermal agitation of electricity in conductors. *Phys. Rev.* 32: 97–109.

21 Moullin, E.B. (1938). *Spontaneous Fluctuations of Voltage due to Brownian Motions of Electricity, Shot Effect and Kindred Phenomena*. Clarendon Press Oxford.

22 Schottky, W. (1918). Über spontane Stromschwankungen in verschiedenen Elektrizitätsleitern. *Ann. Phys.* 57: 541–567.

2

Basic Terms

Average Values

The following general terms can be found in every work that deals with noise problems in a fundamental way, e.g. [1–6].

From Figure 1.1 we can see that the noise is random, but the "strength" of the noise has a defined value. In our case, we have a pure fluctuation phenomenon with a mean value of zero. There are of course cases conceivable where the noise voltage is superimposed on a constant average value. A simple example is a DC power supply. We write in general for the time dependence of the voltage

$$V(t) = \overline{V} + v(t) \tag{2.1}$$

With the pure noise voltage $v(t)$ and the time average

$$\overline{V} = \lim_{T\to\infty} \frac{1}{2T} \int_{-T}^{T} V(t)dt \tag{2.2}$$

Because, by definition. $\overline{v(t)} = 0$.

The oscilloscope shows us the value $v_{RMS} = 2.8$ mV for the strength of the noise. This is the RMS value of the noise voltage. In the case of fluctuating quantities, it is referred to σ as the standard deviation. The square of the voltage variation, also called variance, is the noise power at a resistor $R = 1\ \Omega$.

$$\sigma^2 = \overline{v^2(t)} = \lim_{T\to\infty} \frac{1}{2T} \int_{-T}^{T} v^2(t)dt \tag{2.3}$$

If we have the voltage $V(t)$ (2.1) at a resistor $\mathcal{R}$, then

$$\begin{aligned} \overline{V^2}/R &= Total - Power \\ \overline{V}^2/R &= DC - Power \\ \overline{v^2}/R &= Noise - Power \end{aligned} \tag{2.4}$$

A Guide to Noise in Microwave Circuits: Devices, Circuits, and Measurement, First Edition.
Peter Heymann and Matthias Rudolph.

The linear mean value (2.2) and the root mean square value (2.3) are formed in the time domain. As mentioned at the beginning, an ensemble average value could also be formed. Thus, one does not measure the time dependence of the noise voltage at an amplifier, but the noise voltages at a certain time at an ensemble of completely identical amplifiers. If all amplifiers have warmed up for the same time, the mean values of the fluctuation quantities will be the same. The time averages are equal to the ensemble averages, because this is an ergodic and stationary process. Both views are equivalent. This example is a thought experiment and not very realistic. However, in the derivation of noise from physical elementary processes, we are definitely dealing with ensemble averages. For example, in the treatment of shot noise, the ensemble average is formed by the motion of many electrons.

Amplitude Distribution

We see in Figure 1.1 the distribution of the voltage values around the zero point according to the well-known Gaussian bell-shaped curve. Knowledge of the amplitude distribution, i.e. the Probability distribution function (PDF) is another important parameter of the fluctuation phenomenon in addition to the mean values discussed earlier. For example, to calculate the Bit Error Rate of a digital communication link that is disturbed by noise, it is important to know the probability with which a certain value of the noise voltage is exceeded. Another case is the noise measurement. In this case it is important that the amplifier chain and the display device (Noise Figure Analyzer, Spectrum Analyzer) do not limit the peaks of the noise voltage as a result of driving into non-linearity. So let us look at the properties of the normal distribution of the noise voltage values v from Figure 1.1. It is the Gaussian distribution function (GDF)

$$GDF(v) = \frac{1}{\sqrt{2\pi\sigma^2}} exp\left(-\frac{1}{2}\left(\frac{v}{\sigma}\right)^2\right) \tag{2.5}$$

Figure 2.1 shows the normal distribution for our case with = 2.8 mV. Usually it is shown normalized with $\sigma = 1$. Then the integral = 1, because the probability equals one that each of the voltage values from the ensemble will occur.

In general, statistical thermodynamics leads to the Gaussian distribution. In our case the noise voltage at an ohmic resistance is caused by the thermal motion of the electrons. In the kinetic theory of gases it appears in the form of Maxwell's velocity distribution of the molecules. The other important source of electronic noise is the shot effect. It is caused by a large number of statistically distributed current pulses. Here too, the statistical laws of large numbers apply, the consequence of which is the Gaussian distribution. This distribution is also maintained when passing through linear networks.

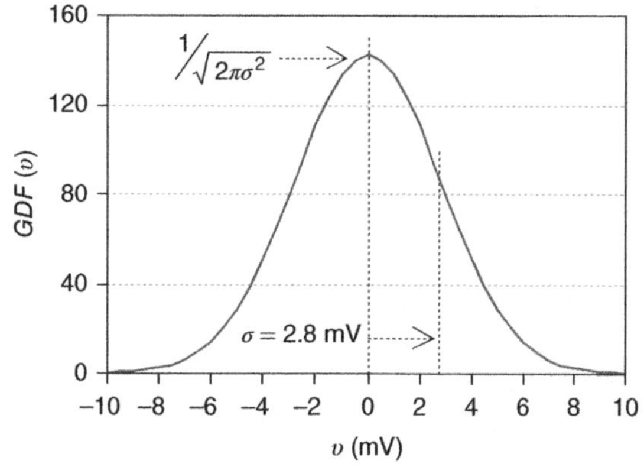

Figure 2.1 Amplitude distribution of the noise voltages v according to the Gaussian bell-shaped curve ($\sigma = v_{RMS} = 2.8$ mV).

The knowledge of the amplitude distribution enables us to calculate the probability with which a certain limit value of the voltage v_1 is exceeded. This is important for the assessment whether the noise can interfere with a data transmission or whether a measuring system is driven into non-linearity and therefore voltage clipping occurs. We must calculate the integral over the Gaussian distribution from this limit value v_1 up to ∞. However, this is not analytically possible. It leads to the normal cumulative distribution function $\Phi(v)$ (Figure 2.2).

$$P(v_1) = 2 \times \int_{v1}^{\infty} GDF(v)dv = 1 - \Phi\left(\frac{v}{\sqrt{2}\sigma}\right) \tag{2.6}$$

It can be assumed that a measuring amplifier does not significantly reduce the noise if it is operated at three times the effective value ($3\sigma \cong 9$ mV). Up to this value it still works linear. The contribution of higher values of the noise voltage is less than 0.1%. This limit value corresponds at the $Z_0 = 50\ \Omega$ input of a microwave amplifier to a power of $P_{IN} = -28$ dBm. With a power Gain of $G_P = 35$ dB,

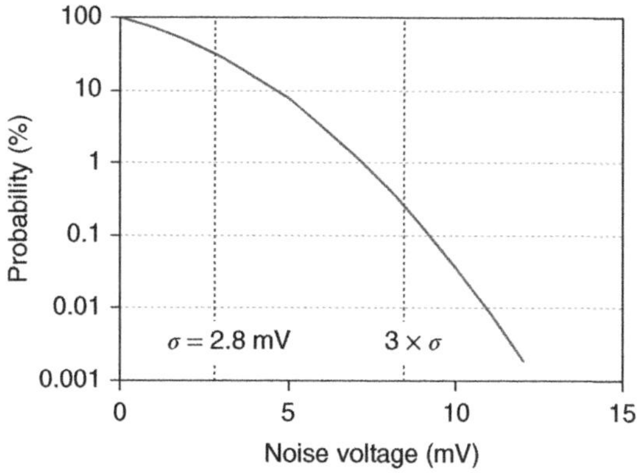

Figure 2.2 Probability $P(v_1)$ of the occurrence of noise voltages $|v| > v_1$ according to (2.6) for the example in Figure 1.1 with $\sigma = 2.8$ mV.

$P_{OUT} = 7$ dBm at the output would already be close to the linearity limit of 10 dBm that is usual specified for small signal amplifiers. It should be noted that this estimation is valid for a given noise signal. In practice, the effective bandwidth of the system also plays an important role, which can significantly change the noise level.

Autocorrelation

If we look at the variance (2.3), we see with the help of Figure 2.3 that this formation of the temporal average is a special case. The voltage curve v(t) multiplied by itself at the same time t_i.

The autocorrelation function (ACF) (2.7) represents a more general form of temporal averaging. We cannot predict the voltage value at a future point in time for noise. But we can use the ACF to obtain a probability assumption about a value at the time $t+\tau$.

$$\rho(\tau) = \overline{v(t)v(t+\tau)} = \lim_{T\to\infty} \frac{1}{2T} \int_{-T}^{T} v(t)v(t+\tau)dt \tag{2.7}$$

One can determine whether the voltage at the time $t+\tau$ in any way depends from the voltage at the time t, for example, whether it has already changed its sign after the time τ or not. The ACF of the noise voltage $v(t)$ in Figure 2.3 is obtained as follows: For a given time difference, the voltages $v(t_1)$ and $v(t_1+\tau)$ are measured and multiplied. Thereby t_1 goes through all values of t. Then the averaging over t is carried out. This gives us a value of the function $\rho(\tau)$ for a certain value of τ. Now τ is changed until $\rho(\tau)$ is known in the area of interest. In practice, this can be done by a correlator, which consists of a splitter, an adjustable delay line (variation of τ) in one branch and a mixer in which the direct and the delayed signal are

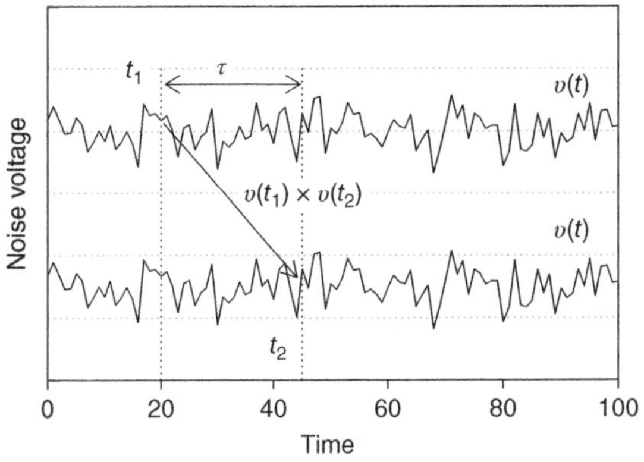

Figure 2.3 Twice the same noise voltage over time. The variance (2.3) is obtained from the product $v(t_1) \times v(t_1)$ for all t. The product $v(t_1) \times v(t_1+\tau)$ of values that were determined for the time difference τ occurrence is more general. It leads to the correlation function.

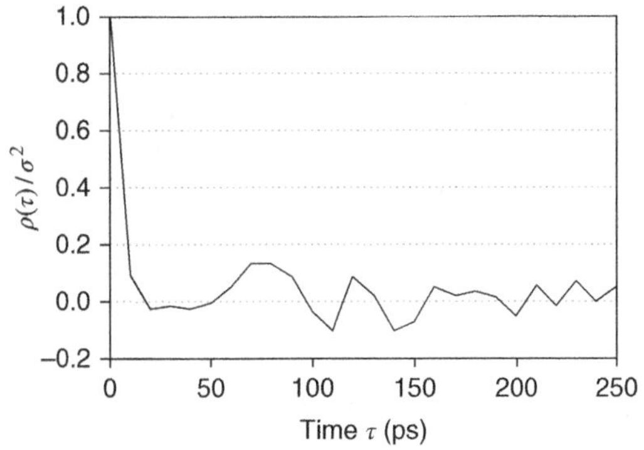

Figure 2.4 Normalized autocorrelation function $\rho(\tau)/\sigma^2$ versus the time shift τ. The fast decay corresponds to the inverse bandwidth of the oscilloscope $^1/_{70\ \text{GHz}} \cong 15$ ps. At $\rho(0)$ one gets the variance σ^2.

multiplied. Afterwards the integration takes place via the time constant of a low pass RC-circuit. Thanks to the availability of fast analog to digital (AD) converters, it is also possible to directly record the variation in time and determine the ACF purely by calculation. This computational method yields the ACF of our signal shown in Figure 2.4 from Figure 1.1.

With ideal white noise, only the value $\rho(0) = \sigma^2$ is different from zero. This means that there is no correlation between the voltage values at different times with the time difference τ. In all practical cases there is a limitation of the frequency band. The transmission system is either a low pass or a band pass. In our examples from Chapter 1, the oscilloscope is a low pass, the spectrum analyzer a bandpass. However, the low-pass character of the oscilloscope is not very pronounced, as it is extremely broadband with $B = 70$ GHz. We can see this in Figure 2.4. The drop from full correlation $\rho(0) = \sigma^2$ to the completely uncorrelated state $\rho(\tau) = 0$ takes place in the extremely short time of $\cong$15 ps. This corresponds to the reciprocal of the bandwidth with $B = 70$ GHz.

In general, the following can be said about the relationship between bandwidth and ACF. If a broadband white noise signal passes through a bandwidth-limiting network, the decay time of the ACF changes inversely to the bandwidth. At the output of a narrow-band filter, white noise is converted into a blurred sine wave. The noise we see on the spectrum analyzer would look like Figure 2.5 in the time domain.

In contrast to the broadband representation of white noise by the oscilloscope, we have a narrow frequency band from the selected resolution bandwidth (RBW) around the center frequency (1 GHz in the example). This of course has an effect on the ACF (Figure 2.6).

The ACF becomes periodic with the center frequency of the filter and decays very slowly. As is well known, there is a close interaction between time domain and frequency domain via the Fourier transform. The settling time of a filter $t_e \cong 1/B$ determines the time step in the display $v(t)$ in Figure 1.1, which we have to do to get

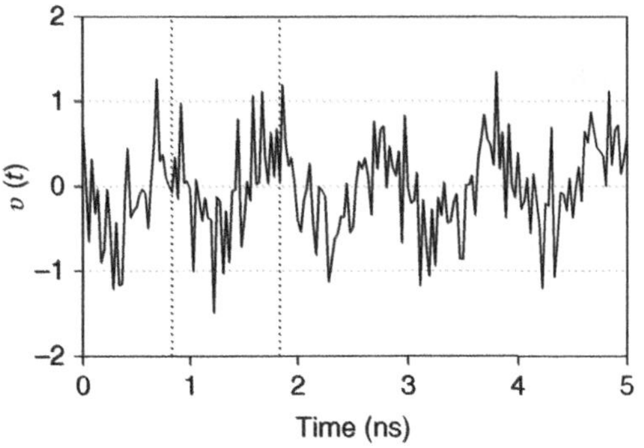

Figure 2.5 White noise after passing through a bandpass filter is a blurred sine wave at the center frequency (1 GHz) in the time domain.

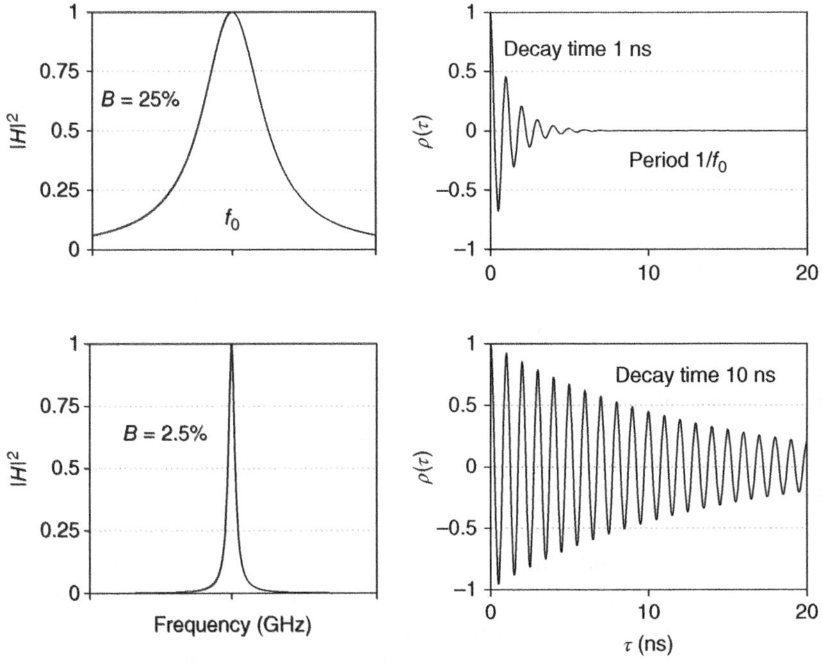

Figure 2.6 ACF $\rho(\tau)$ of band-limited noise. It is periodic with the center frequency of the filter. The correlation time (decay of the envelopes) is given by the bandwidth $\tau_\rho \cong 1/\pi B$.

two voltage values with different signs. If the bandwidth is small t_e large, i.e. the voltage changes occur slowly, we get few zero crossings per time unit, in contrast to the situation in Figure 1.1, where bandwidth is extremely large with $B = 70\,\text{GHz}$, i.e. the zero crossings occur very quickly. So if we have noise where the amplitude changes very slowly, we get a slowly decaying ACF, for fast changing amplitudes the ACF is quickly at zero.

As a simple example, one can determine the ACF of a fully deterministic sinusoidal signal.

$$v(t) = \sqrt{2}v_{RMS}sin(\omega t + \phi) \tag{2.8}$$

If we insert (2.8) in (2.7), we have for the ACF of the sine signal

$$\rho(\tau) = \lim_{T\to\infty} \frac{2v_{RMS}^2}{2T} \int_{-T}^{T} \left\{ \frac{1}{2}cos(\omega\tau) - \frac{1}{2}cos(2\omega t + 2\phi + \omega\tau) \right\} dt \tag{2.9}$$

So it follows:

$$\rho(\tau) = v_{RMS}^2 cos(\omega\tau) - \lim_{T\to\infty} \frac{2v_{RMS}^2}{2T} \int_{-T}^{T} \left\{ \frac{1}{2}cos(2\omega t + 2\phi + \omega\tau) \right\} dt \tag{2.10}$$

Since the time average over the cosine function is zero, we obtain for the ACF of (2.8)

$$\rho(\tau) = v_{RMS}^2 cos(\omega\tau) \tag{2.11}$$

From this result we can deduce two further important properties of ACF.

1. The ACF of a periodic function in t is also a periodic function in τ, with the same frequency.
2. During the formation of the ACF the phase information φ is lost. It is therefore not possible to calculate the original time curve including the phasing from the ACF backwards. Furthermore, as with the noise signal, we see that the ACF for $\tau = 0$ indicates the variance and thus the AC power at $R = 1\,\Omega$.

Cross-Correlation

The ACF is an information about the similarity of a time function to itself. Of course we can also apply this consideration to two different time functions. Are two noise voltages, which are measured at different points of a circuit similar to each other? Do they possibly have the same amplitude but are out of phase? Do they have different amplitudes because one of them has been amplified and an additional noise has been added by the amplifier, so that the original signal is covered but still exists? These considerations are extremely important when superimposing all the components at the output of a network of which a noise model is to be created. This is the case when analyzing the noise contributions of different sources in active networks, e.g. transistors and amplifiers. But it also plays a major role in data transmission and signal identification. Just as sinusoidal voltages of the same frequency are attenuated or amplified by phase-correct superposition, this can also happen with pure noise voltages or with noisy signals.

Let us first consider two different noise voltages $v_1(t)$ and $v_2(t)$ at one and the same time. The variance of the sum of both has the value

$$\overline{(v_1 + v_2)^2} = \overline{v_1^2} + \overline{v_2^2} + \overline{2v_1v_2} = \sigma_1^2 + \sigma_2^2 + 2\rho_{12}\sigma_1\sigma_2 \tag{2.12}$$

In the last term, the correlation coefficient ρ_{12} obviously determines how large the sum of two noise powers is.

$$\rho_{12} = \frac{\overline{v_1v_2}}{\sqrt{\overline{v_1^2}\,\overline{v_2^2}}} = \frac{1}{\sigma_1\sigma_2} \lim_{T\to\infty} \frac{1}{2T} \int_{-T}^{T} v_1(t)v_2(t)dt \tag{2.13}$$

The normalized value is $|\rho_{12}| \leq 1$. We look at the three limiting cases:

1. $\rho_{12} = 0$: Both signals are statistically independent. At one point in time t_1 there is a value $v_1(t_1)$. Completely independent of this value $v_2(t_1)$ take all values from its amplitude range. The total power is equal to the sum of the individual powers.

   $$\overline{(v_1(t) + v_2(t))^2} = \sigma_1^2 + \sigma_2^2 \tag{2.14}$$

   This applies to all noise power from different sources.
2. $\rho_{12} = \pm 1$: Both signals are fully correlated. The superimposed noise power results from the linear addition of the two voltages. Depending on the sign it is maximum or minimum. Both voltages can also cancel each other out if they have the same amplitude and correct phase. However, this extreme case does not occur in practice.
3. $\rho_{12} < 1$: With partial correlation we can make the following consideration for c_{12}:
   We combine the two noise voltages as follows

   $$v_2(t) = av_1(t) + v_{UNC}(t) \tag{2.15}$$

   $v_2(t)$ is proportional to $v_1(t)$, e.g. by amplification a, so fully correlated. To this is added the noise contribution of the amplifier $v_{UNC}(t)$. This comes from a different source and is therefore uncorrelated to $v_1(t)$. We write for the correlation coefficient

   $$\rho_{12} = \frac{\overline{v_1v_2}}{\sqrt{\overline{v_1^2}\,\overline{v_2^2}}} = \frac{a\overline{v_1^2}}{\sqrt{\overline{v_1^2}(a^2\overline{v_1^2} + \overline{v_{UNC}^2})}} \tag{2.16}$$

   and obtain:

   $$\rho_{12} = \frac{1}{\sqrt{1 + \frac{1}{a^2}\frac{\overline{v_{UNC}^2}}{\overline{v_1^2}}}} \tag{2.17}$$

The limiting cases are:

Full correlation: $v_{UNC} \to 0: \quad \rho_{12} = \pm 1$
No correlation: $a \to 0: \quad \rho_{12} = 0$

If we want to express the uncorrelated noise power with the correlation coefficient

$$\overline{v_{UNC}^2} = \overline{v_2^2}(1 - \rho_{12}^2) \tag{2.18}$$

In practice, the superposition of noise quantities in certain frequency bands is of interest. Then of course the phase angle of the signals becomes important, which determines whether the correlation leads to an amplification or a reduction of the sum signal. As an example, let us consider the RC network in Figure 2.7.

As a result of the band limitation there is a phase angle between v and i. We can proceed as with complex alternating current variables. Accordingly, the correlation coefficient becomes complex. Its phase angle results from the original phase shift of v and i and the influence of the complex load.

$$\rho = \frac{\overline{iv^*}}{\sqrt{\overline{i^2 v^2}}} = |c| exp(j\phi_C) \tag{2.19}$$

The noise current square in the load $Y = {}^1\!/\!_R + j\omega C$ results in

$$\overline{i_L^2} = \overline{(i + vY)^2} \tag{2.20}$$

With the introduction of the correlation coefficient (2.19)

$$\overline{i_L^2} = \overline{i^2} + \overline{v^2} YY^* + \sqrt{\overline{i^2 v^2}}(\rho Y^* + \rho^* Y)$$

$$\overline{i_L^2} = \overline{i^2}(1 - |\rho|^2) + \overline{v^2}\left| Y + \rho\sqrt{\frac{\overline{i^2}}{\overline{v^2}}} \right|^2 \tag{2.21}$$

The result (2.21) gives us the variance of the noise current in the complex load Y as a function of the correlation of the two sources v and i. This is significantly determined by the complex correlation coefficient ρ (2.19). The first term gives that part of i, which is not correlated to v.

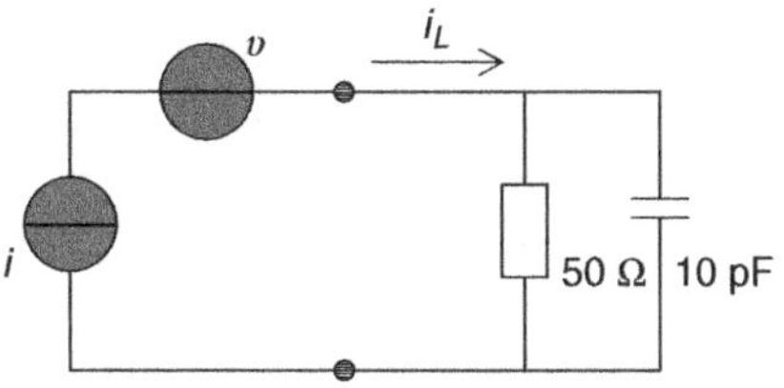

Figure 2.7 RC network with bandlimited noise sources v and i at 1 GHz. The noise current i_L in the load depends decisively on the correlation of v and i.

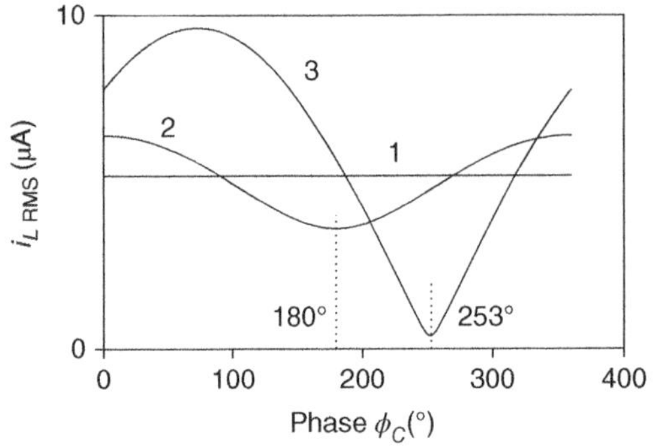

Figure 2.8 Superposition of the two noise currents in the load Y over the phase angle of the correlation coefficient (2.19) Curve 1: $\rho = 0$, no correlation. Curve 2: $|\rho| = 1$, $Y = 1/50\ \Omega$ ($C = 0$). Curve 3: $|\rho| = 1$, $50\,\Omega \,||\, 10\,\text{pF}$, all phase angles of ρ.

If both sources are uncorrelated, the noise current squares in the load Y. Equation (2.21) reduces to:

$$\overline{i_L^2} = \overline{i^2} + \overline{v^2}|Y|^2 \tag{2.22}$$

An example is shown in Figure 2.8. The two noise sources in the circuit Figure 2.7 may have the values: =5 μA, $v = 70$ μV.

For $|\rho| \neq 0$ we see the decisive effect of the phase of the complex ρ on the superposition of noise currents in a load. In the limiting case $|\rho| = 0$ (plot 1), we have the simple addition of the currents and thus the total noise power as the sum of the individual powers (2.22). With $\rho = 1$ there is a strong influence of phase ϕ_C. With a real terminating resistor $Z_L = 50\,\Omega$ (plot 2) we have compensation at $\phi_C = 180°$. No additional phase shift of current and voltage is caused by the circuit. In case Figure 2.7 with $Z_L = 50\,\Omega \,||\, 10\,\text{pF}\ @ f = 1\,\text{GHz}$ (plot 3) we have the compensation at $\phi_C = 253°$.

Noise Spectra

The preceding considerations lead us to the representation of the noise in the frequency domain as displayed on the spectrum analyzer (Figure 1.2). We have already seen in Figures 2.5 and 2.6 the effect of band limiting on the characteristics of noise signals. The communications engineer always has to deal with noise interference in a frequency range relevant to his system. As is well known, time domain and frequency domain are linked with each other via the Fourier transformation. However, due to the statistical nature of the time function, no amplitude spectrum can be specified for it, but only a power spectrum. So far we have considered the mean square value of a fluctuating quantity (2.3). It is a measure for the total power of the noise process. In the frequency domain, we define the power present in a given frequency band around a center frequency. At a bandwidth of

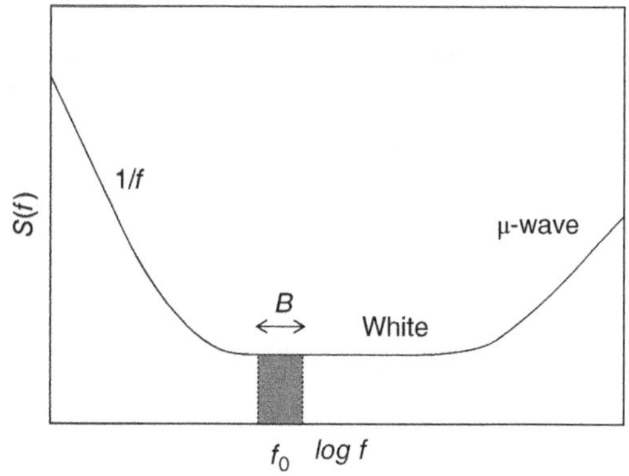

Figure 2.9 Spectral noise power density *S(f)* of a microwave field effect transistor. Below 10 MHz rise by 1/*f* noise, above 10 GHz rise by microwave noise.

$B = 1$ Hz, this is the spectral power density $S(f)$ the noise power at frequency f.

$$\sigma^2 = \int_0^\infty S(f)df \tag{2.23}$$

An example is the typical noise spectrum of a microwave field effect transistor in Figure 2.9 [7]. This curve is only valid up to a cut-off frequency of the transistor, since (2.23) must remain finite for physical reasons. In practice, the noise is considered in the limited frequency range of bandwidth B. This can be chosen so small that $S(f) = S(f_0) = const.$

$$S(f_0)B = \int_B S(f)df = \sigma_B^2 \tag{2.24}$$

σ_B^2 is the square of the fluctuation in this frequency range of bandwidth B. As already shown in Figure 2.5, this narrow band noise can be treated like a signal.

Whichever fluctuation variable is considered, one has the corresponding unit of $S(f)$.

Voltage noise: $[S(f)] = {}^{V^2}/_{Hz}$
Current noise: $[S(f)] = {}^{A^2}/_{Hz}$
Frequency noise: $[S(f)] = {}^{Hz^2}/_{Hz}$
Phase noise: $[S(f)] = {}^{1}/_{Hz}$

In general, one speaks of power density.

Autocorrelation Function and Spectral Power Density

The shape of the power spectrum $S(f)$ depends on the extent to which future amplitude values are predictable. Complete determinacy holds for the sine function. In the frequency spectrum it is represented by a single line, it is "monochromatic".

The other extreme case is the completely stochastic broadband noise signal in Figure 1.1. The future amplitude course is not predictable. In the frequency spectrum it is defined by a constant value of (*f*). It is the "white" spectrum. Here the Wiener–Khintchine theorem applies, because it describes just this fact. From the ACF $\rho(\tau)$ the power spectrum is obtained by Fourier transformation $S(f)$ and vice versa. In general terms, including negative frequencies in the ACF, as is common in signal theory, the complex notation results:

$$\begin{aligned} S(f) &= 2\int_{-\infty}^{+\infty} \rho(\tau)exp(-j2\pi f\tau)d\tau \\ \rho(\tau) &= \frac{1}{2}\int_{-\infty}^{+\infty} S(f)exp(j2\pi f\tau)df \end{aligned} \tag{2.25}$$

Only positive frequencies are technically useful. Therefore a real pair of equations in the limits 0 to ∞.

$$\begin{aligned} G(f) &= 4\int_{0}^{\infty} \rho(\tau)cos(2\pi f\tau)d\tau \\ \rho(\tau) &= \int_{0}^{\infty} G(f)cos(2\pi f\tau)df \end{aligned} \tag{2.26}$$

The total power of the fluctuation process is again

$$\rho(0) = \int_{0}^{+\infty} G(f)df \tag{2.27}$$

In many cases the approximation of (2.26) can be used for low frequencies. For processes whose typical time constant $\tau \cong {}^1\!/_f$ is greater than the decay time of the ACF (correlation time) τ_ρ, applies $\rho(\tau) = 0$. For $\tau < \tau_\rho$ is then $cos(2\pi f\tau) \cong 1$ and (2.26) becomes

$$G_0(f) = 4\int_{0}^{\infty} \rho(\tau)d\tau \quad f \ll \frac{1}{\tau_\rho} \tag{2.28}$$

Band-Limited Noise on the Spectrum Analyzer

In Figure 1.2 we see the Rayleigh distribution of the amplitudes and in Figure 1.3 the reduction of the fluctuation around a constant level by averaging. Which laws underlie this picture? This is important to know when using the SPA for quantitative noise analysis [8, 9]. Figure 1.2 shows white noise within the frequency span of the SPA. The image is obtained, for example, when the RBW of the SPA (B in Figure 2.9) scans over the range of white noise. A portion of the noise power within the bandwidth B around the frequency f_0 is recorded. This is the noisy sine in Figure 2.5, which we can represent as the sum of the real and imaginary parts (I/Q components) in the vector diagram. The SPA measures the magnitude of this signal. It is therefore the vectorial sum of the noise voltages $v_I \cos(\omega_0 t)$ and $v_Q \sin(\omega_0 t)$.

The displayed magnitude of the noise signal in bandwidth B corresponds to the envelope of the I/Q components. The components v_I and v_Q have statistically independent normal distributions of their amplitudes. For their vector sum $v = \sqrt{v_I^2 + v_Q^2}$ results in the Rayleigh distribution (2.29) (Figure 2.10).

$$RDF(v) = \frac{v}{\sigma^2} exp\left(-\frac{1}{2}\frac{v^2}{\sigma^2}\right) \tag{2.29}$$

In the example Figure 1.2 we have the linear display mode of the SPA. The Gaussian distribution of the voltages of the broadband noise at the input changes due to the band limitation by the RBW and the operation of the envelope detector to the Rayleigh distribution. By using video filters or averaging (Figure 1.3) the fluctuation is reduced and the average value is obtained. The average value behind the detector is of course not $\overline{v} = 0$ as with the original Gaussian distribution of the time domain signal but

$$\overline{v} = \int_0^\infty vRDF(v)dv = \sqrt{\frac{\pi}{2}}\sigma \tag{2.30}$$

The noise power at a resistor R calculated with the Rayleigh distribution results to

$$\overline{P} = \frac{1}{R}\int_0^\infty v^2 RDF(v)dv = \frac{2}{R}\sigma^2 \tag{2.31}$$

If we now compare the noise power calculated from the square of the mean value of the voltage (2.30) with (2.31), we get

$$\frac{\overline{v}^2/R}{\overline{P}} = \frac{\sigma^2 \pi/2}{2\sigma^2} = \frac{\pi}{4} \quad 10log\left(\frac{\pi}{4}\right) = -1.05\text{ dB} \tag{2.32}$$

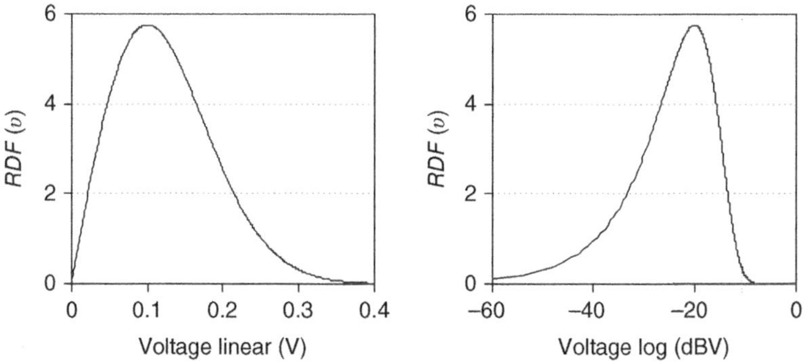

Figure 2.10 Rayleigh distribution *RDF*(*v*) of the envelope of a band-limited noise signal. It results from the uncorrelated I/Q components, each of which has a Gaussian distribution with $\sigma = 0.1$ V have. Left in linear, right in log scale.

The display of the marker on the linear scale of the SPA does not therefore give the correct v_{RMS} of the noise, but a value that is too low by 1.05 dB. With a logarithmic scale, the correction value is even greater. This is 2.5 dB and results from the expected value of the double exponential distribution. Details are dealt with in the chapter on measurement technology.

Thus, the noise power is displayed 2.5 dB too low on the most commonly used dB scale of the SPA. But this is no problem with modern SPAs, because the noise marker corrects this automatically.

References

1 van der Ziel, A. (1954). *Noise*. London: Chapman and Hall.

2 Bendat, J.S. and Piersol, A.G. (1971). *Random Data: Analysis and Measurement*. Wiley Interscience.

3 Freman, J.J. (1958). *Principles of Noise*. Wiley.

4 Pfeifer, H. (1959). *Elektronisches Rauschen*. Leipzig: B.G. Teubner.

5 Meinke, H.H. and Gundlach, F.W. (1992). *Taschenbuch der Hochfrequenztechnik*. Springer Verlag.

6 Müller, P.H. (ed.) (1985). *Lexikon der Stochastik*. Berlin: Akademie Verlag.

7 Müller, R. (1990). *Rauschen*. Springer Verlag.

8 Rauscher, C. (2007). *Grundlagen der Spektrumanalyse*. München: Rohde & Schwarz.

9 Agilent Technologies (2003). Spectrum Analyzer Measurements and Noise. Appl. Note 1303.

3

Noise Sources

Thermal Noise

What noise voltage v does an ohmic resistor generate in an electronic circuit?

The square of variation of the noise voltage $v(t)$ and the spectral power density S_V are given by (2.3) and (2.24), respectively

$$\overline{v^2(t)} = \sigma^2 = S_V B \tag{3.1}$$

The effective value of the noise voltage is

$$v_{RMS} = \sqrt{S_V B} \tag{3.2}$$

To consider thermal noise, we make a simple thought experiment based on the equipartition theorem in thermodynamics [1]: When a gas is in thermal equilibrium, each degree of freedom of particle motion has the same energy

$$E_{th} = \frac{1}{2}kT \tag{3.3}$$

$k = 1.38 \times 10^{-23}$ Ws/K is the Boltzmann constant and T the absolute temperature. According to this the x-component of the velocity u_x of a gas molecule of mass m has the energy

$$\frac{1}{2}mu_x^2 = \frac{1}{2}kT \tag{3.4}$$

The same applies to the free electrons in a plasma or to the electron gas in a solid.

Resistance noise is an expression of the coupling of thermal and electrical fluctuations. The equipartition theorem thus applies not only to the microscopic but also to the macroscopic degrees of freedom of a system. In the case of the circuit Figure 3.1 also for the electrical energy of the capacity C.

$$\frac{1}{2}Cv_C^2 = \frac{1}{2}kT; \quad v_C^2 = \frac{kT}{C} \tag{3.5}$$

Thus we have a relation between noise voltage and temperature.

A Guide to Noise in Microwave Circuits: Devices, Circuits, and Measurement, First Edition.
Peter Heymann and Matthias Rudolph.

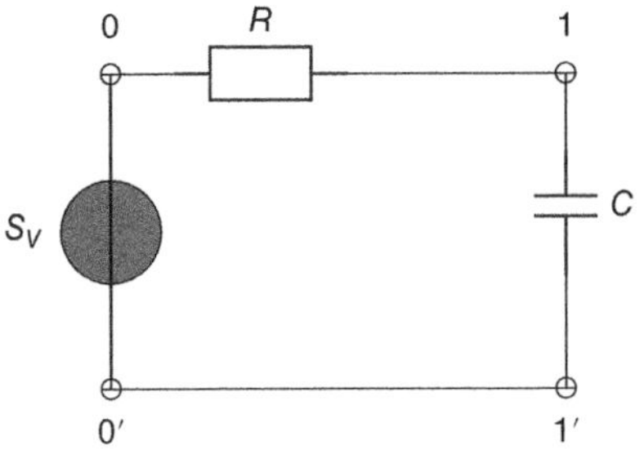

Figure 3.1 Circuit for generating a noise voltage v_C at the capacitance C (1.1′) by the resistance noise of the voltage source (0.0′) with the spectral intensity S_V.

A bandwidth limitation is not considered at first. The power density S_{VC} at the capacitance (1.1′) results from the power density of the voltage source S_V (0,0′) by multiplication with the square of the transfer function $H(f)$. For details see Chapter 4.

$$S_{VC} = S_V|H(f)|^2 = \frac{S_V}{1+(\omega CR)^2} \tag{3.6}$$

The voltage square results from the integration over the power density to

$$v_C^2 = \int_0^\infty S_{VC} df = S_V \int_0^\infty \frac{df}{1+(\omega CR)^2} = \frac{kT}{C} \tag{3.7}$$

The integral has the solution

$$\int_0^\infty \frac{df}{1+(\omega CR)^2} = \frac{1}{2\pi RC} arctan(2\pi fRC)\Big|_0^\infty = \frac{1}{4RC} \tag{3.8}$$

It applies: $arctan(0) = 0$; $arctan(\infty) = {}^{\pi}/_{2}$.

It holds from (3.7)

$$S_V \frac{1}{4RC} = \frac{kT}{C} \tag{3.9}$$

and we obtain the well-known Nyquist formula for the thermal noise of an ohmic resistor [2, 3]:

$$S_V = 4kTR; \quad \overline{v_{TH}^2} = 4kTRB \tag{3.10}$$

Nyquist Formula and Thermal Radiation

It is obvious that (3.10) can only be valid up to a maximum cut-off frequency. The total noise power of a resistor would become infinite if S_V would not approach zero above this cut-off frequency. After the kinetic consideration of the free flight time in the resistance material (Chapter 1), an estimation of this cut-off frequency results in ${}^{1}/_{\tau} \cong {}^{1}/_{10^{-12}}$ s = 1 THz. Strictly physically, the cut-off frequency results from the theory of thermal radiation [1, 4].

So the Nyquist formula is an approximation for "low" frequencies. One can add a Planck factor.

$$\overline{v_{TH}^2} = 4kTRB \times p(f) \tag{3.11}$$

$p(f)$ describes the decrease in noise power at high frequencies due to quantum effects. Planck's formula gives the power of black-body radiation as a function of temperature, which could only be understood after the introduction of the quantum hypothesis.

The dimensionless Planck factor has the following dependence

$$p(f) = \frac{hf}{kT} \frac{1}{exp\left(\frac{hf}{kT}\right) - 1} \tag{3.12}$$

$h = 6.62 \times 10^{-34}$ Ws^2 is Planck's constant (Figure 3.2).

Even at the temperature of liquid nitrogen (77 K), the correction in the microwave range is not important. In the Terahertz range, however, the noise power already decreases.

This consideration also leads us to the direct physical relation between thermal noise of an electronic component and the fundamental phenomenon of thermodynamics: thermal radiation. The Nyquist formula can be understood as a one-dimensional limit case of black-body radiation for low frequencies.

According to Planck, the radiant energy dS is radiated from the unit area of a black body of temperature T within the bandwidth B = 1 Hz into the solid angle $d\Omega$.

$$dS = \frac{2hf^3}{c^2} \frac{d\Omega}{exp\left(\frac{hf}{kT}\right) - 1} \tag{3.13}$$

The unit is Ws/cm^2, $c = 3 \times 10^{10}$ cm/s.

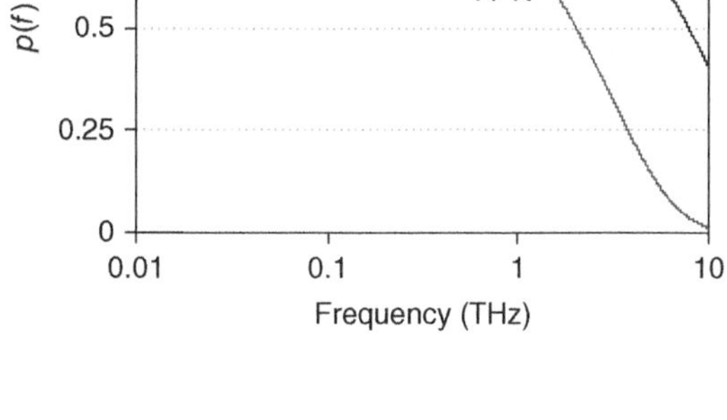

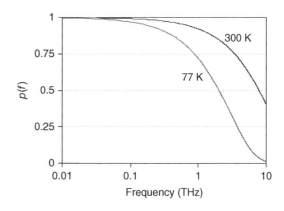

Figure 3.2 Planck factor $p(f)$ for T = 300 K (room temp.) and 77 K (LN_2 temp.) in the terahertz range.

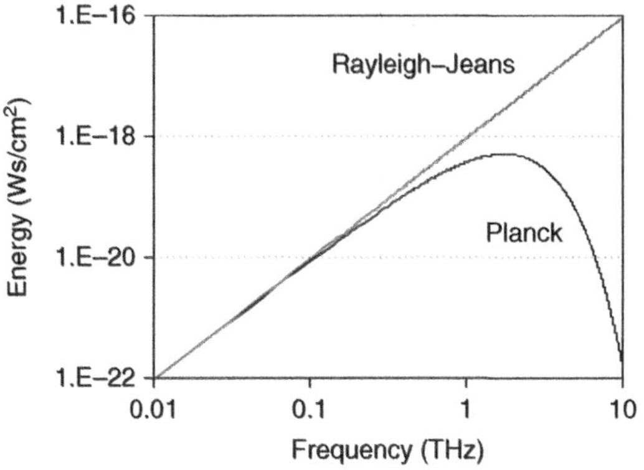

Figure 3.3 Planck's radiation formula for $T = 30$ K (cold outer space) and the Rayleigh Jeans approximation.

This formula is shown in Figure 3.3. In microwave technology and also in radio astronomy, the Rayleigh–Jeans radiation law is usually sufficient. Quantum effects do not play a role. It is the approximation for low frequencies.

$$\frac{hf}{kT} \ll 1$$

For small x applies:

$$exp(x) \cong 1 + x$$

This results in

$$dS = \frac{2kT}{\lambda^2} d\Omega \tag{3.14}$$

It's Rayleigh Jean's law of radiation. A first attempt to explain thermal radiation with classical physics. At high frequencies, such as in the optical range, it leads to the "ultraviolet catastrophe."

Let us imagine an isotropic antenna, which is surrounded on all sides by black-body radiation. It has the effective antenna area

$$A_{EFF} = \frac{\lambda^2}{4\pi} \tag{3.15}$$

λ = wavelength of the radiation in cm (Figure 3.4).

The surface density of the radiant energy (3.14) is integrated over the solid angle of the unit sphere. This results in

$$\int_{\Omega} d\Omega = 4\pi$$

With the total energy

$$\int_{\Omega} A_{EFF} \frac{2kT}{\lambda^2} d\Omega = 2kT \tag{3.16}$$

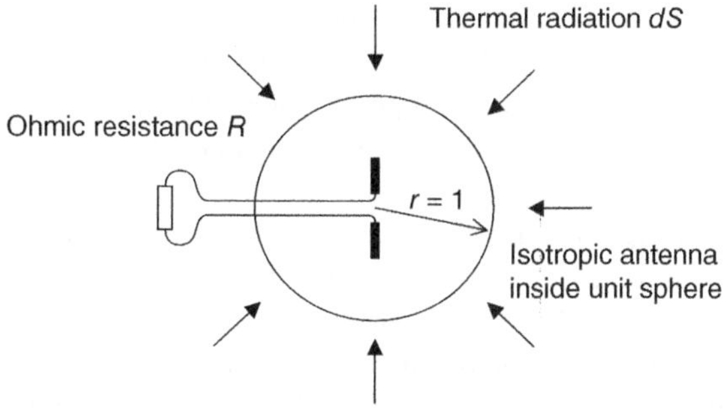

Figure 3.4 Isotropic antenna with matched termination exposed to thermal radiation density dS.

The unit is Ws. From this ½ of the energy arrives in resistance R, since only one direction of polarization is taken up. We have in the bandwidth B the power in the resistor. So we get for the conversion of the thermal energy of the radiation into the electrical energy of the antenna

$$P = kTB \tag{3.17}$$

This is also a form of the Nyquist formula. In contrast to (3.10), the resistor does not occur at all, since it is the available noise power of a resistor R at the temperature T in the bandwidth B. What is the relationship between (3.10) and (3.17)? In Figure 3.1 we replace the capacitance C by a load resistor R_L, which does not generate noise itself. It could for example be at the temperature $T = 0$ K. The spectral power density S_V (3.10) generates a noise current square in the circuit

$$\overline{i^2} = \frac{S_V B}{(R + R_L)^2} \tag{3.18}$$

The noise power in the load resistor is:

$$P = \overline{i^2} R_L = S_V B \frac{R_L}{(R + R_L)^2} \tag{3.19}$$

The available power is the maximum value that is reached during adaptation, i.e. when $R_L = R$

$$P_A = \frac{S_V B}{4R} = kTB \tag{3.20}$$

Validity and Experimental Confirmation of the Nyquist Formula

Our derivation of the Nyquist formula for the thermal noise of resistors from the RC circuit Figure 3.1 is a simple plausibility consideration, not a physically strict procedure. The question of frequency limitation is clarified by the connection with Planck's radiation formula. For the problems of the high-frequency engineer, however, there are usually rather trivial reasons for this, which lie in the frequency limit of the components and circuits. There are a number of experimental confirmations of the Nyquist formula, e.g. on wire resistors up to 2 MΩ in the temperature range 77–380 K. In electronic circuits, the resistors are usually passed through by a direct current. Strictly speaking, the condition of thermodynamic equilibrium is then no longer fulfilled. This does not play a decisive role because the drift velocity is small compared with the thermal velocity. In non-metallic conductors, however, there is additional noise caused by the current flow. This current noise is related to another noise source, which is extremely important in active electronic components. It is the shot noise.

Thermal Noise Under Extreme Conditions

Under extreme conditions we understand here very high frequencies ($f > 100$ GHz) and cryogenic temperatures e.g. LHe ($T = 4$ K). When cooling down to absolute zero, the thermal energy of a system, and thus the fluctuations, does not approach zero. What remains is a quantum physical residual motion, the zero-point energy. In quantum theory one speaks of vacuum energy. So there is a noise contribution of the vacuum fluctuations. Its spectral power density is

$$S_{VAC} = \frac{1}{2}hf \tag{3.21}$$

A comparison with the Nyquist formula (3.10) is shown in Figure 3.5. The zero-point energy can therefore play a role in technical systems [5]. If one works in this environment, one must of course take care that an ohmic resistance at $T = 4$ K has a different value than at room temperature.

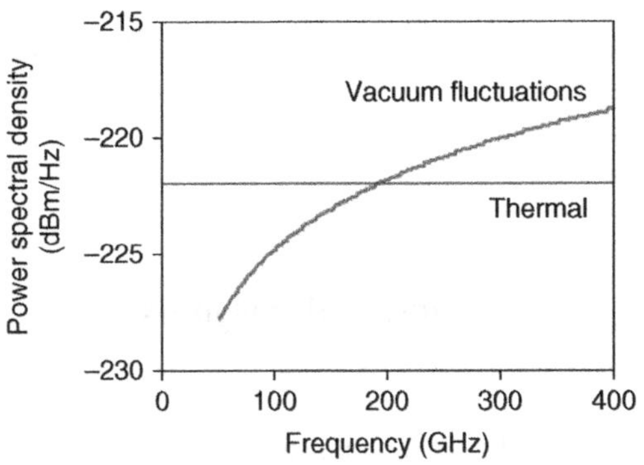

Figure 3.5 Spectral noise power density at $T = 4$ K. The vacuum fluctuations (3.21) exceed the thermal level (3.10) for $f > 200$ GHz.

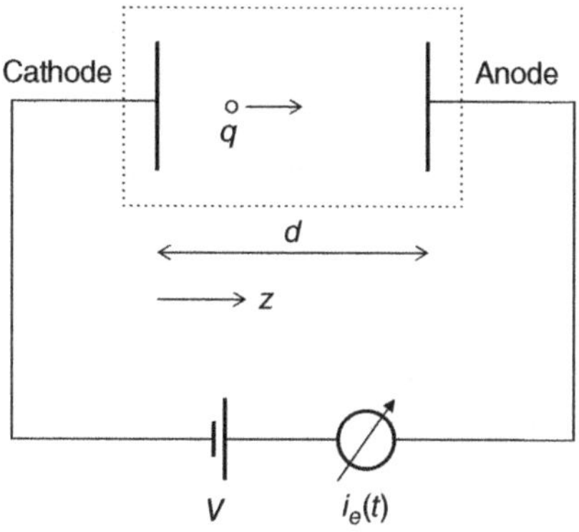

Figure 3.6 A particle with charge q moves in the electric field of a vacuum diode.

Shot Noise

The conduction current consists of transporting the charge of individual electrons. The total current is composed of a large number of current pulses, each of which carries the elementary charge $q = 1.6 \times 10^{-19}$ As. Each electron generates a current impulse in the outer circle of the duration of its transit time. The associated fluctuation is called Shot Noise (Schottky [6]).

Let us first look at the formation of the current pulse in the outer circuit as a result of the motion of the electron. The clearest way to see this is in a vacuum diode. In Figure 3.6 the external voltage V is applied to the vacuum path. All electrons emitted from the cathode are absorbed by the anode without forming a space charge cloud. In this case one speaks of a saturated diode.

The motion of the electron of mass m on the z-axis is according to the basic law of mechanics

$$F = m\ddot{z} \tag{3.22}$$

The force F in the electric field is

$$F = qE = \frac{qV}{d} \tag{3.23}$$

Thus the acceleration

$$\ddot{z} = \frac{qV}{md} \tag{3.24}$$

By integration we obtain the equation of motion for an electron which starts at the time $t = t_0$ with the velocity $u = 0$:

$$u(t) = \dot{z} = \frac{qV}{md}(t - t_0) \tag{3.25}$$

The path is obtained by further integration:

$$z(t) = \frac{qV}{2md}(t - t_0)^2 \tag{3.26}$$

The electron gains kinetic energy $W(t)$ from the acceleration, which comes from the voltage source.

$$W(t) = \frac{1}{2}mu(t)^2 = \frac{q^2V^2}{2md^2}(t - t_0)^2 \tag{3.27}$$

The relationship with the induced current $i_e(t)$ in the outer circle, and the power P is given by

$$P = \frac{dW}{dt} = Vi_e(t) = \frac{q^2V^2}{md^2}(t - t_0) \tag{3.28}$$

We get for the current:

$$i_e(t) = \frac{q^2V}{md^2}(t - t_0) \tag{3.29}$$

In the transit time $\tau_t = t - t_0$ the electron has reached the anode, it has covered the distance d. (3.25) results for d:

$$d = \frac{qV}{2md}\tau_t^2 \tag{3.30}$$

and for the transit time τ_t:

$$\tau_t = d\sqrt{\frac{2m}{qV}} \tag{3.31}$$

thus the induced current pulse of the individual electron during the time of flight $t_0 \leq t \leq t + \tau_t$ is

$$i_e(t) = \frac{2q}{\tau_t^2}(t - t_0) \tag{3.32}$$

and outside that time. $i_e(t) = 0$.

The time dependence of the current is shown in Figure 3.7.

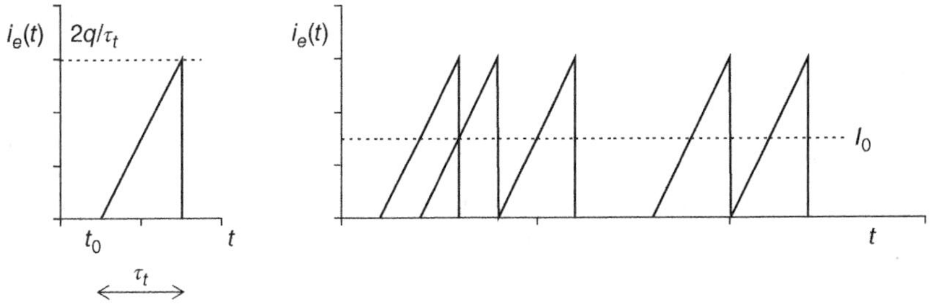

Figure 3.7 The induced current in the outer circuit $i_e(t)$ in the time domain. (a) Single electron. (b) Superposition to the total current I_0. The area of a triangle corresponds to the elementary charge q.

In our example of the high-vacuum diode, the fluctuation is caused by the statistically distributed exit of the electrons from the cathode. After emission, they fly with constant acceleration to the anode, where they disappear from the process after the transit time τ_t. This produces the triangular current pulse (3.32) in the outer circuit.

Technically more interesting is of course the current flow in a semiconductor diode. Although no vacuum path is traversed here, noise generated by the shot effect does occur. It is even the main source of noise in semiconductor electronics. If a stream of electrons or holes flows through a region where the field strength is not constant, the statistically distributed entry times of the particles into this region come into play and generate fluctuations of the current. This is the case both in pn-junctions and in the space charge zones of Schottky diodes. In contrast to the motion in a vacuum, with the constantly increasing speed, the motion in a semiconductor is usually determined by the constant saturation drift velocity. In Silicon this is, e.g. $u_d \approx 10^7$cm/s. The current pulse is therefore not triangular but rectangular. However, the exact shape of the pulse is not important for the following derivation of the shot noise formula.

The total current is the sum of n individual pulses per unit time, each of which transports the elementary charge q.

$$I_0 = qn \tag{3.33}$$

Here we demonstrate the procedure described in Chapter 2, how to obtain the spectral power density from the autocorrelation function according to the Wiener–Khintchine theorem. The ACF of the time function $i(t)$ is (Figure 3.8)

$$\rho_i(\tau) = n \lim_{T\to\infty} \frac{1}{2T} \int_{-T}^{T} i(t)i(t+\tau)dt \tag{3.34}$$

The area under the rectangle $i(t) \times i(t+\tau)$ is the ACF. With (3.32) the result is

$$\rho_i(\tau) = \frac{qI_0}{\tau_t^2}(\tau_t - \tau) \tag{3.35}$$

For low frequencies we use (2.28). We can see that again only the area of the triangle in Figure 3.9 is needed. As integration limit τ_t is sufficient because for longer periods of time only different impulses from the sequence would interact, which are, however, uncorrelated according to the precondition. We obtain the spectral power density $S(f)$ of the noise current:

$$S(f) = 4\int_0^{\infty} \rho_i(\tau)d\tau = \frac{4qI_0}{\tau_t^2}\int_0^{\tau_t}(\tau_t - \tau)d\tau \tag{3.36}$$

Integrated, the Schottky formula for the spectral power density of the shot noise is obtained:

$$S(f) = \frac{4qI_0}{\tau_t^2}\frac{1}{2}\tau_t^2 = 2qI_0 \tag{3.37}$$

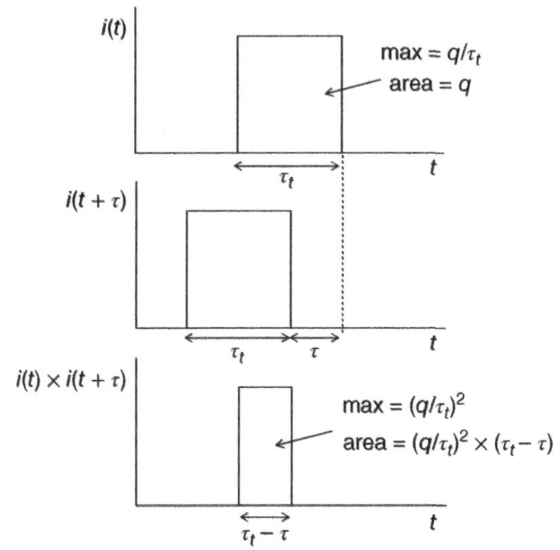

Figure 3.8 Current pulses in a semiconductor. The ACF is calculated from a single pulse by time shift τ.

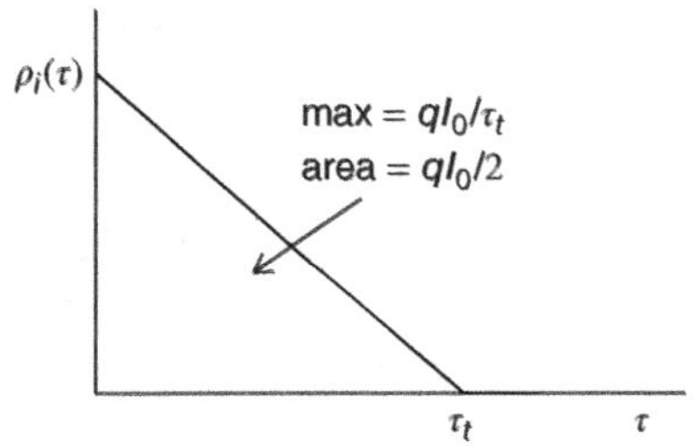

Figure 3.9 ACF $\rho_i(\tau)$ according to (3.34) of the current pulse in Figure 3.8.

Due to our approximation for low frequencies, white noise results. An exact analysis will of course show a decrease for high frequencies, but this is not important for our purposes. There is also a small influence of the pulse shape. The square pulse (Figure 3.8) gives a zero at $f = {}^1\!/_{\tau_t}$. The triangular impulse (Figure 3.7) results in the somewhat more broadband course of the curve (Figure 3.10).

Experiments were carried out early on to verify the Schottky formula. In [7] a saturated vacuum diode was operated as a current source in parallel to an LC resonant circuit. From the fluctuation of the voltage at the resonant circuit, it was possible to confirm (3.36) and to determine the elementary charge q very accurately.

The saturated vacuum diode can be used as a noise source for measurement purposes [8]. It does not require calibration, since its noise power can be directly derived from the physical process of the fluctuating anode direct current I_0. This current can be varied by the heating current of a directly heated hot cathode at constant anode voltage. This results in a defined noise source of adjustable power. However, it only provides white noise in the Megahertz range. According

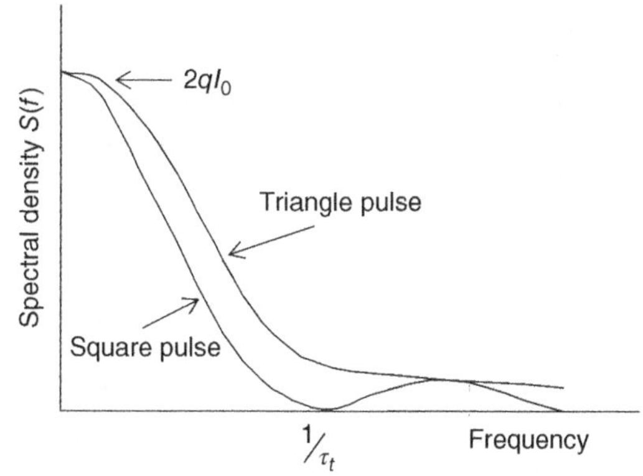

Figure 3.10 Frequency spectrum of the spectral noise current density $S(f)$ of the shot noise for rectangular and triangular current pulses. Typical for a Schottky diode is $^1/_{\tau_t} > 100$ GHz.

to (3.30) the cut-off frequency for $d = 5$ mm and $V = 250$ V is $^1/_{\tau_t} < 500$ MHz. At high frequencies, a resonance is added by lead inductance and cathode–anode capacitance. Today the application is limited to special cases. Calibrated noise generators with semiconductor diodes are almost always used in the reverse voltage range.

Plasma Noise

Closely related to the physics of thermal noise of resistors is the noise of the positive column of a gas discharge plasma. In the pressure range of about 10 mbar, which is typical for fluorescent lamps and gas discharge noise sources, the thermal motion of the electrons is very intense. Energy is constantly supplied to them by the external electric field. The thermalization of their motion is caused by the impacts with the gas molecules. Since these impacts are mainly elastic, they lead to a spatial scattering, but not to a transfer of energy to the gas molecules. The plasma is not isothermal. The electron temperature is much higher than the gas temperature [9]. The electrons determine the electrical properties and thus also the noise. In neon the electron temperature is $T_E \cong 23{,}000$ K. The available noise power according to (3.20) is therefore 80 times the thermal noise power at room temperature. However, this requires suitable measures to extract the power, since the conductivity of the plasma is very low.

$$\kappa = \frac{nq^2\nu}{m(\nu^2 + \omega^2)} \tag{3.38}$$

It means: ν = collision frequency of the electrons with the neutral gas, m = electron mass.

If we look at this plasma in the X-band ($\omega \cong 6 \times 10^{10}$ s^{-1}), the conductivity is $\kappa \cong 6 \times 10^{-4}$ S/m with $\nu \cong 1 \times 10^{10}$ s^{-1}. This corresponds approximately to undoped silicon. A cube of 1 cm edge length would have a resistance of about 160 kΩ. The plasma must therefore be matched by a resonance transformation in a cavity resonator. Broadband matching can also be achieved by inserting the plasma tube at an angle in a waveguide, as is done in gas discharge noise generators [10] (see Chapter 12).

Current Noise of Resistors and Contacts

A prerequisite for the occurrence of shot noise is that the individual processes of charge carrier motion are independent of each other. The electrons emitted by a hot cathode into a vacuum do not influence each other. This also applies to the electrons that pass through potential barriers in a semiconductor. Their total number is subject to noticeable statistical fluctuations. The fluctuations become very small when so many similar elementary events occur that they are no longer independent of each other and can no longer be identified individually. An example is the direct current in a metallic conductor. Due to the high electron density, the current flow is carried only by a very low drift velocity u_D [4, 8].

The current density is

$$J = nq \times u_D \tag{3.39}$$

u_D is the drift velocity of the electrons, and n is the density of the free electrons in the metal, e.g. in Copper $n = 8.5 \times 10^{22}\ \text{cm}^{-3}$. For a current density of $J = 1\ \text{A/cm}^2$ this results in

$$u_D = \frac{1}{8.5 \times 10^{22} 1.6 \times 10^{-19}} = 7 \times 10^{-5}\ \text{cm/s}$$

a very low value, compared with the drift velocity in semiconductor devices of $u_D \cong 10^7$ cm/s. The thermal motion of the particles is superimposed by a very small drift velocity, which has no influence on the fluctuation spectrum. Therefore, there is no current-dependent noise from wire resistors. However, if the real resistor as a circuit element consists of a carbon layer, a very thin metal layer or a semiconductor material, noise voltages above the thermal level can occur when current flows.

Technical Resistors

Technical resistors that are used as components, e.g. SMD resistors show in addition to the thermal noise an "excess noise" in the LF range up to about 20 kHz

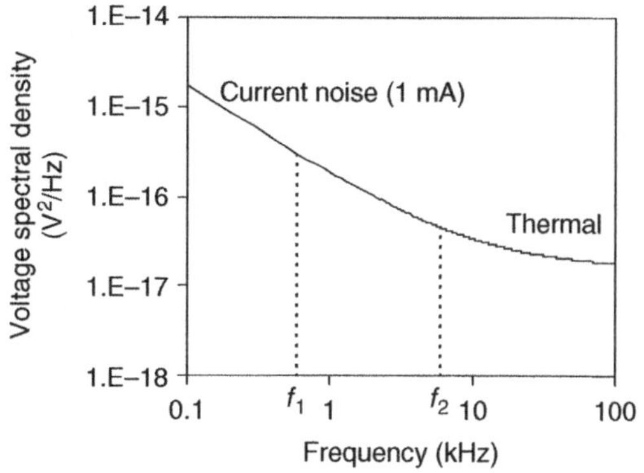

Figure 3.11 Spectral density of the voltage square S_{VC} of a SMD thick film resistor $R = 1\ \text{k}\Omega$, $I = 1$ mA. The $1/f$ noise is for $f < 10$ kHz dominant. f_1, f_2 are the integration limits (1 decade) for (3.42)

with a $1/f$ spectral distribution. An example for the spectral power density S_{VC} of the noise voltage square of a current-carrying SMD thick-film resistor is shown in Figure 3.11.

This noise is caused by the flowing direct current. Its origin can be explained as follows. Thick film resistors consist of a glass frit with ruthenium oxide on a carrier. This is not a homogeneous material and especially at the grain boundaries electrons are captured and released. This is the source of the current noise. In thin film resistors the current noise is about 20 dB lower. They consist of a very homogeneous thin CrNi (Nickel Chromium) layer, which is sputtered onto a ceramic substrate. Remaining noise sources are inhomogeneities, inclusions, and surface imperfections.

The quantitative description is provided by the noise index A. This gives the effective value of the noise voltage v_{EN} at the resistor in microvolts with an applied DC voltage $V_0 = 1$ V. It is clear that a noise voltage can only be specified in connection with a bandwidth. This bandwidth is here a decade, over which the spectral power density in the $1/f$-range must be integrated ($f_2 - f_1$ in Figure 3.11). The noise index A in dB is given by

$$A = 20\,log\left(\frac{v_{EN}}{V_0}\right) \tag{3.40}$$

The units are A in dB, ${}^{v_{EN}}/_{V_0}$ in ${}^{\mu V}/_{V}$. Thus $A = 0$ dB corresponds to an effective value of the noise voltage of $1\ \mu$V, with a current corresponding to $I = {}^{1V}/_{R}$. An example of a manufactorer's data (Panasonic [11]) is shown in Figure 3.12.

The index A is not constant for one size (here 1608), but decreases with increasing volume of the resistive material. This means that low resistance versions have a lower noise index because the current flows more undisturbed in thicker material than in thin layers. With thin-film resistors the overall noise is much lower.

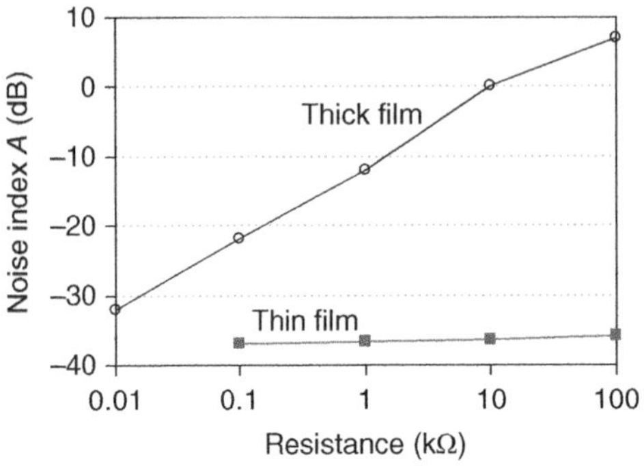

Figure 3.12 Noise index A (dB) of a chip resistor format 1608 (length 1.6 mm, width 0.8 mm).

The noise index A determines the spectral noise power density in the $1/f$-range in Figure 3.11. Its dependence on the DC voltage V_0 is approximately quadratic.

$$S_{VC} = KV_0^2 \times \frac{1}{f} \tag{3.41}$$

K is a device constant that is related to the noise index A as follows.

After (3.2)

$$v_{EN} = \sqrt{S_{VC}B} = \sqrt{\int_{f1}^{f2} KV_0^2 f^{-1} df} \tag{3.42}$$

The bandwidth B is replaced by the integral over the decade from $f_1 \cong 0.6$ kHz to $f_2 \cong 6$ kHz in the $1/f$-range. For the integral with $f_2 = 10 \times f_1$ we obtain:

$$v_{EN} = V_0\sqrt{Kln10} \tag{3.43}$$

With the definition of A (3.37), the relationship between spectral power density, given by K, and noise index A is

$$K = \frac{1}{2.3}\left(\frac{v_{EN}}{V_0}\right)^2 = 0.43 \times \left(10^{\frac{A}{20}} \times 10^{-6}\right)^2 \tag{3.44}$$

Here the different units are to be considered. K refers to V/V, and A refers to μV/V, hence the 10^{-6} in the second term. In [4] one finds K-values, e.g. $K = 1.7 \times 10^{-13}$ for carbon film resistors. This corresponds to a noise index

$$A = 20\,log(\sqrt{2.3K} \times 10^6) = 20\,log(0.625) = -4\text{ dB}$$

Resistors Consisting of Semiconductor Material

In monolithic microwave integrated circuit (MMIC), resistors are often realized by doped semiconductor layers. With ever decreasing dimensions, all components become extremely small. This applies to the transistors as well as to the passive

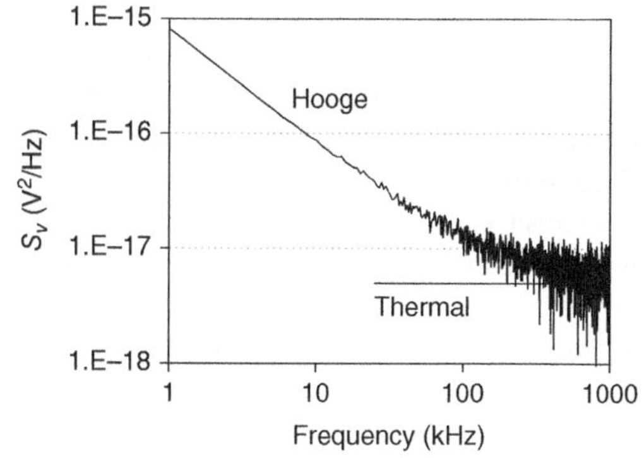

Figure 3.13 Measurement of the spectral density of the noise voltage square of a small resistor made of semiconductor material. (3.42) simulates the measurement with $\alpha = 10^{-5}$.

elements and thus to the ohmic resistors. This leads to the fact that the current is carried by only "few" electrons. If we look at the equivalent circuit diagram of transistors, it can be seen that there is a combination of current or voltage sources and resistors, whereby these parasitic resistors have even smaller dimensions than the circuit elements. This leads to the fact that the current-driven $1/f$ noise of the passive elements can be a significant noise source. This phenomenon is described by the Hooge formula [12].

$$S_{VC} = 4kTR + \frac{\alpha R^2 I^2}{N}\frac{1}{f} \tag{3.45}$$

where α is the dimensionless Hooge constant. Depending on the material its value is 10^{-4}–10^{-5}. N is the total number of electrons in the considered volume (not the density!). The first term gives the always present thermal noise of the resistor R, the second term gives the current noise. R^2I^2 corresponds to the V_0^2 from (3.38). The second term becomes more effective with smaller circuit elements. In Figure 3.13 we show a measurement of the $1/f$ noise on a resistor as it is used in the InGaP/GaAs HBT technology. This resistor consists of the material of the sub-collector. By doping with silicon it contains $n = 5\times10^{18}$ electrons/cm^3, the layer thickness is 0.7 μm. Its lateral dimensions are $w = 5\,\mu$m, $l = 100\,\mu$m. This results in a volume of 3.5×10^{-10} cm^3. The total number of electrons available for current transport, which determine the $1/f$ noise component in (3.42), is $N = 2\times10^9$. The resistance has $R = 300\,\Omega$, the current is $I = 30$ mA.

With a Hooge constant $\alpha = 10^{-5}$, the spectral noise voltage density plotted in Figure 3.13 is S_{VC} according to (3.42) in accordance with a measurement.

Contact Noise

Each resistor material must be connected to the circuit via contacts. There is always a contact resistance. Also in MMIC technology, contact resistances occur in the active and passive components, although here the transitions are metallurgically alloyed. A currentless contact has only thermal noise, which is usually not important. When current flows, LF noise occurs, which usually increases nonlinearly with the current and with the contact resistance. Statistical fluctuations of the contact resistance can be regarded as the cause. The monitoring of contact noise plays a role in quality and lifetime measurement. A strongly noisy contact is of poor quality and will quickly lead to component failure. This can be used in microelectronics as well as in laser technology. Here, for example, inadequate contacts of semiconductor laser diodes are detected by the noise of the operating current.

Again, we are not dealing with white noise here, but with a $1/f$ dependence that is only present at low frequencies. It plays a major role in electroacoustics. It is also important for RF and microwave technology because it can be mixed up into any frequency range by nonlinear processes. Thus it is an important source of the phase noise of an oscillator. It is called pink noise, because in optics a frequency spectrum with a strong low frequency component (red) appears pink.

When calculating the noise voltage according to (3.39), we have to integrate over the frequency, because we do not have a constant spectral power density over the bandwidth B as with white noise.

Generation–Recombination Noise

When analyzing the shot noise, we had assumed that a particle enters a region on one side and after the transit time τ_t disappears on the other side [13]. In semiconductor materials, within this volume the recombination of an electron–hole pair or the temporary capture of electrons in traps can occur. These traps are energy terms in the forbidden band caused by defects. The physical process is not important for statistics. After a dwell time τ_p the electrons are released again and continue to drift. This statistical fluctuation in the number of charge carriers involved in the transport of current is an additional source of noise. We obtain the square of the fluctuation by the following consideration (Figure 3.14).

The mean dwell time of an electron in state 0 is τ_0 in the state 1 it is τ_1. The probability P_0 of finding the electron in state 0:

$$P_0 = \frac{\tau_0}{\tau_0 + \tau_1} \tag{3.46}$$

The total time is $\tau_0 + \tau_1$. The probability to find the electron in state P_1 is $1 - P_0$. The ACF for the fluctuation of this process is given by [13]

$$\rho(\tau) = P_0(1 - P_0)\, exp\left(\frac{-\tau}{\tau_p}\right) \tag{3.47}$$

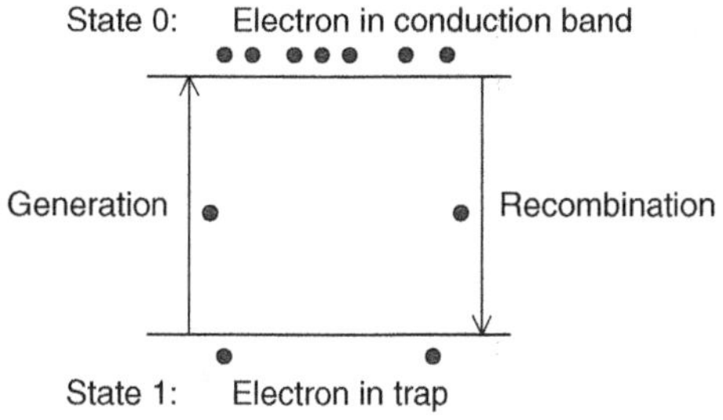

Figure 3.14 Scheme of the exchange of electrons between conduction band and trap.

with the time constant characteristic for this process τ_ρ

$$\frac{1}{\tau_\rho} = \frac{1}{\tau_0} + \frac{1}{\tau_1} \tag{3.48}$$

As is well known, the square of variation is calculated from the ACF for $\tau = 0$ multiplied by the total number of particles (Figure 3.15).

$$\overline{\Delta N^2} = NP_0(1 - P_0) \tag{3.49}$$

The expression applies to both P_0 and P_1. This consideration only applies if the processes run independently of each other. In particular, if there are always free traps in which electrons can be captured. Mostly only conditional transitions are possible in semiconductors, because the possibilities of generation and recombination are not always available for all charge carriers. But the simplest case discussed here already explains the most important effects of the noise behavior. For details see [13].

The transition to the spectral power density $S(f)$ is again made with (2.26).

$$S(f) = 4\int_0^\infty \rho(\tau)\cos(2\pi f\tau)d\tau \tag{3.50}$$

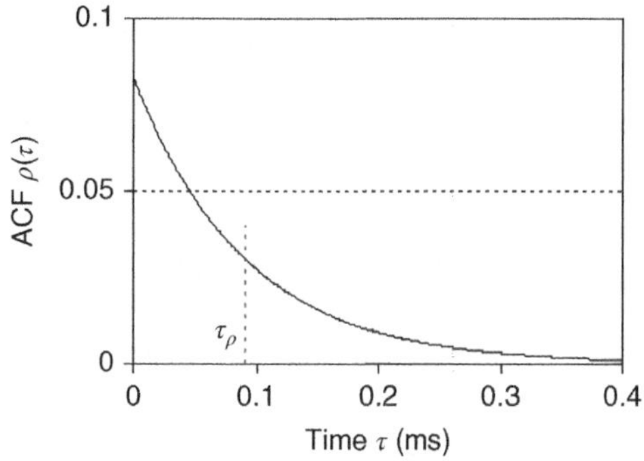

Figure 3.15 Autocorrelation function of generation recombination noise (3.44) for $\tau_0 = 1$ ms and $\tau_1 = 0.1$ ms.

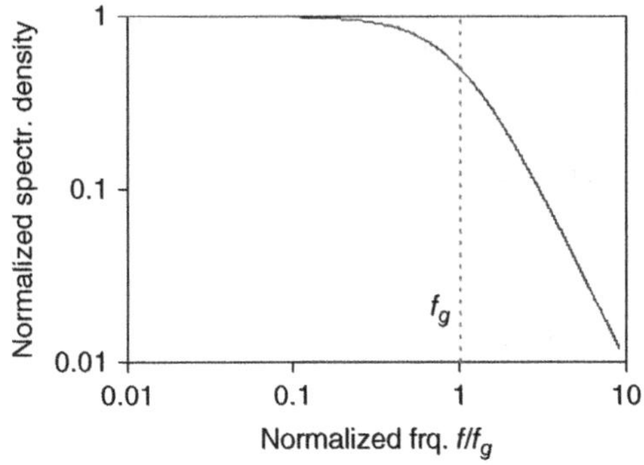

Figure 3.16 Normalized spectral power density of GR noise $S(f)/4\tau_\rho\overline{\Delta N^2}$ after (3.50). The frequency is to f_g normalized.

with (3.44):

$$S(f) = 4NP_0(1 - P_0) \int_0^{\infty} exp\left(-\frac{\tau}{\tau_\rho}\right) cos(2\pi f\tau)d\tau \tag{3.51}$$

The solution of the integral is according to the integral table:

$$\int_0^{\infty} exp(-ax)\,cos(bx)dx = \frac{a}{a^2 + b^2} \tag{3.52}$$

Thus we obtain for the spectral power density of the generation recombination noise a dependence with a typical limiting frequency:

$$S(f) = 4\overline{\Delta N^2}\tau_\rho \frac{1}{1 + \left(\frac{f}{f_g}\right)^2} \tag{3.53}$$

$f_g = {}^1/_{\tau_\rho}$ is the cut-off frequency, which is characteristic for the process.

The GR-noise spectrum has the typical shape of Figure 3.16. At low frequencies it is constant, above the frequency determining the process f_g, there is a steep decay. With the semiconductor material gallium arsenide, this characteristic "shoulder" is often found in the measured LF-noise spectrum. This characteristic suggests a certain contamination of the material or traps at the interface to the passivation layer. If there exists a statistical distribution of the time constants τ_0 and τ_1, a large number of spectra of type (3.50) can overlap to form a $1/f$ spectrum. An example of the superposition is shown in Figure 3.17. This is one way to explain the widespread occurrence of $1/f$ noise.

LF Noise from Transistors

If one measures the spectral density of the noise current square in the output circuit of a microwave transistor of GaAs technology, one obtains a curve as shown in Figure 3.18 [14]. At higher frequencies, the spectrum changes to the

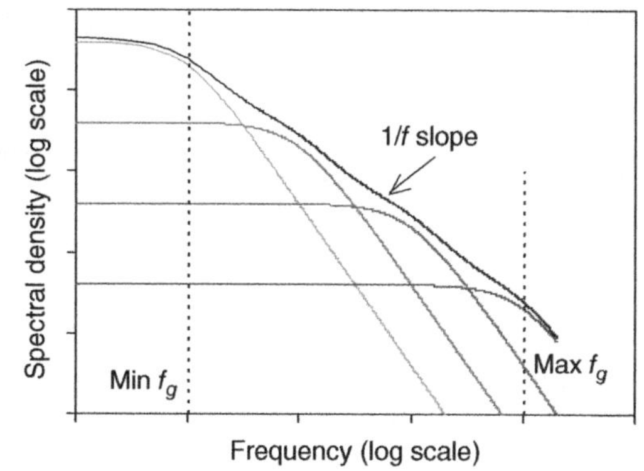

Figure 3.17 Overlay of GR-spectra with different f_g ($\tau_p = 0.1$ s – 0.1 ms) to a $1/f$ spectrum.

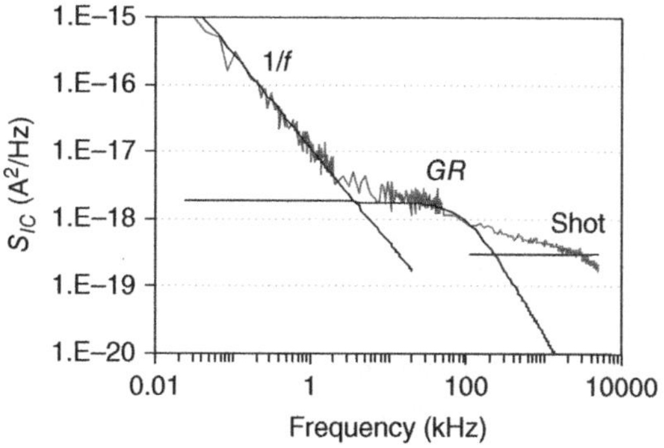

Figure 3.18 Spectral power density S_{IC} of the collector current of a GaInP/GaAs HBT. The three components $1/f$-, *GR*- and shot noise are visible [14].

frequency-independent (white) component of shot or thermal noise. At medium frequencies an additional plateau may appear, the generation recombination noise. At low frequencies there is always a slope with f^{-1}. With Si-BJT the GR component does not appear. The spectrum consists only of the $1/f$ component and the white noise. From this shape stems the definition of the corner frequency f_C which marks the transition between the two ranges. In Figure 3.19 the definition of f_C is schematically represented. The spectrum contains only $1/f$-noise and thermal noise. The amplitude of the $1/f$-noise and thus the frequency f_C up to which it exceeds the thermal noise depends strongly on the technology.

For example, we can specify the following values for f_C of the different technologies (Table 3.1):

The corner frequency is an important parameter for the suitability of different transistor types and materials for certain applications. For oscillator circuits in communications or radar technology, the smallest possible f_C is required to achieve low phase noise. Si-BJT are very good, but often do not have the necessary maximum oscillation frequency for microwave applications. Here SiGe HBT are better suited. GaAs FET are due to the high value of f_C less suitable for oscillator

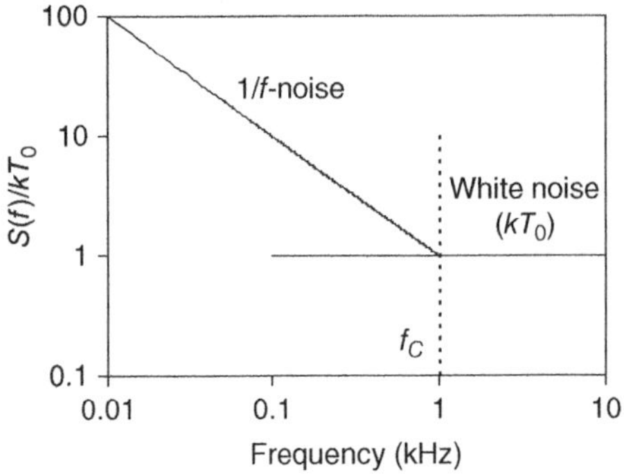

Figure 3.19 Schematic representation of the transition from 1/*f*- to white noise at the corner frequency f_C. The upper limit for Si bipolar transistors is $f_C \cong 1$ kHz.

Table 3.1 Corner frequency f_C of some types of transistors.

Material/device	BJT	HBT	FET
Si	0.1–1 kHz	—	10 kHz–10 MHz
SiGe	—	1–10 kHz	—
GaAs	—	10 kHz–1 MHz	10–100 MHz

circuits, unless phase noise is reduced by special circuit measures. For power applications, LF noise is usually not important.

References

1 Bittel, H. and Storm, L. (1971). *Rauschen*. Springer Verlag.

2 Nyquist, H. (1928). Thermal agitation of electric charge in conductors. *Phys. Rev.* 32: 110–113.

3 Johnson, J.B. (1928). Thermal agitation of electricity in conductors. *Phys. Rev.* 32: 97–109.

4 Landstorfer, F. and Graf, H. (1981). *Rauschprobleme der Nachrichtentechnik*. Oldenbourg Verlag.

5 Kerr, A.R. and Randa, J. (2010). Thermal noise and noise measurements. *IEEE Microwave Mag.* 11: 40–52.

6 Schottky, W. (1918). Über spontane Stromschwankungen in verschiedenen Elektrizitätsleitern. *Ann. Phys.* 57: 541–567.

7 Hull, A.W. and Williams, N.H. (1925). Determination of elementary charge e from measurements of shot-effect. *Phys. Rev.* 25, 147.

8 Meinke, H.H. and Gundlach, F.W. (1992). *Taschenbuch der Hochfrequenztechnik*. Springer Verlag.

9 Mollwo, L. (1958). Elektronentemperatur und Elektronenrauschen in der hochfrequenten Fackelentladung. *Ann. Phys.* 7: 97–129.

10 Mumford, W.W. (1949). A broadband microwave noise source. *Bell Syst. Tech. J.* 28: 608–618.

11 Panasonic (2019). Thin Film Chip Resistors. Datasheet.

12 Hooge, F.N. (1972). Discussion of recent experiments on 1/f-noise. *Physica* 60: 130–144.

13 Müller, R. (1990). *Rauschen*. Springer Verlag.

14 Heymann, P., Rudolph, M., Doerner, R., and Lenk, F. (2001). Modeling of low-frequency noise in GaInp/GaAs hetero-bipolar transistors. In: *IEEE MTT-S, Int. Microwave Symp. Digest*, 1967–1970.

4

Noise and Linear Networks

When passing through a communications circuit or in a measuring system, the input noise is modified. Bandwidth limitations convert white noise into narrowband noise. This leads to a change of the autocorrelation function (Figure 2.6). The amplitude of the noise is increased by amplifiers and decreased by attenuators. The networks add noise components that originate from their own noise. For passive networks (filters) it is the thermal noise of the losses, and for active networks (amplifiers) it is in particular the overthermal noise of the transistors. Assuming linearity of the transmission path, the Gaussian distribution of the amplitudes remains unchanged.

Narrowband Noise

If we filter out a narrow frequency band of width B from white noise, we will see a blurred sine wave in the time domain instead of the statistical fluctuations (Figure 2.5).

In noise modeling of networks one is usually interested in the superposition and interaction of the noise power of different sources within a limited frequency range. As in signal theory, complex pointers can be used for this purpose. The phasor of the narrowband noise signal can be imagined as shown in Figure 4.1.

Frequency and amplitude are not fixed values as with the sinusoidal signal, but fluctuate according to the amplitude noise Δv and the frequency noise determined by the bandwidth of filter B, which is also called phase noise. From the white noise, a sinusoidal-like voltage of amplitude $\bar{v}$ is accentuated.

Calculating with Phasors

Let us repeat here the calculation of the AC power in complex form for the simplest case of sinusoidal signals. In a complex load resistance, current and voltage are out

A Guide to Noise in Microwave Circuits: Devices, Circuits, and Measurement, First Edition.
Peter Heymann and Matthias Rudolph.

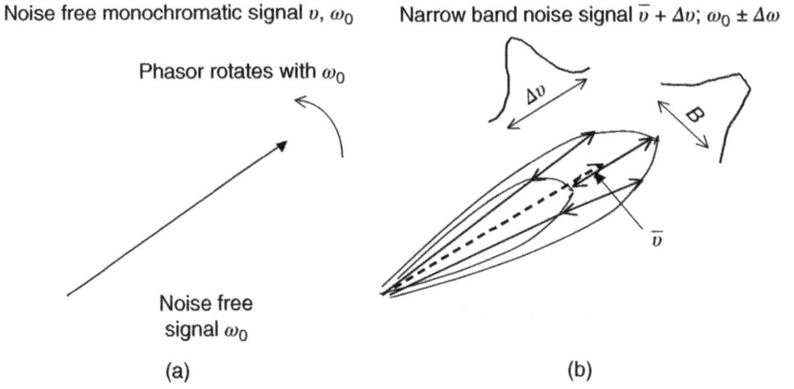

Figure 4.1 (a) Phasor of a monochromatic signal. (b) Phasor of a narrowband noise signal. The voltage is $\bar{v} + \Delta v$, the frequency is $\omega_0 \pm {}^B\!/_2$.

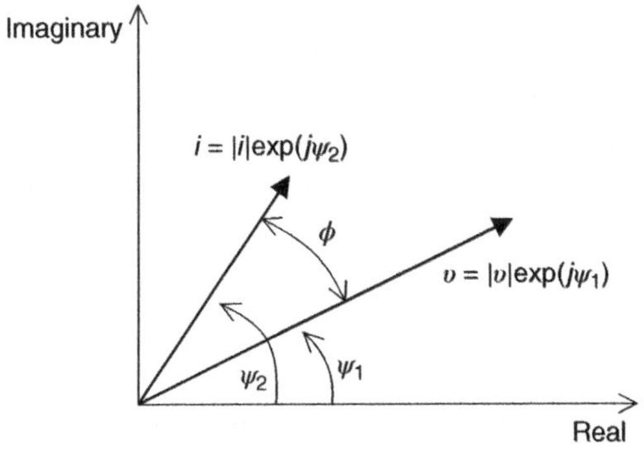

Figure 4.2 Vector diagram of current and voltage in a complex load resistance.

of phase. We represent them by the phasors (4.1) in Figure 4.2.

$$v = |v|\ exp(j\psi_1); \quad i = |i|\ exp(j\psi_2) \tag{4.1}$$

The phase shift between current and voltage is $\phi = \psi_2 - \psi_1$. The instantaneous values are obtained when the phasors rotate with the angular velocity ω_0 and one observes its projection on a reference line, e.g. the real axis.

The time function of the voltage is therefore the real part of the phasor v of Figure 4.2 multiplied by the rotation.

$$v(t) = Re(v\ exp(j\omega_0 t)) \tag{4.2}$$

With the Euler formula

$$exp(jx) = cos\ x + j\ sin\ x$$

$$cos\ x = \frac{1}{2}(\ exp(jx)\ +\ exp(-jx)); \quad sin\ x = \frac{1}{2j}(\ exp(jx)\ -\ exp(-jx)) \tag{4.3}$$

we obtain the well-known time dependence of the sinusoidal voltage

$$v(t) = |v|cos(\omega_0 t + \psi_1) \quad (4.4)$$

In complex notation for voltage and current:

$$v(t) = \frac{1}{2}\{v\, exp(j\omega_0 t) + v^*\, exp(-j\omega_0 t)\}$$
$$i(t) = \frac{1}{2}\{i\, exp(j\omega_0 t) + i^*\, exp(-j\omega_0 t)\} \quad (4.5)$$

The symbol * means the conjugate complex value. For example, $v^* = |v|\; exp\,(-j\psi_1)$.

We can now calculate the instantaneous value of power $p(t)$ in complex notation.

$$p(t) = v(t)i(t) \quad (4.6)$$
$$p(t) = \frac{1}{4}\{vi^* + v^*i + vi\, exp(j2\omega_0 t) + v^*i^*\, exp(-j2\omega_0 t)\} \quad (4.7)$$

For simplification we use the following identities of the complex numbers a and b:

$$a^*b = (ab^*)^*;\;\; a^*b^*\, exp(-j2\omega_0 t) = (ab\, exp(j2\omega_0 t))^*;\;\; a + a^* = 2\, Re(a) \quad (4.8)$$
$$p(t) = \frac{1}{2}Re\{vi^* + vi\, exp(j2\omega_0 t)\} \quad (4.9)$$

The instantaneous power consists of a constant term vi^* and a term oscillating at twice the frequency $vi\, exp(j2\omega_0 t)$.

In context with noise processes we are interested in the time average of the active power

$$P = \overline{p(t)} = \frac{1}{T}\int p(t)dt \quad (4.10)$$

in this case the alternating part vanishes.

$$P = \frac{1}{2}Re(vi^*) \quad (4.11)$$

According to the definition (4.1), v and i are complex phasors whose phase shift must be taken into account. This results in the active power known from alternating current technology to

$$P = \frac{1}{2}|v||i| \times cos\phi \quad (4.12)$$

If the phasors v and i represent effective values, the factor ½ disappears. For sinusoidal alternating current, the relationship between peak value and effective value (RMS value) is $\breve{v} = \sqrt{2}v_{RMS}$.

Let us change from the monochromatic signal to the narrow-band noise signal (Figure 4.1a,b). Now the bandwidth B and the spectral power density $S(f_0)$ become important.

In (2.24) we had defined the spectral power density at the frequency f_0 in general:

$$\sigma^2 = S(f_0)B$$

If $S_i(f_0)$ is the spectral power density of a band-limited noise current, then

$$\overline{i^2} = S_i(f_0)B \tag{4.13}$$

Let's calculate the time average $\overline{i^2}$ from (4.5), then we obtain

$$\overline{i^2} = \frac{1}{2}ii^* \tag{4.14}$$

If i is the RMS value of the current, the average noise power is in a resistor R:

$$P = (ii^*)_{RMS} \times R \tag{4.15}$$

Of course, working with complex noise quantities only makes sense if phasing plays a role. This is at all times the case when two narrowband noise signals of the same center frequency f_0 are superimposed. An example was treated in Chapter 2. Therefore, in complex noise models, one must always work with voltages and currents and not with power, because the phase information is no longer available at the power.

In practice, there exists always band-limited noise. All components of a communication system have a finite bandwidth. In measurement technology, the band is defined by filters. The bandwidth B represents the transmission characteristic of a network. The passband can be described by the transfer function.

This shows the ratio of the complex signals (magnitude and phase) at the output and input of the network. If only voltages or currents are considered, the transfer function is dimensionless. If one relates output voltage to input current, it is a resistance. In general, the following should apply:

Input signal: $x(t) = |x|\ exp\, j(\omega t + \phi_X)$ Output signal: $y(t) = |y|\ exp\, j(\omega t + \phi_Y)$

When passing through a linear network, the frequency does not change. The transfer function $H(j\omega)$ is defined as

$$H(j\omega) = \frac{|y|\ exp\, j\phi_Y}{|x|\ exp\, j\phi_X} \tag{4.16}$$

The amplitudes are amplified or attenuated according to the magnitude of the transfer function

$$|H(j\omega)| = \frac{|y|}{|x|} \tag{4.17}$$

There is also a phase shift

$$arg(H(j\omega)) = \phi(\omega) \tag{4.18}$$

The attenuation of a signal when passing through a passive network is generally given in logarithmic scale (dB):

$$A = -20\ log|H(j\omega)| \tag{4.19}$$

The noise spectrum is proportional to the square of the fluctuation quantity, e.g. the square of the fluctuation current in (4.13), without phase information. When transmitting via a network with the transfer function $H(f)$, the noise spectrum at the output is proportional to the square of the magnitude of $H(f)$.

$$S_{OUT} = |H(f)|^2 S_{IN} \tag{4.20}$$

We have already made use of this in Chapter 3.

Let us consider as an example the transfer through a filter consisting of a parallel resonant circuit. We feed a noise current into the input and consider the noise voltage at the output. The transfer function then has the unit of a resistor. For the sake of completeness we derive the complex resistance of the parallel resonant circuit.

We start with the admittance of the resonant circuit in Figure 4.3.

$$Y = G + j\left\{\omega C - \frac{1}{\omega L}\right\}; \quad G = \frac{1}{R} \tag{4.21}$$

An approximation of the resonance curve is based on the introduction of double detuning. Let us consider the imaginary part.

$$\omega C - \frac{1}{\omega L} = \omega C\sqrt{\frac{L}{L}} - \frac{1}{\omega L}\sqrt{\frac{C}{C}} = \sqrt{\frac{C}{L}}\left\{\omega\sqrt{LC} - \frac{1}{\omega}\frac{1}{\sqrt{LC}}\right\} \tag{4.22}$$

here we can introduce the resonance frequency ω_0

$$\omega_0 = \frac{1}{\sqrt{LC}} \tag{4.23}$$

we obtain for the imaginary part

$$\omega C - \frac{1}{\omega L} = \sqrt{\frac{C}{L}}\left\{\frac{\omega}{\omega_0} - \frac{\omega_0}{\omega}\right\} \tag{4.24}$$

The term in the curly bracket is called double detuning [1].

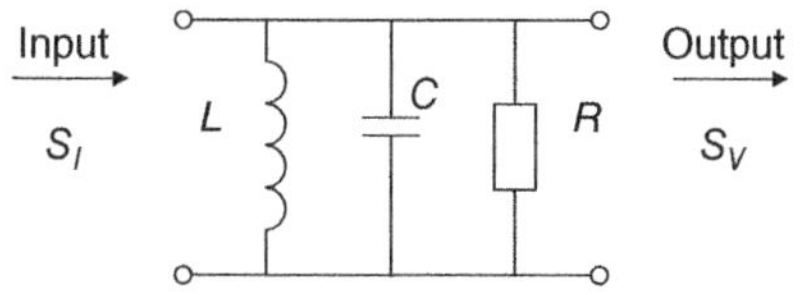

Figure 4.3 Parallel resonant circuit as filter for a noise spectrum. The spectral power density of the noise current S_I at the input is transferred into a spectral power density of the voltage S_V at output.

With the following transformation and the introduction of absolute detuning $\Delta f = f - f_0$

$$\frac{f}{f_0} - \frac{f_0}{f} = \frac{f^2 - f_0^2}{f_0 f} = \frac{(f + f_0)(f - f_0)}{f_0 f} = \frac{(2f_0 + \Delta f)\Delta f}{f_0^2 + f_0 \Delta f} \tag{4.25}$$

we can use for the double detuning close to resonance $\Delta f \ll f_0$:

$$\frac{\omega}{\omega_0} - \frac{\omega_0}{\omega} \cong \frac{2\Delta f}{f_0} \tag{4.26}$$

Furthermore, we introduce the quality factor Q into our admittance formula (4.21)

$$Q = \frac{\omega_0 C}{G} \tag{4.27}$$

and receive:

$$Y = G + j\sqrt{\frac{C}{L}}\frac{2\Delta f}{f_0} \tag{4.28}$$

We can also replace $\sqrt{C/L}$ at the resonance frequency. It is $\sqrt{C/L} = \omega_0 C$.

Thus, the admittance near the resonance is

$$Y = G\left\{1 + jQ\frac{2\Delta f}{f_0}\right\} \tag{4.29}$$

For our approach of the transfer function from an input current to an output voltage, we need the impedance

$$Z = \frac{1}{Y} = \frac{R}{1 + jQ\frac{2\Delta f}{f_0}} \tag{4.30}$$

For the transmission of the noise spectra, we use the square of the magnitude

$$|H|^2 = |Z|^2 = \frac{R^2}{1 + \left(2Q\frac{\Delta f}{f_0}\right)^2} \tag{4.31}$$

and obtain for a white current spectrum $S_I(f)$ fed into the network a bandlimited voltage spectrum $S_V(f)$ at the output:

$$S_V(f) = S_I(f)\frac{R^2}{1 + \left(2Q\frac{\Delta f}{f_0}\right)^2} \tag{4.32}$$

The change of the autocorrelation function was discussed in Chapter 2.

Noise Source with Complex Internal Resistance

The simplest case of a network is the one-port. As usual in circuit theory, we define two cases of noise sources as a one-port. The voltage source with the spectral power density $S_V = 4kTR$ and the current source with the spectral power density $S_I = 4kTG$. Used in more complex noise equivalent circuits, they do not affect the network. The voltage source has the internal resistance "0," the current source has the resistance "∞." Therefore the current source feeds a certain noise current into a circuit, no matter how high its resistance is. In practice they only exist together with an internal resistance R or a conductance G. The equivalent circuit is shown in Figure 4.4 [2].

Both representations are equivalent. If a noise voltage source is given by S_V and R, then the current source results with the same properties:

$$G = \frac{1}{R}; \quad S_I = \frac{S_V}{R^2} \tag{4.33}$$

To consider the noise behavior of a complex resistor we start with the power transfer from a complex source to a complex load. In Figure 4.5 both are connected via an ideal bandpass of bandwidth B to extract a finite power from the white noise spectrum.

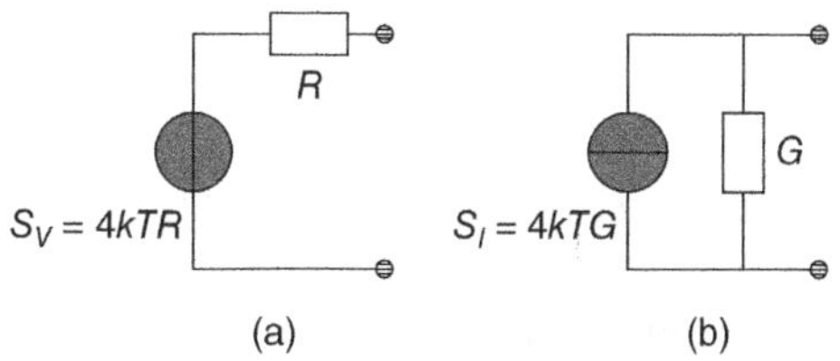

Figure 4.4 Equivalent circuit of a noise voltage source (a) and a noise current source (b). R and G are noise-free.

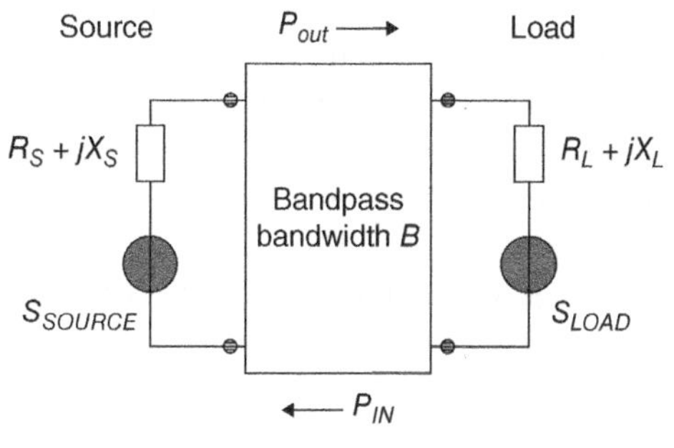

Figure 4.5 Power transfer from a complex source to a complex load over an ideal bandpass.

Let us first assume a sinusoidal source with $S_{SOURCE} = \overline{v_{RMS}^2}$ and set $S_{LOAD} = 0$. The active power transmitted to the complex load $R_L + jX_L$ is

$$P_{OUT} = \frac{\overline{v_{RMS}^2} R_L}{(R_S + R_L)^2 + (X_S + X_L)^2} \tag{4.34}$$

In a thought experiment we can now show that the real part is responsible for the noise of a complex impedance. We assume thermal noise for the source with $S_{SOURCE} = 4kTR_S$. The load is a noisy complex impedance with $Z_L = R_L + jX_L$. The spectral power density S_{LOAD} is to be determined. The outgoing noise power, which we now call N, is according to (4.34):

$$N_{OUT} = \frac{4kTR_S B}{|R_S + Z_L|^2} \times Re(Z_L) \tag{4.35}$$

The incoming power is

$$N_{IN} = \frac{S_{LOAD} B}{|R_S + Z_L|^2} \times R_S \tag{4.36}$$

In thermodynamic equilibrium both powers must be equal.

$$N_{OUT} = N_{IN}; \quad S_{LOAD} R_S = 4kTR_S\, Re(Z_L) \tag{4.37}$$

From this follows for the spectral power density of the noise voltage of a complex impedance Z:

$$S_V = 4kT\, Re(Z) \tag{4.38}$$

The Equivalent Noise Bandwidth

In measurement technology, the noise spectrum is limited by a filter. In spectrum analyzers or noise figure analyzers it is the input filter. It is characterized by the resolution bandwidth (RBW). It is assumed that the input spectrum $S_{IN}(f)$ is frequency independent (white) in the passband of the filter. At the output we measure the square of the fluctuation as an integral over the passband curve according to (2.24). By introducing the equivalent noise bandwidth B_{EQ}, we are independent of the shape of the transmission curve. For this purpose, the real transmission curve $|H(f)|^2$ in Figure 4.6 is replaced by an area equal rectangle of the width B_{EQ} and the height $|H(f)|^2_{MAX}$ [3].

The equivalent noise bandwidth is given by

$$B_{EQ} = \frac{\int_0^\infty |H(f)|^2 df}{|H(f)|^2_{MAX}} \tag{4.39}$$

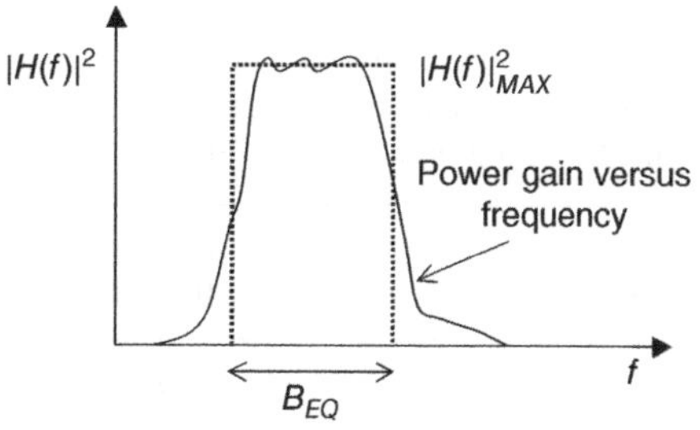

Figure 4.6 The transfer characteristic of an RBW filter is replaced by the equivalent noise bandwidth B_{EQ}.

The fluctuation square at the output, i.e. the measurable noise power, is then corresponding to (2.24) and (4.39).

$$\sigma^2 = S_{IN}|H(f)|^2_{MAX}B_{EQ} \tag{4.40}$$

If we consider, for example, as in Figure 4.3, the noise current at the input and at the output the noise voltage at a load resistor R_L, then the following applies

$$N_N = S_{I,IN}|H(f)|^2_{MAX}B_{EQ} \times \frac{1}{R_L} \tag{4.41}$$

The characterization of a noise measurement system consists in the determination of the quantity $|H(f)|^2_{MAX}B_{EQ}$.

The 3 dB bandwidth usually used for frequency dependencies does not agree with the noise bandwidth. We show this using the example of the parallel resonant circuit. According to (4.39) we have to integrate over the absolute square of the transfer function (4.31).

$$B_{EQ} = \int_0^\infty \frac{df}{1+\left(\frac{2Q}{f_0}\right)^2(\Delta f)^2} = \int_0^\infty \frac{df}{1+\left(\frac{2Q}{f_0}\right)^2(f^2-2ff_0+f_0^2)} \tag{4.42}$$

This integral is of type

$$\int \frac{dx}{X} = \frac{2}{\sqrt{D}}arctan\left(\frac{2ax+b}{\sqrt{D}}\right) \tag{4.43}$$

$$X = ax^2+bx+c; \quad D = 4ac-b^2; \quad a = \frac{4Q^2}{f_0^2}; \quad b = -\frac{8Q^2}{f_0}; \quad c = 4Q^2+1$$

The values of the antiderivative at the integration limits are $-\pi/2$ $(f = 0)$ and $+\pi/2$ $(f = \infty)$ because the zero crossing of the arctan function is at f_0.

This results in

$$B_{EQ} = \frac{2f_0}{4Q} \times \pi = \frac{f_0}{Q} \times \pi/2 = B_{3dB} \times \pi/2 \tag{4.44}$$

with the definition of the 3 dB bandwidth $B_{3dB} = f_0/Q$.

Table 4.1 Ratio of the equivalent noise bandwidth to the 3 dB bandwidth for bandpass filters in spectrum analyzers and noise test receivers.

Filter type	B_{EQ}/B_{3dB} **(linear)**
1-Pole	1.57
4-Poles	1.129
5-Pole	1.114
Gauss digital	1.065

For the multisection filters generally used in spectrum analyzers and test receivers, different ratios of B_{3dB} and B_{EQ} naturally apply. Table 4.1 gives an overview.

As already mentioned, the autocorrelation function changes with the spectrum. One can use a noise time constant τ_ρ, which determines the exponential decay of the ACF. However, the noise bandwidth is not used for this, but B_{3dB}.

$$\tau_\rho = \frac{Q}{\omega_0} = \frac{1}{2\pi B_{3dB}} \tag{4.45}$$

With ideal white noise is $\tau_\rho = 0$. Depending on the bandwidth of the filter, a finite correlation time constant is generated by passing the filter. In a cavity resonator of very high Q, a high quality oscillator signal can be produced from powerful white noise.

$$\rho(\tau) = exp\left(-\frac{\tau}{\tau_\rho}\right) cos(\omega_0\tau) \tag{4.46}$$

Network Components at Different Temperatures

The Nyquist formula is not only an approximation of Planck's radiation law, but also a specification of the fluctuation–dissipation theorem, which was developed from the study of Brownian molecular motion to a fundamental principle of statistical physics. The fluctuations of a particle suspended in a liquid have the same origin as the frictional force that must be overcome to move the particle in a certain direction [2].

Simply said, the following applies to our context. The particle is an electron in a resistive material. The frictional force corresponds to the ohmic resistance. The fluctuations are thermal noise. One consequence is that noise emission and

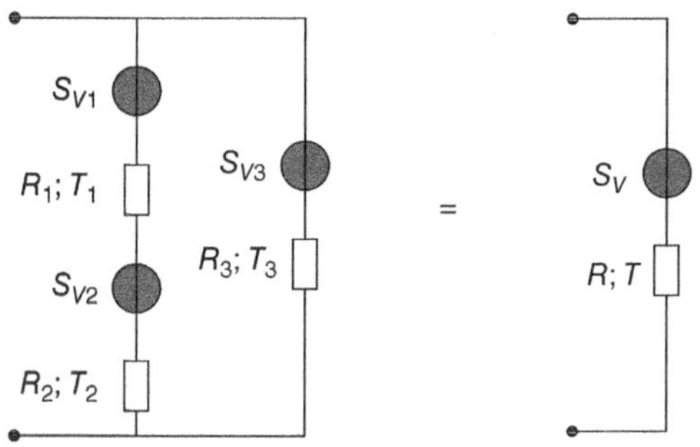

Figure 4.7 R_1, R_2, R_3 at different temperatures forming a noisy one-port. They are replaced by one resistor at temperature T.

absorption refer to the real part of a complex impedance (4.38). Furthermore, for a network with several resistors at different temperatures, this means that the resistor that contributes the most to attenuation also provides the largest part of the noise power.

Let us consider the network (Figure 4.7) with three resistors R_1, R_2, and R_3, which are at different temperatures. It can be replaced by a resistor R, whose spectral noise power density results from the following consideration.

The noise sources are not correlated, so the individual powers add up to the total power. This superimposition must take place at the input plane of the one-port so that the equivalent circuit (right) has the same noise characteristics as the original network (left). For this purpose, the spectral noise voltage densities S_1, S_2, and S_3 can be converted individually to the input plane. According to the procedure with the transfer function (4.20), we calculate the noise voltage square that the source in question produces in the input plane when the other sources are short-circuited. In our case the result is

$$S_{V1}' = S_{V1}\left(\frac{R_3}{R_1 + R_2 + R_3}\right)^2; \quad S_{V2}' = S_{V2}\left(\frac{R_3}{R_1 + R_2 + R_3}\right)^2;$$
$$S_{V3}' = S_{V3}\left(\frac{R_1 + R_2}{R_1 + R_2 + R_3}\right)^2 \tag{4.47}$$

The resulting source is the sum of the transformed sources

$$S_V = S_{V1}' + S_{V2}' + S_{V3}'$$
$$S_V = 4k\frac{T_1R_1R_3^2 + T_2R_2R_3^2 + T_3R_3(R_1 + R_2)^2}{(R_1 + R_2 + R_3)^2} \tag{4.48}$$

In case of equal temperature distribution $T_1 = T_2 = T_3 = T$ the obvious result is that the resulting resistance R has the noise temperature T.

$$S_V = 4kT\frac{R_3(R_1 + R_2)}{R_1 + R_2 + R_3} = 4kTR \tag{4.49}$$

As mentioned earlier, the noise contribution of a resistor in the network corresponds to the portion it would absorb from an input power. Let us look at the resistor R_3 in the network Figure 4.7. With a sinusoidal terminal voltage V the total power P is

$$P = \frac{V^2}{R} = \frac{V^2(R_1 + R_2 + R_3)}{R_3(R_1 + R_2)} \tag{4.50}$$

The power in the resistor R_3 is $P_3 = {}^{V^2}/_{R_3}$. The part of P_3 of the total output is

$$\frac{P_3}{P} = \frac{R_1 + R_2}{R_1 + R_2 + R_3} \tag{4.51}$$

This corresponds to the result in (4.49). At the same temperature R_3 contributes exactly this amount to the total noise.

The relationship between absorption and noise emission becomes particularly clear when a noise signal passes through attenuators or lossy transmission paths, e.g. cables. Here we consider the wave aspect, similar to the passage of a light beam through an optical medium. Figure 4.8 shows the partition of the incoming noise power N_{IN} into three parts: Reflection Re, Absorption Ab, and Transmission Tr. If we normalize to N_{IN}, then

$$Re + Ab + Tr = 1 \tag{4.52}$$

In a passive two-port (Figure 4.8) $N_{OUT} < N_{IN}$ and $Ab > 0$. The signal is attenuated. If the two-port is active, amplification takes place, i.e. power is supplied from another source and (4.52) does not apply. Attenuation and gain are generally specified in dB. The relationship between attenuation (dB) and linear absorption Ab with neglect of reflection ($Re = 0$) is given by

$$N_{OUT} = TrN_{IN} \tag{4.53}$$

because of $Re = 0$, (4.50) becomes $Tr = 1 - Ab$. It follows

$$\alpha(\text{dB}) = 10\, log\left(\frac{1}{1 - Ab}\right) \quad Ab = 1 - 10^{\frac{-\alpha(\text{dB})}{10}} \tag{4.54}$$

Let us consider an amplifier cooled to LN_2 temperature (Figure 4.9). It is located in a cryostat at $T_1 = 77$ K. It is connected by a cable to a noise measuring device.

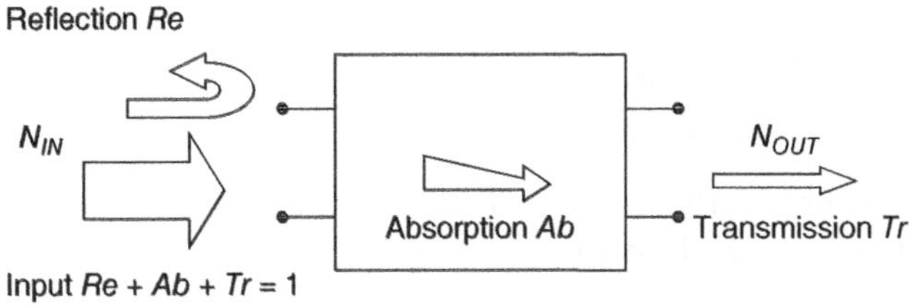

Figure 4.8 A noise power of the value $N_{IN} = 1$ is split into the parts: reflection *Re*, absorption *Ab* and transmission *Tr* when passing through a two-port.

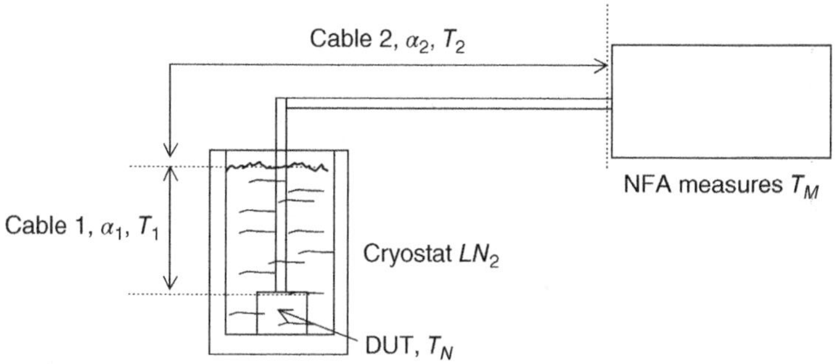

Figure 4.9 Change in noise temperature due to losses at different temperatures. The cable between DUT and NFA is partially immersed in LN_2.

One part of the cable is also cooled, and the other part is at room temperature. The amplifier is well matched and has the characteristic values: gain $G = 10\,\text{dB}$ and noise figure $NF = 0.5\,\text{dB}$. At the input it is terminated with a $50\,\Omega$ resistor at T_1. Its noise power at the output corresponds to a noise temperature T_N.

$$N_{OUT} = kBG(T_1 + T_E); \quad T_E = 290(F - 1); \quad T_N = 1120\,\text{K}$$

The cable has the attenuation values $\alpha_1 = 1$ dB, $\alpha_2 = 2$ dB. Reflection losses are not considered.

According to the dissipation theorem, every lossy component delivers a noise power that corresponds to these losses. The power balance of the setup in Figure 4.9 is therefore as follows:

1 From the available noise power at the output of the DUT $N_{OUT} = kT_N B$ a part is attenuated by the entire cable.

$$(1 - Ab_{1,2}) \times T_N = 0.499 \times 1120\,\text{K} = 559\,\text{K}$$

2 Cable 1 generates noise with T_1, but 2 dB of this is attenuated by cable 2.

$$(1 - Ab_2)Ab_1 \times T_1 = (1 - 0.369) \cdot 0.206 \times 77\,\text{K} = 10\,\text{K}.$$

3 Cable 2 generates noise with T_2.

$$Ab_2 \times T_2 = 0.369 \times 300\,\text{K} = 111\,\text{K}$$

The sum of these results in a noise temperature T_M at the input of the measuring device:

$$T_M = 560\,\text{K} + 10\,\text{K} + 111\,\text{K} = 681\,\text{K}$$

It can be seen that a very careful de-embedding is necessary to deduce the noise temperature of the DUT $T_{DUT} = 1120$ K from the measured noise temperature of $T_M = 681$ K.

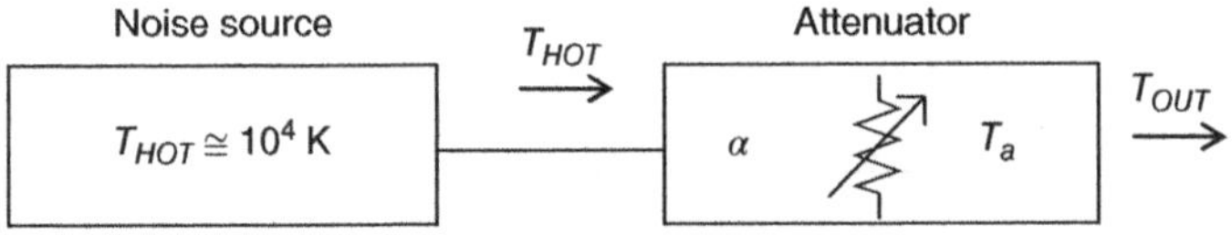

Figure 4.10 Noise source with attenuator. The attenuator reduces the temperature of the source and adds its own noise. The linear attenuation is $\alpha = (1 - Ab)^{-1}$.

Noise Generator and Attenuator

In measurement technology, the configuration is often as shown in Figure 4.10. The power of a noise generator is varied by means of an attenuator. The temperature T_{OUT} appears at the output of the attenuator. It is a combination of the noise generator power and the inherent noise of the attenuator.

The noise temperature of the source is T_{HOT}. In the case of semiconductor noise diodes, for example, $T_{HOT} \cong 10^4$ K. The attenuator is at the ambient temperature T_a. With the considerations of absorption Ab and transmission Tr, we can calculate the output temperature T_{OUT}.

$$T_{OUT} = TrT_{HOT} + AbT_a = (1 - Ab)T_{HOT} + AbT_a \tag{4.55}$$

with (4.54)

$$T_{OUT} = \frac{1}{\alpha}(T_{HOT} - T_a) + T_a \tag{4.56}$$

In terms of attenuation α in dB:

$$T_{OUT} = \frac{T_{HOT} - T_a}{10^{\frac{\alpha(\mathrm{dB})}{10}}} + T_a \tag{4.57}$$

With negligible attenuation ($\alpha = 1$, i.e. 0 dB), we have the full power of the noise generator. With high attenuation we finally have only the noise of the attenuator with its physical temperature T_a.

References

1 Hartnagel, H., Quay, R., Rohde, U.L., and Rudolph, M. (2022). *Fundamentals of RF and Microwave Techniques and Technologies*. Springer Verlag.
2 Schiek, B., Rolfes, I., and Siweris, H.-J. (2006). *Noise in High-Frequency Circuits and Oscillators*. Wiley Interscience.
3 Rauscher, C. (2007). *Grundlagen der Spektrumanalyse*. München: Rohde & Schwarz.

5

Nonlinear Networks

Mixing

A nonlinear network can modify both the frequency distribution and the amplitude distribution of a signal. If there is a band-limited noise spectrum, mixed products outside the passband will be produced at the nonlinear characteristic of a rectifier. The frequencies within the bandwidth B, i.e. $f = {}^{\omega_0}/_{2\pi} \pm {}^{B}/_{2}$ generate sum and difference frequencies near $f = 0$ and $f = 2 \times {}^{\omega_0}/_{2\pi}$. The origin of the triangular spectrum of difference frequencies can be depicted as follows. A vertical arrow in Figure 5.1 represents the spectral power density of a very narrow, "monochromatic noise band." It has the frequency f_1 and mixes with a closely adjacent, equally narrow band f_2 to form the difference frequency $f = f_2 - f_1$. Such pairs are found within the bandwidth B the more, the smaller the frequency difference f is (transformation path 1). If f becomes larger, fewer and fewer pairs are found that fulfill the condition (transformation path 3). The maximum value for f is $f_2 - f_1 = B$. Beyond that there are no more pairs to combine. This is reason for the generation of the triangular LF spectrum from $f = 0$ to the frequency $f = B$ [1].

Band-Limited RF Noise at Input

Before we discuss details of the nonlinear effect of detectors, we want to show how narrowband noise at the center frequency f_0 within the bandwidth $B \lesssim {}^{f_0}/_{5}$ can be treated mathematically in an appropriate way. The basic procedure of describing a non-sinusoidal voltage in the time domain as a Fourier series is generally well known. A common form is the representation with the spectrum of the cosine and sine functions [2].

$$v(t) = V_0 + \sum_{n=1}^{\infty}(a_n \, cos(n\omega t) + b_n \, sin(n\omega t)) \tag{5.1}$$

A Guide to Noise in Microwave Circuits: Devices, Circuits, and Measurement, First Edition.
Peter Heymann and Matthias Rudolph.

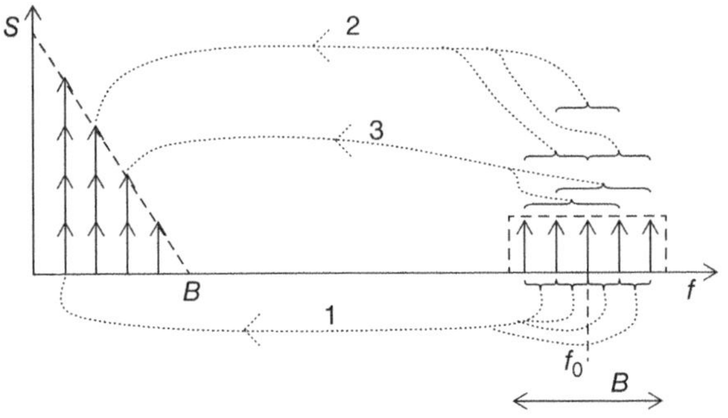

Figure 5.1 "Monochromatic noise bands" (vertical arrows) within B form pairs that are mixed to the difference frequency $f < B$. The result is a triangular LF spectrum below B.

V_0 is the DC component. The Fourier representation is also suitable for stochastic signals although these do not have the defined periodicity $T = {}^{2\pi}/_{\omega}$. For calculating with noise voltages, however, the equivalent representation with the amplitude and phase spectrum (5.1) is advantageous.

$$v(t) = V_0 + \sum_{n=1}^{\infty}(A_n\ cos(n\omega t - \phi_n)) \tag{5.2}$$

it is

$$A_n = \sqrt{a_n + b_n}; \quad \phi_n = arctan\frac{b_n}{a_n}$$

We can transform (5.2) so that a band-limited noise signal is described, which has passed a filter with a center frequency of ω_0 and bandwidth B, at whose input a white noise spectrum is present.

Instead of the harmonic spectrum with the frequencies $n \times \omega$ we introduce the statistically distributed frequencies ω_n within the passband B in its relation relative to the center frequency ω_0.

$$cos(n\omega t - \phi_n) = cos((\omega_n - \omega_0)t + \omega_0 t - \phi_n) \tag{5.3}$$

As we have already seen in Chapter 4, the white noise at the input creates a noisy phasor, which rotates with ω_0. Real and imaginary parts are usually designated I, Q. Real part I for "in phase" and imaginary part Q for "in quadrature." To avoid confusion with the DC-current I, we use $v_I(t)$ and $v_Q(t)$. Both are uncorrelated noise voltages with a Gaussian distribution of the amplitudes.

$$v_I(t) = \sum_{n=1}^{\infty} A_n\ cos[(\omega_n - \omega_0)t - \phi_n]; \quad v_Q(t) = \sum_{n=1}^{\infty} A_n\ sin[(\omega_n - \omega_0)t - \phi_n] \tag{5.4}$$

The phasor of the band-limited noise voltage is then

$$v(t) = v_I(t)cos(\omega_0 t) - v_Q(t)sin(\omega_0 t) \tag{5.5}$$

In analogy to (5.2) we can also write

$$v(t) = R(t)cos(\omega_0 t + \phi(t)) \tag{5.6}$$

with $R^2(t) = v_I^2(t) + v_Q^2(t)$ and $tan\phi = {}^{v_Q}/_{v_I}$. Here $R(t)$ is the fluctuating envelope of the noise signal with the unit of a voltage that is distributed besides ω_0. For the calculation of mean values required in the following concerning this envelope, it must be noted that for a two-dimensional random vector (v) whose components have a Gaussian distribution, the magnitude has a Rayleigh distribution (Figure 5.2a). This results in the following values:

$$\overline{R^2(t)} = 2\overline{v^2(t)}; \quad \overline{R(t)} = \sqrt{\frac{\pi}{2}\overline{v^2(t)}}; \quad \left(\overline{R(t)}\right)^2 = \frac{\pi}{4}\overline{R^2(t)} \tag{5.7}$$

For the mean values of the ${}^{v_I}/_{v_Q}$-components the phase ϕ_n does not play any role, because with noise signals all phases are equally represented and cancel themselves.

$$\overline{v^2(t)} = \overline{v_I^2(t)} = \overline{v_Q^2(t)} = \int_B S(f)df; \quad \overline{v_I(t)v_Q(t)} = 0 \tag{5.8}$$

For the calculations of the detection of a noise signal on a detector characteristic, the generalized mean values are also required. As we have seen in Chapter 4, the correlation coefficients have limited values depending on the filter curve. According to its characteristic the detector responds to the signal (5.6). The linear detector measures $\sqrt{\overline{R^2(t)}}$ the quadratic detector measures $\overline{R^2(t)}$.

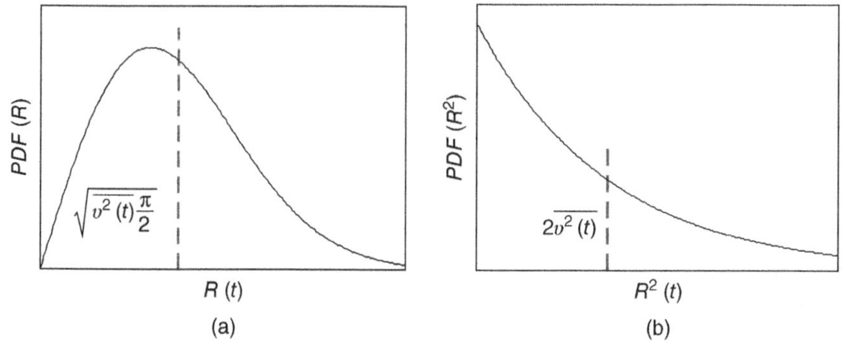

Figure 5.2 Probability distribution of the noise amplitudes behind a detector. (a) Linear characteristic. (b) Quadratic characteristic. These distributions form the basis for averaging. $\overline{R(t)} = \sqrt{\overline{v^2(t)}\,{}^{\pi}/_{2}} = \sigma\sqrt{{}^{\pi}/_{2}}; \overline{R^2(t)} = 2\overline{v^2(t)} = 2\sigma^2$.

Amplitude Clipping

If a network is nonlinear in terms of amplitude, too, it clips the peaks of the transmitted voltage. An example is an amplifier whose transfer characteristic is driven into the nonlinear range by a too strong signal. This can also happen with a highly amplified noise signal. We consider a setup for noise measurement consisting of an amplifier and a quadratic detector. The amplitude distribution, the mean value of which is determined, corresponds to the square of the Rayleigh distribution in Figure 5.2. However, this only applies in the linear range of the amplifier. If the highest amplitudes are clipped, the measured value $\overline{v_m^2}$ is less than the mean value of the Rayleigh distribution (5.7). We calculate the mean values displayed by a square detector. The distribution function Figure 5.2b applies. For the ideal case of the completely linear amplifier, the following applies

$$\overline{v^2} = \frac{1}{2}\int_0^{\infty} R^2\, exp\left(-\frac{R^2}{\overline{R^2}}\right) dR^2 \tag{5.9}$$

All amplitudes from 0 to ∞ contribute to the mean value. Again, R is the envelope of the band-limited noise (5.6).

In case the amplifier cuts off all amplitudes above R_C, the upper integration limit is R_C

$$\overline{v_M^2} = \frac{1}{2}\int_0^{R_C} R^2\, exp\left(-\frac{R^2}{\overline{R^2}}\right) dR^2 \tag{5.10}$$

so that $\overline{v_M^2} < \overline{v^2}$.

The integrals are of the type

$$\int x\, exp(-x)\, dx = -(1+x)\, exp(-x) \tag{5.11}$$

This allows us to estimate up to which voltage the amplifier must be linear if the error of the measurement is less than 1%. Figure 5.3 shows the normalized difference

$$\Delta = \overline{v^2}-\overline{v_M^2}\Big/\overline{R^2}$$

of true value $\overline{v^2}$ and measured value $\overline{v_M^2}$.

We have 1% error ($\Delta = 0.01$) on $R_C^2/\overline{R^2} = 6.6$, i.e. with a clipping of $R_C = 3.6\sqrt{\overline{v^2}}$. This is 3.6 times the effective value of the noise voltage. If one considers the power up to which the amplifier should be linear (5.7), $R_C^2 = 6.6 \times 2\overline{v^2}$, i.e. about 11 dB above the average power of the noise signal.

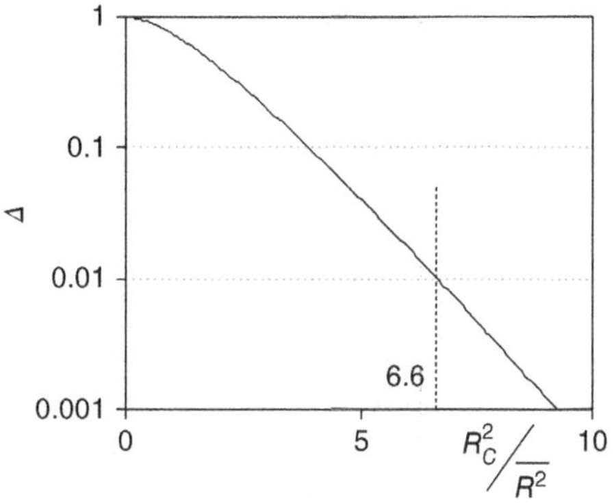

Figure 5.3 Error of a noise measurement due to amplitude clipping with R_C in a measuring system with quadratic detector.

The Detector as a Nonlinear Network

The simplest nonlinear network is the detector diode. Usually the processing of a noisy signal or noise is done by analog data processing with semiconductor diodes in the LF range up to about 20 MHz. In digital data processing the LF signal is digitized with fast AD converters and further processed by computer. Here one has all possibilities of bandwidth selection and subsequent calculation of the quantities of interest. Of particular advantage is the mathematical realization of the Gaussian filter with its optimally defined noise bandwidth (see Chapter 4) [1].

Whether analog or digital, in each case the incoming signal is rectified. In case of noise, the mean square of fluctuation is displayed. As a glance at the display of a spectrum analyzer shows, this mean value is superimposed by a fluctuation. We will deal with this "noise from noise" by using detection with a diode.

Let us first consider the diode as a rectifier of the noise voltage. If we want to measure power, it seems to be particularly suitable, since it is known as a square-law rectifier, at least for small power levels. On the other hand the diode is also used as a linear rectifier. However, if we look at the IV-characteristic of a silicon semiconductor diode, we primarily see neither linear nor quadratic but exponential behavior.

$$I_D = I_S \left\{ exp\left(\frac{V}{2V_{TH}}\right) - 1 \right\} \tag{5.12}$$

I_S is the reverse current (some pA). V is the voltage applied at the pn junction, $V_{TH} = {}^{kT}/_{q}$ is the thermal voltage (26 mV @ 300 K). The 2 in the denominator is typical for Si.

At very low voltage V, the series expansion can be used:

$$exp(x) = 1 + x + \frac{x^2}{2} + \cdots \tag{5.13}$$

The IV characteristic is then determined with good accuracy by

$$I_D \cong I_S \left(\frac{V^2}{2V_{TH}^2} \right) \tag{5.14}$$

It is therefore approximately a quadratic one.

At higher power, the linear behavior is caused by the fact that, due to the voltage drop at the parasitic series resistance R_S of the diode, the externally applied voltage no longer reaches the pn junction completely. The characteristic is therefore sheared at the series resistance and is thus linearized. We have instead of (5.12):

$$I_D = I_S \left\{ exp\left(\frac{V_{IN} - I_D R_S}{2V_{TH}} \right) - 1 \right\} \tag{5.15}$$

after translation

$$ln\left(\frac{I_D}{I_S} + 1 \right) = \frac{1}{2V_{TH}} (V_{IN} - I_D R_S) \tag{5.16}$$

one can see the linearization due to the increasing influence of R_S in the second term

$$V_{IN} = 2V_{TH} ln\left(\frac{I_D}{I_S} \right) + I_D R_S \tag{5.17}$$

For $R_S > 5\ \Omega$ we have practically ohmic behavior in the passband, i.e. a linear IV characteristic.

$$I_D \cong \frac{V_{IN}}{R_S} \tag{5.18}$$

Quadratic and linear rectifications are therefore to be understood as limiting cases of the IV-characteristic (5.12) and (5.15), respectively. Figure 5.4 shows

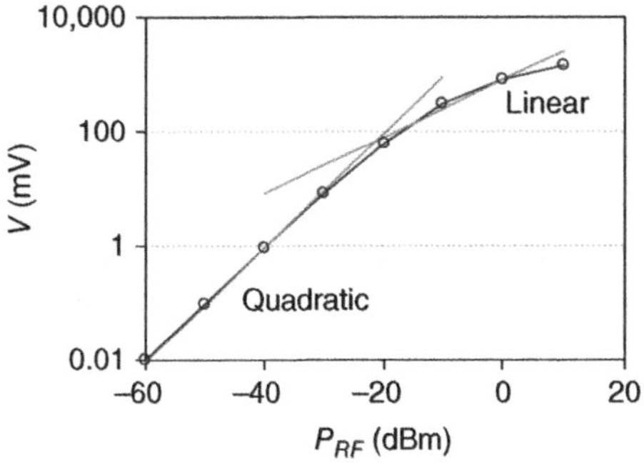

Figure 5.4 Output voltage at the load resistor 1 MΩ versus the RF power of a Si-Schottky diode. Quadratic rectification for $P_{RF} < -30$ dBm, linear for -20 dBm $< P_{RF} < 0$ dBm.

a measuring curve from the data sheet of a typical zero bias Si-Schottky diode (Metelics MSS 20) [3]. The points were measured at $f = 10\,\text{GHz}$. The diode is usually matched at $P_{RF} = -30\,\text{dBm}$. Especially the square part at low input power is very well approximated.

The Noise Spectrum Behind a Quadratic Detector

We look at the quadratic detector in detail because the calculation is clearer than with the linear detector. Moreover it is an example of working with autocorrelation function and frequency spectrum [4–6] (Figure 5.5).

We apply a sinusoidal signal of amplitude V at frequency ω_0 and a superimposed noise signal of bandwidth B symmetrically to ω_0 to the detector.

$$v(t) = (v_I(t) + V)\, cos(\omega_0 t) - v_Q(t)\, sin(\omega_0 t) \tag{5.19}$$

The current of the detector is $i(t) = \kappa v^2(t)$. We set $\kappa = 1AV^{-2}$, but due to this normalization we do not have a correct unit of spectral power density $S(f)$ during the analysis.

The rectified signal is now the current $i(t) = \langle v^2(t)\rangle$. The arithmetical averaging is carried out over a period of ω_0.

$$i(t) = (v_I + V)^2\, cos^2(\omega_0 t) - (v_I + V)v_Q\, sin(2\omega_0 t) + v_Q^2\, sin^2(\omega_0 t) \tag{5.20}$$

the linear term in the center yields no contribution when averaging over a period of ω_0.

$$i(t) = (v_I^2 + 2v_I V + V^2)\, cos^2(\omega_0 t) + v_Q^2 sin^2(\omega_0 t) \tag{5.21}$$

The mean values of $cos^2(\omega_0 t)$ and $sin^2(\omega_0 t)$ are 1/2. So we get

$$i(t) = \frac{1}{2}\left(v_I^2 + v_Q^2 + 2Vv_I + V^2\right) \tag{5.22}$$

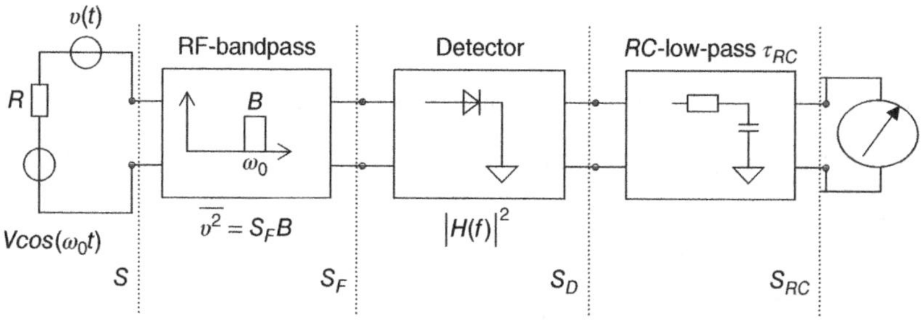

Figure 5.5 Transmission of signal $Vcos(\omega_0 t)$ and noise $v(t)$ via bandpass, detector, and low pass to an indicator. S is the spectral density of white noise at the input. S_F, S_D, and S_{RC} are the modifications along the transmission chain.

For the time average of the current, the following result is obtained

$$\overline{i(t)} = \frac{1}{2}\left(2\overline{v_I^2} + V^2\right) \tag{5.23}$$

We are interested in the noise spectrum of the rectified current $i(t)$. Therefore we need the autocorrelation function $\rho_i(\tau)$ from $i(t)$, which changes into the spectrum after the Fourier transformation.

$$\rho_i(\tau) = \overline{i(t)i(t+\tau)} \tag{5.24}$$

The multiplication gives

$$\rho_i(\tau) = \frac{1}{4}\left\{2\overline{v_I^2(t)v_I^2(t+\tau)} + 4V^2\overline{v_I(t)v_I(t+\tau)} + 4V^2\overline{v_I^2(t)} + 2(\overline{v_I^2(t)})^2 + V^4\right\} \tag{5.25}$$

It is also taken into account that linear terms in v_I and v_Q do not make a contribution and that applies $\overline{v_I^2(t+\tau)} = \overline{v_I^2(t)}$.

Depending on the shape of the filter curve, a correlation coefficient results for the components v_I and v_Q (see Chapter 2).

$$c_I(\tau) = \frac{\overline{v_I(t)v_I(t+\tau)}}{\overline{v_I^2(t)}}; \quad c_Q(\tau) = \frac{\overline{v_Q(t)v_Q(t+\tau)}}{\overline{v_Q^2(t)}}; \quad c_I(\tau) = c_Q(\tau) \tag{5.26}$$

If we subtract the mean value of the detector current (5.23) from the ACF (5.25), only the fluctuating part remains:

$$\rho_I(\tau) - (\overline{i(t)})^2 = \frac{1}{4}\left\{2\overline{v_I^2(t)v_I^2(t+\tau)} + 4V^2\overline{v_I(t)v_I(t+\tau)} - 2(\overline{v_I^2(t)})^2\right\} \tag{5.27}$$

Before we can introduce the correlation coefficient (5.26) in (5.27), we must note that the amplitudes of $v_I(t)$ have a Gaussian distribution. This must be taken into account when calculating quadratic terms like $\overline{v_I^2(t)v_I^2(t+\tau)}$ in (5.27). The following consideration is helpful. Analogous to (2.15) we separate the fluctuating signal into a correlated and an uncorrelated part.

$$v_I(t+\tau) = c_I(\tau)v_I(t) + v_{UNCOR}(\tau) \tag{5.28}$$

The following applies to the mean value of the square

$$\overline{v_I^2(t+\tau)} = \overline{v_I^2(t)} = c_I^2(\tau)\overline{v_I^2(t)} + \overline{v_{UNCOR}^2(\tau)} \tag{5.29}$$

because $\overline{v_I(t)v_{UNCOR}(\tau)} = 0$.

The ACF is obtained by multiplication with $\overline{v_I^2(t)}$.

$$\overline{v_I^2(t)v_I^2(t+\tau)} = c_I^2\overline{v_I^4(t)} + \overline{v_I^2(t)v_{UNCOR}^2(\tau)} \tag{5.30}$$

From (5.29) follows:

$$\overline{v_{UNCOR}^2(\tau)} = \overline{v_I^2(t)}(1 - c_I^2(\tau)) \tag{5.31}$$

Inserted in (5.30):

$$\overline{v_I^2(t)v_I^2(t+\tau)} = c_I^2(\tau)\overline{v_I^4(t)} + (\overline{v_I^2(t)})^2(1 - c_I^2(\tau)) \tag{5.32}$$

Taking into account that for Gaussian distribution $\overline{v_I^4(t)} = 3(\overline{v_I^2(t)})^2$, we have the result

$$\overline{v_I^2(t)v_I^2(t+\tau)} = (\overline{v_I^2(t)})^2\ (1 + 2c_I^2(\tau)) \tag{5.33}$$

If we insert this in (5.27), we obtain, considering (5.26) for the ACF of the detector current:

$$\begin{aligned}\rho_I(\tau) &= \frac{1}{4}\left\{2(\overline{v_I^2(t)})^2 + 4(\overline{v_I^2(t)})^2c_I^2(\tau) + 4V^2\overline{v_I^2(t)}c_I(\tau) - 2(\overline{v_I^2(t)})^2\right\}\\ &= \left(\overline{v_I^2(t)}\right)^2 c_I^2(\tau) + \overline{v_I^2(t)}V^2c_I(\tau)\end{aligned} \tag{5.34}$$

The spectral noise power density of the rectified noise is given by the Wiener–Khinchin theorem to

$$S_D(f) = 4\int_0^{\infty} \rho_I(\tau)\cos(2\pi f\tau)\,d\tau \tag{5.35}$$

Of course we now need the correlation coefficient of the noise voltage $\rho_I(\tau)$ after it has passed the band-limiting filter. As we already mentioned in Chapter 2, this depends on the type of filter curve. The most simple filter curve is the rectangle function. It filters from the available noise spectrum $S(f)$ a narrow rectangular spectrum $S_F(f_0)$ with constant amplitude. The frequency band $B = f_2 - f_1$ is symmetrical relative to the center frequency $f_0 = {}^{(f_1+f_2)}/_2$. The ACF of this noise signal results from the Fourier transformation of the square wave function.

$$\rho(\tau) = \int_{f_1}^{f_2} S_F(f_0)\cos(2\pi f\tau)df \tag{5.36}$$

this will lead to

$$\rho(\tau) = \frac{S_F}{2\pi\tau}\{\sin(2\pi f_2\tau) - \sin(2\pi f_1\tau)\} \tag{5.37}$$

With the use of

$$\sin\alpha - \sin\beta = 2\cos\left(\frac{\alpha+\beta}{2}\right)\sin\left(\frac{\alpha-\beta}{2}\right) \tag{5.38}$$

Results in after the introduction of B and f_0:

$$\rho(\tau) = S_F B\cos(2\pi f_0\tau)\frac{\sin(\pi B\tau)}{\pi B\tau} \tag{5.39}$$

For the integration of (5.35), we need $\rho_I(\tau)$. In comparison with (5.35) we see that holds

$$\rho(\tau) = \rho_I(\tau)\cos(2\pi f_0\tau) \tag{5.40}$$

So we use

$$\rho_I(\tau) = \frac{sin(\pi B\tau)}{\pi B\tau}; \quad \overline{v^2(t)} = S_F B \tag{5.41}$$

The integral (5.35) to be solved is

$$S_D(f) = 4\int_0^\infty \left\{ (\overline{v^2(t)})^2 \frac{sin^2(\pi B\tau)}{(\pi B\tau)^2} + \overline{v^2(t)} V^2 \frac{sin(\pi B\tau)}{\pi B\tau} \right\} cos(2\pi f\tau)\, d\tau \tag{5.42}$$

Let us first look at the noise component on the left in brackets alone, i.e. the case without superimposed sine signal, i.e. $V = 0$.

$$S_{DN}(f) = 4S_F^2 B^2 \int_0^\infty \frac{sin^2(\pi B\tau)}{(\pi B\tau)^2} cos(2\pi f\tau) d\tau \tag{5.43}$$

This integral is of type

$$\begin{aligned} \int_0^\infty \frac{sin^2 x}{x^2} cos(2\,px)\, dx &= \frac{\pi}{2}(1-p) \quad 0 \le p \le 1 \\ &= 0 \qquad\qquad p > 1 \end{aligned} \tag{5.44}$$

with $p = {}^f/_B$ we obtain a triangular LF noise spectrum in the frequency range from $f = 0$ up to the value of the high-frequency bandwidth B (e.g. 1 MHz):

$$S_{DN}(f) = 2S_F^2(B - f) \quad f \le B \tag{5.45}$$

We had already derived this above from plausibility considerations. The spectra $S_F(f)$ behind the input filter and $S_{DN}(f)$ at the output of the detector are shown schematically in Figure 5.6. There are also high-frequency components at $2 \times f_0$, which are not important in this context.

The second term of the integral (5.42) represents the noise component of the rectified signal voltage.

$$S_{DV}(f) = 4S_F BV^2 \int_0^\infty \frac{sin(\pi B\tau)}{\pi B\tau} cos(2\pi f\tau) d\tau \tag{5.46}$$

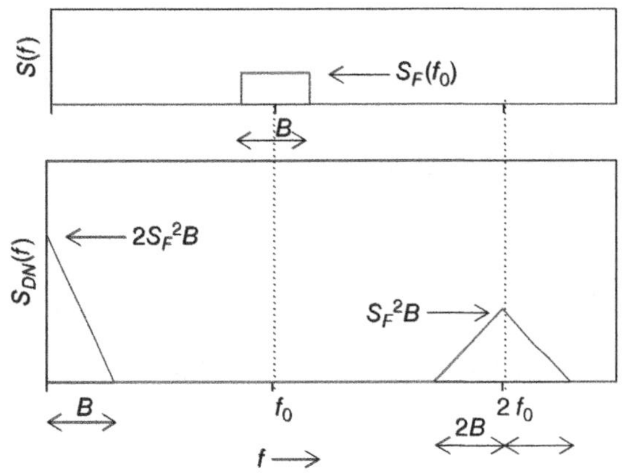

Figure 5.6 Noise without signal. Spectra at quadratic rectification. Top: Narrowband noise at the input. Bottom: Noise components at output with $f < B$ and $2 \times f_0$. LF spectrum according to (5.45).

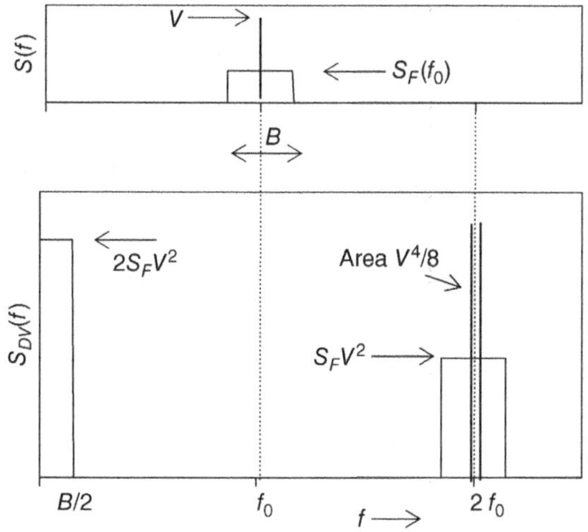

Figure 5.7 Noise with signal. Quadratic detection. Top: Narrowband noise and signal. Bottom: Noise spectra for $f < B/2$ and $2 \times f_0$, signal at $2 \times f_0$. LF spectrum according to (5.48) [6].

This integral is of the type:

$$\int_0^\infty \frac{\sin x}{x}\cos(qx)dx = \frac{\pi}{2} \quad |q| < 1$$
$$= \frac{\pi}{4} \quad |q| = 1$$
$$= 0 \quad |q| > 1 \tag{5.47}$$

With $q = 2f/B$ we obtain a rectangular LF noise spectrum in the frequency range from $f = 0$ to half of the high frequency bandwidth:

$$S_{DV}(f) = 2S_F V^2 \quad f < \frac{B}{2} \tag{5.48}$$

In addition to the LF spectra, noise components in the vicinity of $2 \times f_0$ are also generated. As previously, we are not interested in these here, since they are filtered out by low-pass behavior of the display (Figure 5.7).

The Noise Spectrum Behind a Linear Detector

The voltage generated by the linear detector is directly proportional to the envelope (5.6) of the bandlimited RF signal [4].

$$v_D(t) = R(t) \tag{5.49}$$

As mentioned earlier, $R(t)$ has a Rayleigh distribution of noise amplitudes. The fluctuating part has the autocorrelation function:

$$\rho_R(\tau) = \overline{R(t)R(t+\tau)} - \left(\overline{R(t)}\right)^2 \tag{5.50}$$

The exact solution is more complicated than with the quadratic detector. Equation (5.51) is a good approximation for our rectangular passband curve

$$\rho_R(\tau) \cong \frac{1}{4}(\overline{R})^2 \rho_I^2(\tau) \tag{5.51}$$

The Fourier transformation yields the low-frequency ($f \ll f_0$) noise spectrum of the rectified voltage.

$$S_{DN}(f) = 4\int_0^{\infty} \rho_R(\tau)\cos(2\pi f\tau)d\tau = (\overline{R})^2 \int_0^{\infty} \frac{\sin^2(\pi B\tau)}{(\pi B\tau)^2}\cos(2\pi f\tau)d\tau \tag{5.52}$$

We already had the solution in (5.45).

$$S_{DN}(f) = \frac{(\overline{R})^2}{2B}\left(1 - \frac{f}{B}\right) \tag{5.53}$$

The connection with the input noise $S_F(f_0)$ we find with (5.7):

$$(\overline{R})^2 = \frac{\pi}{2}\overline{v^2}\,\overline{v^2} = S_F(f_0)B \tag{5.54}$$

This results in

$$S_{DN}(f) = \frac{\pi}{4}S_F(f_0)\left(1 - \frac{f}{B}\right) \tag{5.55}$$

So for $f \to 0$ there is no dependence on B, in contrast to the quadratic detector. Because of the approximation (5.51) this solution is not completely accurate. With regard to measuring accuracy, however, it is completely sufficient.

The Sensitivity Limit

In modern spectrum analyzers, there are various ways of evaluating a signal and thus also the noise. For example, one can activate peak, RMS, or sample detectors. Different characteristics of a signal are calculated from these detector signals. Here we consider the limit of the measurement accuracy, which is caused by the noise components that occur behind the detector (5.45) and (5.55). A noise signal at the input produces a mean deflection and a fluctuation in the display, e.g. on the screen of a spectrum analyzer. The average deflection is not a problem since it can easily be compensated. The measurement accuracy is limited by the fluctuation. In all practical cases this fluctuation is reduced by the inertia of the indicator device. With a thermal power meter, its thermal time constant determines the integration. Electronically, the integration time is given by the adjustable cut-off frequency of a low-pass filter. In the spectrum analyzer this is the video bandwidth (VBW).

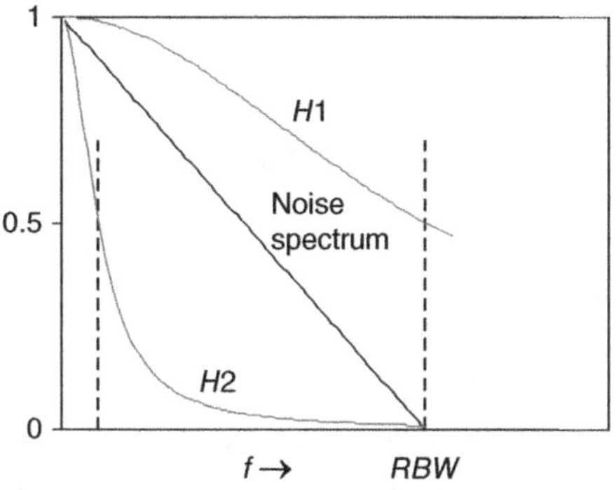

Figure 5.8 The triangular LF noise spectrum before the integrating low pass. *H1* and *H2* are the spectra after integration. *H1*: *VBW* = *RBW*; *H2*: *VBW* = 0.1*RBW*.

Let us first assume a noise without signal. In the input voltage (5.19) let $V = 0$. Let us consider the general case of the free choice of the ratio VBW/RBW on the quadratic detector, as it can easily be set on any spectrum analyzer. Let us follow the noise signal behind the detector (5.45) further on the way to the display. The change in the noise spectrum when passing through a network with the transfer function $H(f)$ is given by (4.20) (Figure 5.8). For the RC low pass this means

$$|H(f)|^2 = \frac{1}{1 + (2\pi f \tau_{RC})^2}; \quad \tau_{RC} = RC \tag{5.56}$$

At the input of the low pass we have the spectrum (5.45). At the output we get

$$S_{RC}(f) = 2S_F^2(f_0)\frac{B - f}{1 + (2\pi f \tau_{RC})^2} \tag{5.57}$$

We remember (2.23) and write for the noise current at the output of the low pass:

$$\overline{i_N^2(t)} = \int_0^{\infty} S_{RC}(f)df \tag{5.58}$$

in our case, that is,

$$\overline{i_N^2(t)} = 2S_F^2(f_0)\int_0^{B} \frac{B - f}{1 + (2\pi f \tau_{RC})^2}df \tag{5.59}$$

This integral has the solution:

$$\overline{i_N^2(t)} = 2S_F^2(f_0)\left\{\frac{B}{2\pi\tau_{RC}}arctan(2\pi B\tau_{RC}) - \frac{1}{8\pi^2\tau_{RC}^2}ln(1 + (2\pi B\tau_{RC})^2)\right\} \tag{5.60}$$

In noise measurement practice it is usually the case that the RF bandwidth B is much larger than the LF bandwidth $f_{RC} = (2\pi\tau_{RC})^{-1}$. In the case of a spectrum analyzer, for example, $B = RBW = 1\,\text{MHz}$, $f_{RC} = VBW = 10\,\text{kHz}$. The following therefore applies $2\pi B\tau_{RC} \gg 1$. This reduces the contribution of the curly bracket in (5.60):

$$\left\{ \frac{B}{2\pi\tau_{RC}} arctan(2\pi B\tau_{RC}) - \frac{1}{8\pi^2\tau_{RC}^2} ln(1 + (2\pi B\tau_{RC})^2) \right\} = \frac{B}{4\tau_{RC}} \tag{5.61}$$

So we get for the fluctuating current behind the integration element at quadratic rectification:

$$\sqrt{\overline{i_N^2(t)}} = S_F(f_0)\sqrt{\frac{B}{2\tau_{RC}}} \tag{5.62}$$

The absolute fluctuation is therefore proportional to the level of the RF input noise $S_F(f_0)$ and the root of the RF bandwidth B. A sufficiently long integration time τ_{RC} reduces the fluctuation arbitralily as long as this technically makes sense.

The analog consideration for the linear detector gives the same integral. In the expression in brackets (5.60), there is only a different factor. Instead of $2S_F^2$ out (5.45) we have ${}^{\pi S_F}/_{4B}$ (5.55). Thus

$$\text{factor linear: } {}^{\pi S_F}/_{4B} \quad \text{factor quadratic: } 2S_F^2 \tag{5.63}$$

What is the value of the relative fluctuation? We have to normalize (5.60) to the direct current value i_{DC} which the detector generates from the applied noise voltage.

In Figure 5.2 the distribution functions of the amplitudes for linear and quadratic rectification are given. The direct currents result from the mean values of the distributions.

With Figure 5.2 we get:

$$i_{LIN} = \sqrt{\overline{v^2}\frac{\pi}{2}}; \quad i_{QUAD} = 2\overline{v^2}; \quad \overline{v^2} = S_F(f_0)B \tag{5.64}$$

The relative fluctuation, i.e. the "noise from noise" is $\sqrt{\overline{i_N^2(t)}/i_{DC}^2}$.

If we denote the expression in brackets in (5.60) with $K(B, \tau_{RC})$, then using (5.63) for linear and quadratic rectification the same relative variation is shown.

$$\sqrt{\frac{\overline{i_N^2(t)}}{i_{DC}^2}} = \sqrt{\frac{1}{2B^2} \times K(B, \tau_{RC})} \tag{5.65}$$

Applied to the spectrum analyzer, this is the known dependence of "noise on noise," i.e. the fluctuation of the display of a mean value:

$$\alpha = \sqrt{\frac{\pi VBW}{4RBW}}; \quad VBW = \frac{1}{2\pi\tau_{RC}}; \quad RBW = B \tag{5.66}$$

For the spectrum analyzer, these are VBW = video bandwidth (LF-range), RBW = resolution bandwidth (RF range).

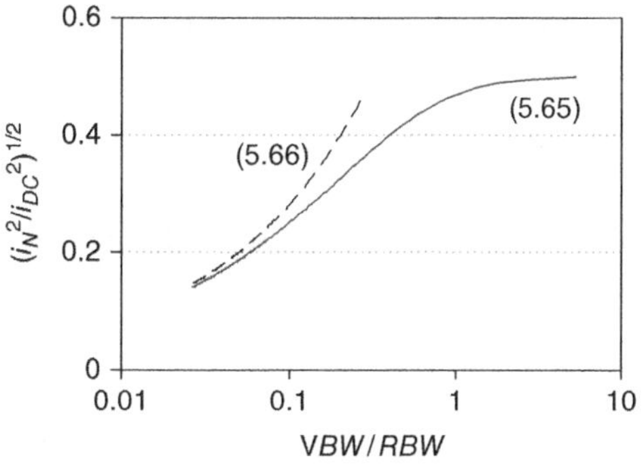

Figure 5.9 Fluctuations of a noise measurement (5.65) versus the video bandwidth normalized to *RBW*.

Without the restriction $VBW \ll RBW$, we must evaluate (5.65). One can see that the fluctuation of the display for $VBW > RBW$ tends to 0.5, independent of the characteristic curve of the detector (Figure 5.9).

Noise with Signal

The input voltage is given by (5.19). The noise spectrum behind the detector is given by (5.45) and (5.48) [6].

$$S_{DN} = 2S_F^2(B-f) \quad f < \frac{B}{2} \tag{5.67}$$

$$S_{DN} = 2S_F^2(B-f) + 2S_F V^2 \quad \frac{B}{2} < f < B \tag{5.68}$$

Due to the presence of the monochromatic signal $V\cos(\omega_0 t)$, an additional noise component with the spectral power density $2S_F V^2$ is generated. When a spectrum analyzer is tuned to an unmodulated signal in narrowband mode, the noise level therefore increases slightly in accordance with (5.48).

What is the smallest voltage V that can be detected in noise with a quadratic detector?

Behind the rectifier we have the current (5.23). We write with (5.18):

$$\overline{i(t)} = \overline{v^2(t)} + \frac{V^2}{2} \tag{5.69}$$

Because of $\kappa = 1AV^{-2}$ the detector current is proportional to the sum of the noise power $\propto \overline{v^2(t)}$ and the signal power $\propto V^2$. Let us define the signal-to-noise ratio of the display to

$$r_0 = \frac{(\overline{i(t)} - \overline{i_0(t)})^2}{\overline{v^2(t)}} \tag{5.70}$$

$\overline{i_0(t)}$ is the current generated by the noise only, i.e. for $V = 0$. Behind the integration low-pass filter the result is

$$r_0 = \frac{V^4}{4\int_0^\infty S_{DN}|H(f)|^2 df} \tag{5.71}$$

Let us assume for the sake of simplicity a rectangular passband up to the LF bandwidth $b \ll B$. This corresponds approximately to the video bandwidth $b \cong VBW$.

$$|H(f)|^2 = 1 \quad f \leq b$$
$$|H(f)|^2 = 0 \quad f > b \tag{5.72}$$

then becomes (5.71):

$$r_0 = \frac{V^4}{8b(S_F^2 B + S_F V^2)} \tag{5.73}$$

With $r_0 = 1$ we assume that the smallest detectable signal V_{min} is equal to the noise floor.

Solved for V_{min} results in:

$$V\sqrt{S_F}\{b + \sqrt{2(b^2 + bB)}\}^{\frac{1}{2}}{}_{min} \tag{5.74}$$

as detection limit of a monochromatic signal in noise. The signal $V\ cos(\omega_0 t)$ is superimposed on noise with the spectral power density S_F. In front of the quadratic rectifier, there is a bandpass with a rectangular passband of center frequency $\omega_0/2\pi$ and bandwidth B. Behind the rectifier there is a low-pass filter as integration element with a rectangular passband curve (no RC low-pass filter) up to frequency b.

The Phase Sensitive Rectifier

The phase sensitive detector (PSD) is often used in physical measurement technology [7]. It is used to measure a weak signal in a strong noise background. Its first application was in the Dicke radiometer of radio astronomy. It was possible to measure very weak antenna noise from space although the level was far below the noise of the receiver. The PSD is the most important component of the lock-in technique. A weak signal is periodically sampled. This can be, for example, a test signal that almost disappears in a very noisy background. If the noise itself is the measured variable, as in radio astronomy, it can be periodically switched on and off. We then have two signals, with the switching frequency, the signal to be measured and the reference signal. Both are multiplied with each other by feeding them to a mixer. The special feature is that the signal and the local oscillator have the same

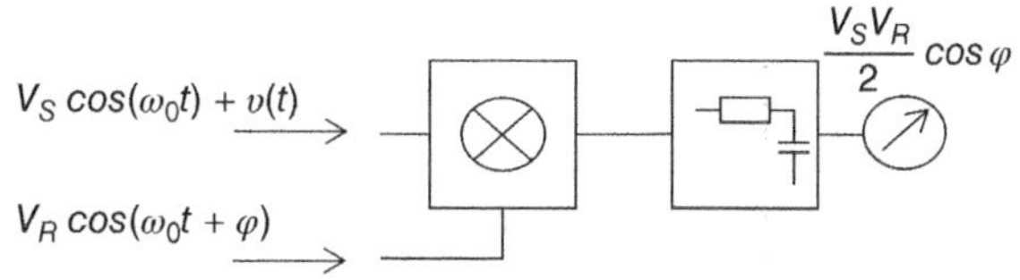

Figure 5.10 Schematic of a phase sensitive detector. Mixed products $\omega \geq \omega_0$ are eliminated by the low pass.

frequency $^{\omega_0}/_{2\pi}$. Let us first look at the mixing process. At the signal input there is signal and noise $V_S \cos(\omega_0 t) + v(t)$. The reference is at LO input of the mixer $V_R \cos(\omega_0 t + \varphi)$. The phase shift φ can be tuned by means of a phase shifter. At the output we have the product of both voltages V_{PSD} (Figure 5.10).

$$V_{PSD} = [V_S \cos(\omega_0 t) + v(t)] V_R \cos(\omega_0 t + \varphi) \tag{5.75}$$

We obtain

$$V_{PSD} = \frac{V_S V_R}{2}[\cos(-\varphi) + \cos(2\omega_0 t + \varphi)] + v(t)\cos(\omega_0 t + \varphi) \tag{5.76}$$

A DC voltage remains at the output, which depends on the phase shift between signal and reference.

$$V_{PSD} = \frac{V_S V_R}{2}\cos(\varphi) \tag{5.77}$$

If both are in phase, then $\cos(\varphi) = 1$ and the output voltage of the PSD is directly proportional to the voltage amplitude V_S of the signal. The constancy of the reference voltage V_R is no problem.

Equation (5.76) shows that the mixer can work as a phase detector. This is the basis for an application in phase noise measurement technology (Chapter 22). If the mixer is operated "in quadrature," i.e. $\varphi = 90\,°$, $\cos(\varphi) = 0$. If the source of signal and reference is an oscillator, the carrier does not appear at the output. However, the fluctuations in phase, caused by the phase noise that is always present, provide a noise spectrum in the baseband that can be measured at the output of the phase detector.

If we set all constants of the equipment to 1, analogous to the quadratic detector, the average detector current is $\overline{i(t)} = V_S$. To calculate its fluctuation spectrum, we need the autocorrelation function.

$$\overline{i(t)i(t+\tau)} = \overline{i^2(t)}\rho_I(\tau) \tag{5.78}$$

and receive the noise spectrum at the detector output with (5.35):

$$S_{DN} = 4\overline{i^2(t)} \int_0^\infty \rho_I(\tau)\cos(2\pi f\tau)d\tau \tag{5.79}$$

For rectangular passband curve at the input with (5.41)

$$S_{DN} = 4S_F B \int_0^\infty \frac{\sin(\pi B\tau)}{\pi B\tau}\cos(2\pi f\tau)d\tau \tag{5.80}$$

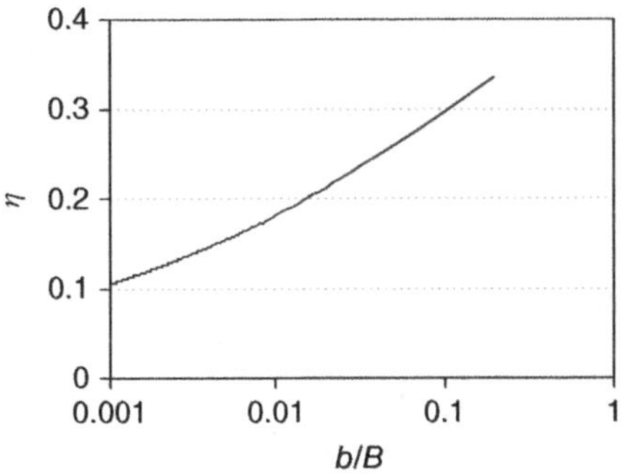

Figure 5.11 Comparison of PSD and quadratic detector for a very weak AC signal. For long integration times ($b/B < 0.01$) the PSD is much more sensitive.

With (5.47) we obtain for the low-frequency noise spectrum

$$S_{DN} = 2S_F \quad f < \frac{B}{2} \tag{5.81}$$

This relation is already known from Figure 5.7. We also want to calculate the signal/noise ratio r_0 of the display for the PSD (5.71).

Without signal, the output voltage of the PSD is 0, i.e. $\overline{i_0(t)} = 0$.

$$r_0 = \frac{V_S^2}{\int_0^\infty S_{DN}|H(f)|^2 df} \tag{5.82}$$

For simplification, we assume a constant passband curve up to bandwidth b for the integration low pass behind the detector.

So that's

$$r_0 = \frac{V_S^2}{2S_F b} \tag{5.83}$$

If we set the detection limit of the signal voltage at $r_0 = 1$ as previously, we obtain

$$V_{min} = \sqrt{2S_F b} \tag{5.84}$$

To compare the sensitivity of the quadratic detector and the PSD, we depict in Figure 5.11 the ratio of the minimum voltages $\eta = {}^{V_{min,PSD}}/_{V_{min,quadr.}}$ in dependence on the ratio of the bandwidths b/B.

Trace Averaging

We have so far achieved the reduction of "noise from noise" with the integration through the RC element. This is the variation of the ratio of RBW to VBW, which can be easily done, especially in the spectrum analyzer. This is also referred to as "video filtering." In real time we see the averaged noise on the display. Each point is the result of averaging with the time constant τ_{RC} of the LF filter.

With "trace averaging," digital signal processing enables a completely different type of averaging. It is based on the theory of observation errors. The individual values of the noise voltage correspond to the random errors of a repeatedly performed measurement. These have a distribution function of the amplitudes characterized by mean value and standard deviation. The aim of the measurement is to determine the mean value accurately. The display should therefore have as little fluctuation as possible. A moving averaging process is applied, so that the value is improved with each run, as more and more measuring points are included.

We have a number of sweeps $n = 1, 2, \ldots N$. With the first data set $n = 1$, the mean value $\overline{v_1}$ is calculated. This is improved with each further sweep according to the formula.

$$\overline{v} = \frac{n-1}{n}\overline{v_{n-1}} + \frac{1}{n}\overline{v_n} \quad n \geq 2 \tag{5.85}$$

As the number of averaging procedures increases, the fluctuation of this mean value naturally decreases. From the definition of the variance

$$\sigma^2 = \frac{1}{N-1}\sum_{n=1}^{N}(v_n - \overline{v})^2 \tag{5.86}$$

When more measuring points are added, i.e. an increase of N, the denominator increases approximately by the square, while the numerator only adds up linearly.

The fluctuation of the mean value is given by

$$\sigma_{\overline{v}} = \sqrt{\frac{1}{N}\frac{1}{N-1}\sum_{n=1}^{N}(v_n - \overline{v})^2} \tag{5.87}$$

The smoothing of the display is therefore improved with $\sqrt{N}$

$$\sigma_{\overline{v}} = \frac{\sigma}{\sqrt{N}} \tag{5.88}$$

It must be noted, however, that the results of video filtering and trace averaging are not always the same. It depends on the representation of the signal and the detector used, of which there is a wide range in modern SPAs. Average detector and sample detector provide the same values as trace averaging when displayed linearly. The two methods do not converge for peak and RMS detectors and for logarithmic display. Video filtering should always be given preference here. If noise-like signals or a mixture of noise and non-stationary signals are involved, a precise analysis of the measurement process is required for their measurement.

References

1 Kummer, M. (1989). *Grundlagen der Mikrowellentechnik*. Berlin: Verlag Technik.
2 Agilent Technologies (2003). Spectrum Analyzer Measurements and Noise. Application Note 1303.
3 Aeroflex/Metelics (2010). Zero Bias Schottky Diode MSS 20. Datasheet, 2010.
4 van der Ziel, A. (1954). *Noise*. London: Chapman and Hall.
5 Lawson, J.L. and Uhlenbeck, G.E. (1965). *Threshold Signals*. Dover Publications Inc.
6 Müller, R. (1990). *Rauschen*. Springer Verlag.
7 Stanford Res. Systems. About Lock-In Amplifiers. Application Note 3.

6

The Noise Factor

Amplifier and Noise Power

The noise power at the output of a receiving system originates from various sources [1]. The main contributions are provided by the antenna, mixer, local oscillator, and amplifier. All contributions are independent of each other and therefore uncorrelated. At the output they add up to the output noise power N_O. Therefore the components of the system can be considered and the noise contributions can be analyzed individually.

We will first consider the simplest case of a linear RF amplifier (Figure 6.1). The input is terminated reflection-free with the resistor R_S. The temperature T_S of this resistor can be varied. At the output there is a matched power meter for measuring N_O.

The amplifier is characterized by its frequency range f_0 and bandwidth B, its gain G_A and its noise figure NF. In our example: $f_0 = 1$ GHz, $B = 1$ MHz, $G_A = 20$ dB; $NF = 3$ dB. G_A and NF are explained and defined below. The thermal noise of R_S increases with temperature according to the Nyquist formula. The measured N_O on the power meter increases linearly with T_S. This relationship is shown in Figure 6.2. N_O consists of the amplified noise of R_S plus the inherent component of the amplifier N_A. The part N_A is generated in every network. In a passive network, thermal noise is generated by the losses, e.g. cable attenuation. In an active network, overthermal noise is added, e.g. shot noise in pn-junctions of transistors.

The definition of gain G_A appropriate to noise considerations is available power gain. This is the ratio of available power at the output of a two-port to the available power of the source. In our example, input and output are matched without reflection, so the exact definition is not yet relevant here.

The dashed curve for the ideal noise-free amplifier, which amplifies with 20 dB only within the bandwidth B and whose contribution $N_A = 0$, represents only the

A Guide to Noise in Microwave Circuits: Devices, Circuits, and Measurement, First Edition.
Peter Heymann and Matthias Rudolph.

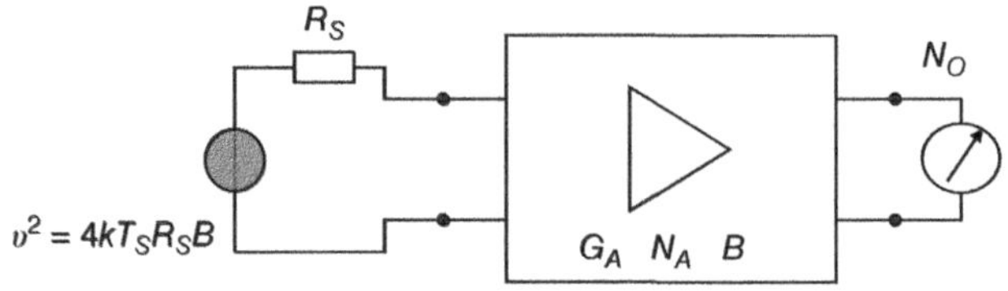

Figure 6.1 Amplifier with gain G_A and bandwidth B. At the input, the resistor R_S is at the temperature T_S. N_O is the noise power at the output.

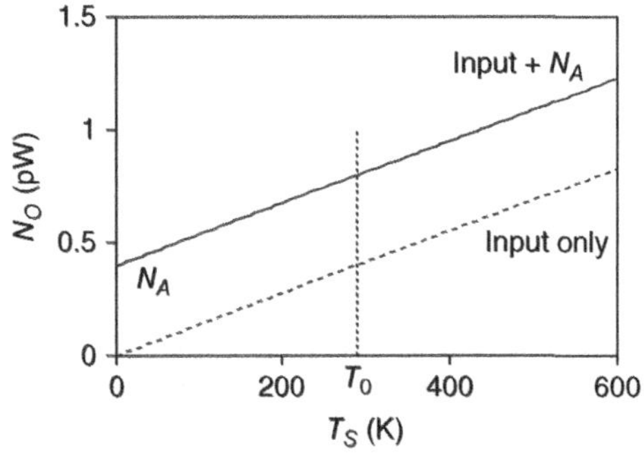

Figure 6.2 Output power N_O versus the temperature T_S of the source resistance. Dashed line: Noise from R_S (6.1). Full line: amplified input noise plus internal noise of amplifier (6.2).

amplified thermal noise of R_S:

$$N_{O1} = kG_ABT_S \tag{6.1}$$

The real amplifier adds its own noise $N_A > 0$ so that the upper straight line corresponds to the normal case.

$$N_O = N_{O1} + N_A = kG_ABT_S + N_A \tag{6.2}$$

The Noise Factor *F*

With these two quantities in Eq. (6.2) the noise factor F can be defined [2]. F is the ratio of the total noise power N_O at the output of a network to the noise power that comes only from the source. However, as Figure 6.3 shows, this is not without unambiguity. Depending on the temperature T_S of the source, a different noise factor of the network is obtained. Part of the definition of F is the assessment of the reference temperature $T_0 = 290\,\text{K}$. This value is not a standardized room temperature in the age of energy saving. It was chosen because in early times of radio relay systems it corresponded to the antenna temperature when the antenna was aligned parallel to the earth's surface. In addition, one has for the widely used expression $^{kT_0}/_q = 0.025\,\text{V}$.

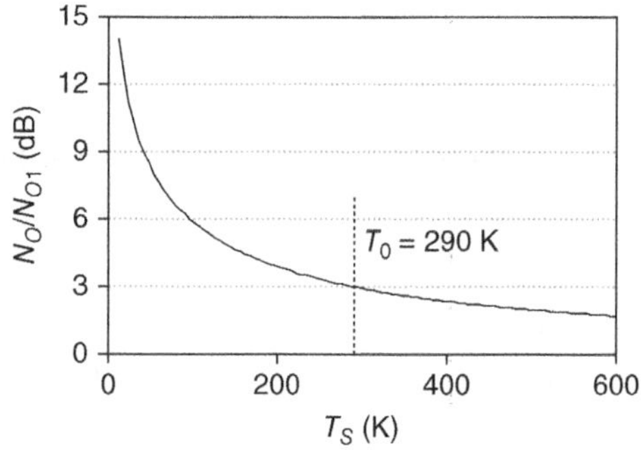

Figure 6.3 Ratio of noise power $10\,log\,(N_O/N_{O1})$ versus the source temperature T_S. The definition of the noise factor F requires the specification $T_0 = 290\,\text{K}$.

The definition of the noise factor is therefore

$$F = \frac{kG_ABT_0 + N_A}{kG_ABT_0} \tag{6.3}$$

Mostly the logarithmic measure, the noise figure NF is used.

$$NF = 10log(F) \tag{6.4}$$

A distinction must therefore be made: For the linear, dimensionless ratio F, the term noise factor F is used. For the logarithmic measure F in dB, NF (noise figure). Occasionally, the term noise measure is used for NF, but this is reserved for a different quantity that includes the gain. See (6.23).

An ideal noise-free amplifier would therefore have the noise factor $F = 1$, i.e. the noise figure $NF = 0\,\text{dB}$. In our example we have $F = 2$, i.e. $NF = 3\,\text{dB}$.

For calculations, it is often useful to consider only the portion of N_A that is caused by the network. It is described by the additional noise figure F_Z:

$$F_Z = \frac{N_A}{kG_ABT_0} \tag{6.5}$$

Comparison of (6.5) with (6.3) shows

$$F_Z = F - 1 \tag{6.6}$$

This observation leads us to the basic principle of the analysis of noise processes in networks. The noise properties are extracted from the network and transferred to the input circuitry. The network is now noise-free and only its transmission properties are considered. In Figure 6.4 we now have a noise voltage source which generates the same output noise power N_O as in Figure 6.1. It combines the noise contributions of source and network. Additionally we have the signal voltage source V_S. So far we have only considered noise sources. In a communication system, however, we always have a signal that is important.

The available noise power of this new source at the input is according to (6.5)

$$\frac{N_A}{G_A} = F_ZkT_0B \tag{6.7}$$

and the effective value of their voltage:

$$v = \sqrt{4kF_Z T_0 R_S B} \tag{6.8}$$

It is now obvious to simply assign the new noise source to the source resistor R_S by increasing its temperature to an effective noise temperature T_e.

$$T_e = F_Z T_0 \tag{6.9}$$

Instead of the reference temperature T_0 of the source resistance, which only affects the source, T_e now appears, which add the noise properties of the network. Of course we keep the reference temperature T_0, since it is only a different definition of the noise factor.

$$F = \frac{T_e + T_0}{T_0} \tag{6.10}$$

Another definition of the noise factor refers to the conditions in a communication system in which signal and noise are always present simultaneously. The goal of designing a transmission system must be to obtain the highest possible signal-to-noise ratio at the output. In Figure 6.4 we have the connection of a two-port with a signal source at the input. A spectrum analyzer would display Figure 6.5.

At the input the signal level is 45 dB above the noise level. At the output, the signal level is amplified by 21 dB, but the signal-to-noise ratio is only 36 dB because the amplifier has added its own noise. This degradation of the signal-to-noise ratio is described by the following definition of the noise factor:

$$F = \frac{S_I/N_I}{S_O/N_O} \tag{6.11}$$

In accordance with the common denotation used in the literature: S_I is the signal power at the input, S_O is the signal power at the output, N_I is the noise power at the input, and N_O is the noise power at the output. Also in this definition it should be noted that thermal noise at the input is assumed with $T_0 = 290\,\text{K}$. The power is therefore $N_I = kT_0B$. Our example applies to a broadband amplifier with $B = 2\,\text{GHz}$.

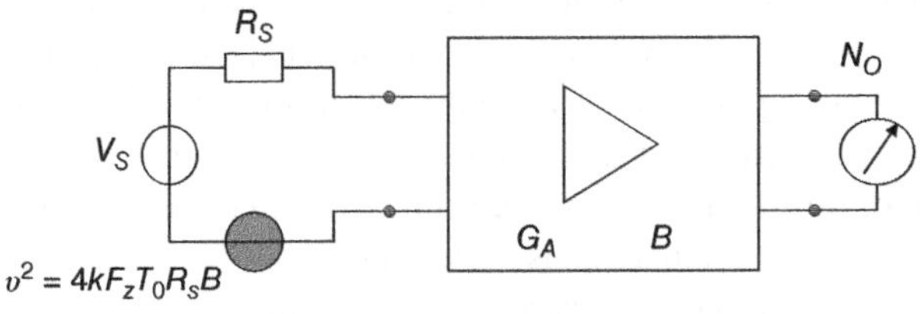

Figure 6.4 Network as in Figure 6.1 but without noise. Noise source and signal source are combined at input.

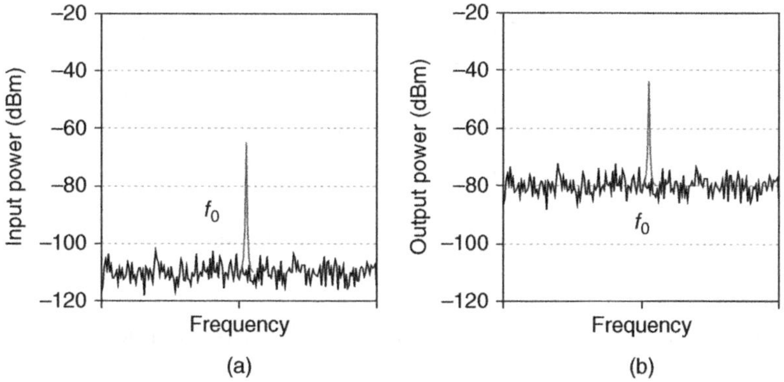

Figure 6.5 (a) Power spectrum with signal at f_0 and noise at amplifier input. (b) At the output. The noise level increases more than the signal level.

In the example Figure 6.5 we have: $S_I = -65\,\text{dBm}$; $S_O = -44\,\text{dBm}$; $N_I = -110\,\text{dBm}$, $N_O = -80\,\text{dBm}$. Since F is a pure ratio, the units are not important. Of course we have to change from the logarithmic dBm to the linear scale mW.

$$F = \frac{10^{\frac{-65}{10}} \Big/ 10^{\frac{-110}{10}}}{10^{-44/10} \Big/ 10^{-80/10}} \cong 8$$

The noise factor of the amplifier in the example Figure 6.5 is therefore $F = 8$, i.e. $NF = 9\,\text{dB}$.

The relation with the noise factor (6.3) defined above, which only referred to noise, is easy to see. The ratio of the signal levels is given by the available power gain.

$$G_A = \frac{S_O}{S_I} \tag{6.12}$$

We obtain from (6.11):

$$F = \frac{1}{G_A}\frac{N_O}{N_I} = \frac{kG_ABT_0 + N_A}{kG_ABT_0} \tag{6.13}$$

as (6.3). In (6.13) we see an important property of the noise factor. It is not a pure characteristic of the two-port itself but depends via G_A on the input circuit.

Cascaded Amplifiers

In a receiving system, components with different noise and gain characteristics are always connected in series [3]. So we need to know how the signal-to-noise ratio

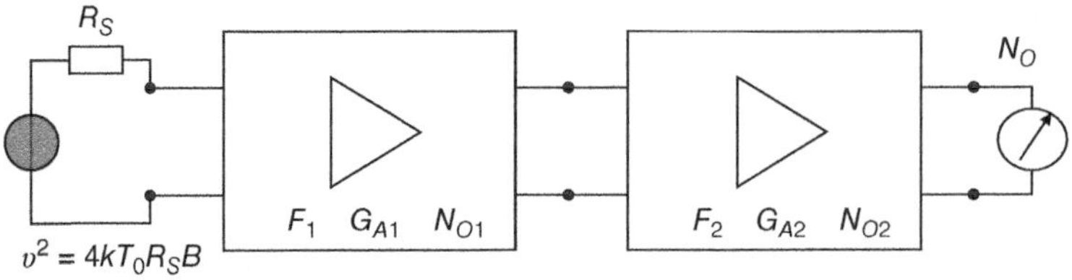

Figure 6.6 Two cascaded networks with kT_0B at the input generate the noise power N_O at the output.

changes when passing through such a chain. We consider two amplifiers with the same bandwidth B in Figure 6.6.

According to the definition of the noise factor (6.13) is the output power:

$$N_O = FG_AN_I \tag{6.14}$$

Let us first consider both amplifiers in Figure 6.6 as a unit with the noise factor F_{12} and the total gain $G_{A1} \times G_{A2}$. At the input there is the thermal noise power $N_I = kT_0B$. At the output the noise power is

$$N_O = F_{12}G_{A1}G_{A2}kT_0B \tag{6.15}$$

At the output of the first stage we have

$$N_{O1} = F_1G_{A1}kT_0B \tag{6.16}$$

This part is amplified by the second stage, so that its contribution to the total noise is

$$N'_{O1} = N_{O1}G_{A2} \tag{6.17}$$

Contribution of network 2 to the total noise N_O is according to (6.5)

$$N'_{O2} = (F_2 - 1)\, G_{A2}kT_0B \tag{6.18}$$

The total noise at the output is the sum of both, since they are uncorrelated.

$$N_O = N'_{O1} + N'_{O2} \tag{6.19}$$

$$N_O = F_1G_{A1}G_{A2}kT_0B + (F_2 - 1)\, G_{A2}kT_0B \tag{6.20}$$

We compare this to (6.15) where we had introduced the overall noise factor F_{12}:

$$F_{12} = F_1 + \frac{F_2 - 1}{G_{A1}} \tag{6.21}$$

In noise temperatures

$$T_{e12} = T_{e1} + \frac{T_{e2}}{G_{A1}} \tag{6.22}$$

This is the important Friis formula. The main statement is: Low noise figure and high gain of the first stage guarantee a low noise figure of the whole system. The subsequent stages only have a small influence on the noise characteristics.

If n networks are cascaded, the total noise factor F is calculated accordingly:

$$F = F_1 + \frac{F_2 - 1}{G_{A1}} + \frac{F_3 - 1}{G_{A1}G_{A2}} + \cdots + \frac{F_n - 1}{G_{A1}G_{A2}\cdots G_{A(n-1)}} \tag{6.23}$$

The Noise Measure *M*

The great importance of the first stage of a chain for the noise of the system is illustrated by the noise measure M, because in addition to the noise factor the gain is included here [4].

$$M = \frac{F - 1}{1 - \frac{1}{G_A}} \tag{6.24}$$

The noise measure is also suitable for estimating the minimum achievable noise factor of an infinite chain of identical amplifiers. All have the same noise factor F and the same gain G_A. We write the Friis formula for this case

$$F_{tot} = F + \frac{F - 1}{G_A} + \frac{F - 1}{G_A^2} + \frac{F - 1}{G_A^3} + \cdots \tag{6.25}$$

$$F_{tot} = F + (F - 1)\left(\frac{1}{G_A} + \frac{1}{G_A^2} + \frac{1}{G_A^3} + \cdots\right) \tag{6.26}$$

With the series expansion

$$\frac{1}{1 - x} = 1 + x + x^2 + x^3 + \cdots \tag{6.27}$$

we set $x = {}^1/_{G_A}$

$$\frac{1}{G_A} + \frac{1}{G_A^2} + \frac{1}{G_A^3} + \cdots = \frac{1}{1 - \frac{1}{G_A}} - 1 \tag{6.28}$$

Inserted in (6.26) results in

$$F_{tot} = 1 + \frac{F - 1}{1 - \frac{1}{G}} = 1 + M \tag{6.29}$$

Definitions of Gain

Here we will look at the different descriptions of gain of an amplifier or a two-port in general and clarify why the Available Power Gain G_A is the appropriate value

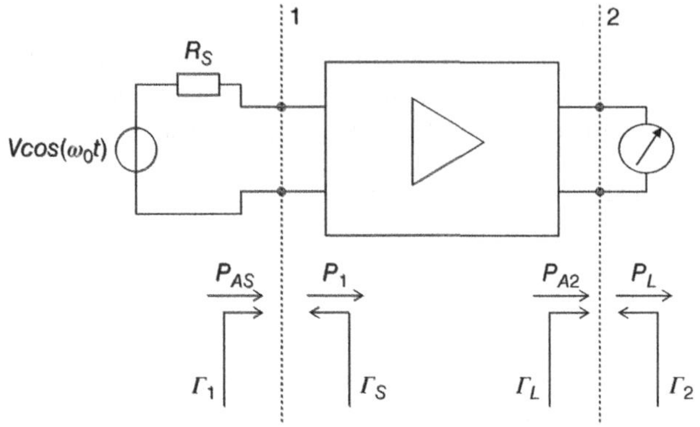

Figure 6.7 Amplifier with signal source at input and power meter at output. Γ are the reflection coefficients at the interfaces.

for noise analysis [5]. In Chapter 3 we introduced the available noise power of an ohmic resistor as the basic element of our considerations. In general, the power transfer from an active one-port with complex internal impedance $Z = R + jX$ on a load resistor is maximum if it is complex conjugated to Z, i.e. it has the value Z^*. This maximum output power of the one-port is called available power P_{AS}. In Figure 6.7 the power available in reference plane 1 of the source is

$$P_{AS} = \frac{V^2}{8R_S} \tag{6.30}$$

It is now just the available power gain that relates the available power at the input and output of a two-port. If we consider in general a signal transfer through the two networks in Figure 6.6, the available individual gains are

$$G_{A1} = \frac{P_{A1}}{P_{AS}}; \quad G_{A2} = \frac{P_O}{P_{A1}} \tag{6.31}$$

and the total gain

$$G = \frac{P_{A1}}{P_{AS}} \frac{P_O}{P_{A1}} = \frac{P_O}{P_{AS}} \tag{6.32}$$

This does not mean that the actual measured output power is P_O. G_A is a pure operand. In practice, the output power is lower due to mismatch at the interfaces. However, this does not matter when considering the signal-to-noise ratio, since the matching situation and thus the losses due to mismatch are the same for signal and noise. However, if the measurement is a noise measurement where the absolute value of the power must be known, a precise knowledge of the complex reflection coefficients at the interface between the networks will be required (see the following text).

Since, when cascaded, the output of the preceding two-port is the source for the following one. The definition G_A of gain for the Friis formula (6.23) is exactly the appropriate one.

As mentioned earlier, G_A is not the gain one is dealing with when using a commercial amplifier in coaxial or waveguide technology. It also does not correspond exactly to the specifications in the data sheet. This is because in a real system there is no matching at input and output. In general, there is neither zero-reflection $\Gamma_1 = \Gamma_S = \Gamma_L = \Gamma_2 = 0$, nor conjugated complex matching $\Gamma_1 = \Gamma_S^*$ and $\Gamma_L = \Gamma_2^*$ realized. As will be shown below, these conditions are often not even desired for the noise optimization of a circuit.

We now analyze the various definitions of gain and the problems of mismatch in the formalism of the waves, i.e. S-parameters and reflection coefficients. The model of impedances, current or voltage sources, and their analysis by equivalent circuitry with short circuit and open circuit are only useful at low frequencies. The wave concept is appropriate above 300 MHz, because here the wavelength is of the same order of magnitude as the circuit dimensions and therefore short circuit and open circuit can no longer be realized in a broad frequency range. Instead of voltage and impedance, power and reflection factor are used. In Figure 6.7 both pictures are combined to explain the relationship. The source (left) is shown as an equivalent voltage source with the internal resistance R_S (the Thevenin equivalent). Two-port and load (right) correspond to the wave picture.

The available power gain which is defined as

$$G_A = \frac{P_{A2}}{P_{AS}} \tag{6.33}$$

is in the formalism of S-parameters and reflection coefficients:

$$G_A = |S_{21}|^2 \frac{1 - |\Gamma_S|^2}{|1 - \Gamma_S S_{11}|^2 \left(1 - |\Gamma_2|^2\right)}; \qquad \Gamma_2 = S_{22} + \frac{S_{12} S_{21} \Gamma_S}{1 - S_{11} \Gamma_S} \tag{6.34}$$

It is a function of the two-port properties represented by the s-matrix $\boldsymbol{S}$ and Γ_S. The maximum value MAG (maximum available gain) is reached when conjugate complex matching is realized at both input and output. It is a sole property of the amplifier itself. If the amplifier is connected between a signal generator whose available power is well defined and a receiver, the transducer power gain G_T, also called operating power gain, is measured.

$$G_T = \frac{P_L}{P_{AS}} \tag{6.35}$$

$$G_T = |S_{21}|^2 \frac{\left(1 - |\Gamma_S|^2\right)\left(1 - |\Gamma_L|^2\right)}{|1 - \Gamma_S S_{11}|^2 |1 - \Gamma_L \Gamma_2|^2} \tag{6.36}$$

It is a function of $\boldsymbol{S}$, Γ_S and Γ_L.

If the output power P_L is not related to the available power of the source but to the power P_1 absorbed by the amplifier input, the power gain G_P is obtained

$$G_P = \frac{P_L}{P_1} \tag{6.37}$$

$$G_P = |S_{21}|^2 \frac{1 - |\Gamma_L|^2}{|1 - \Gamma_L S_{22}|^2 \left(1 - |\Gamma_1|^2\right)} \tag{6.38}$$

It is a function of $\boldsymbol{S}$ and Γ_L.

Often one wants to measure an amplifier in coaxial technology with a commercial noise figure meter. For this purpose the instrument is first calibrated by directly connecting the noise generator. Then the object is connected in between. The displayed gain is the insertion gain G_I

$$G_I = \frac{P_L \text{ with DUT}}{P_L \text{ with source directly connected}} \tag{6.39}$$

It is a function of $\boldsymbol{S}$, Γ_S, and Γ_L. Further definitions, e.g. Unilateral Gain or Maximum Stable Gain are essential for the design of amplifiers, less for noise problems. For illustration, we show a simple example of the values of the different gains as a function of the reflection coefficient of the source $|\Gamma_S|$. All parameters used are real. For microwave amplifiers, a mismatch is usually given at the input and output of $VSWR \leq 2$. For this we set $S_{11} = S_{22} = 0.33$. With a specified gain of 20 dB we have $S_{21} = 10$. The internal feedback is small, we set $S_{12} = 0.035$. This gives the values shown in Figure 6.8. G_P, $|S_{21}|^2$, and MAG do not depend on Γ_S.

Even the simplest circuit has several interfaces where losses occur due to mismatching. In Figure 6.7 there are source and input in reference plane 1 and output and power meter in reference plane 2. With the cascading in Figure 6.6 the interface between the two amplifiers is added. In all these reference planes we have the interaction of a source with a load.

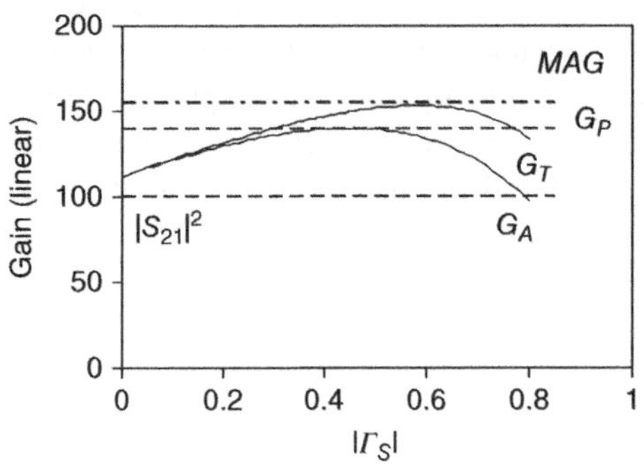

Figure 6.8 Dependence of the different gains on the reflection coefficient of the source.

Source and Load

We therefore want to recall details of the power transmission [6]. In the wave concept, to which the formulas (6.34)–(6.38) belong, one does not work with voltages and resistances but with power waves and reflection factors. These are also the quantities appropriate for noise problems in the microwave range, since most calculations deal with noise power.

A signal flow chart of source and load is shown in Figure 6.9. The generator (left) is characterized by the actual source b_S and the reflection coefficient Γ_G. For the sinusoidal source in Figure 6.7 it corresponds to $V\cos(\omega_0 t)$ and R_S. Here the generator is a one-port with the outgoing wave b_G and the incoming wave a_G. We also consider the connected load (right) as a one-port, since we are only interested in the interaction with the generator (interface 1 in Figure 6.7). For the load a_L is the incoming wave and b_L the outgoing wave. In this picture the components are connected with transmission lines of the characteristic impedance Z_0 (usually $Z_0 = 50\,\Omega$). The waves travel back and forth between the components, even though in reality the distances are short compared to the wavelength and no real wave propagation takes place at all.

What is the relationship between Thevenin equivalent (Figure 6.7), and the wave source. The internal impedance, which generally has the complex value $Z_S = R_S + jX_S$ determines the reflection coefficient Γ_S.

$$\Gamma_S = \frac{Z_S - Z_0}{Z_S + Z_0} \tag{6.40}$$

The source wave, i.e. the actual power generation within the generator, is given by

$$b_S = V\frac{\sqrt{Z_0}}{Z_0 + Z_S} \tag{6.41}$$

A generally complex, normalized wave amplitude with the unit $[b_S] = \sqrt{\mathrm{W}}$.

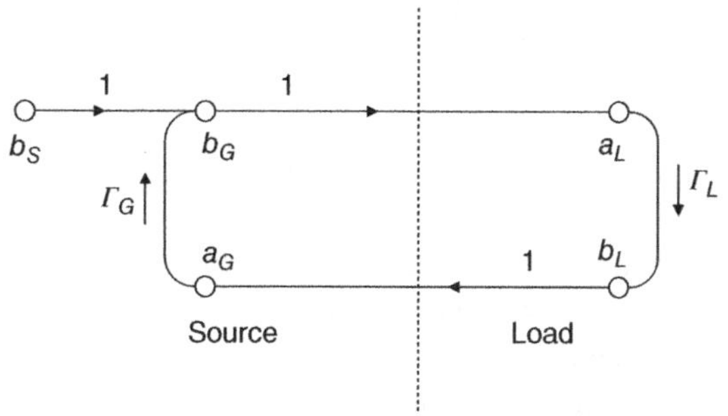

Figure 6.9 Flow chart of a source with complex internal resistance and a load.

The wave running off the generator is composed of the actual source term b_S and the reflected part a_G due to mismatch of the generator.

$$b_G = b_S + a_G \Gamma_S \tag{6.42}$$

If the generator operates to a non-reflecting load ($a_G = 0$), so is b_S the delivered power wave. For example, noise generators are calibrated in the microwave range in this way.

The "1" in diagram Figure 6.9 means that the transmission lines are lossless. The following applies to the waves running on these lines:

$$b_G = a_L; \quad b_L = a_G \tag{6.43}$$

With the definition of the reflection coefficient

$$\Gamma_L = \frac{b_L}{a_L}; \quad \Gamma_G = \frac{b_G}{a_G} \tag{6.44}$$

With (3.40) and (3.41) we obtain for the wave running onto the load

$$a_L = \frac{b_S}{1 - \Gamma_G \Gamma_L} \tag{6.45}$$

and for the wave reflected by the load

$$b_L = \frac{b_S \Gamma_L}{1 - \Gamma_G \Gamma_L} \tag{6.46}$$

The difference is the power absorbed by the load.

How does the power of the waves a and b result from the definition (6.41)?

In (6.30) we have the maximum power of the voltage source. In the simplest case, that all resistors are real and equal, P_{AS} corresponds to the source wave b_S.

We set

$$Z_0 = Z_S = R_S \tag{6.47}$$

thus (6.41)

$$b_S = \frac{V}{2\sqrt{R_S}}; \quad V^2 = 4b_S^2 R_S \tag{6.48}$$

we insert that in (6.30)

$$P_{AS} = \frac{4b_S^2 R_S}{8R_S}; \quad P_{AS} = \frac{1}{2} b_S^2 \tag{6.49}$$

There are two limitations to this result. (i) The source resistance is generally complex. (ii) The reference to the peak value of the voltage V is not common and is not defined for noise processes. If instead one refers to the RMS value $v_{RMS} = \sqrt{2}\,\text{V}$ in the case of the sinusoidal signal, the result for the reflection-free case is ($\Gamma_G = 0$):

$$P_{AS} = |b_S|^2 \tag{6.50}$$

For noise problems, the source is defined by an effective value $\sqrt{\overline{v_{RMS}^2}} = \sqrt{4kTRB}$. This indicates the available power

$$P_{AN} = \frac{\overline{v_{RMS}^2}}{4R} \tag{6.51}$$

Model calculations usually only consider a narrow frequency band, especially when partially correlated noise signals are superimposed in a common load resistor. Broadband consideration is not useful, since different frequencies are uncorrelated in principle. Therefore, noise signals are calculated as if they were monochromatic and the complex formalism can be applied. For the noise voltage in the vicinity of f_0, e.g.

$$\begin{aligned} v(t) &= \sqrt{S_V\left(f_0\right) B}\, exp\, j\left(\omega_0 t + \phi\right) \\ v &= \sqrt{S_V\left(f_0\right) B}\, exp\, (j\phi) \\ \overline{v^2} &= vv^* = S_V\left(f_0\right) B \end{aligned} \tag{6.52}$$

Broadband and Spot Noise Factor

In our considerations we assumed that noise factor F and gain G_A are constant within the considered bandwidth B. This is also obvious in a transmission system, where we assume the presence of a signal of frequency f_0, which lies in the center of a narrow frequency band of width B. Where $B \ll f_0$ applies. However, if we look at a wider frequency band, e.g. the entire transmission band of a broadband amplifier in Figure 6.10, this condition is no longer fulfilled. Here too, a noise factor can be defined by relation of the noise power at the input and output, as in definitions (6.3) and (6.11). This is called the integrated or broadband noise factor, as opposed to the spot noise factor at a defined frequency. Almost always, one is dealing with the spot noise factor without emphasizing it. Very important is the integration over the entire passband for applications in the MHz range or in acoustics. Here the noise voltage at the output is of interest, to which all frequency ranges contribute through integration. This can also be a strong part of the $1/f$ noise.

Starting from the definition of the noise factor (6.11) we can write

$$\frac{N_O}{S_O} = F\frac{N_I}{S_I} \tag{6.53}$$

With the frequency-dependent variables $G_A(f)$ and $F(f)$ applies:

$$N_O = FG_A N_I = \int_0^{\infty} F(f) G_A(f) k T_0 df \tag{6.54}$$

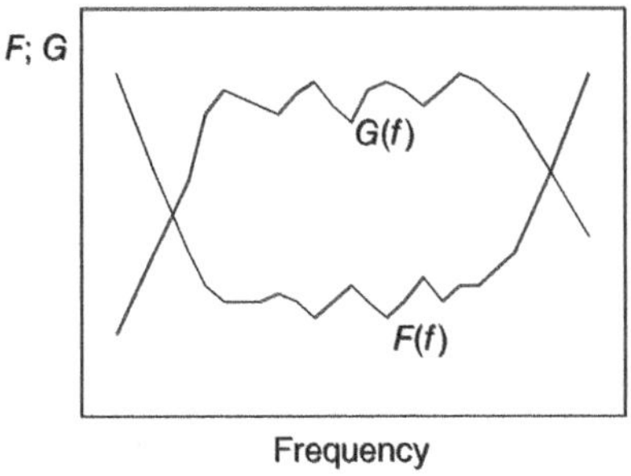

Figure 6.10 Fine structure of noise factor F and gain G of an amplifier versus frequency.

and

$$G_A N_I = \int_0^\infty G_A(f) k T_0 df \tag{6.55}$$

Thus we obtain for the integrated noise factor

$$\overline{F} = \frac{\int_0^\infty F(f) G_A(f) df}{\int_0^\infty G_A(f) df} \tag{6.56}$$

For a non-uniformly shaped transfer curve as shown in Figure 6.10, an equivalent noise bandwidth B_{eq} is defined:

$$B_{eq} = \frac{1}{G_{A,MAX}} \int_0^\infty G(f) df \tag{6.57}$$

The relationship between output noise power and integrated noise factor is then:

$$N_O = \overline{F} k T_0 B_{eq} G_{A,MAX} \tag{6.58}$$

This can be used, for example, to calculate the noise voltage at the output of a measuring amplifier.

Noise Factor of a Passive Network

One can assign a noise figure to attenuators and cables just like to amplifiers [7]. We have seen in Chapter 4 with an example that the inherent noise of these passive two-ports can be understood with the dissipation theorem. Here we want to derive the additional noise factor F_Z according to (6.7). The passive network shall be at the temperature T_a. The available noise power at the output is the numerator of F in (6.59), the part which comes only from the source is the denominator. We use the normalized power balance (4.52)

$$Ab + Tr + Re = 1$$

These linear quantities (not dB) stand for the absorption Ab, the transmission Tr, and the reflection Re. Here $Re = 0$.

From the source the part $kT_0B \times Tr$ attains the output. According to the dissipation theorem, the noise contribution of the network is $kT_aB \times Ab$. Thus we have

$$F = \frac{kT_0BTr + kT_aBAb}{kT_0BTr} \tag{6.59}$$

With $Ab = 1 - Tr$ we obtain

$$F = 1 + \frac{T_a}{T_0}\left(\frac{1}{Tr} - 1\right) \tag{6.60}$$

With the attenuation introduced in Chapter 4 one usually writes

$$\alpha = \frac{1}{Tr} = \frac{1}{1 - Ab} \tag{6.61}$$

$$F = 1 + \frac{T_a}{T_0}(\alpha - 1) \tag{6.62}$$

If the cable or attenuator is at ambient temperature ($T_a \cong T_0$) so simply

$$F = \alpha \tag{6.63}$$

A 3 dB attenuator therefore has the noise figure $NF = 3$ dB.

The equivalent noise temperature T_E (not to be confused with T_a) is by definition

$$T_E = (\alpha - 1)T_0 \tag{6.64}$$

Antenna Temperature

Each antenna receives not only the wanted signal but also interference signals [8]. The interfering signals can be "Man Made" or of natural origin. They limit the sensitivity of a receiving system. The internal noise sources of the system can be influenced by noise matching, optimally designed input stages, and minimization of cable losses. In the transmission path and the receiving antenna, the limits are determined by noise originating from terrestrial and cosmic sources. In 1933 [9] Karl G. Jansky was the first to systematically scan the night sky for radio radiation. He discovered that it has a noise character and that a maximum of the radiation comes from the center of the Milky Way. Jansky is regarded as the founder of radio astronomy. We consider summarily the natural sources whose noise contributions are received by the antenna as electromagnetic waves. An overview of the brightness temperatures T_B (definition see (6.66)) as a function of frequency is shown in Figure 6.11.

Below 100 MHz, noise generated by technology prevails (man-made noise). In addition there is galactic noise, which is dominant up to 1 GHz. The latter comes from the center of the Milky Way and from radio stars. Thermal sources

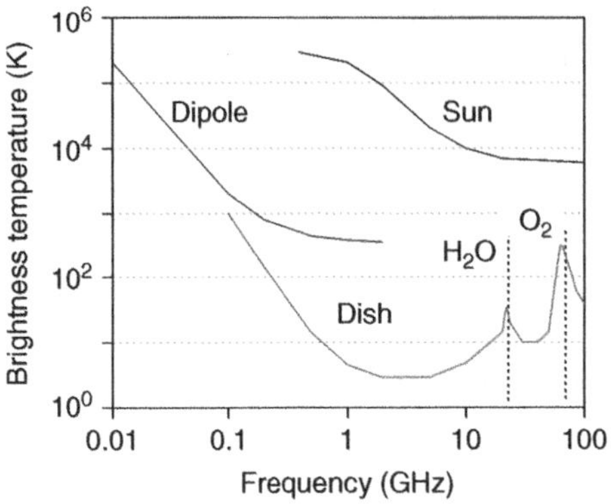

Figure 6.11 Brightness temperature T_B versus frequency. The dipole receives from all directions, the parabolic antenna looks into cold space.

are the earth's surface and the atmosphere. This is particularly effective above 10 GHz. Between 1 and 10 GHz there is a minimum of cosmic noise. Here the isotropic residual radiation with a temperature of 2.7 K remains, which is of great importance for space physics and cosmology. The relative maxima at 22 and 63 GHz are not cosmic in nature. They result from the absorption of water vapor and the O_2 molecule present in the atmosphere. As we have seen above that increased absorption also leads to increased emission.

All external contributions can be combined to one noise source in the antenna. For this purpose the antenna temperature T_A was introduced. The antenna delivers an available noise power

$$N = kB \times T_A \tag{6.65}$$

This is the definition of the antenna temperature. It is the fictitious temperature at which the real part of the antenna impedance is assumed. Its effect is that the same noise power occurs as in the real antenna exposed to the disturbing radiation of the sky. The antenna temperature has nothing to do with the physical ambient temperature that one would measure with a thermometer. For a strongly focused parabolic antenna directed into cold space, T_A can be a few 10 K, for an antenna directed toward the sun a few 10^4 K.

As already mentioned, the noise radiation interfering on the antenna is composed very inhomogeneously. There are area sources and point sources. The classification depends on the directivity of the antenna. For the antennas commonly used in communication systems, the earth's surface, the atmosphere, and outer space are area sources. The outer space actually also consists of point sources, but these can only be resolved with the extreme antenna systems of radio astronomy. A typical point source is the sun.

The sources are characterized by a brightness temperature T_B (brightness temperature). What is the relationship between brightness temperature and antenna temperature?

In Chapter 3 we had introduced the Planck formula (3.13) for the radiation of the black body. The sources of radio radiation from space, including the quiet sun, are very similar to the black body. For us, the Rayleigh–Jeans approximation (3.14) is sufficient, which was derived with the methods of classical electrodynamics even before the foundation of quantum theory. We first define the relationship between the temperature of brightness and the temperature of the black body. The brightness temperature is defined from the observer's point of view, i.e. in our case from the antenna. The temperature of the black body is a property of the source. The brightness temperature T_B is the temperature of the black body that it must have to produce the radiation intensity present at the antenna. The surface of the black body emits the intensity

$$I = \frac{2kT_B}{\lambda^2} \quad [I] = \mathrm{W/m^2\,Hz\,sr} \tag{6.66}$$

Various forms of the radiation formula are in use. In (3.14) we had the form that contained the solid angle $d\Omega$. Ours is adapted to the batwing radiator. One can see from the unit that it is the power in Watts radiated from the unit area 1 m^2 in the bandwidth 1 Hz into the unit solid angle 1 sr. The unit of extremely low spectral power flux density is called Jansky Jy.

$$1\,\mathrm{Jy} = 10^{-26}\,\mathrm{W/m^2\,Hz}$$

An antenna directed to the sky sees a section of the radiating surface, the size of which depends on the directivity. A measure of the size is the solid angle at which the area appears from the earth. A bounded source appears under the solid angle Ω_S (sun in the optical range $\Omega_S = 7\times10^{-5}$ sr). The solid angle Ω_A that the antenna captured depends on its size. The gain of an antenna is defined by this solid angle. G is the antenna gain compared with the monopole, i.e. the isotropic omnidirectional radiator. This has $G = 1$ (3.15).

$$\frac{4\pi}{\Omega_A} = G \tag{6.67}$$

The isotropic omnidirectional radiator exists only as an operand. It emits uniformly into the entire solid angle 4π. Each technical antenna has a radiation pattern and therefore receives preferably from the direction of the antenna lobe. The ratio (6.67) is a kind of gain. The logarithmic measure is called dBi. As an indication of the relation to the isotropic radiator.

$$G(\mathrm{dBi}) = 10\,log(G)$$

For parabolic mirrors the following relation is given:

$$G = \frac{4\pi A_{EFF}}{\lambda^2} \cong 10\frac{D^2}{\lambda^2}\eta \tag{6.68}$$

A_{EFF} is the effective antenna area. It is smaller than the geometric circular area of the parabolic mirror with diameter D. This reduction is measured by the antenna efficiency $\eta = 0.6$–0.8. The total power of the receiving antenna comes from the solid angle Ω_A and not from the full solid angle 4π as with the isotropic radiator. Therefore, with (6.67)

$$\Omega_A = \frac{\lambda^2}{A_{EFF}} \tag{6.69}$$

If the area in the sky radiating with intensity I (6.66) is opposite an absorbing receiving area of the size A_{EFF}, it absorbs the noise power N_A.

$$N_A = \frac{1}{2} A_{EFF} I \Omega_A B \tag{6.70}$$

The factor ½ takes into account that the antenna only receives one polarization direction. According to the definition of the antenna temperature, N_A is the power corresponding to the resistance on the temperature T_A.

$$N_A = kT_A B = \frac{1}{2} A_{EFF} I \Omega_A B \tag{6.71}$$

With the radiation intensity (6.66) the following is obtained

$$kT_A B = \frac{A_{EFF} \Omega_A B k T_B}{\lambda^2} \tag{6.72}$$

as $A_{EFF}\Omega_A/\lambda^2$ we get for batwing radiators that are larger than the spot of the antenna:

$$T_A = T_B \tag{6.73}$$

This also applies to sources of small dimensions to which a strongly focusing antenna is directed, e.g. the sun in Figure 6.12b. If the antenna lobe is directed at the earth's surface with grazing incidence, the average is $T_B \cong 290\,\text{K}$. The specification of the reference temperature $T_0 = 290\,\text{K}$ for the noise factor F has its origin in this. From an airplane or satellite, there is a fine structure of T_B, which can provide information about surface structure, agricultural data, and other important information. This method is called passive radar. Another form of passive radar uses the tracks that a flying object produces in the electromagnetic waves of telecommunications over densely populated areas. However, this is not the subject of our considerations.

Seen from the observer, the source appears under the solid angle Ω_S. For a point source, is $\Omega_S < \Omega_A$, for the batwing radiator, $\Omega_S > \Omega_A$, (Figure 6.12b). For the batwing radiator, the gain of the antenna is not important. As the gain, i.e. A_{EFF}, increases, Ω_A decreases, so that the product remains constant. With the batwing radiator, the antenna temperature is independent of the gain of the antenna. For the point source, however, the following applies

$$T_A = \frac{\Omega_S}{\Omega_A} T_B \tag{6.74}$$

If we direct a parabolic antenna toward the sun, the antenna temperature is composed proportionally (6.74) of the brightness temperature of the sun and

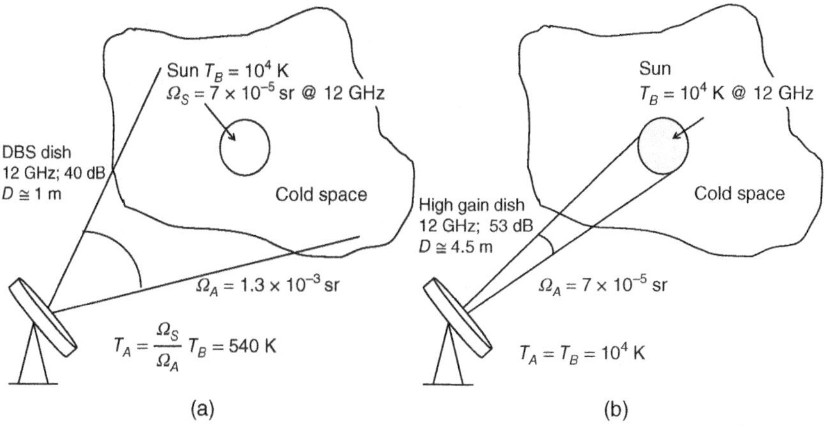

Figure 6.12 Parabolic antenna pointed at the sun. (a) Weak directionality, cold space predominates (T_A = 540 K). (b) Strong directionality on the sun ($T_A = 10^4$ K).

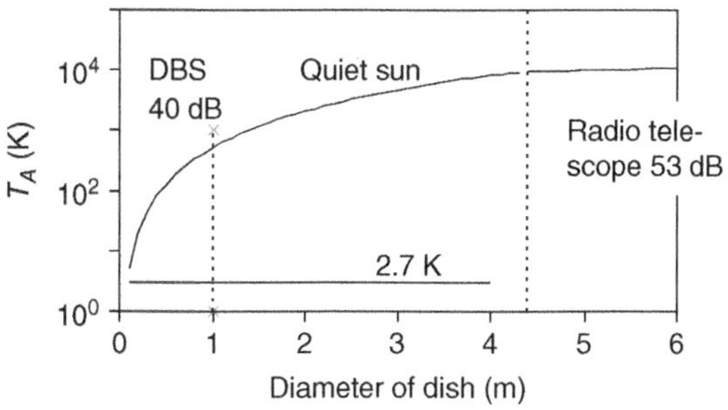

Figure 6.13 12 GHz parabolic antenna pointed at the quiet sun. T_A increases with the diameter until $D \cong 4.4$ m. There the antenna spot is equal to the sun's surface.

that of cold space. Figure 6.13 shows the increase of T_A with the antenna size, i.e. with increasing directivity. For a typical DBS antenna with 40 dBi gain T_A = 540 K. A large part of the cold environment of the sun is still detected (Figure 6.12a). A rather small radio astronomical antenna with D = 4.4 m has the aperture angle $\Omega_A = 7 \times 10^{-5}$ sr and thus just detects the solar disk. Even with stronger directionality the antenna temperature does not rise above $T_A = 10^4$ K (Figure 6.12b).

If we consider not only the two extreme cases of the area source $\Omega_S \gg \Omega_A$ and the point source $\Omega_S \ll \Omega_A$, but also a structure of the source $T_B(\theta, \phi)$ and an antenna lobe $G(\theta, \phi)$ then T_A results from the integral over the two characteristics:

$$T_A = \frac{1}{4\pi} \int_\Omega G(\theta, \phi) T_B(\theta, \phi) d\Omega \tag{6.75}$$

An antenna itself always has a contribution T_L to the antenna temperature, which results from the ohmic losses of the construction parts. Sources are mainly the finite conductivity of the metal surface, connecting cables and rotary joints. These losses generate noise with the physical temperature T_a of these construction parts. In (4.56) we derived the temperature of a noise generator with an attenuator. We had assumed a high value $T_{HOT} \cong 10^4$ K.

$$T_{OUT} = \frac{1}{\alpha}\left(T_{HOT} - T_a\right) + T_a \tag{6.76}$$

Here α is the attenuation (linear) due to the losses at the ambient temperature T_a. If we apply this to T_L, in order to estimate the contribution of the losses to the antenna temperature T_A, we set $T_{HOT} = 0$ K. Then we get:

$$T_L = T_a\left(1 - \frac{1}{\alpha}\right) \tag{6.77}$$

Antenna losses of 1 dB ($\alpha = 1.26$) on $T_a = 300$ K contribute $T_L \cong 62$ K to the antenna temperature.

The Reference Temperature $T_0 = 290$ K

In the context with the antenna temperature, it is instructive to reconsider the definition of the reference temperature for the noise factor $T_0 = 290$ K [10, 11]. This originates from the middle of the twentieth century, when the terrestrial radio-relay links were developed. The receiver input stages were diode mixers, with noise figures of $NF = 6 \ldots 10$ dB. The antennas were aligned parallel to the earth surface and therefore had a $T_A \cong 290$ K.

Today, in space-based communication systems, we have $T_A = 10 \ldots 30$ K and, thanks to advances in semiconductor technology, $NF \cong 1$ dB in the input stage. An evaluation of the noise power behind the antenna, i.e. at the input of the first stage with the noise figure based on T_0 can lead to misleading results. In (6.7) we had

$$N_A = F_Z k T_0 B \tag{6.78}$$

The noise generated by the amplifier is given by the additional noise factor $F_Z = F - 1$. Related to the input, i.e. without amplification, the N_A is according to (6.78). In addition, there is the noise contribution N from the antenna (6.70).

$$N_{IN} = N_A + N \tag{6.79}$$

$$N_{IN} = kT_0B(F - 1) + kT_AB \tag{6.80}$$

Figure 6.14 shows an example of the noise level (in dBm) at the input of the first stage for two different antenna temperatures depending on the noise figure of the receiver. At $NF < 5$ dB the true noise level is much lower than if T_0 would be expected. It can be seen that for space systems, calculating with temperatures

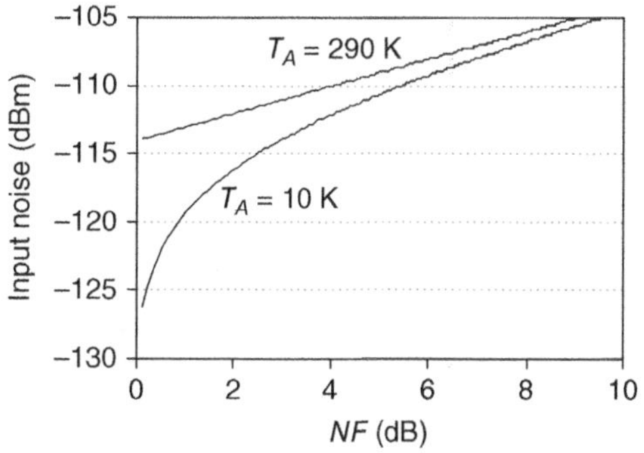

Figure 6.14 Calculated noise level N_{IN} (dBm) of a receiver (6.81) for the reference temperatures $T_A = 10$ and 290 K versus the noise figure of the preamplifier $B = 1$ MHz.

is better than with noise figure. In the early terrestrial systems mentioned above ($NF > 6$ dB) the difference did not play a role. The reference to the input impedance, which is noisy with $T_0 = 290$ K, leads to a too high noise level, because the real part of the antenna impedance does not generate noise with T_0, but with the lower temperature T_A. In logarithmic measure with N_{IN} (dBm), NF (dB), and B (MHz) is (6.80):

$$N_{IN}(\text{dBm}) = 10\,log\left(10^{\frac{NF(\text{dB})}{10}} + \frac{T_A}{T_0} - 1\right) - 114.5\left(\frac{\text{dBm}}{\text{MHz}}\right) \tag{6.81}$$

Noise Factor and Detection Limit

It is easy to estimate the detection limit of a system on the basis of the noise factor. In the case the signal disappears in the noise, the system is useless. If signal and noise level are equal, one defines the sensitivity limit. The decisive quantity here is the noise factor F of the receiver. We had a definition of F in (6.11). There is S_i the available signal power of the voltage source v_S

$$S_i = \frac{v_S^2}{4R} \tag{6.82}$$

The available noise power from the internal resistance R of the source is

$$N_i = kTB \tag{6.83}$$

Other noise sources are not present. The definition of sensitivity limit is $S_O/N_O = 1$. The signal just disappears in noise, i.e. signal and noise voltage at the output are equal.

Thus we have $S_i = FN_i$ and

$$v_{MIN} = \sqrt{F4kTBR} \tag{6.84}$$

In Figure 6.15 the smallest detectable voltages according to (6.84) are shown for two examples.

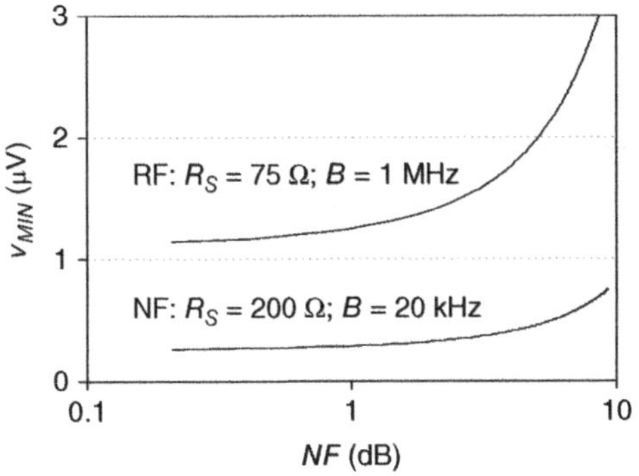

Figure 6.15 Sensitivity limit (6.84) for RF ($B = 1$ MHz) and LF amplifiers ($B = 20$ kHz) versus the noise figure.

References

1 Rothe, H. and Dahlke, W. (1956). Theory of noisy fourpoles. *Proc. IRE* 44: 811–818.

2 Hewlett Packard (2004) Fundamentals of RF- and Microwave Noise Figure Measurements. Application Note 57-1.

3 Friis, H.T. (1944). Noise figures of radio receivers. *Proc. IRE* 32: 419–422.

4 Haus, H.A. and Adler, R.B. (1959). *Circuit Theory of Linear Noisy Networks.* Wiley.

5 Meinke, H.H. and Gundlach, F.W. (1992). *Taschenbuch der Hochfrequenztechnik.* Springer Verlag.

6 Kummer, M. (1989). *Grundlagen der Mikrowellentechnik.* Berlin: Verlag Technik.

7 Schiek, B., Rolfes, I., and Siweris, H.-J. (2006). *Noise in High-Frequency Circuits and Oscillators.* Wiley Interscience.

8 Bokulic, R.S. (1991). Use basic concepts to determine antenna noise temperature. *Microwaves&RF* 30: 107–115.

9 Jansky, K.G. (1933). Radio waves from outside the solar system. *Nature* 132: 66.

10 North, D.O. (1942). The absolute sensitivity of radio receivers. *RCA Rev.* 6: 332–343.

11 Harper, T. and Grebenkemper, J. (1984). Where lies the thermal noise floor? *Microwaves&RF* 23: 115–117.

7

Noise of Linear Two-Ports

Representation of Two-Ports

We will now look at the noise behavior of two-port devices in more general terms, especially with regard to measurement technology and noise modeling. Since in practice these are usually input stages of a communication system or a noise receiver, the signal level will be small or even non-existent. The two-port is therefore linear. As is well known, there are many possibilities for linear two-port systems to calculate the combination of input and output variables [1, 2]. For noise modeling purposes, the representations in admittance (Y matrix) and impedance (Z matrix) are suitable. The chain form (A-matrix) is particularly suitable for measurement purposes. The hybrid form (H-matrix) is suitable for transistor modeling.

It is important to note that the following considerations apply to a specific frequency. For noise considerations, this is not a monochromatic signal but a narrow frequency band of width B, symmetrical around the frequency f_0.

Let us first look at the Y-form. The following applies

$$i_1 = v_1 Y_{11} + v_2 Y_{12} \quad i_2 = v_1 Y_{21} + v_2 Y_{22} \tag{7.1}$$

Of the four signal variables v_1, v_2, i_1, and i_2, only two are independent of each other. This is also the reason why the noise characteristics of a two-port are clearly described by two independent noise sources. In the Y-mode, these are the two currents i_1 and i_2 and for the Z-mode, the two voltages v_1 and v_2. They are mixed at the A-mode and at the H-mode (Figure 7.1). In these two equivalent sources all noise sources within the two-port are combined. In a passive two-port it is the thermal noise of ohmic resistors or line losses. In an active two-port, there is also the over-thermal noise of the transistors or other components. This also means that the two sources are generally correlated with each other, as they combine common physical causes. In this representation, the remaining two-port is noise-free and has only its network properties, which are given by the corresponding matrices

A Guide to Noise in Microwave Circuits: Devices, Circuits, and Measurement, First Edition.
Peter Heymann and Matthias Rudolph.

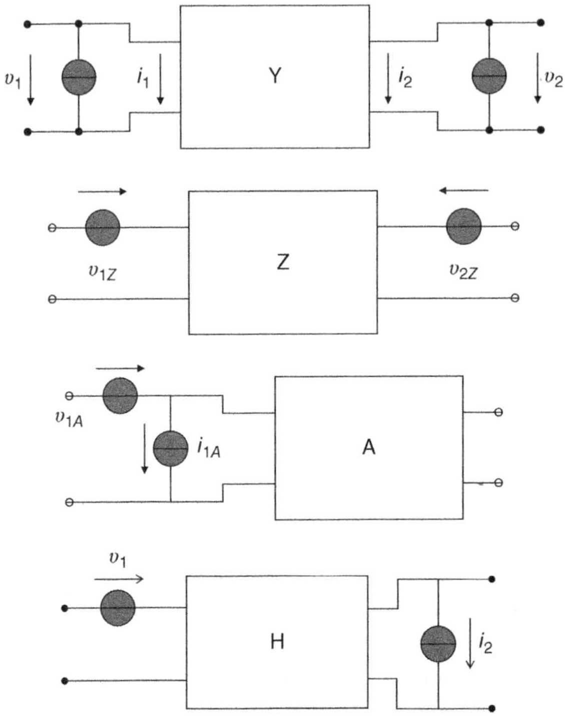

Figure 7.1 Placement of current and voltage sources in the different types of the two-port.

(Y, Z, A, H). The modes of the two-port representation can be converted into each other using the methods of circuit algebra. Admittance, impedance, chain, and hybrid representation are thus equivalent, i.e. their effect on connected networks is identical. For example, the sources of the chain mode result from the admittance form by

$$v_{1A} = -\frac{i_2}{Y_{21}}; \quad i_{1A} = i_1 + v_{1A}Y_{11} = i_1 - i_2\frac{Y_{11}}{Y_{21}} \tag{7.2}$$

Both v_{1A} and i_{1A} include parts of i_2 and are therefore partially correlated with each other. A entity of four quantities is therefore required for complete noise characterization of the two-port. The two sources and the complex correlation coefficient. We have already experience with the principle of noise analysis that all sources are placed at the input. Therefore, the representation in the form of the chain matrix is particularly suitable for our purposes.

Noise Modeling Using the Chain Matrix

The transmission of current and voltage of a monochromatic signal in a linear regime is given in the formalism of the chain parameters by [3] (Figure 7.2)

$$\begin{aligned} v_2 &= Av_3 + Bi_3 \\ i_2 &= Cv_3 + Di_3 \end{aligned} \tag{7.3}$$

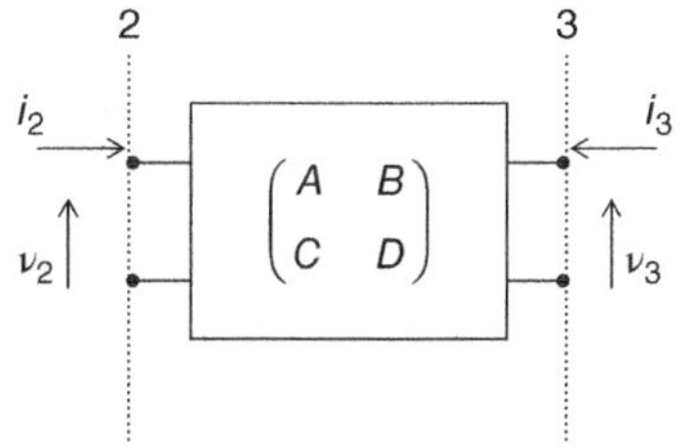

Figure 7.2 Schematic of the currents and voltages in the chain matrix.

Adding the noise sources v_N and i_N at the input, we get

$$v_2 = Av_3 + Bi_3 + v_N$$
$$i_2 = Cv_3 + Di_3 + i_N \tag{7.4}$$

The noise sources v_N and i_N are correlated. We consider the correlation by splitting the current into a component that is fully correlated with the voltage source i_C and into the uncorrelated part i_U.

$$i_N = i_C + i_U \tag{7.5}$$

The quantitative measure of correlation is the complex correlation admittance Y_{COR} (Figure 7.3).

$$Y_{COR} = \frac{i_C}{v_N} \tag{7.6}$$

The current at the input is now

$$i_1 = i_N + i_2 = i_U + Y_{COR}v_N + i_2 \tag{7.7}$$

and the voltage

$$v_1 = v_N + v_2 \tag{7.8}$$

This corresponds to the following equivalent circuit [4] (Figure 7.4).

One obtains for i_1

$$i_1 = i_2 + i_U + v_1Y_{COR} - v_2Y_{COR} \tag{7.9}$$

With connection of the noise source at the input i_S, Y_S, we obtain the short-circuit current in reference plane 2 ($-Y_{COR}$ is short-circuited)

$$i_{TOT} = i_S + i_U + v_N(Y_S + Y_{COR}) \tag{7.10}$$

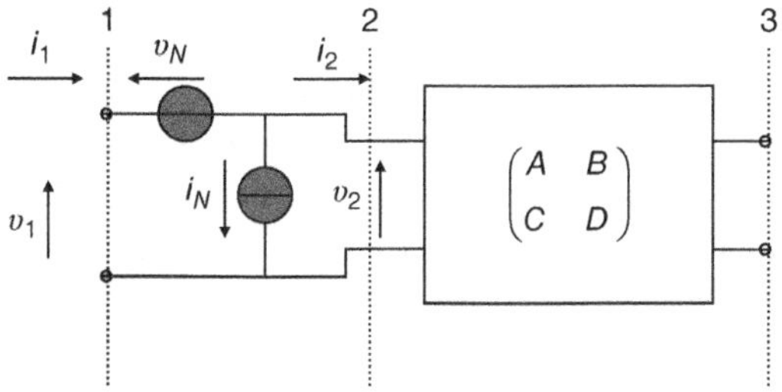

Figure 7.3 Network in chain mode with correlated noise sources at the input.

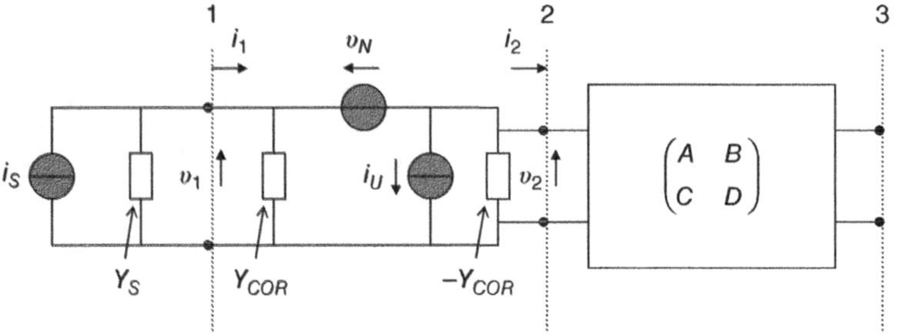

Figure 7.4 Equivalent circuit with source i_S, Y_S, and correlation coefficient Y_{COR}. Because Y_{COR} not relevant to the circuit, $-Y_{COR}$ is added.

Since all sources are uncorrelated in this representation, the square of fluctuation is

$$\overline{|i_{TOT}|^2} = \overline{|i_S|^2} + \overline{|i_U|^2} + \overline{|v_N|^2}|Y_S + Y_{COR}|^2 \tag{7.11}$$

With regard to the noise properties, we only need to look at reference plane 2. The network between planes 2 and 3 is by definition noise-free. It amplifies noise and signal power equally. The signal-to-noise ratio remains unchanged.

We write for the noise factor in reference plane 2

$$F = \frac{\overline{|i_{TOT}|^2}}{\overline{|i_S|^2}} = 1 + \frac{\overline{|i_U|^2} + \overline{|v_N|^2}|Y_S + Y_{COR}|^2}{\overline{|i_S|^2}} \tag{7.12}$$

This formula comprises the three noise sources i_S, i_U, and v_N. While i_S is the thermal noise of the generator internal resistance, the sources extracted from the two-port i_U and v_N include both thermal and overthermal components. A uniform description is obtained if all three are assumed to be noisy resistors at the reference temperature T_0.

$$\begin{aligned} \overline{|i_S|^2} &= 4kT_0G_SB \\ \overline{|i_U|^2} &= 4kT_0G_UB \\ \overline{|v_N|^2} &= 4kT_0R_NB \end{aligned} \tag{7.13}$$

R_N is the noise resistance of the two-port, G_U the noise conductance.

Inserting (7.13), the following results for F

$$F = 1 + \frac{G_U + R_N|Y_S + Y_{COR}|^2}{G_S} \tag{7.14}$$

It becomes obvious that the noise factor of a two-port is determined not only by the characteristics of the two-port itself, but also by the input circuit through $Y_S = G_S + jB_S$. This is also descriptive clear, since the two sources i_U and v_N feed their noise currents not only to the right into the two-port, but also to the left into the generator. There they are partially reflected and, depending on the phase position, are superimposed to form a total current at the two-port input, which is then amplified. The following questions arise: What is the minimum possible value of the noise factor F? What is its value and at what generator admittance Y_{OPT} is it achieved? It is also important: what is the gain of the two-port in this circuit?

Real and imaginary parts of Y_S must be optimized to obtain F_{MIN}. We first transform (7.14) to

$$F = 1 + \frac{G_U}{G_S} + \frac{R_N}{G_S}[(G_S + G_{COR})^2 + (B_S + B_{COR})^2] \tag{7.15}$$

With respect to the imaginary part, F is minimal if the following applies

$$(B_S + B_{COR})^2 = 0; \quad B_{OPT} = -B_{COR} \tag{7.16}$$

The value of F, for the case of circuit optimization with respect to B_S, is also called tuned noise factor (F tuned, F_{TUN}). In this case we calculate the derivative of the remaining term

$$\frac{\partial F_{TUN}}{\partial G_S} = 0$$

$$\frac{\partial F_{TUN}}{\partial G_S} = R_N - \frac{G_U}{G_S^2} - R_N\left(\frac{G_{COR}}{G_S}\right)^2 = 0 \tag{7.17}$$

This results in the following for the optimum source conductance

$$G_{OPT} = \sqrt{\frac{G_U}{R_N} + G_{COR}^2} \tag{7.18}$$

For tuning the generator to Y_{OPT}, the minimum noise factor of the two-port results in

$$F_{MIN} = 1 + 2R_N(G_{COR} + G_{OPT}) \tag{7.19}$$

This is now a characteristic of the two-port and independent of the input circuit. Using F_{MIN}, the noise factor for any source admittance Y_S is obtained:

$$F = F_{MIN} + \frac{R_N}{G_S}|Y_S - Y_{OPT}|^2 \tag{7.20}$$

For Y_{OPT} we receive

$$Y_{OPT} = G_{OPT} + jB_{OPT} = \sqrt{\frac{G_U}{R_N} + G_{COR}^2} - jB_{COR} \tag{7.21}$$

Let us still consider the case of negligible correlation between v_N and i_U, i.e. $Y_{COR} = 0$.

$$G_{OPT} = \sqrt{\frac{G_U}{R_N}}; \quad B_{OPT} = 0 \tag{7.22}$$

$$F_{MIN} = 1 + 2\sqrt{R_N G_U} = 1 + \frac{1}{2kT_0B}\sqrt{|v_N|^2|i_U|^2} \tag{7.23}$$

Derived from the equivalent circuit diagrams with two sources, we have shown that the noise characteristics of a two-port are determined by four quantities. Their measurement is called complete noise characterization. The quantities are derived directly from the network considerations:

$$R_N, G_U, G_{COR}, \text{and } B_{COR}.$$

In the practice of circuit design and measurement technology, the variables derived from this are preferred.

$$F_{MIN}, R_N, G_{OPT}, \text{and } B_{OPT}$$

In the wave representation of RF technology, reflection factors are used instead of admittances.

$$\Gamma_{OPT} = \frac{1-y_{OPT}}{1+y_{OPT}}; \quad y_{OPT} = Y_{OPT}Z_0$$
$$\Gamma_S = \frac{1-y_S}{1+y_S}; \quad y_S = Y_S Z_0 \tag{7.24}$$

The expression for the noise factor F in this notation is

$$F = F_{MIN} + \frac{4R_N}{Z_0}\frac{|\Gamma_{OPT}-\Gamma_S|^2}{|1+\Gamma_{OPT}|^2(1-|\Gamma_S|^2)} \tag{7.25}$$

As mentioned earlier, the noise factor F is a "spot noise factor." The values can be different for other frequencies. For transistors the manufacturer provides a table. In Table 7.1 are the data of a Si-bipolar transistor. This is an example of a mismatched two-port, as can be seen from the S-parameters in Figure 7.5.

The dependence of the effective noise factor F on the four noise parameters and the source reflection factor can be clearly shown by circles of constant noise figure in the Smith chart (Figure 7.6). We transform (7.25) into a circular equation.

$$\frac{F-F_{MIN}}{4r_N}|1+\Gamma_{OPT}|^2 = \frac{|\Gamma_S-\Gamma_{OPT}|^2}{1-|\Gamma_S|^2} \tag{7.26}$$

Table 7.1 Manufacturer's specifications of the noise parameters: Infineon Technologies BFP405 $V_{CE} = 1.0\,\text{V}$, $I_C = 5.0\,\text{mA}$; common emitter circuit.

f (GHz)	NF_{MIN} (dB)	$\lvert\Gamma_{OPT}\rvert$	$\angle\Gamma_{OPT}$ (°)	$R_N/50\,\Omega$
1.8	1.49	0.28	18	0.31
2.4	1.64	0.24	26	0.31
3.0	1.66	0.21	44	0.29
4.0	1.83	0.1	66	0.25
5.0	3.19	0.05	139	0.23

Source: Based on Infineon Technologies [5].

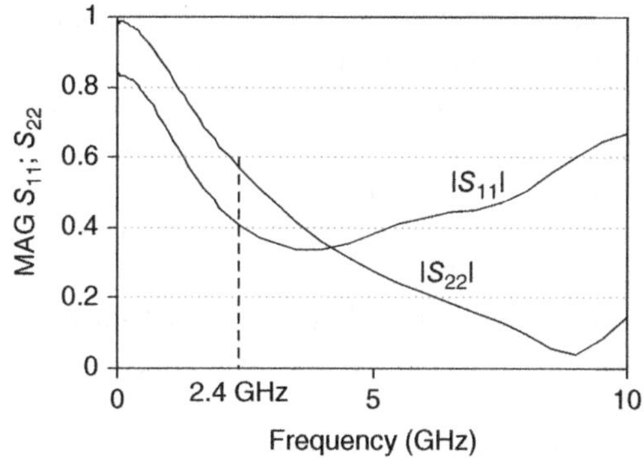

Figure 7.5 Example for mismatching of the BJT Infineon BFP405 at input ($|S_{11}|$) and output ($|S_{22}|$).

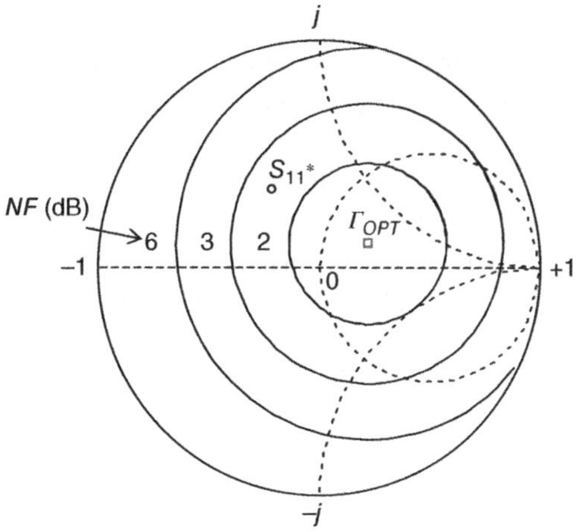

Figure 7.6 Smith chart of the source reflection coefficient Γ_S with circles of constant noise figure, which increases with distance from Γ_{OPT}. Data according to Table 7.1.

We summarize the noise quantities in N

$$N = \frac{F - F_{MIN}}{4r_N}|1 + \Gamma_{OPT}|^2 \tag{7.27}$$

with $r_N = {}^{R_N}/_{Z_0}$.

After some arithmetics one obtains a circular equation:

$$\left|\Gamma_S - \frac{\Gamma_{OPT}}{1+N}\right|^2 = \left\{\frac{1}{1+N}\sqrt{N^2 + N(1 - |\Gamma_{OPT}|^2)}\right\}^2 \tag{7.28}$$

The center of the circle is

$$\frac{\Gamma_{OPT}}{1+N} \tag{7.29}$$

and the radius

$$\frac{1}{1+N}\sqrt{N^2 + N(1 - |\Gamma_{OPT}|^2)} \tag{7.30}$$

Figure 7.6 shows the circles of constant noise figure of the example transistor BFP 405 in the Smith chart of the complex reflection factor Γ_S for the frequency $f = 2.4\,\text{GHz}$. In the case of noise matching ($\Gamma_{OPT} = 0.24/26°$) we have the minimum noise figure $NF = NF_{MIN} = 1.64\,\text{dB}$. If we move away from this point, the noise figure increases. A measure for this increase is given by the noise resistance R_N. For example, in the case of power matching ($\Gamma_S = S_{11}^*$), the noise figure has increased to $NF = 2.3\,\text{dB}$.

When designing a low-noise input stage, compromises between noise matching ($\Gamma_S \cong \Gamma_{OPT}$) and power matching ($\Gamma_S \cong S_{11}^*$) are unavoidable. This must also be optimized over the required bandwidth. The selection of a transistor with low noise resistance R_N is advantageous here, as this reduces the problem. It should also be noted that each transformation circuit is lossy, which in turn increases the overall noise figure. To get a first overview of the frequency dependence, it is useful to analyze the circuits at the lowest and at the highest frequency together with the frequency response curve of S_{11}^*. A CAD program should then be used for optimization.

If it is a system with antenna and cable at the input, it is usually not possible to work in noise matching. To ensure broadband capability, the amplifier must be matched to the characteristic impedance of the cable. A solution is offered by "active antennas," where the receiving antenna and amplifier are designed as one module and the amplifier is connected directly to the base point of the antenna [6]. Since the connecting cable is not required, noise matching can be ensured.

References

1 Feldtkeller, R. (1962). *Vierpoltheorie*. Stuttgart: Hirzel Verlag.

2 Schiek, B. (1984). *Messsysteme der Hochfrequenztechnik*. Hüthig Verlag.

3 Rothe, H. and Dahlke, W. (1956). Theory of noisy fourpoles. *Proc. IRE* 44: 811–818.

4 Müller, R. (1990). *Rauschen*. Springer Verlag.

5 Infineon Technologies (2019). BFP 405 NPN Silicon RF-Transistor. Datenblatt.

6 Landstorfer, F. and Graf, H. (1981). *Rauschprobleme der Nachrichtentechnik*. Oldenbourg Verlag.

8

Calculation Methods for Noise Quantities

Noise Voltages, Currents, and Spectra

When analyzing signal transmission passing a network, complex sinusoidal variables are used. Complex because amplitudes and phasing must be considered. Here we follow the presentation in [1].

Ohm's law for alternating current is

$$v(t) = Z(\omega)i(t) \tag{8.1}$$

Here $v(t)$ the complex voltage

$$v(t) = \sqrt{2}V_{eff}\ exp(j\omega t + \phi) \tag{8.2}$$

$i(t)$ the complex current

$$i(t) = \sqrt{2}I_{eff}\ exp(j\omega t) \tag{8.3}$$

and $Z(\omega)$ a complex, frequency-dependent impedance.

$$Z(\omega) = |Z(\omega)|\ exp(j\phi) \tag{8.4}$$

For fluctuating quantities, one works with squares of fluctuation that correspond to the noise power. Instead of the effective values, the variance σ^2 as a time average is used.

$$V_{eff}^2 = \overline{v(t)^2} = \sigma_V^2 \text{ and } I_{eff}^2 = \overline{i(t)^2} = \sigma_I^2 \tag{8.5}$$

For the squares of the effective values, the following result is obtained from (8.1)

$$V_{eff}^2 = |Z(\omega)|^2 I_{eff}^2 \tag{8.6}$$

For noise variables, the RMS value is obtained by integrating the spectral power density, here the one-sided spectral power density $G(\omega)$ over the frequency.

A Guide to Noise in Microwave Circuits: Devices, Circuits, and Measurement, First Edition.
Peter Heymann and Matthias Rudolph.

$G_V(\omega)d\omega$ is a differentially small contribution to the mean square of the voltage $\overline{v(t)^2}$. In a limited frequency band $B = f_2 - f_1$ and with $f = \omega/2\pi$ is the relation

$$V_{eff}^2 = \overline{v(t)^2} = \int_{f1}^{f2} G_V(f)df \tag{8.7}$$

The spectra of current and voltage are interrelated by the square of the impedance

$$G_V(f) = |Z(f)|^2 G_I(f) \tag{8.8}$$

Working with noise spectra and with signal quantities is therefore done using a $|Z(f)|^2$ corresponding quantity. Instead of the impedance $Z(f)$, also a conductance can be used if one wants to get from the voltage spectrum to the current spectrum. The phase relationship is not important, since the noise sources are characterized by quadratic mean values.

In the literature the designation of the power spectra is not uniform, which can lead to some confusion. Here a reminder: For the double-sided power spectrum, we use $S(f)$ for the one-sided $G(f)$. $S(f)$ is mathematically oriented and corresponds to the Fourier transform

Spectrum ↔ correlation function

and thus directly to the Wiener–Khintchine theorem. The integration is performed in the frequency range $-\infty < f < +\infty$. The one-sided power spectrum $G(f)$ is technically oriented. Here the integration is performed in the frequency range $0 \leq f < +\infty$, since there are no negative frequencies in practice. The relation is simple $G(f) = 2 \times S(f)$. For the thermal noise of a resistor R and the shot noise of a direct current I_{DC}, the following applies

$$S = 2kTR \quad G = 4kTR$$

$$S = qI_{DC} \quad G = 2qI_{DC} \tag{8.9}$$

The distinction between $S(f)$ and $G(f)$ is not always used consistently, so we should always make sure which form is used in a specific case.

Let us consider as an example the inherent noise of a lossy resonant circuit in Figure 8.1. The one-sided spectral power density $G_2 = 4kTR$ of the white noise of the loss resistance R in the reference plane 2, 2′ appears at the output modified by the LC-circuit with the transfer function $H(f)$. There is no more white noise. The frequency response in the reference plane 1, 1′ corresponds to the resonance curve.

$$G_1(f) = |H(f)|^2 G_2 \tag{8.10}$$

Using the resonant frequency $f_0 = 1/2\pi\sqrt{LC}$, we can write (see also (4.31))

$$G_1(f) = \frac{1}{(2\pi fCR)^2 + \left[1 - \left(\frac{f}{f_0}\right)^2\right]^2} \times 4kTR \tag{8.11}$$

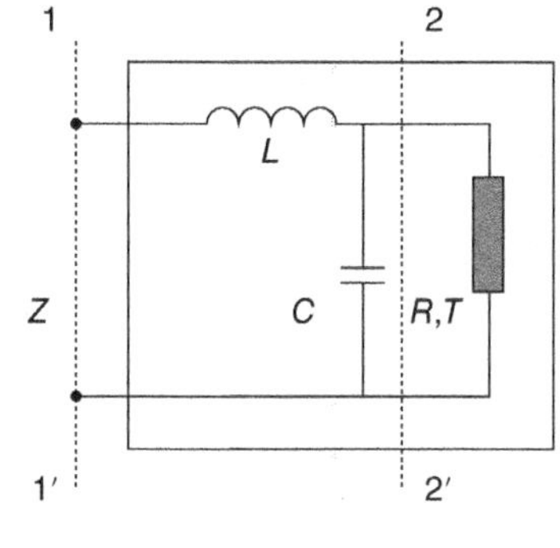

Figure 8.1 Noise of a lossy resonant circuit. Example data: $R = 200\,\Omega$ at $T = 300\,\text{K}$, $C = 0.1\,\mu\text{F}$, $L = 4\,\text{mH}$ with $f_0 = 8\,\text{kHz}$.

To calculate the total noise power, weighted over all frequencies, in the reference plane 1, 1′ we must (8.11) integrate in the frequency range $0 < f \leq \infty$. In case of the two-sided spectral power density, we would have in the reference plane 2, 2′: $S_2 = 2kTR$. Then we would also have to take into account the negative frequencies and integrate in the range $-\infty < f < +\infty$. Both results are shown in Figure 8.2.

This result is obtained as follows. We had already integrated a resonance curve in the frequency domain in (4.42). But there we used the approximation of the double detuning, which is only valid in the vicinity of the resonance frequency ($\Delta f \ll f_0$). Here we have the whole frequency range, so the integral is more complicated. It is of type [2]:

$$\int \frac{dx}{a + bx^2 + cx^4} = \frac{1}{4cq^3 \sin\alpha} \left\{ \sin\frac{\alpha}{2} \ln\frac{x^2 + D}{x^2 - D} + 2\cos\frac{\alpha}{2} \arctan\frac{x^2 - q^2}{2qx\sin\frac{\alpha}{2}} \right\} \tag{8.12}$$

with the abbreviations:

$$a = 1, b = (2\pi RC)^2 - 2f_0^{-2}, c = f_0^{-4}, q = f_0, \alpha = \arccos\left(-\frac{b}{2\sqrt{c}}\right),$$

$$D = 2f_0 x \cos\frac{\alpha}{2} + f_0^2$$

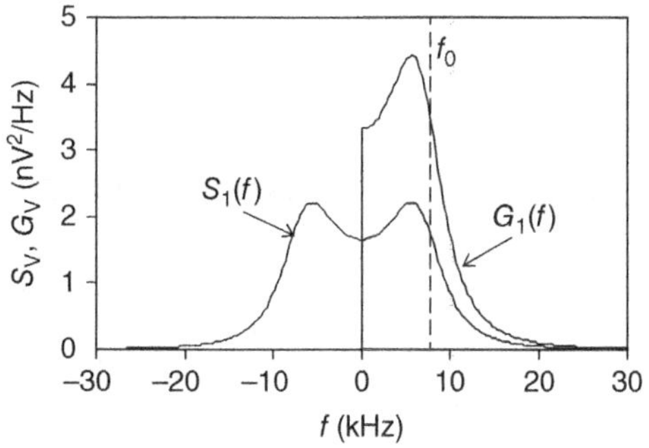

Figure 8.2 Spectral noise power density at the input plane 1, 1′ of the circuit Figure 8.1.

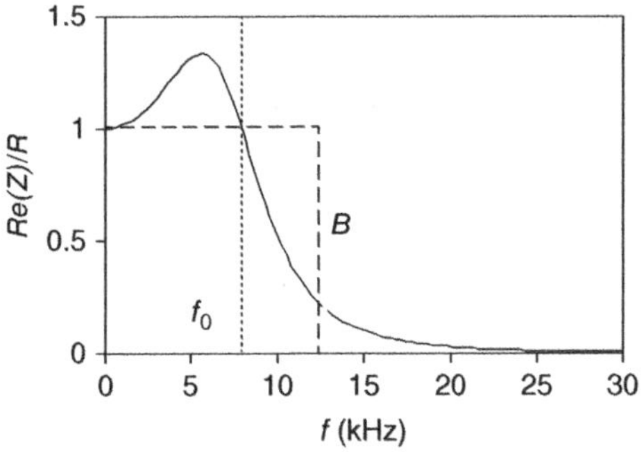

Figure 8.3 Spectral noise power density $G_1(f)$ in the input reference plane of Figure 8.1. From the white noise in plane 2, 2′ this frequency response appears in plane 1, 1′.

When setting the frequency limits for calculation of the area, it must be noted that the arctan term at $x = 0$ makes a jump by π. One obtains the following results for $f = +\infty$: 6.25×10^3 and for $f = 0$: -6.25×10^3 (Figure 8.3).

The value of the integral for the one-sided spectral power density G_1 thus corresponds to an equivalent rectangular bandwidth $B = 12.5\,\text{kHz}$. That means, if the noise power of the 200 Ω resistor across the network in Figure 8.1 would be measured with a broadband voltmeter (e.g. DC – 50 kHz), the following noise voltage would be obtained

$$v_N = \sqrt{4 \times 1.38 \times 10^{-23} \times 300 \times 200 \times 1.25 \times 10^4} = 0.2\,\mu\text{V}$$

Of course, this can only be done with a suitable preamplifier.

For the integration over S_1 in the range $-\infty$ to $+\infty$, one obtains the double value of the integral and must therefore be start from a spectral noise voltage density of the resistor of $S_2 = 2kTR$.

The expression (8.13)

$$Re(Z) = \frac{R}{(2\pi fCR)^2 + \left\{1 - \left(\frac{f}{f_0}\right)^2\right\}^2} \tag{8.13}$$

is also the real part of the network impedance in the input reference plane 1, 1′. Using this, we can formulate the generalized Nyquist theorem:

$$G = 4kT\,Re(Z) \tag{8.14}$$

Calculating with Current, Voltage, and Noise Waves

In linear circuit analyses, voltages and currents are interrelated using suitable matrices. A simple case is the calculation of input and output current (I_1, I_2) in a two-port whose properties, which are described by the admittance matrix. Accordingly, at input and output, the two voltage sources (V_1, V_2) can be connected and the impedance matrix can be used. Basically, the following applies: Two alternating voltages of the same frequency can cancel or amplify each other when superimposed in a load resistor, as is the case with optical interference. When noise power is superimposed in a certain frequency band, amplification or attenuation can also occur, depending on the correlation. In Chapter 2 we have shown by way of example what effect the correlation of the signals can have when superimposed to generate the total noise power. If one wants to create a formalism that allows the noise analysis in a comparable way as the network analysis, one arrives at the noise correlation matrices [3].

In RF and microwave technology, the wave model is the appropriate description for the signal flow. The circuit components are described by scattering matrices. In this system the network analyzer operates and provides the corresponding complex s-parameters. In the world of lower frequencies, the equivalent circuit with lumped elements (R, L, C) is common. For modeling, the application of the equivalent circuit is extended to the highest frequencies, so that the wave picture and the lumped elements merge into each other. Well known are the transformation formulas of the s-parameters into the usual two-port parameters Y, Z, etc. This consideration also applies to the noise parameters. The description of the noise, which is usually based on lumped elements, can also be done in the wave model. This results, as mentioned earlier, in the very powerful method of noise correlation matrices.

We will first introduce the noise waves and explain their relationship with the noise currents in the admittance representation. Here we follow the presentation in [4].

In the equivalent circuit with lumped elements (Figure 8.4), one works with voltages and currents, in the wave model with the complex wave amplitudes. The noise characteristics of the two-port are determined in the Y equivalent circuit by adding the two noise current sources I_{N1} and I_{N2}. These are simply added to the current vector. We use capital letters here because we will normalize later.

$$\begin{aligned} I_1 &= Y_{11}V_1 + Y_{12}V_2 + I_{N1} \\ I_2 &= Y_{21}V_1 + Y_{22}V_2 + I_{N2} \end{aligned} \tag{8.15}$$

in matrix form

$$I = \boldsymbol{Y} \cdot \boldsymbol{V} + I_N \tag{8.16}$$

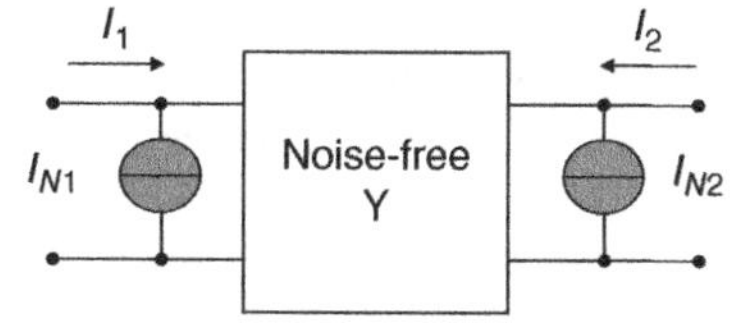

Figure 8.4 Two-port in Y representation with additional noise current sources I_{N1} and I_{N2}.

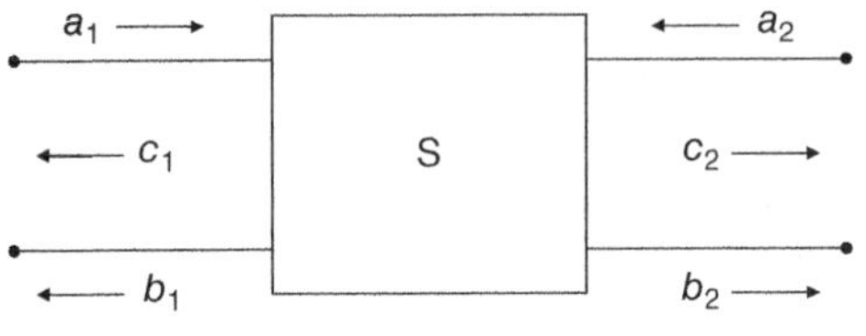

Figure 8.5 Two-port in s-parameter form with incoming waves $a_{1,2}$, outgoing waves $b_{1,2}$, and noise waves $c_{1,2}$.

In the wave picture (Figure 8.5) incoming and outgoing waves are characterized by the wave amplitudes a and b. For the two-port the s-parameters apply. The noise power generated in the two-port is described by the outgoing vector c.

The power transported with the incoming wave is

$$P_{1,in} = a_1 a_1^* = |a_1|^2 \tag{8.17}$$

where a and b refer to effective values. The same applies to the outgoing wave. The unit of the wave amplitudes is $[a] = \sqrt{W}$. The values in Figure 8.5 are connected via the s-matrix. Analogue to (8.15) applies

$$\begin{aligned} b_1 &= S_{11}a_1 + S_{12}a_2 + c_1 \\ b_2 &= S_{21}a_1 + S_{22}a_2 + c_2 \end{aligned} \tag{8.18}$$

in matrix form

$$\boldsymbol{b} = \boldsymbol{S} \cdot \boldsymbol{a} + \boldsymbol{c} \tag{8.19}$$

Before we examine the relationship between the noise wave vector c and the noise current vector I_N, we consider the wave vectors a and b. They represent purely propagating waves. They are related to the effective values of current and voltage of the incoming and outgoing wave.

$$a_1 = \frac{V_{1,in}}{\sqrt{Z_0}} = I_{1,in}\sqrt{Z_0}; \qquad b_1 = \frac{V_{1,out}}{\sqrt{Z_0}} = I_{1,out}\sqrt{Z_0} \tag{8.20}$$

Current and voltage at a specific position on the line (I_1, V_1), i.e. here at port 1, result from the superposition of the two waves running in opposite directions.

$$\frac{V_1}{\sqrt{Z_0}} = \frac{V_{1,in} + V_{1,out}}{\sqrt{Z_0}} = a_1 + b_1; \qquad I_1\sqrt{Z_0} = (I_{1,in} - I_{1,out})\sqrt{Z_0} = a_1 - b_1 \tag{8.21}$$

Conversely, the wave variables can be represented by current and voltage. For this purpose, the normalization is used.

$$v_1 = \frac{V_1}{\sqrt{Z_0}}; \qquad i_1 = I_1\sqrt{Z_0} \tag{8.22}$$

From (8.21) we get

$$a_1 = \frac{1}{2}(v_1 + i_1); \qquad b_1 = \frac{1}{2}(v_1 - i_1) \tag{8.23}$$

For the noise waves, we also use the normalized quantities. We look to gate 1 (input).

$$v_{N,1} = \frac{V_{N,1}}{\sqrt{Z_0}}; \qquad i_{N,1} = I_{N,1}\sqrt{Z_0} \tag{8.24}$$

The elements of the Y-matrix are also normalized.

$$y_{i,k} = Y_{i,k}\sqrt{Z_0} \tag{8.25}$$

The matrix Eq. (8.2) is then transformed into

$$\boldsymbol{i} = \boldsymbol{y} \cdot \boldsymbol{v} + \boldsymbol{i}_N \tag{8.26}$$

$\boldsymbol{i}$, $\boldsymbol{v}$, $\boldsymbol{i}_N$ are column vectors, and $\boldsymbol{y}$ is the normalized admittance matrix. Inserting (8.21) results in

$$\boldsymbol{a} - \boldsymbol{b} = \boldsymbol{y} \cdot (\boldsymbol{a} + \boldsymbol{b}) + \boldsymbol{i}_N = \boldsymbol{y} \cdot \boldsymbol{a} + \boldsymbol{y} \cdot \boldsymbol{b} + \boldsymbol{i}_N \tag{8.27}$$

transformed:

$$\boldsymbol{b} + \boldsymbol{y} \cdot \boldsymbol{b} = \boldsymbol{a} - \boldsymbol{y} \cdot \boldsymbol{a} - \boldsymbol{i}_N \tag{8.28}$$

further transformation with the unit matrix [**1**]:

$$([\mathbf{1}] + \boldsymbol{y}) \cdot \boldsymbol{b} = ([\mathbf{1}] - \boldsymbol{y}) \cdot \boldsymbol{a} - \boldsymbol{i}_N \tag{8.29}$$

Resolving to $\boldsymbol{b}$ allows the comparison with the matrix of the waveform (8.5)

$$\boldsymbol{b} = ([\mathbf{1}] + \boldsymbol{y})^{-1} \cdot ([\mathbf{1}] - \boldsymbol{y}) \cdot \boldsymbol{a} - ([\mathbf{1}] + \boldsymbol{y})^{-1} \cdot \boldsymbol{i}_N \tag{8.30}$$

This comparison yields the quested correlation of the noise currents $\boldsymbol{i}_N$ in the Y representation with the noise wave vector $\boldsymbol{c}$.

$$\boldsymbol{c} = -([\mathbf{1}] + \boldsymbol{y})^{-1} \cdot \boldsymbol{i}_N \tag{8.31}$$

The s-matrix is derived from the Y-matrix by the following operation.

$$\boldsymbol{S} = ([\mathbf{1}] + \boldsymbol{y})^{-1} \cdot ([\mathbf{1}] - \boldsymbol{y}) \tag{8.32}$$

This corresponds to the known transformation equation $Y \rightarrow S$ for the array elements.

The Noise Correlation Matrix

The goal of noise analysis of a communication system is to calculate the level of fluctuations in a receiver. These are, for example, fluctuating currents in a load resistor, or fluctuating voltages at this resistor. They consist generally of the interaction of various noise sources. Some interfere on the input from outside, others are located in the system itself. When superimposing the noise to the output level, both the levels of the individual sources and their correlation must be taken into account. Although the fluctuating quantities are not deterministic signals, we can perform network analysis as with signal quantities if we work with the spectral power densities in a certain frequency range. Because of correlation, we need both auto spectral densities and cross spectral densities. These are known to be the Fourier transforms of the autocorrelation function and the cross-correlation function, respectively. The integral of the auto spectral density over frequency is the variance of the fluctuating signal, its noise power [3].

We have compared the equivalent circuit model and wave model earlier, for the circuit model in normalized form (8.26), for the wave model (8.19).

The components i_{N1}; i_{N2} of the vector $\boldsymbol{i}_N$ can be correlated. This information must also be present in the wave model. We look at the power that is transported with the noise wave, e.g. at the output with c_2 according to Figure 8.5

$$N = c_2 c_2^* = |c_2|^2 \tag{8.33}$$

The powers of the single waves and their correlation are shown in the noise correlation matrix $\boldsymbol{C}$, which is determined by the tensor product (outer product) of the two vectors $\mathbf{c} = \begin{pmatrix} c_1 \\ c_2 \end{pmatrix}$ and $\mathbf{c}^* = \begin{pmatrix} c_1^* \\ c_2^* \end{pmatrix}$

$$\mathbf{c} \otimes \mathbf{c}^* = \mathbf{c} \cdot (\mathbf{c}^*)^T = \mathbf{c} \cdot \mathbf{c}^H = \begin{pmatrix} |c_1|^2 & c_1 c_2^* \\ c_2 c_1^* & |c_2|^2 \end{pmatrix} = \boldsymbol{C} \tag{8.34}$$

Mathematically it is the multiplication of a column vector with a row vector. $\mathbf{c}^H$ means the Hermitian vector to $\mathbf{c}$. The column vector becomes a row vector and the elements are replaced by their conjugate complex values. For matrices, the following applies: rows and columns are swapped and the elements are conjugated complex. The result is a matrix. The two noise waves c_1 (input) and c_2 (output) are combined to the vector $\mathbf{c}$. In general, the noise power will be frequency dependent. The elements of the correlation matrix $C_{I,K}(f)$ are called correlation spectra. They are the Fourier transforms of the autocorrelation function of the noise quantity. The elements of the main diagonal ($i = k$) are the autocorrelation spectra, and the elements of the secondary diagonal are the cross correlation spectra. From the definition of the Fourier transformation follows

$$C_{i,k}(f) = C_{k,i}^*(f) \tag{8.35}$$

The elements of the main diagonals are real and symmetrical, and the cross correlation spectra are complex. The real part is symmetrical, and the imaginary part is asymmetrical.

The variance σ^2, i.e. the noise power, results from the integration of the spectral power density over the entire frequency range.

$$\int_{-\infty}^{\infty} C_{i,i}(f) df = \sigma_i^2 \tag{8.36}$$

If we look at the two-sided spectral power density $S(f)$ of white noise integrated in the frequency interval df, which is mathematically correct, the result is

$$\sigma^2 = 2C_{i,i}df \tag{8.37}$$

As discussed earlier, the factor 2 does not occur when the technically oriented one-sided spectral power density $G(f)$ is used.

When analyzing the noise properties of a network, one can now work with the scattering matrices or the corresponding matrices of the equivalent circuit for the signal properties and with the correlation matrices for the noise properties. The values of the correlation matrices of passive elements can be derived from thermodynamic considerations. For transistors and other active elements there are a number of noise models. For complicated networks with active elements, e.g. amplifiers, measurements are usually required.

The Correlation Matrix of Passive Components

The simplest example is an ohmic source resistor R_S, which is connected to a matched load R_L via a transmission line with the characteristic impedance Z_0. This can be a 50 Ω load resistor. Both are at temperature T. According to Figure 8.5 the s-matrix consists of S_{22} only, which is equal to the reflection coefficient $\Gamma = (R_S - Z_0)/(R_S + Z_0)$ of the source (Figure 8.6).

From (8.19) results the scalar equation

$$b_2 = \Gamma a_2 + c_2 \tag{8.38}$$

The wave a_2 is emitted by the load resistor, the wave b_2 by the source resistor R_S. Both are uncorrelated. By squaring we get the power.

$$|b_2|^2 = |\Gamma|^2 |a|_2{}^2 + |c_2|^2 \tag{8.39}$$

The spectral power density of the wave coming from the load resistor is calculated according to the Nyquist formula $|a_2|^2 = kT$. Since the system is in thermodynamic equilibrium, the following must apply $|a_2|^2 = |b_2|^2$. It results for the power of the noise wave in the bandwidth 1 Hz:

$$|c_2|^2 = kT(1 - |\Gamma|^2) \tag{8.40}$$

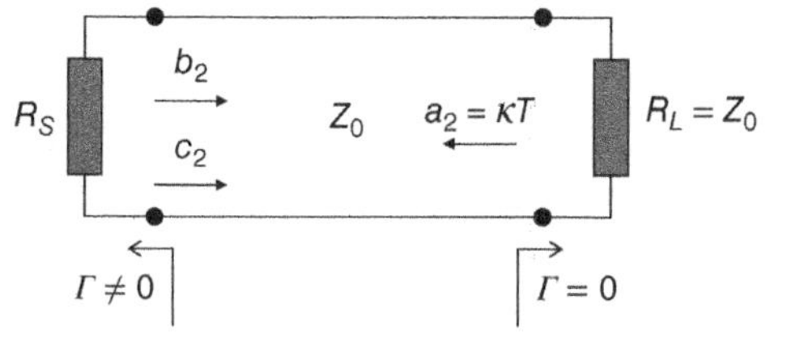

Figure 8.6 R_S represents the mismatched source, R_L the matched load.

This corresponds to the known expression for the available power of a source with the reflection coefficient Γ:

$$P_{AV} = \frac{|b|^2}{1 - |\Gamma|^2} \tag{8.41}$$

The power of the noise wave $|c_2|^2$ corresponds to the signal power $|b|^2$ and kT is the available power of the resistor, according to Nyquist's formula.

In this simple case the correlation matrix is reduced to one parameter $C_{22} = |c|^2$.

A general rule for deriving the correlation matrix of passive networks is given by the Bosma theorem [5]. In thermal equilibrium we obtain the correlation matrix in s-parameter form from the s-matrix of the passive network:

$$\boldsymbol{C}^S = kT([\mathbf{1}] - \boldsymbol{S} \cdot \boldsymbol{S}^H) \tag{8.42}$$

$\boldsymbol{S}^H$ is the Hermitic matrix to $\boldsymbol{S}$.

The simplest example is the series resistance. However, it is easy to see that for these simple networks the s-Matrix formalism is complicated and inappropriate (Figure 8.7).

The s-matrix of the series resistance is $z = {}^{R}/_{Z_0}$

$$\boldsymbol{S} = \begin{pmatrix} \frac{z}{2+z} & \frac{2}{2+z} \\ \frac{2}{2+z} & \frac{z}{2+z} \end{pmatrix} = \frac{1}{2+z}\begin{pmatrix} z & 2 \\ 2 & z \end{pmatrix} = \boldsymbol{S}^H \tag{8.43}$$

Transition to the Hermitian Matrix does not change anything, because z is real and rows and columns are equal. Used in Bosma's theorem it follows

$$\begin{aligned} \boldsymbol{S} \cdot \boldsymbol{S}^H &= \frac{Z_0^2}{(2Z_0 + R)^2}\begin{pmatrix} \left(\frac{R}{Z_0}\right)^2 + 4 & \frac{4R}{Z_0} \\ \frac{4R}{Z_0} & \left(\frac{R}{Z_0}\right)^2 + 4 \end{pmatrix} \\ &= \frac{4Z_0R}{(2Z_0 + R)^2}\begin{pmatrix} \frac{R}{4Z_0} + \frac{Z_0}{R} & 1 \\ 1 & \frac{R}{4Z_0} + \frac{Z_0}{R} \end{pmatrix} \end{aligned} \tag{8.44}$$

The difference to the unit matrix $[\mathbf{1}] - \boldsymbol{S} \cdot \boldsymbol{S}^H$ is

$$[\mathbf{1}] - \boldsymbol{S} \cdot \boldsymbol{S}^H = \frac{4RZ_0}{(2Z_0 + R)^2}\begin{pmatrix} 1 & -1 \\ -1 & 1 \end{pmatrix} \tag{8.45}$$

Thus we derived the noise correlation matrix of the ohmic series resistance R in a transmission line system with the characteristic impedance Z_0 in the S-parameter formalism.

$$\boldsymbol{C}^S = kT \times \frac{4RZ_0}{(2Z_0 + R)^2}\begin{pmatrix} 1 & -1 \\ -1 & 1 \end{pmatrix} \tag{8.46}$$

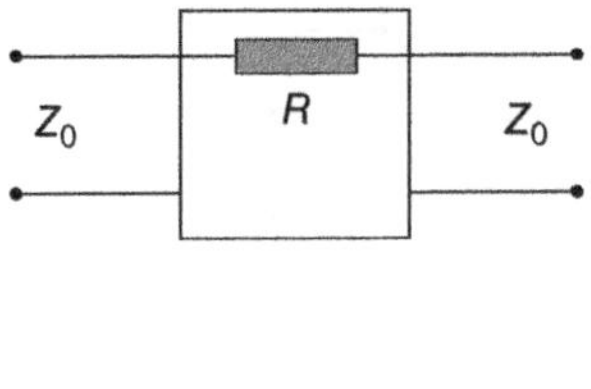

Figure 8.7 Derivation of the correlation matrix of a series resistor.

For the parallel resistor the following results analogously

$$C^S = kT \times \frac{4RZ_0}{(2Z_0 + R)^2}\begin{pmatrix} 1 & 1 \\ 1 & 1 \end{pmatrix} \tag{8.47}$$

The Noise of Simple Passive Networks

Example 8.1 ***Parallel Connection of Two Resistors***

The simplest case: We only have two ohmic resistors that generate thermal noise. What are the noise parameters? These are examples of how to deal with the noise sources, as a basis for the application of correlation matrices. The network is terminated at the input with a generator resistor $R_g = {}^1\!/_{G_g}$ and short-circuited at the output. We know that the circuit on the output has no influence on the noise parameters. To calculate the noise factor, we use the definition with the short-circuit currents. With no special remark all noise parameters are in $B = 1$ Hz bandwidth. Figure 8.8 shows the current source equivalent circuit.

The noise factor is given by $F = \overline{i_{TOT}^2}/\overline{i_g^2}$. The short-circuit currents in the output generated by the conductance values G_1, G_2, and G_g are

$$i_1 = \sqrt{4kTG_1}; \qquad i_2 = \sqrt{4kTG_2}; \qquad i_g = \sqrt{4kTG_g} \tag{8.48}$$

Since all sources are uncorrelated, the following applies to the power

$$\overline{i_{TOT}^2} = \overline{i_1^2} + \overline{i_2^2} + \overline{i_g^2} \tag{8.49}$$

This means for the noise factor

$$F = 1 + \frac{\overline{i_1^2} + \overline{i_2^2}}{\overline{i_g^2}} = 1 + R_g(G_1 + G_2) \tag{8.50}$$

We can see immediately that the minimum noise factor $F_{MIN} = 1$ ($NF_{MIN} = 0$ dB) occurs for the short-circuited input $R_g \to 0$. The dependence on R_g is depicted in Figure 8.10.

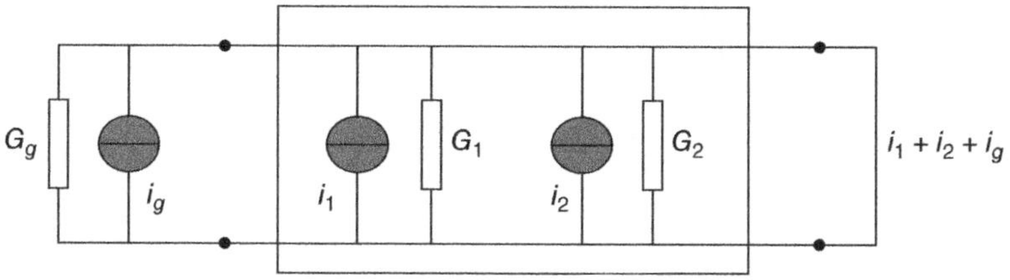

Figure 8.8 Two parallel conductances with noise current sources. At the input the generator is the source. At the output there is a short circuit.

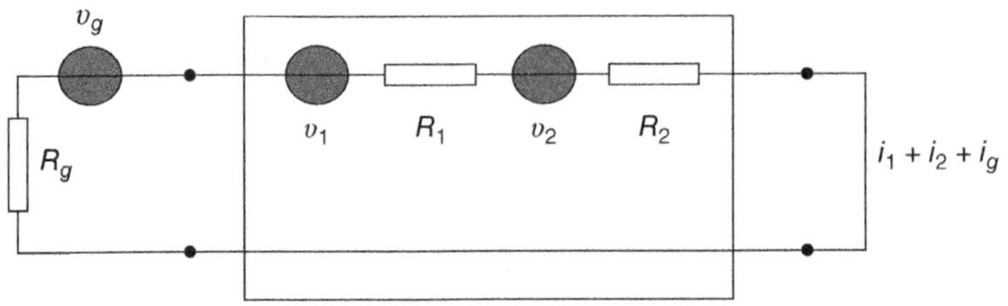

Figure 8.9 Two resistors in series with voltage sources. At the input the generator resistance, the output is short-circuited.

Example 8.2 ***Series Connection***

We now connect both resistors in series and calculate the short circuit currents i_1, i_2, and i_g from the equivalent circuit Figure 8.9 with the voltage sources $v_k = \sqrt{4kTR_k}$ $k = 1{,}2, g$:

$$i_1 = \frac{\sqrt{4kTR_1}}{R_1 + R_2 + R_g}; \qquad i_2 = \frac{\sqrt{4kTR_2}}{R_1 + R_2 + R_g}; \qquad i_g = \frac{\sqrt{4kTR_g}}{R_1 + R_2 + R_g} \tag{8.51}$$

In this case the noise factor is

$$F = \frac{\overline{v_g^2} + \overline{v_1^2} + \overline{v_2^2}}{\overline{v_g^2}} = 1 + \frac{1}{R_g}(R_1 + R_2) \tag{8.52}$$

Here we can see that $F_{MIN} = 1$ (0 dB) for open input $R_g \to \infty$ occurs. The dependence is also illustrated in Figure 8.10.

The first case of parallel connection is a typical example of working with admittances, both for the network and for the noise correlation matrices.

For the second case, the series connection, the impedances are the adequate formalism.

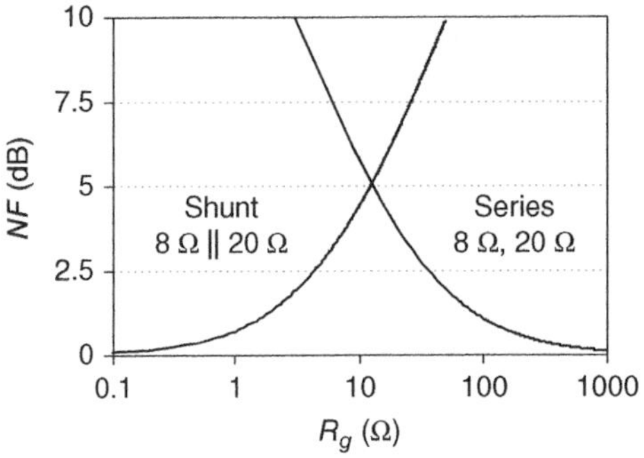

Figure 8.10 Noise figure for parallel and series connection of two resistors versus the generator resistance.

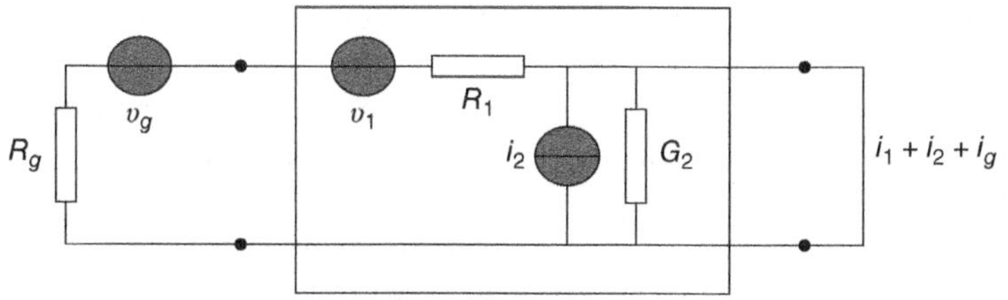

Figure 8.11 Two resistors in combination. Input and output as in Figure 8.10. $R_1 = 8\ \Omega$, $1/G_2 = 20\ \Omega$.

Example 8.3 *Combination Series – Parallel*

A combination of the two limit cases is 8 Ω in series and 20 Ω as shunt (Figure 8.11).

Here, too, we start from the definition of F based upon the power of the short-circuit currents in the output.

$$F = \frac{1}{\frac{\overline{v_g^2}}{(R_1+R_g)^2}}\left\{\frac{\overline{v_g^2}}{(R_1+R_g)^2} + \frac{\overline{v_1^2}}{(R_1+R_g)^2} + \overline{i_2^2}\right\} \tag{8.53}$$

$$F = 1 + \frac{1}{R_g}\left[R_1 + G_2\left(R_1+R_g\right)^2\right] \tag{8.54}$$

The noise figure NF is plotted versus R_g in Figure 8.12. In contrast to the previous examples, the network generates noise for both short circuit and open circuit at the input. There is an optimal source resistance R_{OPT} for which the noise factor is minimal. We obtain this by deriving F with respect to R_g.

$$\frac{\partial F}{\partial R_g} = -\frac{R_1}{R_g^2} - \frac{R_1^2 G_2}{R_g^2} + G_2 = 0 \tag{8.55}$$

$$R_{OPT} = \sqrt{R_1 R_2 + R_1^2} \tag{8.56}$$

Figure 8.12 Noise figure of the circuit Figure 8.11 versus the generator resistance.

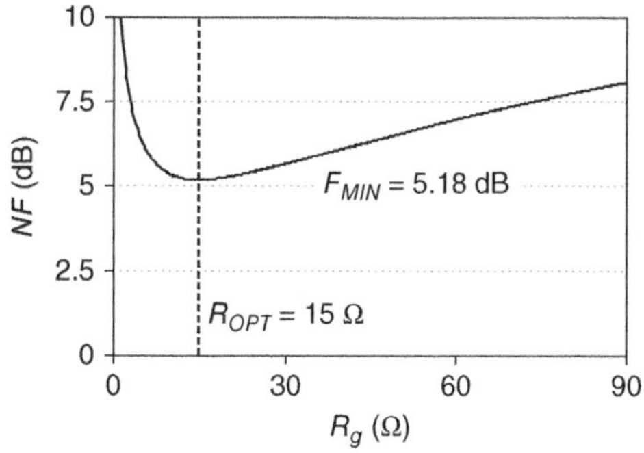

In this example, the minimum noise figure $NF_{MIN} = 5.18$ dB at the source resistance $R_{OPT} = 15\,\Omega$.

The consideration carried out is only useful for simple special cases and should show the principles of noise modeling. Every real network will behave like Example 8.3. It will deliver noise power for both short-circuited and open circuit inputs. This corresponds to the formalism already discussed (Chapter 7) of the chain matrix representation of the network with the two noise sources at the input. For the limiting cases 1 (Series R) and 2 (Shunt R) one single noise source is sufficient. In the case of the series connection of R_1 and R_2, we have the voltage source $v_N = v_1 + v_2$ and thus for the noise resistance (Eq. (7.13))

$$R_N = \frac{\overline{v_N^2}}{4kT} = R_1 + R_2 \tag{8.57}$$

The current source i_U, being uncorrelated with v_N is not required

$$G_U = \frac{\overline{i_U^2}}{4kT} = 0 \tag{8.58}$$

In the case of parallel connection of R_1 and R_2, the noise currents add up to $i_U = i_1 + i_2$.

$$G_U = \frac{1}{4kT}\left(\overline{i_1^2} + \overline{i_2^2}\right) = G_1 + G_2 \tag{8.59}$$

The voltage source v_N is not required. These simple results are only valid because the thermal noise powers of R_1 and R_2 are not correlated.

We now want for case 3 transform the internal noise sources and v_1 i_2 to the input. We use the notations given in Figure 8.13.

We calculate the contributions of the internal sources to the short-circuit current in the output and transfer it to the input using the network matrix. In our case, if

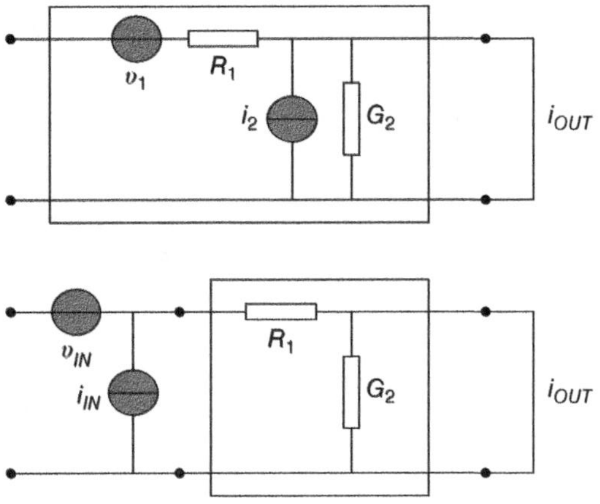

Figure 8.13 The noise contributions of the intrinsic elements R_1, G_2 transformed to the input are converted to v_{IN}, i_{IN}.

the input is open, the contribution is $i_{OUT} = i_2$. The voltage v_1 does not contribute to the output current because the voltage source is open. It must be added, however, to the voltage that is transformed to the input across the network. The output voltage is $v_O = 0$ because it is short-circuited.

Input and output variables are connected by the admittance matrix.

$$\begin{pmatrix} i_i \\ i_O \end{pmatrix} = \begin{pmatrix} G_1 & -G_1 \\ -G_1 & G_1 + G_2 \end{pmatrix} \begin{pmatrix} v_i \\ v_O \end{pmatrix} \tag{8.60}$$

It remains as a contribution to v_{IN} (without consideration of the arrows) $i_{OUT} = Y_{21} v_i$

$$\frac{\overline{v_i^2}}{4kT} = \frac{\overline{i_{OUT}^2}}{4kTY_{21}^2} = \frac{G_2}{G_1^2} \tag{8.61}$$

Thus we obtain for the noise resistance R_N, characterizing the voltage source $v_{IN} = v_i + v_1$:

$$R_N = \frac{G_2}{G_1^2} + R_1 \tag{8.62}$$

In our example with $R_1 = 8\ \Omega$ and $R_2 = 20\ \Omega$ the noise resistance becomes $R_N = 11.2\ \Omega$. The current source i_{IN} we also get from the Y-matrix

$$i_i = Y_{11} v_i = \frac{Y_{11}}{Y_{21}} i_{OUT} \tag{8.63}$$

Because $|Y_{11}| = |Y_{21}|$ and $i_{OUT} = i_2$ we get for the noise conductance G_N of the current source i_{IN} $G_N = {}^1/_{R_2}$, in our example $G_N = 50$ mS. This is in agreement with the result for R_{OPT} obtained above. This is the ratio of the two noise sources

$$R_{OPT} = \sqrt{\frac{\overline{v_{IN}^2}}{\overline{i_{IN}^2}}} = \sqrt{\frac{R_N}{G_N}} \tag{8.64}$$

In our example $R_{OPT} = \sqrt{{}^{11.2\ \Omega}/_{0.05\ \mathrm{S}}} = 15\ \Omega$.

Example 8.4 ***The Correlation Matrices of* R *Series and* R *Parallel***

In the literature one finds for the two basic circuits

$$\begin{aligned} \boldsymbol{C}^Z &= 2kT\Delta fRe\{\boldsymbol{Z}\} \\ \boldsymbol{C}^Y &= 2kT\Delta fRe\{\boldsymbol{Y}\} \end{aligned} \tag{8.65}$$

The bandwidth is here again set to $\Delta f = B = 1$ Hz. The factor 2 comes from the integration of the two-sided spectral power density S in the frequency range $-\infty < f < +\infty$. This corresponds to the mathematically correct procedure, but does not matter in the end, because in the transition to the IEEE noise quantities these factors cancel out. No matter whether $2kT\Delta f$ or $4kT\Delta f$ is used here.

The definition of the noise correlation matrices by the spectral power densities of the fluctuating quantities leads in the case of currents i_1, i_2 or voltages v_1, v_2 at the two-port to the mean power $\overline{i_k i_n^*}$ and $\overline{v_k v_n^*}$ with $k, n = 1, 2$. The power refers to a 1 Ω resistor. The time averaging is used here, e.g. $\overline{i_k i_n^*}$. The mathematically more correct use of the ensemble average is often found, e.g. $\langle i_k i_n^* \rangle$. Practically it does not matter, because we assume ergodic signals. The elements of the main diagonal represent the auto power spectral density, and the elements of the secondary diagonal represent the cross power spectral density and thus the correlation. Thus we have

$$\boldsymbol{C}^Z = \begin{pmatrix} \overline{v_1 v_1^*} & \overline{v_1 v_2^*} \\ \overline{v_2 v_1^*} & \overline{v_2 v_2^*} \end{pmatrix}; \qquad \boldsymbol{C}^Y = \begin{pmatrix} \overline{i_1 i_1^*} & \overline{i_1 i_2^*} \\ \overline{i_2 i_1^*} & \overline{i_2 i_2^*} \end{pmatrix} \tag{8.66}$$

We want to establish the connection with the network matrices. These are for series R

$$\boldsymbol{Z} = \infty; \qquad \boldsymbol{Y} = \frac{1}{R}\begin{pmatrix} 1 & -1 \\ -1 & 1 \end{pmatrix} \tag{8.67}$$

and for Parallel-R

$$\boldsymbol{Z} = R\begin{pmatrix} 1 & 1 \\ 1 & 1 \end{pmatrix}; \qquad \boldsymbol{Y} = \infty \tag{8.68}$$

There is only one form at a time.

With the Series-R (Figure 8.9) we have the two noise voltage sources v_1 and v_2 and the ohmic resistors R_1 and R_2. The short-circuit currents at the input and output are

$$i_1 = i_2 = \frac{v_1 + v_2}{R_1 + R_2} \tag{8.69}$$

The elements of the correlation matrix are all the same in this simple case. Furthermore v_1 and v_2 are uncorrelated, since they represent the thermal noise of two different ohmic resistors.

$$\overline{i_1 i_1^*} = \frac{\overline{(v_1 + v_2)^2}}{(R_1 + R_2)^2} = \frac{\overline{v_1^2} + \overline{v_2^2}}{(R_1 + R_2)^2} = 4kT\frac{1}{R_1 + R_2} \tag{8.70}$$

We obtain for the noise correlation matrix of the series resistance $\boldsymbol{C}^Y = 4kT\, Re(\boldsymbol{Y})$.

For the reasons mentioned earlier, this corresponds to the known approach except for a factor of 2. As long as we do not mix up both factors in a calculation, this does not matter. Analogously, for parallel R holds $\boldsymbol{C}^Z = 4kT\, Re(\boldsymbol{Z})$. In general: The elements of the noise correlation matrix of passive networks (even if they are complicated) are obtained from the real part of the elements of the network matrix multiplied by $4kT$ where T is the physical temperature of the network.

Table 8.1 summarizes the matrix representations of the simple circuit of series and parallel resistor.

Table 8.1 Matrices of simple networks.

Network	Series impedance	Shunt impedance
Symbol	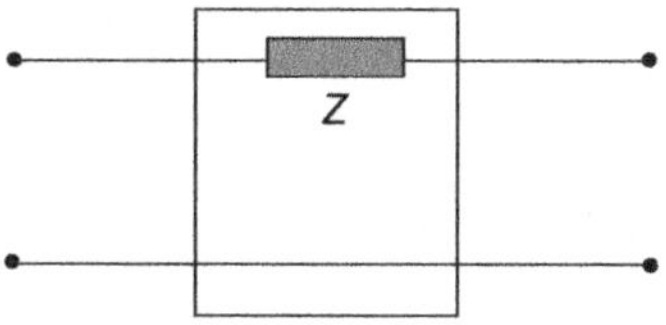	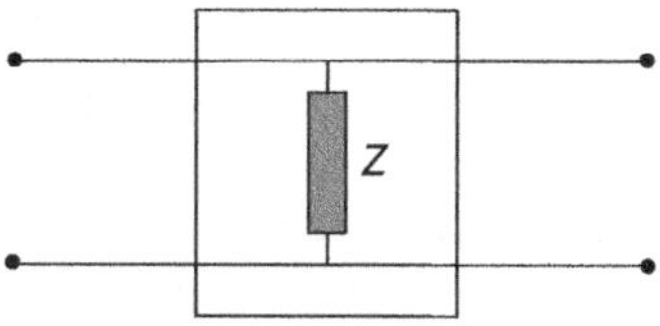
Network matrix	$\boldsymbol{Y} = \frac{1}{Z}\begin{pmatrix} 1 & -1 \\ -1 & 1 \end{pmatrix}$	$\boldsymbol{Z} = Z\begin{pmatrix} 1 & 1 \\ 1 & 1 \end{pmatrix}$
Y/Z correlation matrix	$\boldsymbol{C}^Y = \frac{4kT}{Re(Z)}\begin{pmatrix} 1 & -1 \\ -1 & 1 \end{pmatrix}$	$\boldsymbol{C}^Z = 4kTRe(Z)\begin{pmatrix} 1 & 1 \\ 1 & 1 \end{pmatrix}$
A correlation matrix	$\boldsymbol{C}^A = 4kTRe(Z)\begin{pmatrix} 1 & 0 \\ 0 & 0 \end{pmatrix}$	$\boldsymbol{C}^A = \frac{4kT}{Re(Z)}\begin{pmatrix} 0 & 0 \\ 0 & 1 \end{pmatrix}$

Example 8.5 ***Example 8.3 with the Admittance Matrix***

The elements of the noise correlation matrix thus correspond to the elements of the network matrix. In this formalism we can thus calculate the noise parameters for Example 8.3. The admittance matrix is

$$\boldsymbol{Y} = \frac{1}{R_1R_2}\begin{pmatrix} R_2 & -R_2 \\ -R_2 & R_1 + R_2 \end{pmatrix} \tag{8.71}$$

We leave out the factor $4kT$ or $2kT$, because it is canceled down in the further calculation or is not important for the understanding. Besides, all resistances are real. Thus the elements of the Y-correlation matrix are for our example:

$$\boldsymbol{C}^Y = \begin{pmatrix} \frac{R_2}{R_1R_2} & -\frac{R_2}{R_1R_2} \\ -\frac{R_2}{R_1R_2} & \frac{R_1+R_2}{R_1R_2} \end{pmatrix} = \begin{pmatrix} 0.125 & -0.125 \\ -0.125 & 0.175 \end{pmatrix} \tag{8.72}$$

The calculation of IEEE noise parameters from the elements of the Y correlation matrix is not a general standard. In most cases, the correlation matrix in chain form is used. Here we give the formulas with the numerical values for our example:

$$R_N = \frac{C_{22}^Y}{|Y_{21}|^2} = \frac{0.175}{(0.125)^2} = 11.2\,\Omega \tag{8.73}$$

$$G_{OPT} = \sqrt{\begin{matrix} \frac{1}{C_{22}^Y}(C_{11}^Y|Y_{21}|^2 - C_{21}^Y Y_{11} Y_{21}^* - C_{12}^Y Y_{11}^* Y_{21}) \\ +|Y_{11}|^2 - Im\left(Y_{11}^* - \frac{C_{21}^Y}{C_{22}^Y} Y_{21}^*\right)^2 \end{matrix}} = 0.067\,\text{S} \tag{8.74}$$

$$B_{OPT} = -Im\left(Y_{11} - Y_{21}\frac{C_{Y12}}{C_{Y22}}\right) = 0 \tag{8.75}$$

$$F_{MIN} = 1 + 2\frac{C^Y_{22}}{|Y_{21}|^2}\left(G_{OPT} + Re\left[Y^*_{11} - Y^*_{21}\frac{C^Y_{21}}{C^Y_{22}}\right]\right) = 3.3 = 5.18\ \text{dB} \tag{8.76}$$

We have thus confirmed the results of the aforementioned network analysis.

Example 8.6 ***Correlation by Transformation***

In this treatment of Example 8.3, the correlation of the noise sources transformed to the input is included indirectly, but is not explicitly obvious. Although the two intrinsic noise sources v_1 and i_2 are not correlated, since they originate from two different thermal noise generating resistors, after the transformation $v_1, i_1 \rightarrow v_{IN}$, i_{IN} these two are partially correlated. One can see this, for example, in the formula for the noise resistance (8.62) $R_N = {}^{G_2}/_{G_1^2} + R_1$. The part ${}^{G_2}/_{G_1^2}$ contains with G_2 a contribution from the current source i_2 to the voltage source v_{IN}.

We want to analyze the correlation of v_{IN} and i_{IN} in this example [6, 7]. As shown in Chapter 7, the correlation coefficient Y_{COR} is introduced to this purpose. Through it flows the part i_C of the current i_{IN} that is fully correlated with v_{IN}

Therefore applies

$$i_C = Y_{COR}v_{IN} \tag{8.77}$$

Quantitatively, this separation is realized by the correlation coefficient

$$\rho = \frac{\overline{v_{IN}i^*_{IN}}}{\sqrt{\overline{|v_{IN}|^2|i_{IN}|^2}}} \tag{8.78}$$

With the numerator we have the simplification that in our example no complex quantities occur. The voltage source is $v_{IN} = \sqrt{4kT\,R_N}$ the current source is $i_{IN} = \sqrt{4kT\,G_2}$.

For the sake of clarity, we will replace the expression $\sqrt{4kT}$ by K since it is canceled down at the end anyway, but cannot be simply omitted for dimensional reasons. In the numerator we have the voltage with the correlated part v_C and with uncorrelated part v_U.

$$v_{IN} = v_C + v_U, \qquad i_{IN} = i_2 \tag{8.79}$$

The numerator is

$$\overline{v_{IN}i_2} = \overline{v_Ci_2} + \overline{v_Ui_2} \tag{8.80}$$

The second term disappears because the mean value of two uncorrelated fluctuation quantities is zero. It follows

$$\overline{v_{IN}i_2} = K\frac{R_1}{\sqrt{R_2}}K\frac{1}{\sqrt{R_2}} = K^2\frac{R_1}{R_2} \tag{8.81}$$

In the denominator we have

$$\overline{|v_{IN}|^2} = K^2\left(R_1 + \frac{R_1^2}{R_2}\right) \tag{8.82}$$

and

$$\overline{|i_{IN}|^2} = K^2 \,{}^1\!/_{R_2} \tag{8.83}$$

The correlation coefficient is therefore

$$\rho = \frac{R_1}{\sqrt{R_1 R_2 + R_1^2}} \tag{8.84}$$

In our numerical example $\rho = 0.535$. The correlation admittance is calculated from the correlation coefficient

$$Y_{COR} = \rho\sqrt{\frac{\overline{i_{IN}^2}}{\overline{v_{IN}^2}}} \tag{8.85}$$

In our example:

$$G_{COR} = \rho\sqrt{\frac{1}{G_2 R_N}} = 0.036\ \mathrm{S} \tag{8.86}$$

With this consideration we have the partition of the input noise current i_{IN} into the fully correlated part i_C and the uncorrelated part i_U with respect to v_{IN}. Since the spectral power densities are always of interest, we write the squares of the fluctuating currents:

$$\overline{i_C^2} = Y_{COR}^2\overline{v_{IN}^2}; \qquad \overline{i_U^2} = \overline{i_{IN}^2}(1 - |\rho|^2) \tag{8.87}$$

Since it is common practice to use the noise resistance R_N and the noise conductance G_N instead of the fluctuation squares, we write:

$$R_N = \frac{\overline{|v_{IN}|^2}}{4kT_0B}; G_N = \frac{\overline{|i_U|^2}}{4kT_0B} \tag{8.88}$$

We had set $B = 1$ Hz as previously.

Now we have the quantities G_{COR}, R_N, G_N, and with (7.18) and (7.19) we can determine the minimum noise factor F_{MIN}. Initially G_{OPT}

$$G_{OPT} = \sqrt{\frac{G_N}{R_N} + G_{COR}^2} = 0.067\ \mathrm{S} \tag{8.89}$$

$$F_{MIN} = 1 + 2R_N(G_{COR} + G_{OPT}) = 3.3 \tag{8.90}$$

We have thus calculated the noise properties of the two-port starting from the intrinsic physical quantities, the current and voltage sources, by transformation to the input. This procedure is useful for understanding transformation and correlation, but can only be performed in simple cases. In practical calculations with more complicated circuits, correlation matrices are the method of choice.

Transformation of Noise Sources in Different Network Representations

In the examples we have seen the importance of the different configurations of the noise sources of a two-port and their appropriate combination. The two-port equations known from network theory have the advantage that they are independent of input and output circuitry. In a systematic consideration we have to consider the directional arrows of current and voltage. The equivalent circuit in Figure 8.14 is in admittance form. It applies in matrix notation for the relation of currents and voltages [1]:

$$\begin{pmatrix} i_1 \\ i_2 \end{pmatrix} = \boldsymbol{Y} \begin{pmatrix} v_1 \\ v_2 \end{pmatrix} \tag{8.91}$$

This model of the network describes the transfer properties. However, it is not a noise equivalent circuit. This becomes easily clear by the following consideration: If input and output are short-circuited, the voltages $v_1 = v_2 = 0$ also the currents $i_1 = i_2 = 0$ because the voltage controlled current sources $v_2 Y_{12}$ and $v_1 Y_{21}$ can no longer be activated. In a real two-port, however, short-circuit noise currents still flow. Every passive network is lossy, every active network contains resistors and transistors as internal noise sources. We obtain a generally valid noise equivalent circuit if the two noise current sources i_{N1} and i_{N2} are added. This is done in Figure 8.15. The two-port equation is now

$$\begin{pmatrix} i_1 \\ i_2 \end{pmatrix} = \boldsymbol{Y} \begin{pmatrix} v_1 \\ v_2 \end{pmatrix} + \begin{pmatrix} i_{N1} \\ i_{N2} \end{pmatrix} \tag{8.92}$$

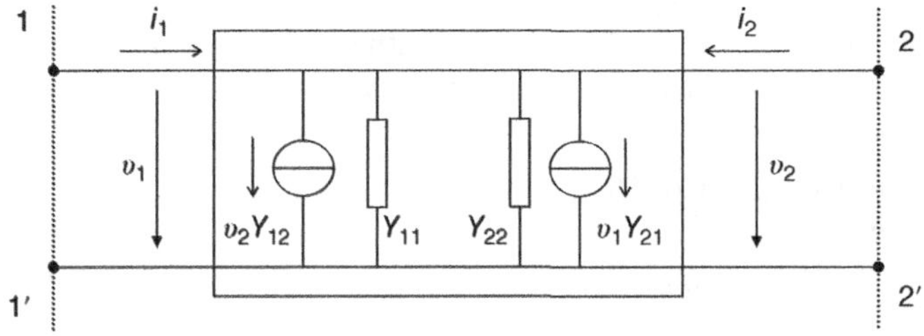

Figure 8.14 Two port in admittance representation.

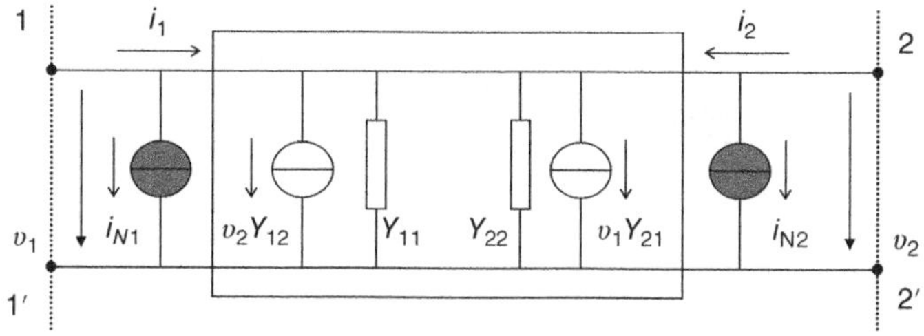

Figure 8.15 Two-port in admittance representation with noise current sources.

or in detail

$$i_1 = Y_{11}v_1 + Y_{12}v_2 + i_{N1}$$
$$i_2 = Y_{21}v_1 + Y_{22}v_2 + i_{N2} \tag{8.93}$$

As we have already seen in the simple Example 8.3, the external sources are mostly correlated, since they originate by transforming internal sources to the ports. Therefore we have to know the power spectra of the sources themselves as well as their cross spectra. The connection of the noise current sources to the Y-matrix is, as we have already shown in Chapter 7, only one possibility. Another is to connect voltage sources to the impedance matrix or the mixed circuit at the input of the chain matrix (ABCD). Also important for the noise analysis of transistors is the hybrid matrix (H), with a noise voltage source at the input and a noise current source at the output (see Figure 7.1) (Figure 8.16).

The table shows the different representations and their transformation.

The transformation equations (Table 8.2 column 3), by which the noise currents i_{Y1} and i_{Y2} of the admittance form can be converted into the noise quantities of the other forms, can be derived according to the following scheme. Based on the network equations (column 1), the currents i_1, i_2 are replaced by the noise currents i_{Y1}, i_{Y2}. Taking into account the directional arrows, the new noise quantities are added to the right side. Then the terms with the voltages v_1 and v_2 are deleted. Here is an explicit example for the transformation of the Y-form into the ABCD-form.

$$A_{11}v_2 + A_{12}i_{Y2} = v_1 - v_A$$
$$A_{21}v_2 + A_{22}i_{Y2} = i_{Y1} - i_A \tag{8.94}$$

after deleting $A_{11}v_2$, $A_{21}v_2$, and v_1 remains

$$v_A = -A_{12}i_{y2}$$
$$i_A = i_{Y1} - A_{22}i_{Y2} \tag{8.95}$$

The scheme applies not only to the Y-representation with its noise currents i_{Y1}, i_{Y2}, but also for other initial configurations. Let us consider as an example

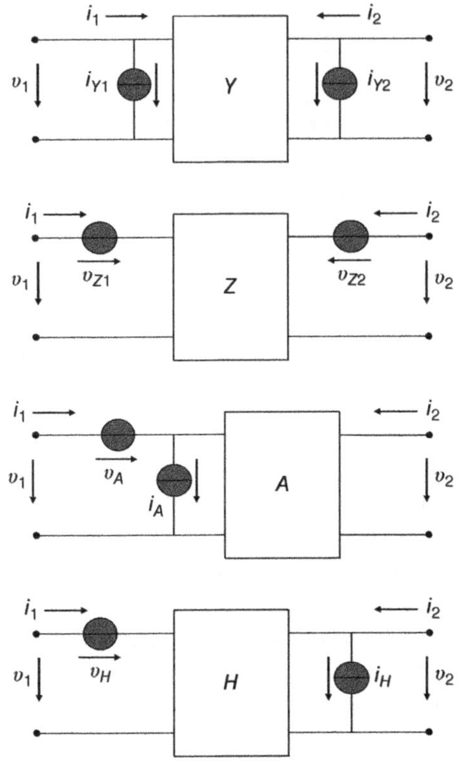

Figure 8.16 Two-port with noise sources in the different representations.

the transformation of the Z-representation with its noise voltages $v_{Z1,}v_{Z2}$ into the ABCD form

$$A_{11}v_{Z2} + A_{12}i_2 = v_{Z1} - v_A$$
$$A_{21}v_{Z2} + A_{22}i_2 = i_1 - i_A \tag{8.96}$$

After the deletion of $A_{12}i_2, A_{22}i_2$, and i_1 remains

$$v_A = v_{Z1} - A_{11}v_{Z2}$$
$$i_A = -A_{21}v_{Z2} \tag{8.97}$$

It can be seen that during the transformation both quantities change and previously uncorrelated quantities can become correlated in the new form. In the last example, the two uncorrelated noise voltages v_{Z1} and v_{Z2} are transformed into the two partially correlated quantities v_A and i_A. They are partially correlated because v_A in the term $A_{11}v_{Z2}$ includes a voltage, which also determines the current i_A (8.97). We had already seen this in Example 8.3.

Table 8.2 Two-port equations with noise sources and transformation equations for these sources [1].

Network equation	Noise sources	Transformation equations
Y		$v_Z \rightarrow i_Y$
$Y_{11}v_1 + Y_{12}v_2 = i_1$	$-i_{Y1}$	$i_{Y1} = -Y_{11}v_{Z1} - Y_{12}v_{Z2}$
$Y_{21}v_1 + Y_{22}v_2 = i_2$	$-i_{Y2}$	$i_{Y2} = -Y_{21}v_{Z1} - Y_{22}v_{Z2}$
Z		$i_Y \rightarrow v_Z$
$Z_{11}i_1 + Z_{12}i_2 = v_1$	$-v_{Z1}$	$v_{Z1} = -Z_{11}i_{Y1} - Z_{12}i_{Y2}$
$Z_{12}i_1 + Z_{22}i_2 = v_2$	$-v_{Z2}$	$v_{Z2} = -Z_{21}i_{Y1} - Z_{22}i_{Y2}$
ABCD		$i_Y \rightarrow i_A,\ v_A$
$A_{11}v_2 + A_{12}i_2 = v_1$	$-v_A$	$v_A = -A_{12}i_{Y2}$
$A_{21}v_2 + A_{22}i_2 = i_1$	$-i_A$	$i_A = i_{Y1} - A_{22}i_{Y2}$
H		$i_Y \rightarrow i_H,\ v_H$
$H_{11}i_1 + H_{12}v_2 = v_1$	$-v_H$	$v_H = -H_{11}i_{Y1}$
$H_{21}i_1 + H_{22}v_2 = i_2$	$-i_H$	$i_H = -H_{21}i_{Y1} + i_{Y2}$

Correlation Matrix and IEEE Elements

The noise correlation matrices are available in a suitable form for each network description, as are the different networks used in signal analysis. For the noise analysis of the two-port, the chain form had proved to be optimal, since here the two noise sources are at the input. This corresponds to the basic idea that all internal noise sources are transformed to the input and the remaining network is noise-free [8].

The noise correlation matrix in chain form with the notations of Figure 8.17:

$$\mathbf{C}^A = \begin{pmatrix} C_{11}^A & C_{12}^A \\ C_{21}^A & C_{22}^A \end{pmatrix} = \begin{pmatrix} \overline{v_N v_N^*} & \overline{v_N i_N^*} \\ \overline{v_N^* i_N} & \overline{i_N i_N^*} \end{pmatrix} \tag{8.98}$$

As is well known, the noise characteristics of the two-port depend on its input circuit Y_S. They are therefore described by the four noise parameters F_{MIN}, R_N, G_{OPT}, and B_{OPT}. Here we want to derive the relationship of these IEEE parameters with the elements of the noise correlation matrix.

The analysis is analogous to that already shown in Chapter 7, but here it leads to formulas for the four noise parameters in the formalism of the correlation matrices.

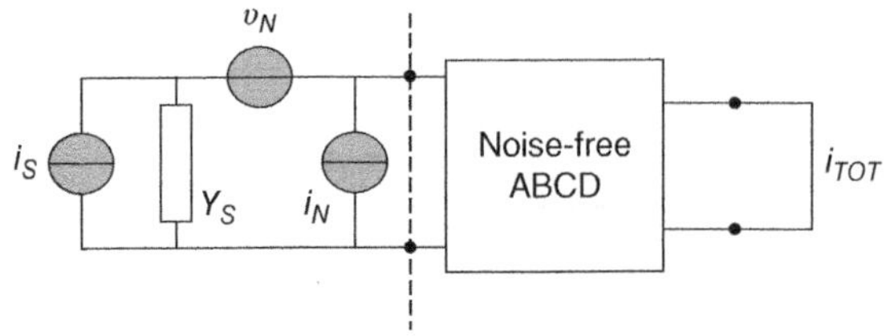

Figure 8.17 Two-port as chain matrix with noise sources at the input.

The input is connected to a source. This is, e.g. a noise generator with the current source i_S and the input admittance $Y_S = G_S + jB_S$. The network includes the voltage source v_N and the current source i_N. Both are correlated. The source i_S is uncorrelated to both.

The noise factor F is defined by the ratio of available noise power at the output (inherent noise plus amplified power of the source) to amplified power of the source alone. Written with the currents in the output circuit, as already used:

$$F = \frac{\overline{i_{TOT} i^*_{TOT}}}{\overline{i_S i^*_S}} \tag{8.99}$$

with

$$i_{TOT} = i_S + i_N + v_N Y_S \tag{8.100}$$

The gain is canceled out because it affects both currents equally.

With i_{TOT} in (8.99) the result is:

$$F = \frac{1}{\overline{i_S i^*_S}} \overline{(i_S + i_N + v_N Y_S)(i_S + i_N + v_N Y_S)^*} \tag{8.101}$$

or expanded:

$$F = \frac{1}{\overline{i_S i^*_S}} (\overline{i_S i^*_S} + \overline{i_S i^*_N} + \overline{i_S v^*_N Y^*_S} + \overline{i_N i^*_S} + \overline{i_N i^*_N} + \overline{i_N v^*_N Y^*_S} + \overline{v_N Y_S i^*_S} + \overline{v_N Y_S i^*_N} + \overline{v_N v^*_N |Y_S|^2}) \tag{8.102}$$

Here, several terms are omitted, since the quantities are not correlated and the mean value of the product therefore disappears.

$$\overline{i_S i^*_N} = \overline{i_S v^*_N Y^*_S} = \overline{i_N i^*_S} = \overline{v_N Y_S i^*_S} = 0 \tag{8.103}$$

it remains

$$F = \frac{(\overline{i_S i^*_S} + \overline{i_N i^*_N} + \overline{i_N v^*_N Y^*_S} + \overline{v_N Y_S i^*_N} + \overline{v_N v^*_N |Y_S|^2}}{\overline{i_S i^*_S}} = 1 + \frac{\overline{i_N i^*_N} + \overline{i_N v^*_N Y^*_S} + \overline{v_N Y_S i^*_N} + \overline{v_N v^*_N |Y_S|^2}}{\overline{i_S i^*_S}} \tag{8.104}$$

We recognize in (8.104) the elements of the noise correlation matrix.

$$\overline{v_N v_N^*} = C_{11}^A; \quad \overline{i_n i_n^*} = C_{22}^A; \quad \overline{v_N i_N^*} = C_{12}^A; \quad \overline{v_N^* i_N} = C_{21}^A = C_{12}^{A*} \tag{8.105}$$

For the thermal noise of the real part of the source admittance, the Nyquist formula applies.

$$\overline{i_S i_S^*} = 4kT\, Re(Y_S) \tag{8.106}$$

Inserted in (8.104) the noise factor F is

$$F = 1 + \frac{C_{22}^A + C_{11}^A |Y_S|^2 + C_{12}^A Y_S + C_{12}^{A*} Y_S^*}{2kTG_S} = 1 + \frac{C_{22}^A + C_{11}^A |Y|^2 + 2Re(C_{12}^A Y_S)}{2kTG_S} \tag{8.107}$$

Using $Y_S = G_S + jB_S$

$$F = 1 + \frac{1}{2kTG_S}\{C_{22}^A + Re(C_{11}^A)G_S^2 + Re(C_{11}^A)B_S^2 + j\, Im(C_{11}^A)G_S^2 + j\, Im(C_{11}^A)B_S^2 +$$
$$+ 2(Re(C_{12}^A)G_S - Im(C_{12}^A)B_S)\} \tag{8.108}$$

The minimum noise factor F_{MIN} is known to occur for the source admittance $Y_S = Y_{OPT}$. In order to find them, we are looking for the minimum of $F(Y_S)$. For this purpose, we set one after the other

$$\frac{\partial F}{\partial B_S} = 0; \frac{\partial F}{\partial G_S} = 0$$

(8.108) is very well suited for this purpose.

$$\frac{\partial F}{\partial B_S} = 2Re(C_{11}^A)B_S + 2\, Im(C_{11}^A)B_S - 2\, Im(C_{12}^A) = 0 \tag{8.109}$$

So we have for B_{OPT}

$$B_{OPT} = \frac{Im(C_{12}^A)}{C_{11}^A} \tag{8.110}$$

The derivation of (8.108) with respect to G_S results in

$$\frac{\partial F}{\partial G_S} = -C_{22}^A + C_{11}^A G_S^2 - C_{11}^A B_S^2 + 2Im(C_{12}^A)B_S = 0 \tag{8.111}$$

Inserting (8.110) for B_{OPT} and solving to G_{OPT} gives

$$G_{OPT} = \frac{1}{C_{11}^A}\sqrt{C_{11}^A C_{22}^A - [Im(C_{12}^A)]^2} \tag{8.112}$$

If we insert G_{OPT} (8.112) and B_{OPT} (8.110) in the expression (8.108) for F, we get F_{MIN}.

$$F_{MIN} = 1 + \frac{1}{2kT}\{Re(C_{12}^A) + \sqrt{C_{11}^A C_{22}^A - [Im(C_{12}^A)]^2}\} \tag{8.113}$$

The noise resistance R_N describes the power of the voltage source v_N in 1 Hz bandwidth.

$$\overline{v_N v_N^*} = 4kTR_N = C_{11}^A; R_N = \frac{C_{11}^A}{4kT} \tag{8.114}$$

With this we have established the connection between the four noise parameters F_{MIN}, R_N; $G_{OPT} + jB_{opt}$, which the RF-engineer knows and the noise correlation matrix (in $\boldsymbol{C}^A$-form), which is more familiar to the software community.

We still need the solution, for the $C_{i,k}^A$ if we want to derive the noise correlation matrix from measured values that are usually obtained in IEEE form.

C_{11}^A is already known from (8.114).

With (8.110) we get

$$Im(C_{12}^A) = 4kTR_N B_{OPT} \tag{8.115}$$

We can use these two formulas in (8.112) and obtain C_{22}^A.

$$C_{11}^A C_{22}^A = G_{OPT}^2 C_{11}^{A2} + [Im(C_{12}^A)]^2 \tag{8.116}$$

$$C_{22}^A = 4kTR_N(G_{OPT}^2 + B_{OPT}^2) = 4kTR_N |Y_{OPT}|^2 \tag{8.117}$$

The remaining $Re(C_{12}^A)$ we calculate using F_{MIN} (8.113).

$$\begin{aligned} 2kT(F_{MIN} - 1) &= Re(C_{12}^A) + \sqrt{C_{11}^A C_{22}^A - [Im(C_{12}^A)]^2} \\ &= Re(C_{12}^A) + 4kTR_N G_{OPT} \end{aligned}$$

$$Re(C_{12}^A) = 4kT\left(\frac{F_{MIN} - 1}{2} - R_N G_{OPT}\right) \tag{8.118}$$

Thus we obtain the elements of the noise correlation matrix in chain form ($\boldsymbol{C}^A$) with the four noise parameters accessible to the measurement F_{MIN}, R_N; $G_{OPT} + jB_{opt}$.

$$\boldsymbol{C}^A = 4kT\begin{pmatrix} R_N & \frac{F_{MIN}-1}{2} - R_N Y_{OPT}^* \\ \frac{F_{MIN}-1}{2} - R_N Y_{OPT} & R_N |Y_{OPT}|^2 \end{pmatrix} \tag{8.119}$$

FET-Like Network with the Y-Correlation Matrix

We will discuss an example that is very similar to a field effect transistor, but contains only a minimum of elements. It corresponds to the approach for the intrinsic part of an FET chip in the Pospieszalski model [9]. In contrast to the previous passive circuits we have here a voltage controlled current source that converts a noise voltage at the capacitance into a noise current in the output circuit. In analogy to the FET we call the capacitance C_{gs} and the transconductance (gain) of the current source as g_m. The output current of the current source is therefore $g_m v_{gs}$. We do not consider the physical noise sources of the FET: shot and diffusion noise in the channel and the induced gate noise. Here only the thermal noise of the resistors R_g and R_d (correspond to gate or drain resistance of the FET) are considered. DC operating voltages are also of no interest in this context. Input and output are short circuited. The short-circuit noise currents are i_1 and i_2. This corresponds to the admittance representation of the network (Figure 8.18).

The noise sources are

$$v_g = \sqrt{4kT_gBR_g}; i_d = \sqrt{4kT_dBG_d} \tag{8.120}$$

with $G_d = {}^1/_{R_d}$. T_g and T_d are fictitious temperatures in the Pospieszalski model, we set $T_g = 300$ K and with $B = 1$ Hz bandwidth. The noise voltage v_g generates a short-circuit current in the input circuit

$$i_1 = \frac{v_g}{Z_{in}} = \frac{v_g}{R_g + \frac{1}{j\omega C_{gs}}} \tag{8.121}$$

and the control voltage at the gate-source capacitance:

$$v_{gs} = \frac{i_1}{j\omega C_{gs}} \tag{8.122}$$

We must first calculate the noise currents and can only proceed to the spectral power densities once we know the correlation of the contributions that form the noise power at the output. Explicit is the short-circuit current in the input circuit:

$$i_1 = \frac{v_g \omega C_{gs}}{1 + (\omega C_{gs} R_g)^2} (\omega C_{gs} R_g + j1) \tag{8.123}$$

and the control voltage at the gate-source capacitance

$$v_{gs} = \frac{v_g}{1 + (\omega C_{gs} R_g)^2} (1 - j\omega C_{gs} R_g) \tag{8.124}$$

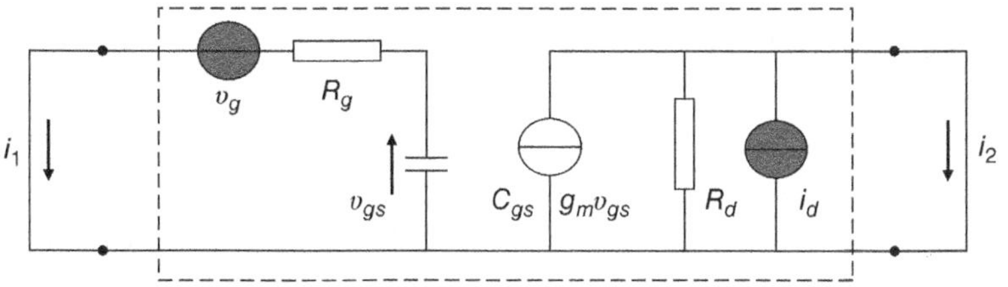

Figure 8.18 Noise equivalent circuit in admittance representation with the most important elements of an intrinsic FET.

This drives by $v_{gs}g_m$ a part of the output current i_2,which is fully correlated to i_1. The second part of i_2 is the thermal noise current

$$i_d = \sqrt{4kT_dBG_d} \tag{8.125}$$

This is uncorrelated with i_1, since both come from different resistors. So we have in the output circuit the current i_2 that is partially correlated to i_1. We had already analyzed the case in Chapter 2. We write the general form again here and follow the description in van der Ziel [10].

The partial correlation of two fluctuating quantities x and y is

$$y = ax + z \text{ with } \overline{y} = \overline{x} = \overline{z} = 0 \tag{8.126}$$

a is a constant and z is a fluctuating quantity that is independent of x.

The following applies

$$\overline{xy} = a\overline{x^2} \tag{8.127}$$

Because $\overline{xz} = 0$.

The correlation coefficient is

$$\rho = \frac{\overline{xy}}{\sqrt{\overline{|x|^2}\,\overline{|y|^2}}} \tag{8.128}$$

We want to explicitly write down this partition into a correlated part and an uncorrelated one for our example: $i_2 = ai_1 + i_d$ with a as a dimensionless constant. i_d is independent of i_1. In our case

$$a = \frac{g_m}{j\omega C_{gs}} \tag{8.129}$$

If a current correlated with a voltage is interrelated in this way, as in Chapter 7, a has the dimension of a conductance and represents the correlation conductance Y_{COR}.

The short-circuit current in the output circuit is therefore

$$i_2 = \left[\frac{v_g g_m}{1 + (\omega C_{gs}R_g)^2} + i_d\right] - j\frac{v_g g_m \omega C_{gs} R_g}{1 + (\omega C_{gs}R_g)^2} \tag{8.130}$$

The correlation coefficient is

$$\rho_Y = \frac{\overline{i_1^* i_2}}{\sqrt{\overline{|i_1|^2}\,\overline{|i_2|^2}}} \tag{8.131}$$

When calculating the products we have to consider that $\overline{v_g i_d} = 0$ is because they are both uncorrelated. Terms containing this product will disappear. But this can only be seen when explicitly calculated. A Math Program that returns products, squares, or magnitudes cannot reveal this.

Taking this fact into account we receive:

$$\overline{i_1^* i_2} = -\frac{g_m v_g}{Z_{in}(1+\omega^2 C_{gs}^2 R_g^2)}(1+j\omega C_{gs} R_g) \tag{8.132}$$

and

$$\overline{|i_2|^2} = \frac{g_m^2 v_g^2}{1+\omega^2 C_{gs}^2 R_g^2} + i_d^2 \tag{8.133}$$

Because i_1 does not contain any mixed terms, the calculation of the magnitude is unproblematic. This is the result for the correlation coefficient:

$$\rho_Y = -\frac{g_m v_g}{\sqrt{\overline{|i_1|^2}\ \overline{|i_2|^2}} Z_{in}(1+\omega^2 C_{gs}^2 R_g^2)}(1+j\omega C_{gs} R_g) \tag{8.134}$$

Figure 8.19 shows the complex ρ_Y for three temperatures of the drain resistance R_d with the circuit elements listed below. With increasing T_d the noise contribution from i_2 that is not correlated to i_1 increases. Therefore, the magnitude of the correlation coefficient becomes smaller. On the contrary, it would be lower for $T_d \to 0$ up to $|\rho_Y| \to 1$ and finally the uncorrelated part disappears.

We arrive at the usual noise parameters (IEEE parameters): F_{MIN}; R_N; and Y_{OPT} using the admittance matrix $\boldsymbol{Y}$ of the network Figure 8.18 and the noise correlation matrix $\boldsymbol{C^Y}$ in admittance form.

Using (8.123) and (8.130) the matrices are

$$\boldsymbol{Y} = \begin{pmatrix} \frac{1}{Z_{in}} & 0 \\ \frac{g_m}{j\omega C_{gs} Z_{in}} & G_d \end{pmatrix}; \boldsymbol{C^Y} = \begin{pmatrix} \frac{v_g^2 \omega^2 C_{gs}^2}{1+\omega^2 C_{gs}^2 R_g^2} & -j\frac{v_g^2 g_m \omega C_{gs}}{1+\omega^2 C_{gs}^2 R_g^2} \\ j\frac{v_g^2 g_m \omega C_{gs}}{1+\omega^2 C_{gs}^2 R_g^2} & \frac{v_g^2 g_m^2}{1+\omega^2 C_{gs}^2 R_g^2} + i_d^2 \end{pmatrix} \tag{8.135}$$

The four noise parameters are obtained from the formulas (8.73, 8.74, 8.75, 8.76). In our example with the numerical values $R_g = 5\ \Omega$; $R_d = 300\ \Omega$; $C_{gs} = 0.5$ pF; and

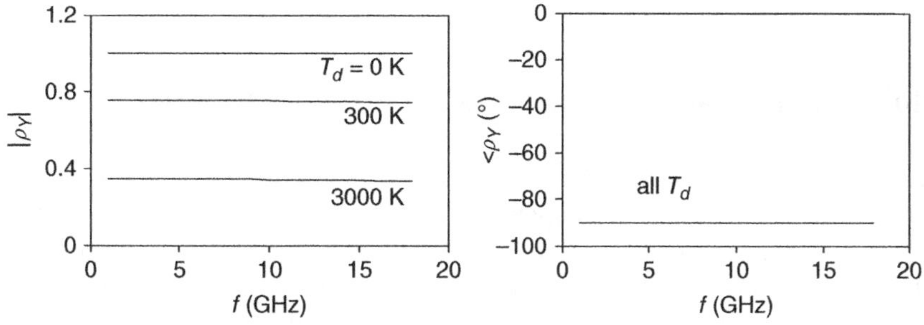

Figure 8.19 Correlation coefficient in admittance representation according to (8.134) for different drain temperatures T_d.

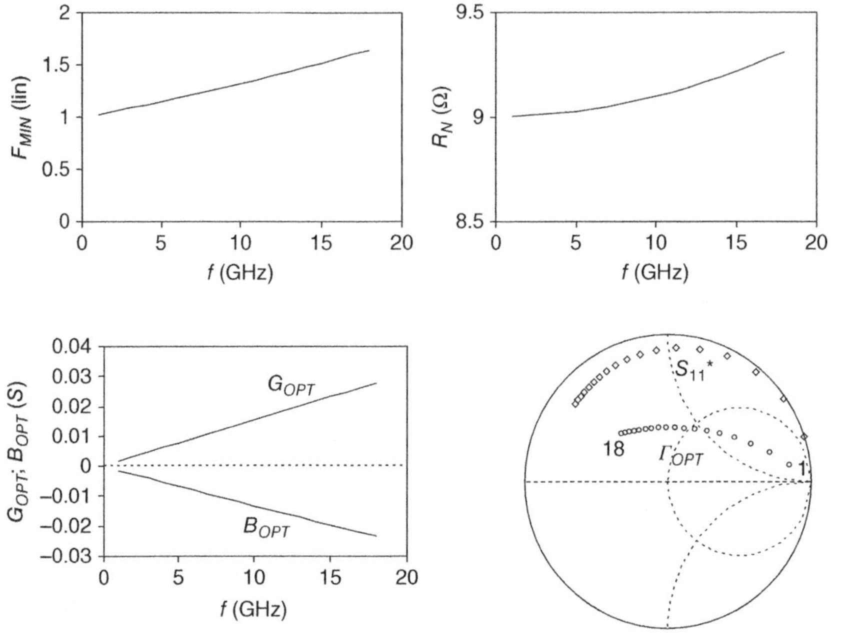

Figure 8.20 The four noise parameters of the circuit Figure 8.18. In the Smith chart S_{11}^* and Γ_{OPT} from 1 to 18 GHz.

$g_m = 30$ mS and the temperatures $T_g = T_d = 300$ K the noise parameter values are very similar to a GaAs FET. They are shown in Figure 8.20. In microwave technology, reflection factors are used instead of admittances. Therefore we plot S_{11}^* and Γ_{OPT} in the Smith chart. Normalized to $Z_0 = 50\ \Omega$ and with $y_{OPT} = Y_{OPT}Z_0$ the relation is

$$\Gamma_{OPT} = \frac{1 - y_{OPT}}{1 + y_{OPT}} \tag{8.136}$$

Noise Sources at Input with ABCD Correlation Matrix

In Example 8.4, we see how the noise currents are discussed in the Y representation. This is often the starting point for noise models of transistors, since the intrinsic physical processes can be described well in this way, e.g. [9, 11]. The noise of a more general two-port, e.g. a complete transistor or amplifier, is, as we know, described with two noise sources at the input. We want to apply this to our circuit Figure 8.18. The sources are transformed in such a way that they generate the same short-circuit current in the output circuit (Figure 8.21).

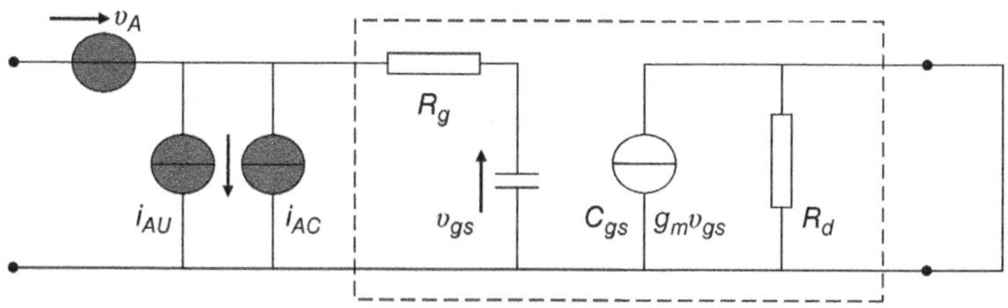

Figure 8.21 Noise equivalent circuit in chain representation with the main elements of an intrinsic FET.

The table of transformations (Table 8.2) shows the following for the transformation of the already known currents in Y representation i_1 and i_2 into the sources of the ABCD representation: v_A and $i_A = i_{AU} + i_{AC}$. The current consists of a part i_{AU} that is uncorrelated to v_A and a part i_{AC} that is fully correlated to v_A.

$$v_A = -\frac{i_2}{Y_{21}}; \qquad i_A = i_1 - i_2\frac{Y_{11}}{Y_{21}} \tag{8.137}$$

The partial correlation of both variables is obvious, since both v_A as well as i_A comprise i_2. We receive

$$v_A = v_g - \frac{i_d}{g_m}(1 + j\omega C_{gs}R_g); \qquad i_A = -j\frac{i_d}{g_m}\omega C_{gs} \tag{8.138}$$

To calculate the correlation admittance Y_{COR} we need the correlation coefficient

$$\rho_A = \frac{\overline{v_A^* i_A}}{\sqrt{\overline{|v_A|^2}\ \overline{|i_A|^2}}} \tag{8.139}$$

Its numerator is explicit:

$$\overline{v_A^* i_A} = \frac{i_d^2}{g_m^2}(\omega^2 C_{gs}^2 R_g + j\omega C_{gs}) \tag{8.140}$$

and the denominator

$$\overline{|v_A|^2} = v_g^2 + \frac{i_d^2}{g_m^2}(1 + \omega^2 C_{gs}^2 R_g^2); \overline{|i_A|^2} = \frac{i_d^2}{g_m^2}\omega^2 C_{gs}^2 \tag{8.141}$$

The frequency dependence of the complex correlation coefficient of v_A and i_A for three effective drain temperatures T_d is shown in Figure 8.22. It can be seen that for $T_d \to 0$ also the current source i_A disappears (8.138) and with it the correlation.

With knowledge of ρ_A we can calculate the correlation admittance Y_{COR}.

$$Y_{COR} = \rho_A \sqrt{\frac{\overline{|i_A|^2}}{\overline{|v_A|^2}}} = G_{COR} + jB_{COR} \tag{8.142}$$

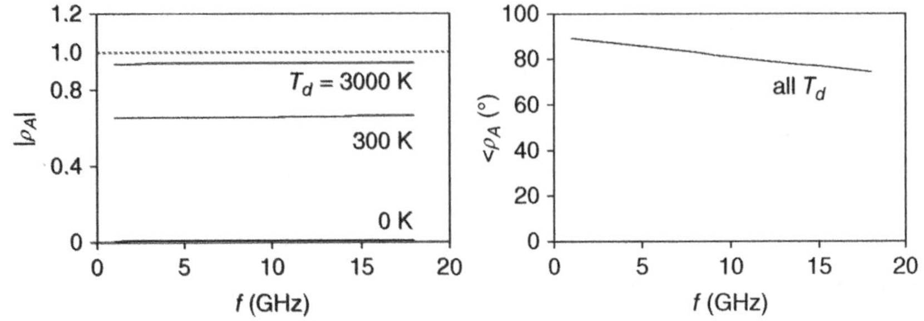

Figure 8.22 Correlation coefficient for the chain representation Figure 8.19 according to (8.140) and (8.141) for three temperatures T_d.

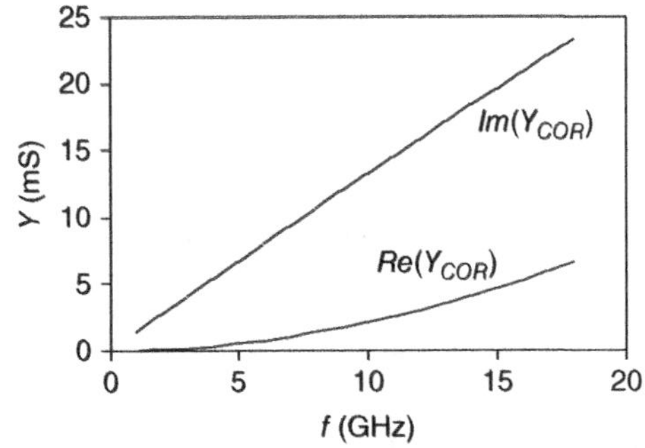

Figure 8.23 Complex correlation admittance for the circuit Figure 8.19.

The result is shown in Figure 8.23.

In the usual formalism one uses instead of the noise generators v_A and i_A the noise resistance R_N and the noise conductance G_N. R_N is given by the voltage source v_A. For the bandwidth B we have generally $B = 1$ Hz.

$$R_N = \frac{\overline{|v_A|^2}}{4kT_0B} \tag{8.143}$$

The uncorrelated part of i_A determines the noise conductance G_N.

The current partition is obtained via

$$\overline{|i_{AU}|^2} = \overline{|i_A|^2}(1 - |\rho_A|^2) \tag{8.144}$$

and thus

$$G_N = \frac{\overline{|i_{AU}|^2}}{4kT_0B} \tag{8.145}$$

Figure 8.24 shows R_N and G_N for three effective temperatures of the drain resistance.

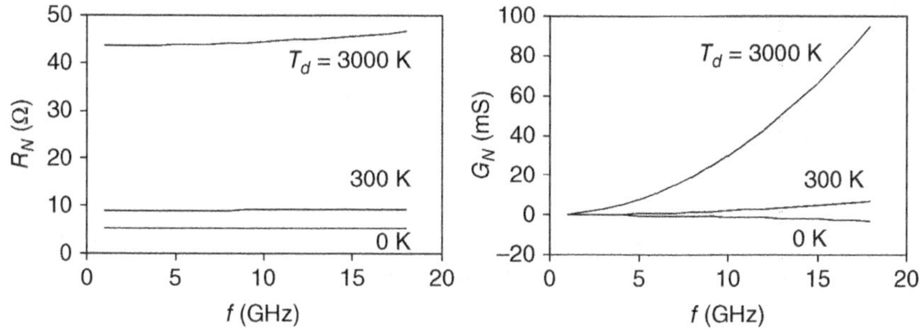

Figure 8.24 Noise resistance R_N and conductance of the uncorrelated current G_N for three temperatures T_d. Circuit according to Figure 8.19.

As already shown in Chapter 7 we can, with the knowledge of G_{COR}, B_{COR}, R_N, and G_N calculate the four IEEE noise parameters.

$$F_{MIN} = 1 + 2R_N G_{COR} + 2\sqrt{R_N G_N + R_N^2 G_{COR}^2}$$

$$G_{OPT} = \sqrt{\frac{G_N}{R_N} + G_{COR}^2}; \qquad B_{OPT} = -B_{COR} \tag{8.146}$$

Usually, noise modeling is not performed in this way, but using the noise correlation matrix in chain representation $\boldsymbol{C}^A$. The great advantage of this is the possibility to combine any complex network of passive and active components in this way.

For our example, the matrix (8.98) using (8.140) and (8.141) is

$$\boldsymbol{C}^A = \begin{pmatrix} v_g^2 + \frac{i_d^2}{g_m^2}(1 + \omega^2 C_{gs}^2 R_g^2); & \frac{i_d^2}{g_m^2}(\omega^2 C_{gs}^2 R_g - j\omega C_{gs}) \\ \frac{i_d^2}{g_m^2}(\omega^2 C_{gs}^2 R_g + j\omega C_{gs}); & \frac{i_d^2}{g_m^2}\omega^2 C_{gs}^2 \end{pmatrix} \tag{8.147}$$

From this, the four IEEE noise parameters are calculated as

$$R_N = \frac{C_{11}^A}{4kT_0}; G_{OPT} = \sqrt{\frac{C_{22}^A}{C_{11}^A} - Im\left(\frac{C_{12}^A}{C_{11}^A}\right)^2}; B_{OPT} = Im\left(\frac{C_{12}^A}{C_{11}^A}\right)$$

$$F_{MIN} = 1 + \frac{Re(C_{12}^A) + C_{11}^A G_{OPT}}{2kT_0} \tag{8.148}$$

Examples of working with the noise correlation matrices are given in the following chapters, where models of important components are discussed, transformation of the different representations into each other and cascading the chain matrices, e.g. in Chapter 9.

References

1 Bittel, H. and Storm, L. (1971). *Rauschen*. Springer Verlag.
2 Gradstein, J.S. and Ryshik, J.M. (1981). *Tables of Series, Products and Integrals*. Verlag H. Deutsch.
3 Hillbrandt, H. and Russer, P. (Apr. 1976). Efficient method for computer aided analysis of linear amplifier networks. *IEEE Trans. Circuits Syst.* 23: 235–238.
4 Schiek, B. (1999). *Grundlagen der Hochfrequenz-Messtechnik*. Berlin, Heidelberg: Springer Verlag.
5 Bosma, H. (1967). On the theory of linear noisy systems. *Philips Res. Rep.* (Supply 10).
6 Müller, R. (1990). *Rauschen*. Springer Verlag.
7 Meinke, H.H. and Gundlach, F.W. (1992). *Taschenbuch der Hochfrequenztechnik*. Springer Verlag.
8 Gronau, G. (1992). *Rauschparameter und Streuparameter Messtechnik*. Verlag Nellisen-Wolff.
9 Pospieszalski, M.W. (Sept. 1989). Modeling of noise parameters of MESFETs and MODFETs and their frequency and temperature dependence. *IEEE Trans. Microwave Theory Tech.* 37: 1340–1350.
10 van der Ziel, A. (1954). *Noise*. London: Chapman and Hall.
11 Rudolph, M. (2006). *Introduction to Modeling HBTs*. Artech House.

9

Diodes and Bipolar Transistors

Semiconductor Diode

We look at the semiconductor diode, which consists of a pn-junction. Current flow is caused by diffusion of charge carriers from regions of high particle density to volumes of lower density and by drift of the particles in the electric field. For the noise properties, that current components are important, which dominate depending on the voltage applied [1].

In the reverse direction, the space-charge layer is so wide that no diffusion currents flow. In the electric field, only few electrons from the p-region drift to the positive electrode at the n-region, and in opposite direction a few holes. They form the currents I_{R1} and I_{R2}, whose sum is the saturation current I_S. When electron–hole pairs are generated in the space charge layer, the current I_{R3} is added. They are shown schematically in Figure 9.1.

Forward biased we have the diffusion currents I_{F1} and I_{F2} and in case of recombination in the space charge layer I_{F3}. Due to their thermal energy, more and more majority carriers move to the other region and reach the counter electrode, even if there is still a retarding potential. This leads to the typical current–voltage characteristic of the diode (9.1) (Figure 9.2).

$$I = I_S \left[exp\left(\frac{qV}{kT}\right) - 1 \right] \tag{9.1}$$

This is the simplest form without recombination. With recombination, a parameter is $m \cong 2$ added that reduces the slope.

$$I = I_S \left[exp\left(\frac{qV}{mkT}\right) - 1 \right] \tag{9.2}$$

All currents are shot noise sources that are not correlated with each other. In the case of reverse biasing we have $I_S = I_{R1} + I_{R2}$ and thus for the noise

$$\overline{i_R^2} = \overline{i_{R1}^2} + \overline{i_{R2}^2} = 2q\left(I_{R1} + I_{R2}\right) B = 2qI_S B \tag{9.3}$$

A Guide to Noise in Microwave Circuits: Devices, Circuits, and Measurement, First Edition.
Peter Heymann and Matthias Rudolph.

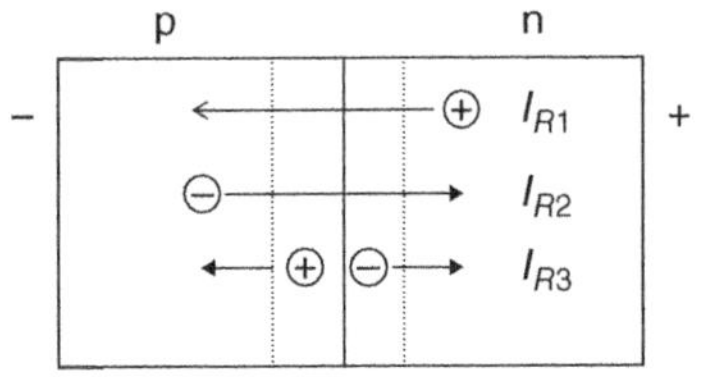

Figure 9.1 The pn-junction in reverse direction. n-region positively biased.

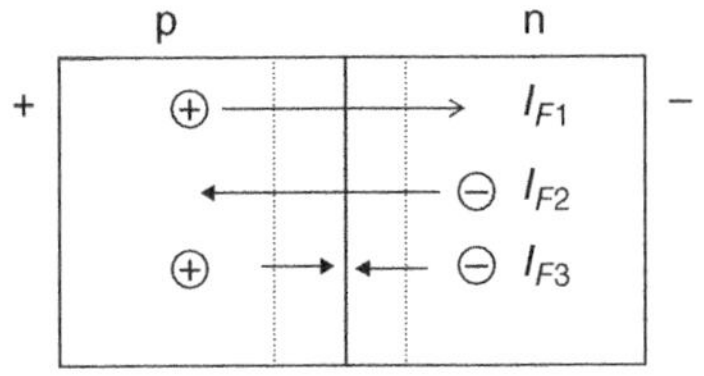

Figure 9.2 Forward biased pn-junction. n-region is negatively biased.

If the reverse voltage is reduced the current increases accordingly (9.1). The reverse current continues to flow anyway, since the voltage is still negative.

$$I = I_{F1} + I_{F2} - I_{R1} - I_{R2} = I_{F1} + I_{F2} - I_S \tag{9.4}$$

The part of diffusion currents is

$$I_{F1} + I_{F2} = I + I_S \tag{9.5}$$

The noise contribution

$$\overline{i_F^2} = 2q\left(I + I_S\right)B \tag{9.6}$$

The sum of the noise of all forward and reverse currents is therefore

$$\overline{i^2} = \overline{i_F^2} + \overline{i_R^2} = 2q\left(I + 2I_S\right)B \tag{9.7}$$

Let us consider the following three special cases:

$$\begin{array}{llll} 1. & \text{strong forward bias} & I \gg I_S & \overline{i^2} = 2qIB \\ 2. & \text{zero current} & V = 0;\ I = 0 & \overline{i^2} = 4qI_SB \\ 3. & \text{reverse bias} & I = -I_S & \overline{i^2} = 2qI_SB \end{array} \tag{9.8}$$

Instead of the *IV*-characteristic (9.1) the differential conductance of the diode $g(V)$ can be used

$$g(V) = \frac{dI}{dV} = \frac{q}{kT}\left(I + I_S\right) \tag{9.9}$$

In the currentless case this conductance must be $g(0) = g_0$. There is thermal noise generation according to the Nyquist formula.

$$\overline{i^2} = 4kTg_0B = 4qI_SB \tag{9.10}$$

Figure 9.3 Normalized noise temperature of the pn-junction according to (9.14) as a function of the bias voltage. $I(V)$ characteristic after (9.1).

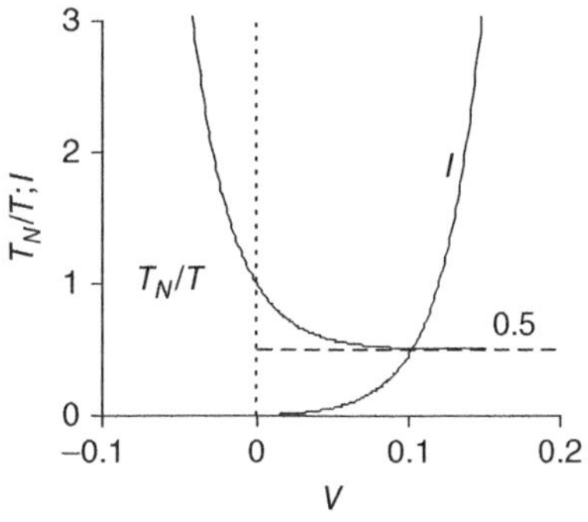

We can apply this description as thermal noise to the entire diode characteristic by introducing an equivalent noise temperature of diode T_N, which is dependent on the bias point.

$$\overline{i^2} = 4kT_N g(V) B \tag{9.11}$$

We set (9.7) and (9.11) equal and solve for T_N

$$4kT_N g(V) B = 2q\left(I + 2I_S\right)B \tag{9.12}$$

With (9.9) results in

$$\frac{T_N}{T} = \frac{1}{I + I_S}\left(\frac{I}{2} + I_S\right) \tag{9.13}$$

If we add (9.1) for I, we get

$$\frac{T_N}{T} = \frac{1}{2}\left\{1 + exp\left(-\frac{eV}{kT}\right)\right\} \tag{9.14}$$

The plot of the normalized noise temperature T_N/T (9.14) is shown in Figure 9.3 together with the diode characteristic curve (9.1).

This comparison of the shot noise of the pn-diode with the thermal noise of its differential conductance (9.9) shows that the currentless diode ($V = 0$) generates thermal noise. Forward biased it has the noise temperature $T_N/T = 1/2$ (1.8 dB) and in the reverse direction it increases strongly. The diode therefore generates shot noise corresponding to the current flowing in both the forward and reverse direction. In the currentless state it also generates noise, but in this case it is thermal noise corresponding to its conductance.

Bipolar Transistor

In high-frequency technology, npn-transistors in grounded emitter circuit are common practice. Before we examine the noise properties, let us briefly recall the principle of operation. Here we will see which currents occur and what is the shot noise contribution due to their passage through pn-junctions.

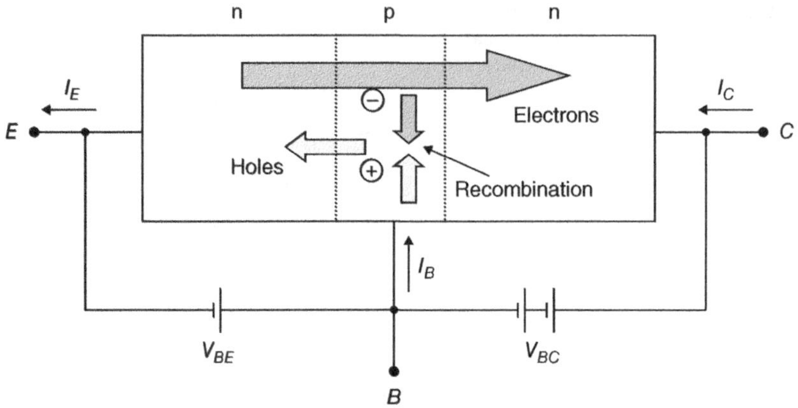

Figure 9.4 Schematic diagram of the npn-bipolar transistor with currents of the charge carriers.

The bipolar transistor can be treated as consisting of two diodes, which both have a common semiconductor region. This is the base, a very thin, p-doped layer. As with the diode, we have drift and diffusion currents through the two pn-junctions. A schematic picture shows Figure 9.4.

The emitter–base diode is biased in the forward direction, the collector–base diode in the reverse direction. We must distinguish the direction of the particle currents inside (electrons flow from − to +) from the direction of the technical currents in the outer circuit (from + to −). In the base–emitter diode, which is forward biased, the electrons flow from the n-doped emitter volume into the base. However, only a small part of them attain the base-electrode. Rather, the effect is that the small volume of the base is flooded with electrons, almost all of which enter the blocked pn-junction of the collector–base diode. This makes the diode conductive and the strongly positive collector electrode absorbs the electrons. Some electrons recombine with the holes in the base. A small part of the emitter current is carried by the holes that enter the emitter from the base. All these currents produce shot noise.

The Ebers–Moll model [2] is a simple approach for the currents in the transistor. As shown in Figure 9.5 for the npn-transistor, it consists of two diodes completed by the transfer currents, which describe the coupling of the two diodes via the common base.

The currents at the contacts are obtained to:

$$\begin{aligned} I_E &= I_F - \alpha_R I_R \\ I_C &= \alpha_F I_F - I_R \\ I_B &= I_E - I_C \end{aligned} \tag{9.15}$$

Figure 9.5 Equivalent circuit of the npn-BJT according to the Ebers-Moll model.

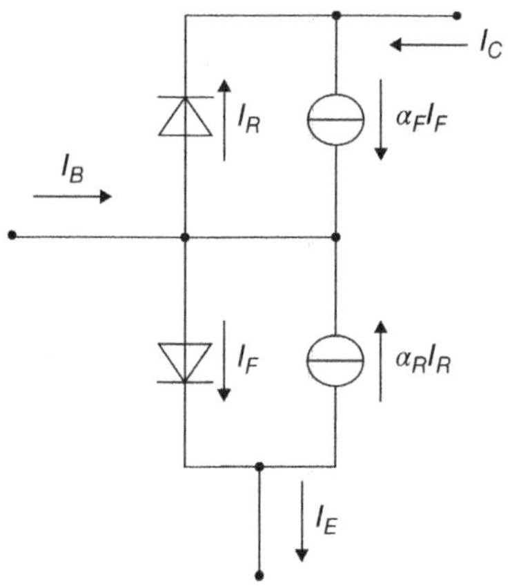

$\alpha_F \approx 0.95 \ldots 0.99$ is the forward current gain, $\alpha_R \approx 0.1 \ldots 0.5$ the reverse current gain in a common-base circuit.

The Ebers–Moll equations of this model based on the diode current (9.1) are

$$I_E = I_{ES}\left\{exp\left(\frac{V_{BE}}{V_T}\right) - 1\right\} - \alpha_R I_{CS}\left\{exp\left(\frac{V_{BC}}{V_T}\right) - 1\right\}$$
$$I_C = \alpha_F I_{ES}\left\{exp\left(\frac{V_{BE}}{V_T}\right) - 1\right\} - I_{CS}\left\{exp\left(\frac{V_{BC}}{V_T}\right) - 1\right\}$$
$$I_B = I_E - I_C \tag{9.16}$$

The thermal voltage is used $V_T = {}^{kT}/_{q} \cong 26$ mV (at $T = 300$ K). I_{ES} and I_{CS} are the saturation currents of both diodes.

For the output characteristics in common-emitter circuit $I_C = f(V_{EC})$ with I_B as parameter it follows

$$I_C = \frac{I_B\left\{\alpha_F - \frac{I_{CS}}{I_{ES}} exp\left(\frac{V_{EC}}{V_T}\right)\right\} + I_{CS}\left\{1 - exp\left(\frac{V_{EC}}{V_T}\right)\right\} + \alpha_R I_{CS}\left\{exp\left(\frac{V_{EC}}{V_T}\right) - 1\right\}}{1 - \alpha_F + \frac{I_{CS}}{I_{ES}}\left(1 - \alpha_R\right) exp\left(\frac{V_{EC}}{V_T}\right)} \tag{9.17}$$

An example is shown in Figure 9.6.

Small-Signal Equivalent Circuit

The equivalent circuit with the diodes (Figure 9.5) is not suitable for noise analysis. The transistor is operated at a defined bias point in the linear range and the diodes are replaced by their resistance at this operating point. We select therefore at a certain point in the characteristic Figure 9.6. In addition, one must include reactive

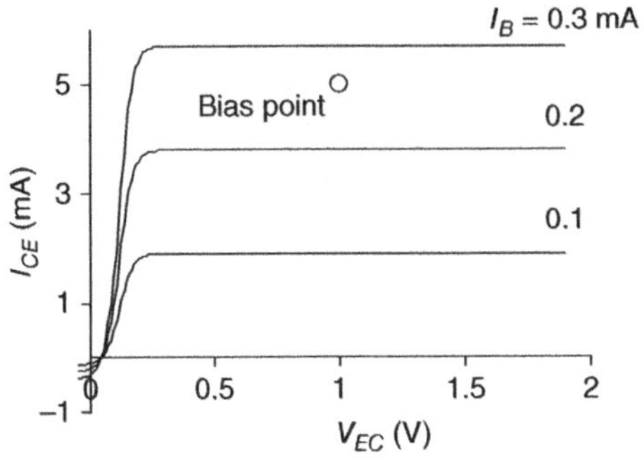

Figure 9.6 *VI*-characteristic according to (9.17); $\alpha_F = 0.95$; $\alpha_R = 0.2$; $I_{ES} = 10^{-12}$ mA. Parameter is the base current I_B. A favorable operating point in the linear range is indicated.

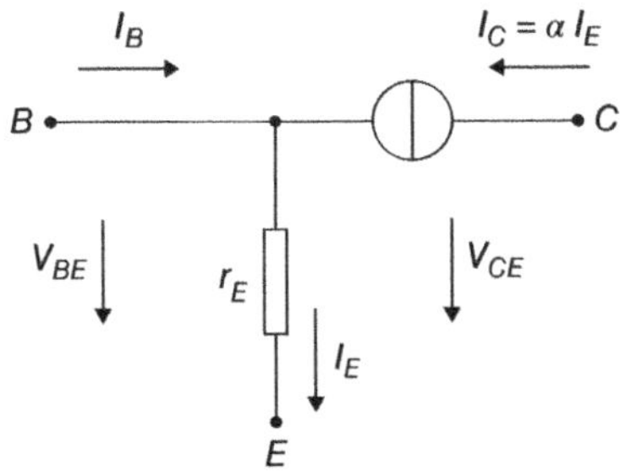

Figure 9.7 Simplest small signal equivalent circuit of a BJT.

elements in order to model the frequency dependence of the noise parameters. The noise analysis is performed with the small-signal equivalent circuit. Here, of course, a complexity that increases with the frequency range is required.

In the simplest T-configuration in Figure 9.7 we first see the transition from the diode characteristic to the equivalent resistance corresponding (9.9) for a certain bias point. r_E is the resistance of the base–emitter diode when looking into the emitter (forward direction).

$$r_E = \frac{V_T}{I_E} \tag{9.18}$$

The parasitic ohmic resistances of the n- or p-doped access layers are denoted by capital letters, e.g. $\mathcal{R}_B$. The resistance r_E corresponds to the derivative of the diode characteristic (9.1) and is denoted with lower case letters. The completion of Figure 9.7 to a simple noise equivalent circuit without reactive elements leads to the Nielsen noise model of the BJT [3].

Hawkins BJT Noise Model

Figure 9.8 shows the somewhat more complicated model by Hawkins [4], which contains the capacitance of the base–emitter diode C_E as reactive element and a complex source resistance $\mathcal{Z}_S$.

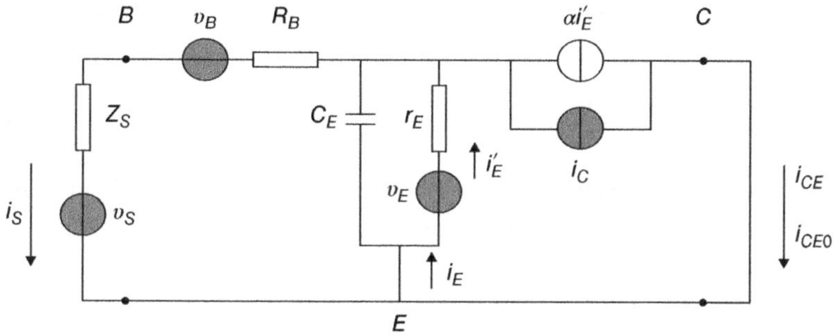

Figure 9.8 Noise model of the BJT according to Hawkins.

In Figure 9.8 the current directions are different from those in the circuits Figures 9.5 and 9.7. This is in itself irrelevant, as they are noise currents, only i_E and $\alpha i'_E$ have the same direction. Figure 9.8 also describes the Nielsen model if we make the following simplifications. The source is real: $Z_S = R_S$ the capacity is neglected: $C_E = 0$. Thus $i'_E = i_E$. The noise voltage source of the resistor r_E is derived from the shot noise of the emitter current.

$$S_{i,E} = 2qI_E \tag{9.19}$$

we replace $\mathcal{I}_E$ accordingly (9.16)

$$r_E = \frac{kT}{qI_E}; \quad I_E = \frac{kT}{qr_E} \tag{9.20}$$

$\mathcal{I}_E$ inserted in (9.19) gives

$$S_{i,E} = 2kT\frac{1}{r_E}; \quad S_{v,E} = \overline{v_E^2} = 2kTr_E; \quad B = 1\,\text{Hz} \tag{9.21}$$

We have already seen above (Figure 9.3) that the forward biased diode generates only ½ of thermal noise.

The resistors $R_S = Re(Z_S)$ and $\mathcal{R}_B$ generate thermal noise.

$$S_{v,S} = 4kTR_S; \quad S_{v,R} = 4kTR_B \tag{9.22}$$

By means of the introduction

$$\alpha = \frac{\alpha_0}{1 + jf/f_b} \tag{9.23}$$

a frequency dependence is achieved even without reactive elements. The cut-off frequency f_b is determined by the transist time τ_B in the base layer

$$f_b = \frac{1}{2\pi\tau_B} \tag{9.24}$$

The approach for the shot noise of the collector current also generates a frequency dependence (derivation see the following).

$$S_{i,C} = 2qI_C\left(1 - \frac{|\alpha|^2}{\alpha_0}\right) \tag{9.25}$$

For the Nielsen model, the noise factor F is obtained as above carried out from the noise current squares in the short-circuited output circuit (collector-emitter):

$$F = 1 + \frac{R_B}{R_S} + \frac{r_E}{2R_S} + \left\{1 - \alpha_0 + \left(\frac{f}{f_b}\right)^2\right\}\frac{\left(R_S + R_B + r_E\right)^2}{2\alpha_0 R_S r_E} \tag{9.26}$$

This is a first approximation for the noise factor of a bipolar transistor.

The Hawkins model better describes the reality of RF-technology. It considers a complex source impedance $Z_S = R_S + jX_S$ and the junction capacitance of the base emitter diode C_E. Both are already shown in Figure 9.8. Here we want to perform the derivation of the noise factor again starting from the definition.

$$F = \frac{\overline{i_{CE}^2}}{\overline{i_{CE0}^2}} \tag{9.27}$$

$\overline{i_{CE}^2}$ is the noise current square in the shorted collector-emitter circuit with the contributions of the transistor sources and the input circuit. The contributions of the transistor are $\overline{v_B^2}$, $\overline{v_E^2}$, and $\overline{i_C^2}$. In the denominator $\overline{i_{CE0}^2}$ represents the noise current square if only the noise of the source is effective and the transistor does not generate noise but only amplifies. It is determined by $\overline{v_S^2}$. For the calculation of i_{CE} we follow [5].

The short-circuit current in the output is composed of

$$i_{CE} = \alpha i'_E + i_C; \quad i'_E = \frac{i_{CE} - i_C}{\alpha} \tag{9.28}$$

In the emitter node the current balance is

$$\begin{aligned} &i_E - i'_E - j i'_E \omega C_E r_E + j\omega C_E v_E = 0 \\ &i_E = i'_E\left(1 + j\omega C_E r_E\right) - j\omega C_E v_E \end{aligned} \tag{9.29}$$

The following also applies to the emitter node

$$\begin{aligned} &i_S + i_{CE} - i_E = 0 \\ &i_S = i_E - i_{CE} \end{aligned} \tag{9.30}$$

In the input circuit the voltage balance is

$$i_S\left(Z_S + R_B\right) + i'_E r_E = v_S + v_B + v_E \tag{9.31}$$

From (9.28)–(9.30) we obtain

$$i_S = \frac{i_{CE} - i_C}{\alpha}\left(1 + j\omega C_E r_E\right) - j\omega C_E v_E - i_{CE} \tag{9.32}$$

We now combine this equation the voltage balance (9.31)

$$\left\{\frac{i_{CE} - i_C}{\alpha}\left(1 + j\omega C_E r_E\right) - j\omega C_E v_E - i_{CE}\right\}\left(Z_S + R_B\right) + \frac{i_{CE} - i_C}{\alpha} r_E$$
$$= v_S + v_B + v_E \tag{9.33}$$

Now we just have to solve to i_{CE}.

$$i_{CE} = \alpha \frac{v_S + v_B + v_E\left(1 + j\omega C_E\left[Z_S + R_B\right]\right) + \frac{i_C}{\alpha}\left\{\left(1 + j\omega C_E r_E\right)\left(Z_S + R_B\right) + r_E\right\}}{\left(1 - \alpha + j\omega C_E r_E\right)\left(Z_S + R_B\right) + r_E} \tag{9.34}$$

now we still need the output short-circuit current i_{CE0} which is produced by the source alone. v_B, v_E, and i_C are equal to zero. It results in:

$$i_{CE0} = \frac{\alpha v_S}{\left(1 - \alpha + j\omega C_E r_E\right)\left(Z_S + R_B\right) + r_E} \tag{9.35}$$

According to (9.27) we need the mean values of the current squares to obtain the noise factor F:

$$F = 1 + \frac{R_B}{R_S} + \frac{r_E}{2R_S} + \left\{\frac{\alpha_0}{|\alpha|^2} - 1\right\}\frac{\left(R_S + R_B + r_E\right)^2 + X_S^2}{2r_E R_S} +$$
$$+ \frac{\alpha_0}{|\alpha|^2}\frac{r_E}{2R_S}\left\{\left(\omega C_E X_S\right)^2 - 2\omega C_E X_S + \omega^2 C_E^2\left(R_S + R_B\right)^2\right\} \tag{9.36}$$

The optimum value of the source impedance $\mathcal{Z}_S = \mathcal{Z}_{OPT}$, for which $\mathcal{F} = \mathcal{F}_{MIN}$, is obtained by partial differentiation of (9.36). The calculation becomes a little clearer by introducing the following coefficients:

$$a = \left\{1 - \frac{|\alpha|^2}{\alpha_0} + \left(\omega C_E r_E\right)^2\right\}\frac{\alpha_0}{|\alpha|^2}$$
$$A_1 = a\frac{\left(R_S + R_B\right)^2}{2r_E R_S} + \frac{\alpha_0}{|\alpha|^2}\left\{1 + \frac{R_B}{R_S} + \frac{r_E}{2R_S}\right\}$$
$$B_1 = -\frac{\alpha_0 \omega C_E r_E}{|\alpha|^2 R_S}; \quad C_1 = \frac{a}{2r_E R_S} \tag{9.37}$$

Eq. (9.36) now has the simple structure

$$F = A_1 + B_1 X_S + C_1 X_S^2 \tag{9.38}$$

The derivation with respect to $\mathcal{X}_S$

$$\frac{\partial F}{\partial X_S} = B + 2CX_{OPT} = 0 \tag{9.39}$$

this results in

$$X_{OPT} = \frac{\alpha_0 \omega C_E r_E^2}{|\alpha|^2 a} \tag{9.40}$$

We insert this into (9.36) and obtain

$$F_X = a\frac{R_S + R_B}{2r_E R_S} + \frac{\alpha_0}{|\alpha|^2}\left\{1 + \frac{R_B}{R_S} + \frac{r_E}{2R_S}\right\} - a\frac{X_{OPT}^2}{2r_E R_S} \tag{9.41}$$

We define new coefficients to simplify the structure of (9.41).

$$A_2 = a\frac{R_B}{r_E} + \frac{\alpha_0}{|\alpha|^2}$$

$$B_2 = a\frac{R_B^2 - X_{OPT}^2}{2r_E} + \frac{\alpha_0}{|\alpha|^2}\left\{R_B + \frac{r_E}{2}\right\}; \quad C_2 = \frac{a}{2r_E} \tag{9.42}$$

This will change (9.41) to

$$F_X = A_2 + \frac{B_2}{R_S} + C_2 R_S \tag{9.43}$$

Derivation with respect to R_S provides the minimum relative to the real part of $\mathcal{Z}_S$.

$$\frac{\partial F_X}{\partial R_S} = -\frac{B_2}{R_{OPT}^2} + C_2 = 0 \tag{9.44}$$

this results in

$$\frac{B_2}{R_{OPT}} = C_2 R_{OPT} \tag{9.45}$$

If we insert this in (9.43), we get for the minimum noise factor

$$F_{MIN} = A + 2C_2 R_{OPT} = a\frac{R_B + R_{OPT}}{r_E} + \frac{\alpha_0}{|\alpha|^2} \tag{9.46}$$

We use (9.23) in the expression for α (9.37) and introduce the cut-off frequency f_e of the base–emitter diode

$$f_e = \frac{1}{2\pi\, r_E C_E} = \frac{1}{\tau_E} \tag{9.47}$$

We arrive at:

$$a = \frac{1}{\alpha_0}\left\{\left[1 + \left(\frac{f}{f_b}\right)^2\right]\left[1 + \left(\frac{f}{f_e}\right)^2\right] - \alpha_0\right\} \tag{9.48}$$

By comparing the formula (7.20) for $\mathcal{F}$, which contains the noise resistance $\mathcal{R}_{\mathcal{N}}$, with (9.36) the noise resistance can be calculated.

The achievable minimum noise figure can be limited by the emitter cut-off frequency (9.36) as well as by the base cut-off frequency (9.22). For Silicon microwave BJT for the lower GHz range we have the emitter limited case. For example, for the AT-41400 [6] $f_B = 23$ GHz. It is far above the range of application of approx. 6 GHz. We write here a summary of the four noise parameters

$$X_{OPT} = \frac{\alpha_0}{a|\alpha|^2}\omega C_E r_E^2; \quad R_{OPT} = \sqrt{R_B^2 - X_{OPT}^2 + \frac{\alpha_0 r_E}{a|\alpha|^2}\left(2R_B + r_E\right)}$$

$$R_N = \frac{R_B}{\alpha_0} + \frac{r_E}{2}\left\{1 + \left(\frac{R_B}{r_E}\right)^2\left(\frac{f}{f_B}\right)^2\right\}; \quad F_{MIN} = \frac{a}{r_E}\left(R_B + R_{OPT}\right) + \frac{\alpha_0}{|\alpha|^2} \tag{9.49}$$

We calculate the noise values for the AT-41400 at the bias point $V_{CE} = 3\,\text{V}$; $I_C = 10\,\text{mA}$. The relevant elements of the equivalent circuit diagram are: $R_B = 6.8\,\Omega$; $r_E = 2.6\,\Omega$; $C_E = 2.1$ pF; $\alpha_0 = 0.99$; $f_B = 23$ GHz.

The measured values of the four noise parameters are taken from the data sheet. Z_{OPT} was converted to Γ_{OPT}, which is common in RF-technology (Figure 9.9).

The simple model provides for $f < 1$ GHz even for Γ_{OPT} surprisingly good results. NF_{MIN} and R_N are modeled very well in the entire frequency range.

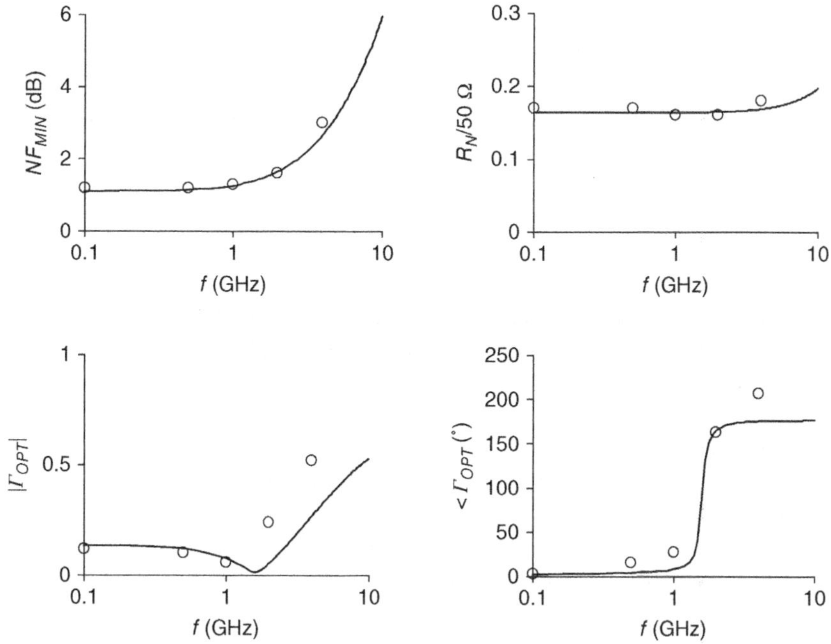

Figure 9.9 Noise parameters of the BJT AT-41400 according to data sheet [6] (circles). Calculation according to (9.49) (lines). Source: Data from Agilent Technologies [6].

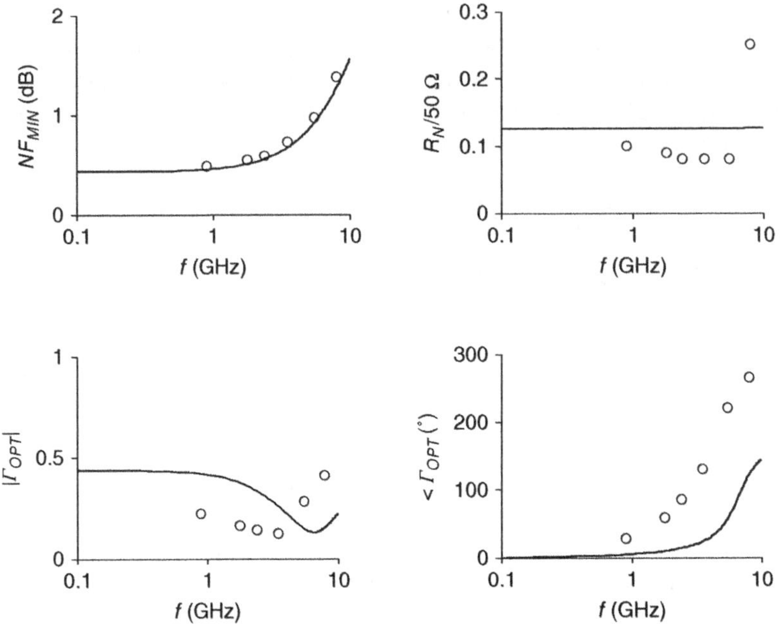

Figure 9.10 Noise parameters according to the Hawkins model for the SiGe-BJT BFP 760 [7] (lines) compared with manufacturer's specifications (circles).

However, for typical microwave transistors, such as the SiGe-BJT BFP 760, the simple model is no longer adequate. The device can be used up to 10 GHz, because up to this frequency the maximum power gain is $G_{MAX}(f) \geq 10$ dB. In Figure 9.10 we show the noise modeling in common-emitter circuit for the bias point $V_{CE} = 3$ V; $I_C = 10$ mA. The elements are $R_B = 5\,\Omega$; $r_E = 2.6\,\Omega$; $C_E = 0.5$ pF; $\alpha_0 = 0.998$; $f_B = 79$ GHz; $f_E = 103$ GHz.

NF_{MIN} is reproduced well over the entire frequency range. With the other quantities the inadequacy of the model is clearly visible. Neglecting the parasitic elements of the transistor and the correlation in the equivalent circuit (Figure 9.7) leads to unusable results.

In addition to the frequency dependence, the current dependence is also important and should be well reproduced by a noise model. Figure 9.11 shows an example at $f = 2.4$ GHz.

Above $I_c > 10$ mA the agreement is good. However, the simple model cannot reproduce the important effect of increasing NF_{MIN} with decreasing current because it does not include the decrease in gain.

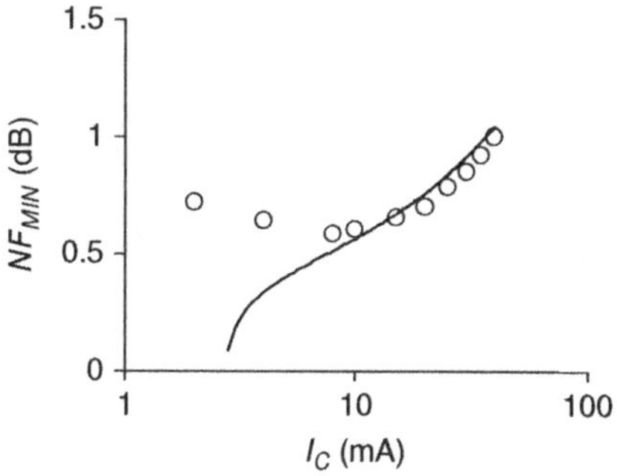

Figure 9.11 Current dependence of the noise figure of the BFP 760 at $f = 2.4\,\text{GHz}$ [7].

Two Approaches for the Collector Noise Current Source

It is useful for the understanding of noise modeling to analyze the approach for the collector source spectral noise current density (9.25). We follow the presentation in [8].

The shot effect of the emitter current gives the spectral noise power density $S_{i,E}$. We write (9.21) with the conductance $g_E = {}^1/_{r_E}$

$$S_{i,E} = 2kTg_E \tag{9.50}$$

The identity

$$S_{i,E} = 2S_{i,E} - S_{i,E} = 4kTg_E - 2qI_E \tag{9.51}$$

will be useful in the further calculation. The basic idea is that almost all charge carriers that have passed the base–emitter junction also reach the base–collector transition. The following applies to the noise spectrum

$$S_{i,C} = 2qI_C = 2q\alpha_0 I_E \tag{9.52}$$

Since the noise current of the base–emitter junction contributes with the part α to the noise current of the collector, it is obvious that $S_{i,E}$ and $S_{i,C}$ are strongly correlated. The cross spectrum results from

$$S_{i,EC} = i_E^* i_C = -i_E^* \alpha i_E = -\alpha |i_E|^2 = -\alpha 2qI_E = -2kT\alpha g_E \tag{9.53}$$

The minus sign results from the consideration that in the short-circuited input circuit (without Z_S and R_B) the noise currents i_E and i_C are in opposite direction.

This correlation of the two sources is not beneficial. It can be avoided by combining the noise currents αi_E and i_C in the collector source.

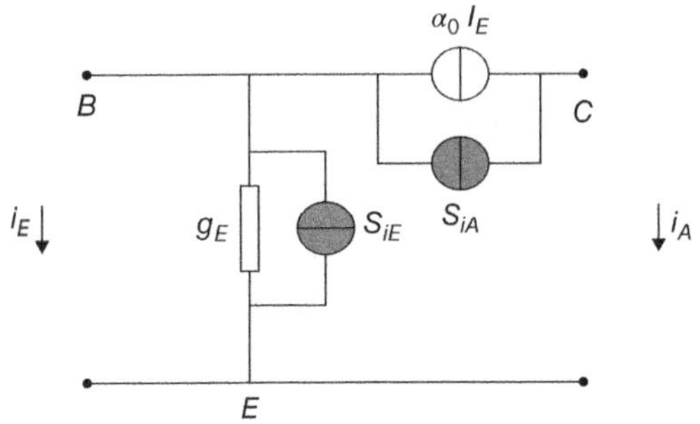

Figure 9.12 Noise model of the BJT with the sources in the emitter and collector circuit S_{iE} and S_{iA} to analyze the correlation.

We denote the combined noise current source in parallel with the collector current source with i_A.

$$i_A = \alpha i_E + i_C \tag{9.54}$$

with the spectral power density

$$S_{i,A} = \left(\alpha i_E + i_C\right)^* \left(\alpha i_E + i_C\right) \tag{9.55}$$

when multiplied, the result is

$$S_{i,A} = |\alpha|^2 \left|i_E\right|^2 + \left|i_C\right|^2 + \alpha^* i_E^* i_C + \alpha i_E i_C^* \tag{9.56}$$

In the first two terms, the contributions of the amplified emitter shot current and the collector shot current can be seen. The last two terms are the cross-correlation, which we have already calculated (9.53). We can write with (9.51) and (9.54)

$$S_{i,A} = |\alpha|^2 \left(4kTg_E - 2qI_E\right) + 2qI_C - 4|\alpha|^2 kTg_E = 2q\left(I_C - |\alpha|^2 I_E\right) \tag{9.57}$$

This gives us the spectral power density of the collector current source, as used in the Hawkins model (9.23) (Figure 9.12)

$$S_{i,A} = 2qI_E\left(\alpha_0 - |\alpha|^2\right) = 2qI_C\left(1 - \frac{|\alpha|^2}{\alpha_0}\right) \tag{9.58}$$

Let us now look at the correlation between the two sources

$$S_{i,EA} = i_E^*\left(\alpha i_E + i_C\right) = \alpha S_{i,E} + S_{i,EC} \tag{9.59}$$

This leads to

$$S_{i,EA} = \alpha\left(4kTG_{EB} - 2qI_E\right) - \alpha 2kTG_{EB} = 0 \tag{9.60}$$

and shows that the correlation disappears. With this approach, to combine the noise currents αi_E and i_C in the collector source, the noise analysis is therefore much easier.

BJT Noise Model with Correlation Matrices

The models discussed earlier fast reach their limits at higher frequencies. For transistors with cut-off frequencies above 100 GHz, much more elements of the equivalent circuit, especially parasitic ones, have to be considered. The analysis based on network formulas as in the Hawkins model would be an unsolvable task. The method of correlation matrices is favorably used. It allows the step-by-step composition of a complicated circuit from simple elements and is ideally suited to computer applications.

Here is a summary of the calculation rules. The networks 1 and 2 are combined, which are characterized by their noise correlation matrices.

$$\begin{aligned}&\text{Parallel:} && \boldsymbol{C}^{Y} = \boldsymbol{C}_1^{Y} + \boldsymbol{C}_2^{Y}\\ &\text{Series:} && \boldsymbol{C}^{Z} = \boldsymbol{C}_1^{Z} + \boldsymbol{C}_2^{Z}\\ &\text{Cascade:} && \boldsymbol{C}^{A} = A_1 \boldsymbol{C}_2^{A} A_1^{H} + \boldsymbol{C}_1^{A}\end{aligned} \tag{9.61}$$

$\boldsymbol{A}_1$ is the chain matrix of network 1, $\boldsymbol{C}_1^{A}$ the noise correlation matrix of network 1 in chain representation. Transformation matrices $\boldsymbol{T}$ for the conversion of the three forms can be taken from the table. The conversion is performed according to the formula

$$\boldsymbol{C}^{i} = \boldsymbol{T}\boldsymbol{C}^{j}\boldsymbol{T}^{H} \tag{9.62}$$

with $i, j = Y, Z, A$.

The Π-Model

We will first demonstrate the application of correlation matrices on the simplest equivalent circuit of the BJT, the hybrid-Π model. As shown in Figure 9.13, the noise-free transistor consists only of the resistance of the base–emitter diode r_π with the parallel junction capacity C_E and the voltage controlled current source $g_m v_{BE}$. At the ports the two noise current sources i_1 i_2 are added. Using this we can define the Y-correlation matrix $\boldsymbol{C}^{Y}$ and continue calculating with the admittance matrix $\boldsymbol{Y}$ of the circuit. The two noise sources represent the shot noise of base and collector current ($I_B; I_C$). Both are not correlated. First of all, we write the following for the direct current values with the current gain β_0

$$r_\pi = \frac{V_T \beta_0}{I_C}; \quad g_{mo} = \frac{\beta_0}{r_\pi}; \quad \beta_0 = \frac{I_C}{I_B} = \frac{\alpha_0}{1-\alpha_0} \tag{9.63}$$

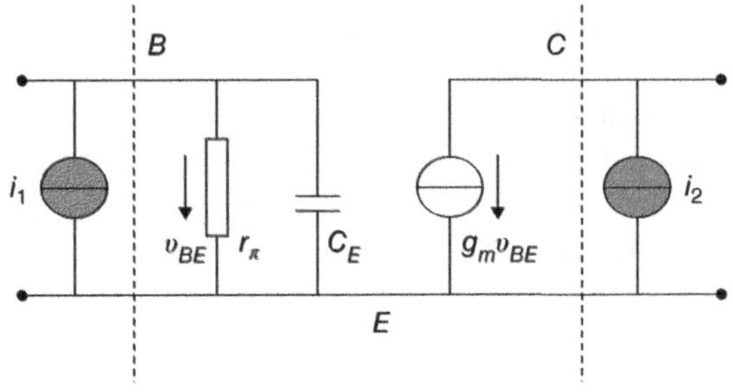

Figure 9.13 Π-model of the BJT with noise sources in Y-representation.

The frequency dependence of the gain is taken into account by introducing the base time constant τ_B in g_m

$$g_m = \frac{g_{m0}}{1 + j\omega\tau_B} \tag{9.64}$$

In $B = 1$ Hz bandwidth we have

$$\mathbf{C}^Y = 2q\begin{pmatrix} I_B & 0 \\ 0 & I_C \end{pmatrix}; \quad \mathbf{Y} = \begin{pmatrix} \frac{1}{r_\pi} + j\omega C_E & 0 \\ g_m & 0 \end{pmatrix} \tag{9.65}$$

With knowledge of these elements and the relations (8.73)–(8.76) we can determine the four noise parameters F_{MIN}; $|\Gamma_{OPT}|$; $arg(\Gamma_{OPT})$ and R_N. The result cannot satisfy, because it accounts only for the simplest model of the intrinsic transistor. The analysis has to be extended for a real transistor by additional elements. For example, the base resistance R_B is already included in the Hawkins model (Figure 9.7) and is absent here.

The combination of cascaded elements is performed by the chain matrix. We first convert (9.65) into the chain form $\boldsymbol{Y} \rightarrow \boldsymbol{A}$

$$\boldsymbol{A} = -\frac{1}{Y_{21}}\begin{pmatrix} Y_{22} & 1 \\ det(Y) & Y_{11} \end{pmatrix} = -\frac{1}{g_m}\begin{pmatrix} 0 & 1 \\ 0 & \frac{1}{r_\pi} + j\omega C_E \end{pmatrix} \tag{9.66}$$

According to Table 9.1, the transformation matrix and its Hermitian form are as follows

$$\boldsymbol{T} = \begin{pmatrix} 0 & A_{12} \\ 1 & A_{22} \end{pmatrix} = \begin{pmatrix} 0 & -\frac{1}{g_m} \\ 1 & -\frac{1}{g_m}\left(\frac{1}{r_\pi} + j\omega C_E\right) \end{pmatrix}$$

$$\boldsymbol{T}^H = \begin{pmatrix} 0 & 1 \\ -\frac{1}{g_m^*} & -\frac{1}{g_m^*}\left(\frac{1}{r_\pi} - j\omega C_E\right) \end{pmatrix} \tag{9.67}$$

Table 9.1 Transformation matrices **T** for converting the correlation matrices according to (9.62) [9].

		Original representation			
		C^Y	C^Z	C^A	C^S
Resulting representation	$\boldsymbol{C}^Y$	$\begin{pmatrix} 1 & 0 \\ 0 & 1 \end{pmatrix}$	$\begin{pmatrix} y_{11} & y_{12} \\ y_{21} & y_{22} \end{pmatrix}$	$\begin{pmatrix} -y_{11} & 1 \\ -y_{21} & 0 \end{pmatrix}$	$\begin{pmatrix} 1+y_{11} & y_{12} \\ y_{21} & 1+y_{22} \end{pmatrix}$
	$\boldsymbol{C}^Z$	$\begin{pmatrix} z_{11} & z_{12} \\ z_{21} & z_{22} \end{pmatrix}$	$\begin{pmatrix} 1 & 0 \\ 0 & 1 \end{pmatrix}$	$\begin{pmatrix} 1 & -z_{11} \\ 0 & -z_{21} \end{pmatrix}$	$\begin{pmatrix} 1+z_{11} & z_{12} \\ z_{21} & 1+z_{22} \end{pmatrix}$
	$\boldsymbol{C}^A$	$\begin{pmatrix} 0 & a_{12} \\ 1 & a_{22} \end{pmatrix}$	$\begin{pmatrix} 1 & -a_{11} \\ 0 & -a_{21} \end{pmatrix}$	$\begin{pmatrix} 1 & 0 \\ 0 & 1 \end{pmatrix}$	—
	$\boldsymbol{C}^S$	$\frac{1}{4}\begin{pmatrix} 1+s_{11} & s_{12} \\ s_{21} & 1+s_{22} \end{pmatrix}$	$\frac{1}{4}\begin{pmatrix} 1-s_{11} & -s_{12} \\ -s_{21} & 1-s_{22} \end{pmatrix}$	—	$\begin{pmatrix} 1 & 0 \\ 0 & 1 \end{pmatrix}$

The relation of $\boldsymbol{C}^S$ and $\boldsymbol{C}^A$ is not easy to represent.

The rule for the conversion of correlation matrices is $\boldsymbol{C}^A = \boldsymbol{T}\boldsymbol{C}^Y\boldsymbol{T}^H$. Thus after two matrix multiplications the noise correlation matrix is obtained in chain form for the equivalent circuit in Figure 9.13

$$\boldsymbol{C}^A = 2q \begin{pmatrix} \frac{I_C}{|g_m|^2} & \frac{I_C}{|g_m|^2}\left(\frac{1}{r_\pi} - j\omega C_E\right) \\ \frac{I_C}{|g_m|^2}\left(\frac{1}{r_\pi} + j\omega C_E\right) & I_B + \frac{I_C}{|g_m|^2}\left(\frac{1}{r_\pi^2} + \omega^2 C_E^2\right) \end{pmatrix} \tag{9.68}$$

This is the basis for extending the model with noisy parasitic resistors and reactive elements. To explain the procedure, we add the base path resistance R_B as the first step of completion (Figure 9.14).

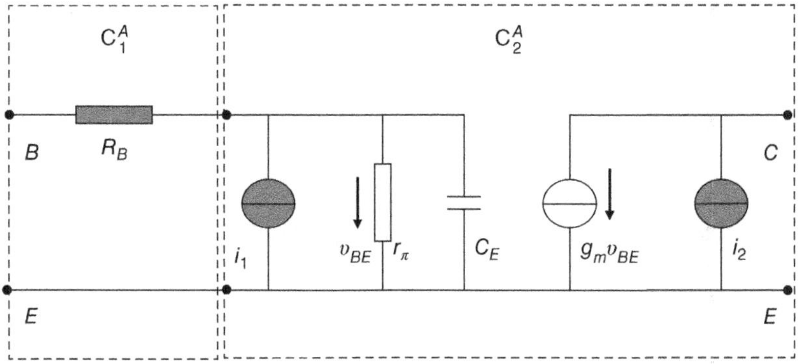

Figure 9.14 Base resistance R_B and intrinsic transistor are two cascaded networks. They are described by matrices in chain form.

The thermal noise generating resistor R_B is cascaded with the intrinsic transistor. We arrive at the noise correlation matrix of the combination $\boldsymbol{C}_1^A$ and $\boldsymbol{C}_2^A$ by applying the above transformation (9.61)

$$\boldsymbol{C}^A = \boldsymbol{A}_1 \boldsymbol{C}_2^A \boldsymbol{A}_1^H + \boldsymbol{C}_1^A \tag{9.69}$$

$\boldsymbol{C}_2^A$ is already known, it is the matrix of the intrinsic transistor (9.68). $\boldsymbol{A}_1$ is the chain matrix of the series resistor R_B. $\boldsymbol{C}_1^A$ is the noise correlation matrix of this resistor. According to (8.65) it results for passive elements from the real part of the network matrix. However, this applies only to the Y- and Z-matrices, not to the chain matrix. This must be obtained by conversion. We start from the $\boldsymbol{Y}$-matrix of the series resistance R_B (Figure 9.14 left). We need the $\boldsymbol{A}$-matrix for conversion.

$$\boldsymbol{Y} = \frac{1}{R_B}\begin{pmatrix} 1 & -1 \\ -1 & 1 \end{pmatrix}; \quad \boldsymbol{A}_1 = \begin{pmatrix} 1 & R_B \\ 0 & 1 \end{pmatrix}; \quad \boldsymbol{A}_1^H = \begin{pmatrix} 1 & 0 \\ R_B & 1 \end{pmatrix}$$

$$\boldsymbol{C}^Y = \frac{4kT}{R_B}\begin{pmatrix} 1 & -1 \\ -1 & 1 \end{pmatrix} \tag{9.70}$$

The transformation to $\boldsymbol{C}^A$-matrix is carried out according to (9.62) with $\boldsymbol{C}^A = \boldsymbol{T}\boldsymbol{C}^Y\boldsymbol{T}^H$. According to Table 9.1, $\boldsymbol{T}$ in this case is

$$\boldsymbol{T} = \begin{pmatrix} 0 & A_{12} \\ 1 & A_{22} \end{pmatrix} = \begin{pmatrix} 0 & R_B \\ 1 & 1 \end{pmatrix}; \quad \boldsymbol{T}^H = \begin{pmatrix} 0 & 1 \\ R_B & 1 \end{pmatrix} \tag{9.71}$$

Multiplying the three matrices according to (9.62) gives the chain correlation matrix of the series resistance (basic path resistance R_B in Figure 9.14) to

$$\boldsymbol{C}^A = 4kT\begin{pmatrix} R_B & 0 \\ 0 & 0 \end{pmatrix} \tag{9.72}$$

Using (9.69) we can now calculate the chain correlation matrix of the network in Figure 9.14. $\boldsymbol{A}_1$ and $\boldsymbol{A}_1^H$ we have in (9.60), $\boldsymbol{C}_2^A$ is the matrix of the intrinsic transistor (9.68) and $\boldsymbol{C}_1^A$ is given from (9.72).

We set

$$K1 = \frac{1}{r_\pi} + j\omega C_E; \quad K2 = \frac{1}{r_\pi^2} + \omega^2 C_E^2 \tag{9.73}$$

The result for the matrix elements is

$$C_{11}^A = 2q\frac{I_C}{|g_m|^2}\left[1 + R_B\left\{K1 + R_B\left(K_1^* + R_B\frac{I_B}{I_C}|g_m|^2 + K2\right)\right\}\right] + 4kTR_B$$

$$C_{22}^A = 2q\frac{I_C}{|g_m|^2}\left(\frac{I_B}{I_C}|g_m|^2 + K2\right)$$

$$C_{21}^A = 2q\frac{I_C}{|g_m|^2}\left\{K1 + R_B\left(\frac{I_B}{I_C}|g_m|^2 + K2\right)\right\}; \quad C_{12}^A = C_{21}^{A*} \tag{9.74}$$

A numerical example provides a slightly different frequency response of the noise parameters compared with the Hawkins model for $f > 1$ GHz. This is due to the arbitrarily introduced frequency dependence of the gain, $\alpha(f)$ in (9.21) and $g_m(f)$ in (9.64). They do not have the same effect. One can easily show this effect by converting $g_m = {}^{\alpha}/_{r_\pi(1-\alpha)}$.

The T-Model with Correlation Matrices

Figure 9.15 shows on the left the simplest noise model of the intrinsic transistor in T-form.

The input admittance is determined by the emitter current, which is practically given by the amplified base current:

$$i_E = i_B(1 + \beta) \tag{9.75}$$

This results for Y_{11} in

$$Y_{11} = (1 - \alpha)\left(\frac{1}{r_E} + j\omega C_E\right) \tag{9.76}$$

The admittance matrix of the T-model in common-emitter circuit is therefore

$$\boldsymbol{Y} = \begin{pmatrix} (1-\alpha)\left(\frac{1}{r_E} + j\omega C_E\right) & 0 \\ \alpha\left(\frac{1}{r_E} + j\omega C_E\right) & 0 \end{pmatrix} \tag{9.77}$$

Noise modeling is always based on the physical noise sources, which are positioned at a suitable location in the equivalent circuit. If possible, the circuit is reduced to the extent that only the intrinsic transistor is used as the starting basis. All external elements can be added in the formalism of the correlation matrices. In the beginning there is usually the correlation matrix of the intrinsic transistor of the BJT in $\mathbf{C}^\mathrm{Y}$-form. The location of the two uncorrelated sources in Figure 9.15a

$$\overline{v_r^2} = 2kTr_E = 2qI_E r_E^2; \quad \overline{i_C^2} = 2qI_B \tag{9.78}$$

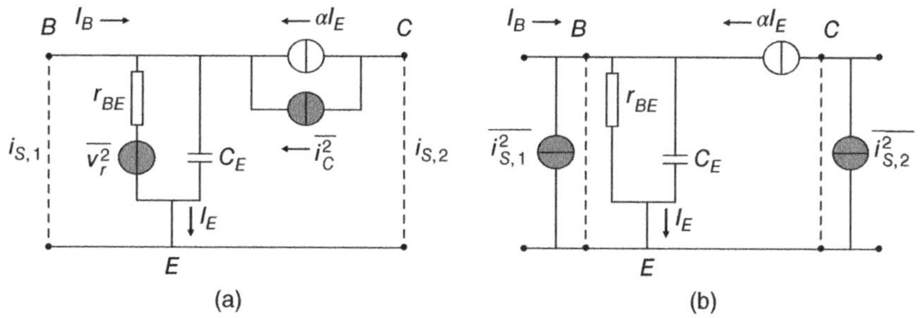

Figure 9.15 Intrinsic noise sources of the network (a) are converted to external noise (b).

of course does not correspond to the two short-circuit currents $i_{S,1}$ and $i_{S,2}$which we need for the $\boldsymbol{C}^{\mathrm{Y}}$-matrix. In (9.78) we have a voltage source at the input and a current source between base and collector. This arrangement based on the considerations by van der Ziel [10]. At first one is astonished why in the current source not the collector direct current $\mathcal{I}_C$ appears but the much smaller base current $\mathcal{I}_B$. The main part of the noise power in the output circuit comes from the shot effect of the emitter current when passing through the emitter–base junction, whose power is amplified. $\overline{i_C^2}$ adds up the shot effect of a series of very small currents. These are: (i) Electrons recombining in the base. (ii) Holes injected into the emitter from the base. (iii) The reverse saturation currents of the emitter–base diode and the reverse-biased base–collector diode.

The intrinsic noise sources must be transformed to the input resp. output. We intentionally do not write here the current squares $\overline{i_S^2}$ corresponding to the spectral power density, since we need the currents directly to calculate the correlation that we expect. From the circuit Figure 9.15a we attain to Figure 9.15b by applying the superposition theorem for networks with several sources and linear resistances. As is well known, the required currents are calculated by successively considering only the effect of one source at a time and finally adding all the currents calculated in this way. The effect of the other sources is canceled by short-circuiting voltage sources and leaving out current sources.

As a first step (Figure 9.16) we short-out the voltage source v_r and calculate the currents i_1' and i_2' which are supplied by the collector current source i_C. One can see that the same short-circuit current flows through the input and output.

$$i_1' = i_2' = \sqrt{2qI_B} \tag{9.79}$$

In the second step (Figure 9.17), we delete the collector noise source and consider the noise currents i_1'' and i_2'', that are generated by v_r. In the input loop the

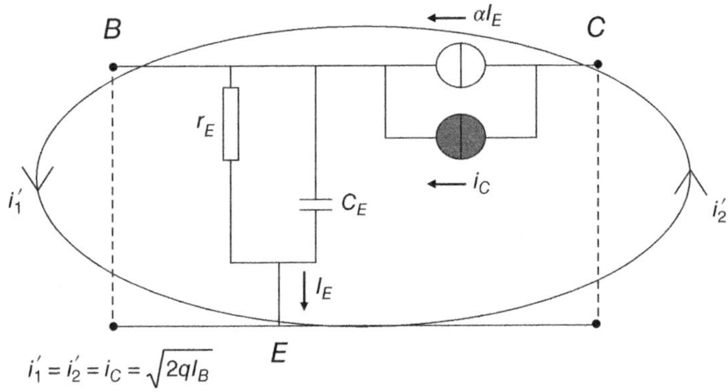

Figure 9.16 First step: The voltage source is ineffective (short circuited).

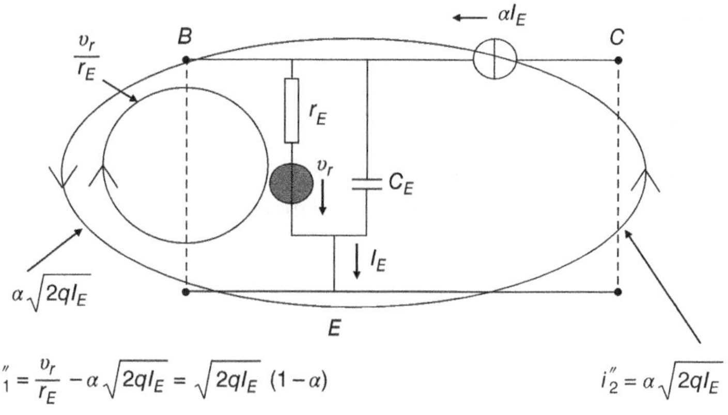

Figure 9.17 Second step: The collector noise current source is ineffective (removed).

voltage v_r and the current generated by the controlled source act together but in opposite directions.

$$i_1'' = \frac{v_r}{r_E} - \alpha\sqrt{2qI_E} = \sqrt{2qI_E}(1-\alpha) \tag{9.80}$$

The short-circuit current in the output circuit is

$$i_2'' = \alpha\sqrt{2qI_E} \tag{9.81}$$

Now we add both partial currents to the total short circuit current. It must be regarded that i_1' and $\alpha\sqrt{2qI_E}$ have the same direction. Therefore i_1' has a negative sign:

$$i_{S,1} = -i_1' + i_1''; \quad i_{S,2} = i_2' + i_2'' \tag{9.82}$$

$$i_{S,1} = -\sqrt{2qI_B} + (1-\alpha)\sqrt{2qI_E} \tag{9.83}$$

$$i_{S,2} = \sqrt{2qI_B} + \alpha\sqrt{2qI_E} \tag{9.84}$$

As elements of the $\mathbf{C}^{\mathrm{Y}}$-correlation matrix, we need the spectral power densities $\overline{i_{S,1}i_{S,1}^*}$ and $\overline{i_{S,2}i_{S,2}^*}$ of the noise current sources in Figure 9.15b and their cross spectral density $\overline{i_{S,1}i_{S,2}^*}$.

$$\begin{aligned} C_{11}^Y &= \overline{i_{S,1}i_{S,1}^*} = 2q\left[-\sqrt{I_B} + (1-\alpha)\sqrt{I_E}\right]\left[-\sqrt{I_B} + (1-\alpha^*)\sqrt{I_E}\right] \\ &= 2q\left[I_B - (1-\alpha^*)\sqrt{I_BI_E} - (1-\alpha)\sqrt{I_BI_E} + (1-\alpha)(1-\alpha^*)I_E\right] \end{aligned} \tag{9.85}$$

The mixed products with $\sqrt{I_BI_E}$ disappear because the two sources in Figure 9.15a are not correlated. It remains:

$$C_{11}^Y = 2q\left[I_B + |1-\alpha|^2 I_E\right] \tag{9.86}$$

For the output circuit we receive

$$\begin{aligned} C_{22}^{Y} &= \overline{i_{S,2} i_{S,2}^{*}} = 2q\left[\sqrt{I_B} + \alpha\sqrt{I_E}\right]\left[\sqrt{I_B} + \alpha^{*}\sqrt{I_E}\right] \\ &= 2q\left[I_B + \alpha^{*}\sqrt{I_B I_E} + \alpha\sqrt{I_B I_E} + |\alpha|^2 I_E\right] = 2q\left[I_B + |\alpha|^2 I_E\right] \end{aligned} \tag{9.87}$$

After the transformation, two new sources occur, each of which contains parts of the two original sources. Therefore, their correlation does not disappear, but rather results to

$$\begin{aligned} C_{12}^{Y} &= \overline{i_{S,1} i_{S,2}^{*}} = 2q\left[-\sqrt{I_B} + (1-\alpha)\sqrt{I_E}\right]\left[\sqrt{I_B} + \alpha^{*}\sqrt{I_E}\right] \\ &= 2q\left[-I_B - \alpha^{*}\sqrt{I_B I_E} + (1-\alpha)\sqrt{I_B I_E} + (1-\alpha)\alpha^{*} I_E\right] \end{aligned} \tag{9.88}$$

It remains

$$C_{12}^{Y} = 2q\left[\left(\alpha^{*} - |\alpha|^2\right) I_E - I_B\right] \tag{9.89}$$

With this analysis for the intrinsic transistor, we have the basis for the step-by-step extension by passive elements, up to the complete noise model. For high microwave frequencies, the inclusion of the complete small-signal equivalent circuit is indispensable. But this is pure computer calculation based on the formalism of the noise correlation matrices. The next step in our example is to add the base path resistor R_B. For this purpose the transition from the Y-form to the A-form $\mathbf{C}^Y \rightarrow \mathbf{C}^A$ and cascading with R_B must be performed. This has already been shown for the hybrid model above.

The Y-form with the two noise short circuit currents at input and output (Figure 9.15b) is in any case the best starting basis for modeling. However, the A-form, with its two sources at the input and the subsequent noise-free network, is known to be better suited to the actual problem.

We want to attain the representation Figure 9.18b from the representation Figure 9.18a by the well-known methods of network analysis.

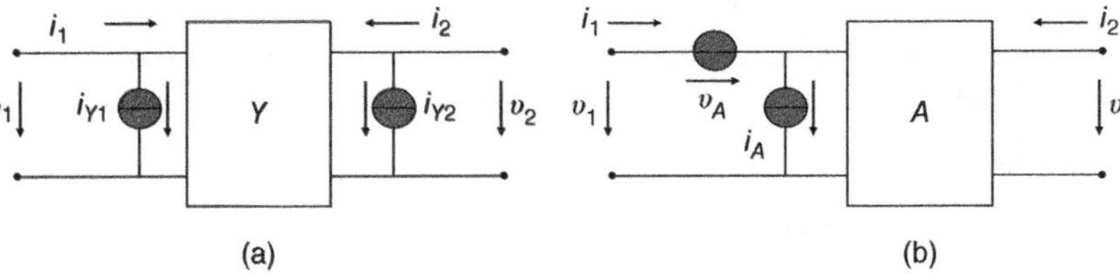

Figure 9.18 Transition from admittance (***Y***-matrix) a to chain representation (***A***-matrix) b.

Transformation of the Y-Sources to the Input

Table 8.2 shows the formulae for the transformation of the short-circuit currents i_{Y1} and i_{Y2} of the Y-form into the two input sources v_A and i_A the A-form

$$\begin{aligned} v_A &= -A_{12} i_{Y2} \\ i_A &= i_{Y1} - A_{22} i_{Y2} \end{aligned} \tag{9.90}$$

In our case $i_{Y1} = i_{S1}$; $i_{Y2} = i_{S2}$. This results in (9.83) and (9.84)

$$v_A = -\frac{\sqrt{2q}}{Y_{21}}\left(\sqrt{I_B} + \alpha\sqrt{I_E}\right) \tag{9.91}$$

$$i_A = \sqrt{2q}\left(-\sqrt{I_B} + (1-\alpha)\sqrt{I_E} - \frac{Y_{11}}{Y_{21}}\left[\sqrt{I_B} + \alpha\sqrt{I_E}\right]\right) \tag{9.92}$$

According to the procedure described earlier, we must first split the noise current i_A into a part which is correlated to v_A and an uncorrelated part. We recall (Chapter 2) that the general form of correlation of two fluctuating quantities, x and y, can be written as follows:

$$y = ax + z; \quad \overline{x} = \overline{y} = \overline{z} = 0 \tag{9.93}$$

a is a constant and z is a fluctuating quantity independent of x. In our case y and z are two noise currents and x is a noise voltage. Thus a has the dimension of a conductance, it is the correlation conductance Y_{COR}. The current i_A (9.92) thus consists of a current fully correlated with v_A, i.e. i_{AC} and an uncorrelated part i_{AU}. (9.93) thus reads

$$i_A = Y_{COR} v_A + i_{AU}; \quad i_{AC} = Y_{COR} v_A \tag{9.94}$$

For the correlation coefficient in our case

$$\rho = \frac{\overline{v_A i_A}}{\sqrt{\overline{v_A^2}\,\overline{i_A^2}}} \tag{9.95}$$

With its use one gets Y_{COR}. We multiply the left part of (9.94) by v_A and write the mean value

$$\overline{v_A i_A} = Y_{COR} \overline{v_A^2} \tag{9.96}$$

Inserting $\overline{v_A i_A}$ from the correlation coefficient (9.95)

$$Y_{COR} = \rho \frac{\sqrt{\overline{v_A^2}\,\overline{i_A^2}}}{\overline{v_A^2}} = \rho\sqrt{\frac{\overline{i_A^2}}{\overline{v_A^2}}} \tag{9.97}$$

for the uncorrelated current one obtains

$$\overline{i_{AU}^2} = \overline{i_A^2}\left(1 - |\rho|^2\right) \tag{9.98}$$

We must calculate the correlation coefficient (9.55) with (9.91) and (9.92). Again, it is important that mixed products, i.e. those that $I_B \times I_E$ contain, do not contribute to the mean value because the original noise currents are not correlated. We obtain for the power of the voltage source

$$\overline{|v_A|^2} = \frac{2q}{|Y_{21}|^2}\left(I_B + |\alpha|^2 I_E\right) \tag{9.99}$$

And for the power of the current source

$$\overline{|i_A|^2} = 2q\left[I_B + (1-\alpha)(1-\alpha^*)I_E\right] - 2q\frac{Y_{11}^*}{Y_{21}^*}\left[-I_B + (1-\alpha)\alpha^* I_E\right] - \\ -2q\frac{Y_{11}}{Y_{21}}\left[-I_B + (1-\alpha^*)\alpha I_E\right] + 2q\frac{|Y_{11}|^2}{|Y_{21}|^2}\left[I_B + |\alpha|^2 I_E\right] \tag{9.100}$$

The numerator of (9.95) results in

$$\overline{v_A i_A} = -\frac{2q}{Y_{21}}\left[-I_B + (1-\alpha^*)\alpha I_E\right] + 2q\frac{Y_{11}^*}{|Y_{21}|^2}\left[I_B + |\alpha|^2 I_E\right] \tag{9.101}$$

This allows us to calculate the correlation coefficient for the AT 41400 (Figure 9.19) and, as the next step, the correlation conductance Y_{COR} according to (9.97).

For the further calculation we also need the uncorrelated part of the current (9.98). In Figure 9.20 we have $\overline{|i_A|^2}$ as spectral power density of the total current in 1 Hz bandwidth and the uncorrelated part. Because of the low correlation the difference is only small.

We can now calculate the correlation conductance (9.97). In Figure 9.21 the magnitude is shown, the angle is equal to the angle of ρ.

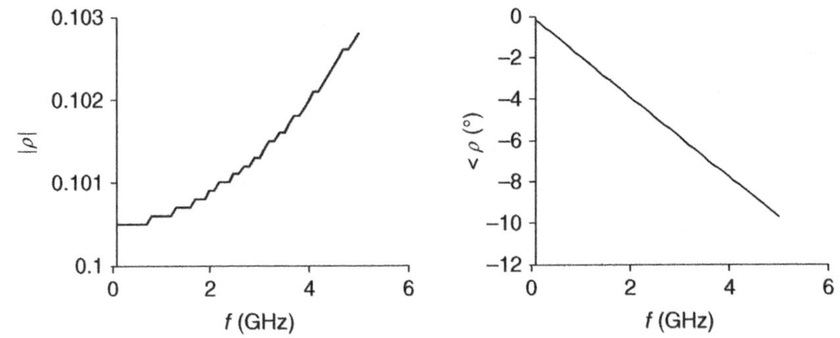

Figure 9.19 Values of the correlation coefficient for the AT 41400 according to (9.101).

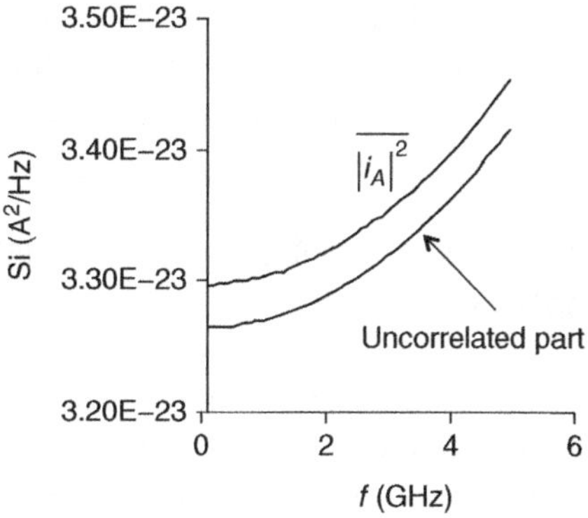

Figure 9.20 Spectral power density of the current source i_A separated into total current and uncorrelated part.

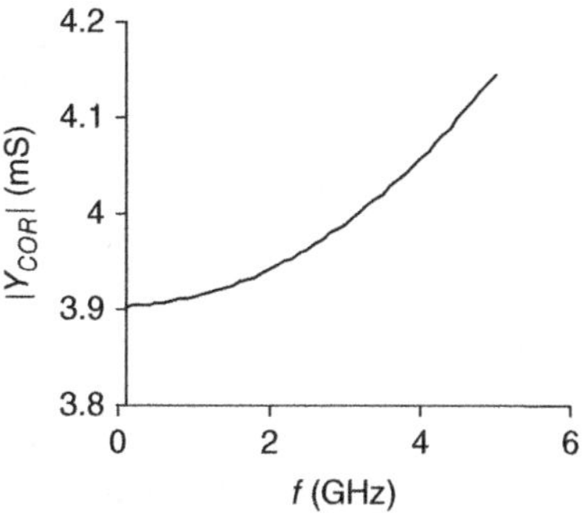

Figure 9.21 Correlation conductance for the AT 41400.

Now we have all the quantities together to move on to the IEEE noise parameters. We had already introduced the formulas in (7.13).

The noise resistance is

$$R_N = \overline{|v_A|^2}\frac{1}{4kTB} \tag{9.102}$$

The noise conductance G_U corresponds to the uncorrelated part of the current source. The correlated part is accounted for by the correlation conductance.

$$G_U = \overline{|i_U|^2}\frac{1}{4kTB} \tag{9.103}$$

Since we have done the example calculations for 1 Hz bandwidth, B is also 1 Hz. With these quantities we get

$$G_{OPT} = \sqrt{\frac{G_U}{R_N} + G_{COR}^2}; \quad B_{OPT} = -B_{COR} \tag{9.104}$$

$$F_{MIN} = 1 + 2R_N G_{COR} + 2\sqrt{R_N G_U + R_N^2 G_{COR}^2} \tag{9.105}$$

This means that we have reached our goal for the example. However, it does not correspond to the practice of the real transistor, since the base path resistance

R_B is not included. For the extension by all external elements the application of correlation matrices is the method of choice. Here the aim was to demonstrate the details of calculating with noise sources.

Modeling of a Microwave Transistor with Correlation Matrices

We now want to perform noise modeling for a microwave HBT. It is an unpackaged chip in InGaP/GaAs technology. The elements of the small-signal equivalent circuit were determined by on-wafer S-parameter measurements at different bias points. Here we follow the elaboration of Rudolph [11]. The step-by-step composition of the model from relatively clear networks impressively demonstrates the efficiency of the method of correlation matrices. Let us first look at the small-signal equivalent circuit. It is a T-model of the intrinsic transistor in common-emitter circuit. Included are the base path resistance R_{B2} and parasitic capacitances. This intrinsic transistor (dashed box) is embedded in the lead structures at the three electrodes. The capacitances of the contact pads on the wafer are also included (Figure 9.22).

The values of the equivalent circuit elements are summarized in Table 9.2.

The intrinsic transistor contains the shot noise sources, whose position in the circuit and their correlation is the subject of physical considerations. The voltage source v_{B2} belongs to the base path resistance R_{B2}. In any case it is uncorrelated with the shot noise currents i_B and i_C. These noise sources inside the intrinsic transistor generate short-circuit currents between the terminals of the two-port, which we need for further considerations. They are i_{S1} between the terminals $(B_I - E_I)$ at the input and i_{S2} between the terminals $(C_I - E_I)$ at the output (Figure 9.23).

We describe the noise current sources at the base and collector with a correlation approach, which has already led to the understanding of the Hawkins model

$$\begin{aligned}
\overline{i_B^2} &= 2q\left\{I_B + |1 - exp(-j\omega\tau)|^2 I_C\right\} \\
\overline{i_C^2} &= 2qI_C \\
\overline{i_B i_C} &= 2q\left\{I_C exp(j\omega\tau) - 1\right\}
\end{aligned} \tag{9.106}$$

A simplified form [12] neglects the correlation by $\tau = 0$. It reads:

$$\overline{i_B^2} = 2qI_B; \quad \overline{i_C^2} = 2qI_C; \quad \overline{i_B i_C} = 0 \tag{9.107}$$

Here we want to compare these two approaches using practical examples. The elements of the noise correlation matrix of the intrinsic transistor in Y-form are determined by the short-circuit currents i_{S1} and i_{S2}. For their calculation we need

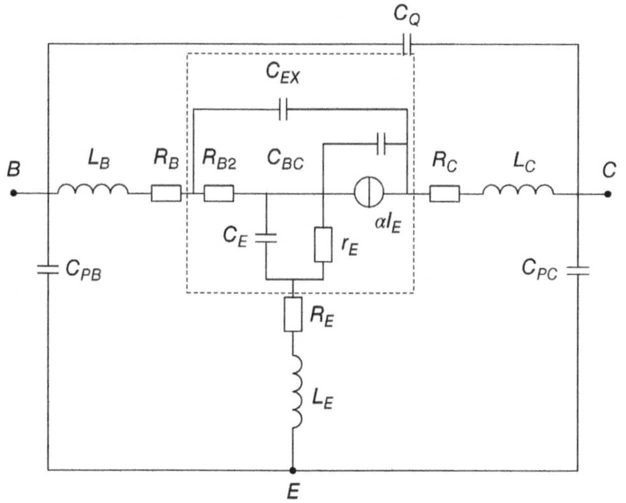

Figure 9.22 Small signal equivalent circuit of a microwave HBT. The intrinsic transistor (dashed) is gradually extended by the parasitic elements.

Table 9.2 Values of the elements in Figure 9.22 derived from S-parameter measurements in active and passive state.

	Resistors (Ω)	Capacitances (fF)	Inductances (pH)
Intrinsic	r_E: 9.6	C_E: 400	
	R_{B2}: 5.0	C_{EX}: 36	
		C_{BC}: 9	
Parasitic	R_B: 2.7	C_{PB}: 20	L_B: 38
	R_C: 3.5	C_{PC}: 22	L_C: 1
	R_E: 1.6		
		C_Q: 20	
	α_0: 0.989	τ: 3 ps	ω_α: $2\pi \times 36$ GHz

$3 \times 30\,\mu\text{m}^2$ chip; $V_{CE} = 4\,\text{V}$; $I_C = 6\,\text{mA}$.

the following network parameters:

$$A = \frac{1}{1 + R_{B2}\left\{(1-\alpha)Y_{BE} + Y_{BC}\right\}}; \quad B = \frac{(1-\alpha)Y_{BE} + Y_{BC}}{1 + R_{B2}\left\{(1-\alpha)Y_{BE} + Y_{BC}\right\}}$$
$$C = \frac{R_{B2}\left(Y_{BC} - \alpha Y_{BE}\right)}{1 + R_{B2}\left\{(1-\alpha)Y_{BE} + Y_{BC}\right\}}; \quad D = \frac{\alpha Y_{BE} - Y_{BC}}{1 + R_{B2}\left\{(1-\alpha)Y_{BE} + Y_{BC}\right\}} \tag{9.108}$$

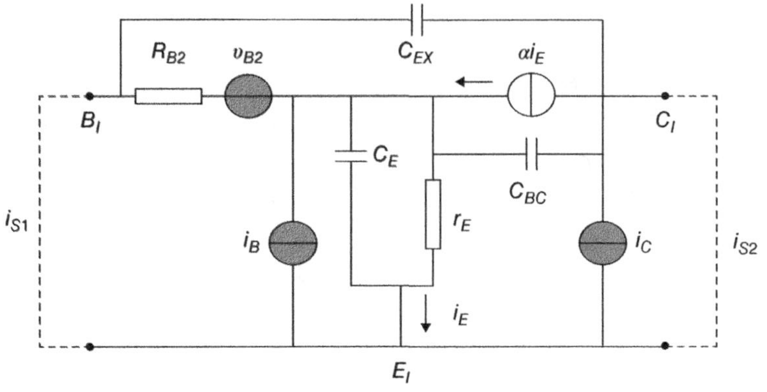

Figure 9.23 Intrinsic transistor from Figure 9.22 with noise sources.

with

$$\alpha = \frac{\alpha_0 exp(-j\omega\tau)}{1+j\frac{\omega}{\omega_\alpha}}; \quad Y_{BE} = \frac{1}{r_E} + j\omega C_E; \quad Y_{BC} = j\omega C_{BC} \tag{9.109}$$

We receive

$$\begin{aligned}
\overline{i_{S1}^2} &= |A|^2\overline{i_B^2} + |B|^2\overline{v_{B2}^2} \\
\overline{i_{S2}^2} &= \overline{i_C^2} + |C|^2\overline{i_B^2} + |D|^2\overline{v_{B2}^2} + 2Re\left(C\overline{i_B i_C}\right) \\
\overline{i_{S2}i_{S1}^*} &= (CA^*)\,\overline{i_B^2} + A^*\overline{i_B i_C} + (DB^*)\,\overline{v_{B2}^2} \\
\overline{i_{S1}i_{S2}^*} &= \left(\overline{i_{S2}i_{S1}^*}\right)^*; \quad \overline{v_{B2}^2} = 4kTR_{B2}
\end{aligned} \tag{9.110}$$

Thus we have the **Y**-correlation matrix of the intrinsic transistor, which also contains the thermal part of R_{B2}.

$$\mathbf{C_I^Y} = \begin{pmatrix} \overline{i_{S1}^2} & \overline{i_{S1}i_{S2}^*} \\ \overline{i_{S2}i_{S1}^*} & \overline{i_{S2}^2} \end{pmatrix} \tag{9.111}$$

For arithmetic operations with the correlation matrices, the network matrices are also required. The elements of the $\boldsymbol{Y}$-matrix of the intrinsic transistor are

$$\begin{aligned}
Y_{11} &= Y_{EX} + \frac{Y_{BC} + (1-\alpha)Y_{BE}}{DnY}; \quad Y_{12} = -Y_{EX} - \frac{Y_{BC}}{DnY} \\
Y_{21} &= -Y_{EX} + \frac{\alpha Y_{BE} - Y_{BC}}{DnY}; \quad Y_{22} = Y_{EX} + \frac{Y_{BC}\left(1 + Y_{BE}R_{B2}\right)}{DnY}
\end{aligned} \tag{9.112}$$

with

$$Y_{EX} = j\omega C_{EX} \quad DnY = 1 + R_{B2}\left\{(1-\alpha)Y_{BE} + Y_{BC}\right\} \tag{9.113}$$

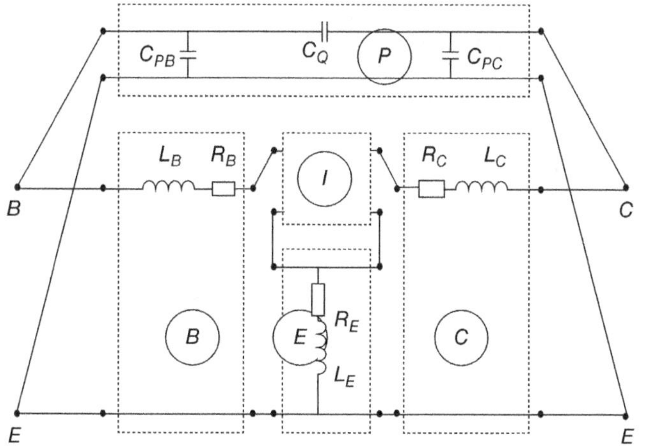

Figure 9.24 Schematic structure of the noise equivalent circuit from individual networks.

The other networks are arranged around the intrinsic transistor. From these the noise equivalent circuit of the entire transistor is developed step by step. The schematic is shown in Figure 9.24. The networks I (intrinsic transistor Figure 9.23) and E (emitter) are connected in series. The **Z** matrices must be added to their combination. The resulting network IE is cascaded between the base network (B) and the collector network (C). To combine them, the chain matrices must be multiplied by each other. The network of pad capacities (P) is in parallel to this resulting combination. Consequently, the **Y** matrices must be added.

The first step is to calculate the $\boldsymbol{Z}$-matrix of the intrinsic transistor $\mathbf{Z}_I$

$$\boldsymbol{Z}_I = \frac{1}{DY}\begin{pmatrix} Y_{22} & -Y_{12} \\ -Y_{21} & Y_{11} \end{pmatrix}; \quad DY = Y_{11}Y_{22} - Y_{12}Y_{21} \tag{9.114}$$

We also need the $\boldsymbol{Z}$-Matrix of the emitter network E. This is $\mathbf{Z}_E$. The correlation matrix simply results from the real part of $\mathbf{Z}_E$

$$\boldsymbol{Z}_E = \left(R_E + j\omega L_E\right)\begin{pmatrix} 1 & 1 \\ 1 & 1 \end{pmatrix}; \quad \boldsymbol{C}_E^Z = 4kTRe\left(\boldsymbol{Z}_E\right) \tag{9.115}$$

The sum is the $\boldsymbol{Z}$-Matrix $\mathbf{Z}_{IE}$, which includes the intrinsic transistor and the emitter network. The same applies to the correlation matrices. For this we have to convert $\mathbf{C}_\mathbf{I}^\mathrm{Y}$ into $\mathbf{C}_\mathbf{I}^\mathbf{Z}$.

$$\boldsymbol{Z}_{IE} = \boldsymbol{Z}_I + \boldsymbol{Z}_E; \quad \boldsymbol{C}_{IE}^Z = \boldsymbol{C}_I^Z + \boldsymbol{C}_E^Z \tag{9.116}$$

The transformation $\mathbf{C}_\mathbf{I}^\mathrm{Y} \rightarrow \mathbf{C}_\mathbf{I}^\mathbf{Z}$ is carried out according to rule (9.62) with the transformation matrix $\mathbf{T}_1$ from Table 9.1.

$$\boldsymbol{C}_I^Z = \boldsymbol{T}_1 \times \boldsymbol{C}_I^Y \times \boldsymbol{T}_1^H; \quad \boldsymbol{T}_1 = \boldsymbol{Z}_I; \quad \boldsymbol{T}_1^H = \boldsymbol{Z}_I^H \tag{9.117}$$

Now follows the cascading of networks B, IE and C. The **Z**-matrix of B and C does not exist, so we start from the **Y**-matrix. Analogous to (9.115), we write for B:

$$\boldsymbol{Y_B} = \frac{1}{R_B + j\omega L_B}\begin{pmatrix} 1 & -1 \\ -1 & 1 \end{pmatrix}; \quad \boldsymbol{C_B^Y} = 4kTRe\left(\boldsymbol{Y_B}\right) \tag{9.118}$$

And for the collector network

$$\boldsymbol{Y_C} = \frac{1}{R_C + j\omega L_C}\begin{pmatrix} 1 & -1 \\ -1 & 1 \end{pmatrix}; \quad \boldsymbol{C_C^Y} = 4kTRe\left(\boldsymbol{Y_C}\right) \tag{9.119}$$

For the last step we need the **Y**-matrix of pad capacities, which has no real part.

$$\boldsymbol{Y_P} = j\omega\begin{pmatrix} C_{PB} + C_Q & -C_Q \\ -C_Q & C_{PC} + C_Q \end{pmatrix} \tag{9.120}$$

To cascade *B*, *IE*, and *C*, we have to convert network and correlation matrices into chain form. This requires several steps.

(1) Intrinsic transistor and emitter: The impedance matrix of the combination of intrinsic transistor and emitter network (Z_{IE}) must be transformed into the chain matrix $\mathbf{Z_{IE}} \rightarrow \mathbf{A_{IE}}$

$$\boldsymbol{A_{IE}} = \frac{1}{\left(Z_{IE}\right)_{21}}\begin{pmatrix} \left(Z_{IE}\right)_{11} & DZ_{IE} \\ 1 & \left(Z_{IE}\right)_{22} \end{pmatrix}$$
$$DZ_{IE} = \left(Z_{IE}\right)_{11}\left(Z_{IE}\right)_{22} - \left(Z_{IE}\right)_{12}\left(Z_{IE}\right)_{21} \tag{9.121}$$

Transformation of the correlation matrix in impedance form (9.116) into the chain form according to (9.62) and Table 9.1

$$\boldsymbol{C_{IE}^A} = \boldsymbol{T_2} \times \boldsymbol{C_{IE}^Z} \times \boldsymbol{T_2^H}$$
$$\boldsymbol{T_2} = \begin{pmatrix} 1 & -\left(A_{IE}\right)_{11} \\ 0 & -\left(A_{IE}\right)_{21} \end{pmatrix}; \quad \boldsymbol{T_2^H} = \begin{pmatrix} 1 & 0 \\ -\left(A_{IE}\right)_{11}^{*} & -\left(A_{IE}\right)_{21}^{*} \end{pmatrix} \tag{9.122}$$

(2) Base: The admittance matrix of the base network must be transformed into the chain matrix $\boldsymbol{Y_B} \rightarrow \boldsymbol{A_B}$

$$\boldsymbol{A_B} = \frac{1}{\left(Y_B\right)_{21}}\begin{pmatrix} -\left(Y_B\right)_{22} & -1 \\ -DY_B & -\left(Y_B\right)_{11} \end{pmatrix}$$
$$DY_B = \left(Y_B\right)_{11}\left(Y_B\right)_{22} - \left(Y_B\right)_{12}\left(Y_B\right)_{21} \tag{9.123}$$

Transformation of the correlation matrix in admittance form (9.118) into the chain form according to (9.62) and Table 9.1

$$\boldsymbol{C_B^A} = \boldsymbol{T_3} \times \boldsymbol{C_B^Y} \times \boldsymbol{T_3^H}$$
$$\boldsymbol{T_3} = \begin{pmatrix} 0 & \left(A_B\right)_{12} \\ 1 & \left(A_B\right)_{22} \end{pmatrix}; \quad \boldsymbol{T_3^H} = \begin{pmatrix} 0 & 1 \\ \left(A_B\right)_{12}^{*} & \left(A_B\right)_{22}^{*} \end{pmatrix} \tag{9.124}$$

(3) Collector: The admittance matrix of the collector network into the chain matrix $\boldsymbol{Y}_C \rightarrow \boldsymbol{A}_C$

$$\boldsymbol{A_C} = \frac{1}{(Y_C)_{21}} \begin{pmatrix} -(Y_C)_{22} & -1 \\ -DY_C & -(Y_C)_{11} \end{pmatrix}$$

$$DY_C = (Y_C)_{11}(Y_C)_{22} - (Y_C)_{12}(Y_C)_{21} \tag{9.125}$$

Transformation of the correlation matrix in admittance form (9.119) into the chain form according to (9.62) and Table 9.1

$$\boldsymbol{C_C^A} = \boldsymbol{T}_4 \times \boldsymbol{C_C^Y} \times \boldsymbol{T_4^H}$$

$$\boldsymbol{T}_4 = \begin{pmatrix} 0 & (A_C)_{12} \\ 1 & (A_C)_{22} \end{pmatrix}; \quad \boldsymbol{T_4^H} = \begin{pmatrix} 0 & 1 \\ (A_C)_{12}^* & (A_C)_{22}^* \end{pmatrix} \tag{9.126}$$

Now we can calculate the correlation matrix $\boldsymbol{C^A}$ as the result of cascading the three networks Basis: $\boldsymbol{C_B^A}$ Intrinsic + Emitter: $\boldsymbol{C_{IE}^A}$ and collector: $\boldsymbol{C_C^A}$ (9.61).

$$\boldsymbol{C^A} = \boldsymbol{A_B} \times (\boldsymbol{A_{IE}} \times \boldsymbol{C_C^A} \times \boldsymbol{A_{IE}^H} + \boldsymbol{C_{IE}^A}) \times \boldsymbol{A_B^H} + \boldsymbol{C_B^A} \tag{9.127}$$

In (9.127) there are two steps included. First, the collector and intrinsic transistor plus emitter networks are combined (in brackets), and then the resulting network is combined with the collector network. The resulting correlation matrix $\boldsymbol{C^A}$ describes the complete transistor, since the matrix of pad capacitances $\boldsymbol{Y_P}$ has no real part and thus does not contribute to the noise.

As a result of our considerations we obtain both the noise parameters and the S-parameters. These are naturally influenced by the pad capacitance. Therefore the matrix $\boldsymbol{Y_P}$ must be taken into account. First of all, the chain matrix $\boldsymbol{A_N}$, which corresponds to $\boldsymbol{C^A}$ will be calculated. It results simply as the product

$$\boldsymbol{A_N} = \boldsymbol{A_B} \times \boldsymbol{A_{IE}} \times \boldsymbol{A_C} \tag{9.128}$$

The pad network P is in parallel to this network, so that the $\boldsymbol{Y}$-matrices must be added. First the transformation $\boldsymbol{A}_N \rightarrow \boldsymbol{Y}_N$

$$\boldsymbol{Y_N} = \frac{1}{(A_N)_{12}} \begin{pmatrix} (A_N)_{22} & -DA_N \\ -1 & (A_N)_{11} \end{pmatrix}$$

$$DA_N = (A_N)_{11}(A_N)_{22} - (A_N)_{12}(A_N)_{21} \tag{9.129}$$

The $\boldsymbol{Y}$-matrix of the entire transistor (Figure 9.22) is simply the sum of

$$\boldsymbol{Y} = \boldsymbol{Y_N} + \boldsymbol{Y_P} \tag{9.130}$$

The S-parameters are given by the matrix equation

$$\boldsymbol{S} = (\boldsymbol{I} - 50\boldsymbol{Y}) \times (\boldsymbol{I} + 50\boldsymbol{Y})^{-1} \tag{9.131}$$

$\boldsymbol{I}$ is the unit matrix.

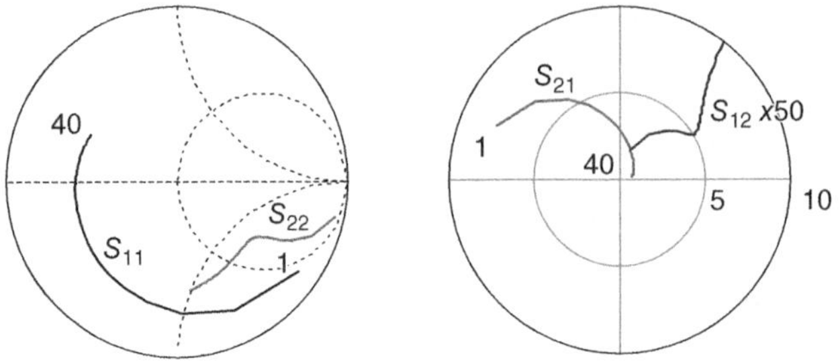

Figure 9.25 S-parameter of the InGaP/GaAs chip in Figure 9.22 calculated according to (9.131). $f = 1\text{–}40$ GHz.

We can now calculate the S-parameters and the four noise parameters (Figure 9.25).

$$R_N = \frac{C_{11}^A}{4kT}; \quad F_{MIN} = 1 + \frac{C_{12}^A + C_{11}^A Y_{OPT}}{2kT}$$

$$Y_{OPT} = \sqrt{\frac{C_{22}^A}{C_{11}^A} - \left(Im\frac{C_{12}^A}{C_{11}^A}\right)^2} + jIm\left(\frac{C_{12}^A}{C_{11}^A}\right)$$

$$y_{OPT} = 50Y_{OPT}; \quad \Gamma_{OPT} = \frac{1 - y_{OPT}}{1 + y_{OPT}} \tag{9.132}$$

Figures 9.26 shows the noise parameters [11] (Figure 9.27).

Further details on measuring and modeling microwave HBTs can be found in [13–15].

Simplest Π-Model

We now return to the simple model in the LF-range and analyze its performance in comparison with the information from a data sheet [1]. We are interested in how far we can get with this minimum effort. The noise model with the resistor R_S at the input and with a shorted output is shown in Figure 9.28.

To calculate the noise factor we again use the ratio of the currents i_S and i_{S0} in the short-circuited output. Since these are noise currents and we calculate the power, we need $\overline{i_S^2}$ and $\overline{i_{S0}^2}$. The quantity $\overline{i_S^2}$ is generated by all noise sources present in Figure 9.28. The quantity $\overline{i_{S0}^2}$ is only generated by $\overline{v_{RS}^2}$, i.e. the transistor only amplifies and does not generate noise itself ($\overline{v_{RS}^2} = \overline{i_B^2} = \overline{i_C^2} = 0$). Through network considerations we arrive at the following formulas.

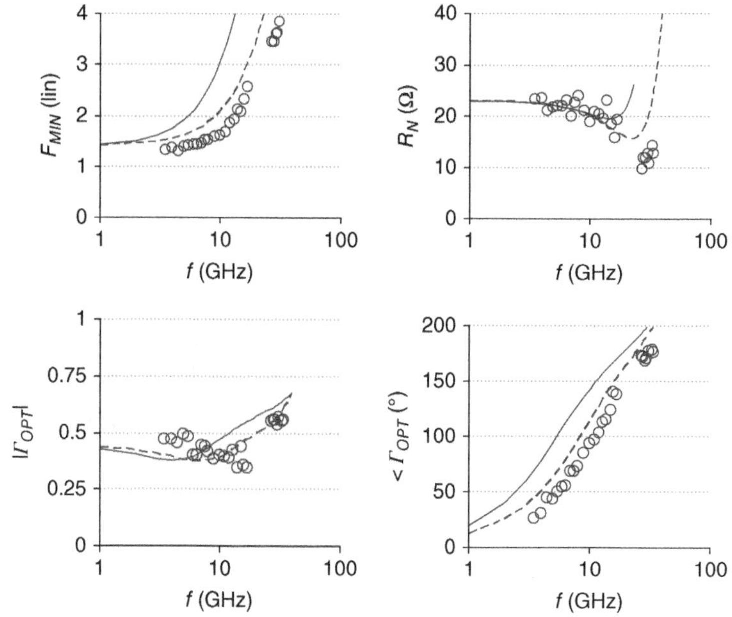

Figure 9.26 Noise parameters of the chip. Circles: Measurement; dashed: noise model; Full lines: without correlation.

Figure 9.27 Comparison of S_{11}^* and Γ_{OPT}.

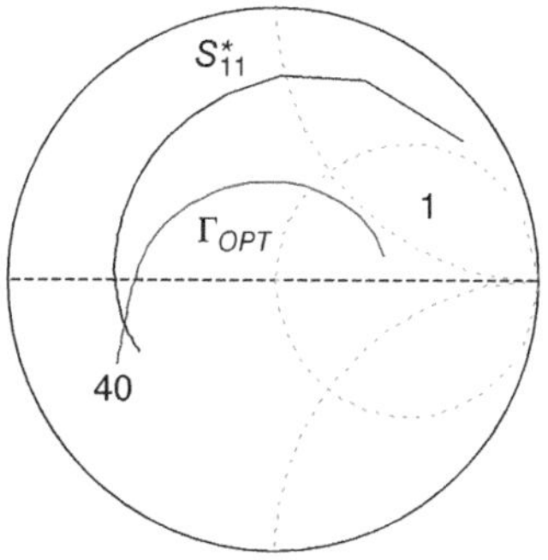

With all sources:

$$\overline{i_S^2} = \overline{i_C^2} + \frac{g_m^2 r_\pi^2}{\left(R_S + R_B + r_\pi\right)^2} \left\{ \overline{i_B^2}\left(R_S + R_B\right)^2 + \overline{v_{RB}^2} + \overline{v_{RS}^2} \right\} \tag{9.133}$$

Only the generator source

$$\overline{i_{S0}^2} = \frac{g_m^2 \overline{v_S^2} r_\pi^2}{\left(R_S + R_B + r_\pi\right)^2} \tag{9.134}$$

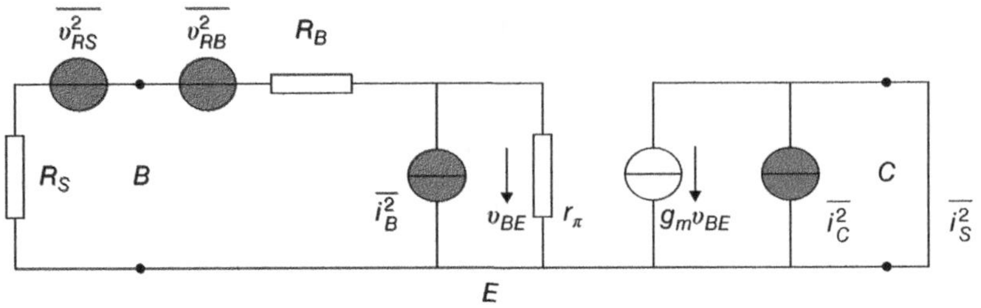

Figure 9.28 Noise equivalent circuit of a BJT in a simple model.

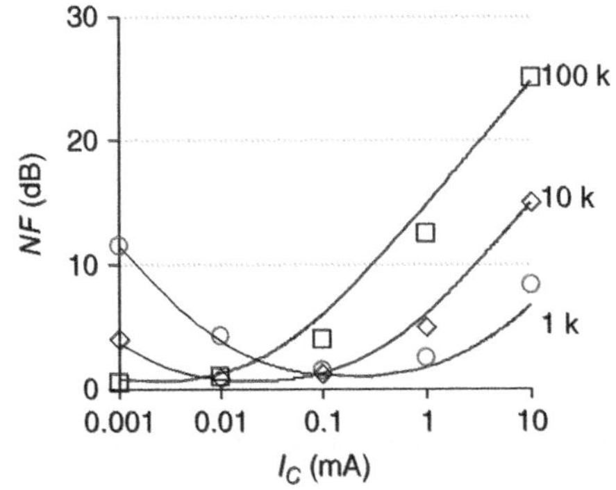

Figure 9.29 Noise figure of the BCW 60 at $f = 1$ kHz versus the collector current. Symbols: Manufacturer's data; lines: Result of Π-model. Parameter is the source resistance R_S.

The noise sources in $B = 1$ Hz bandwidth are given by

$$\overline{i^2_B} = 2qI_B; \quad \overline{i^2_C} = 2qI_C; \quad \overline{v^2_S} = 4kTR_S; \quad \overline{v^2_{RB}} = 4kTR_B \tag{9.135}$$

With

$$F = \frac{\overline{i^2_S}}{\overline{i^2_{S0}}} \tag{9.136}$$

we can test whether we can verify the dependencies of the noise factor F on the generator resistance R_S and the collector current I_C given in the data sheet. Of course, this is only possible for the low frequency range at which the reactive elements of the transistor do not play any role, here $f = 1$ kHz.

The Infineon data sheet BCW 60 [16] depicts the diagram Figure 9.29. It shows the dependence $F(I_C)$ with R_S as parameter for $f = 1$ kHz. The calculation with (9.133)–(9.136) with the transistor data $\alpha = 0.985$; $R_b = 100\,\Omega$ gives good results for the simple noise model.

Also the variation of R_S with I_C as parameter gives good results shown in Figure 9.30.

With the simple Π-noise model of the BJT one can work well in the frequency range of acoustics without knowing the exact equivalent circuit. Collector current

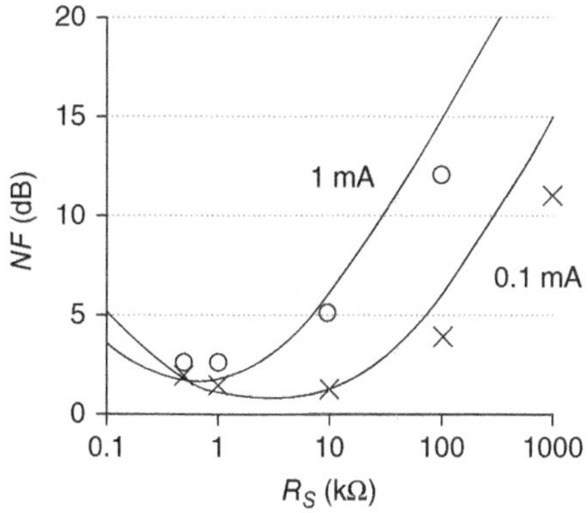

Figure 9.30 Noise figure of the BCW 60 at $f = 1$ kHz versus the source resistance R_S. Symbols: Manufacturer's data; lines: Result of Π-model. Parameter is the collector current.

and current gain are sufficient. For R_B an estimated value is sufficient, which is of little influence.

Contour Diagram

One can find information about the noise behavior of transistors, which refers to two sources at the input. This corresponds to the theory of noisy two-ports in chain matrix representation, as we discussed above. In addition to this, a contour diagram is usually also given, in which one can see at a glance the noise behavior as a function of R_S and I_C. In general, a contour diagram (Muscheldiagramm) is a graphical representation of the characteristics of a technical system when operating parameters are varied. Here, in Figure 9.34, the property is the noise figure, the operating parameters are R_S and I_C.

We take the NPN-BJT 2N3392 as an example. In the data sheet we find for $f = 1\,\text{kHz}$ the information about the noise sources in the equivalent circuit Figure 9.31.

The positioning of the two sources in front of the noise-free transistor corresponds to the usual approach in the theory of noisy two-ports. Both sources are not correlated (Figure 9.32).

To calculate the four noise parameters from the data $v_N(I_C)$ and $i_N(I_C)$ in Figure 9.31 we can use the formalism of noise correlation matrices. Since we have only real quantities and v_N and i_N are not correlated, the **A**-correlation matrix looks very simple.

$$\mathbf{C}^A = \begin{pmatrix} \overline{v_N^2} & 0 \\ 0 & \overline{i_N^2} \end{pmatrix} \tag{9.137}$$

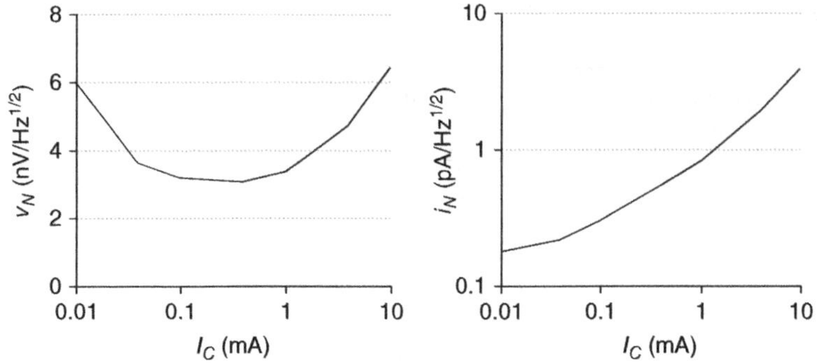

Figure 9.31 Noise sources of the BJT 2N3392 [19] according to manufacturer's specifications for $f = 1$ kHz.

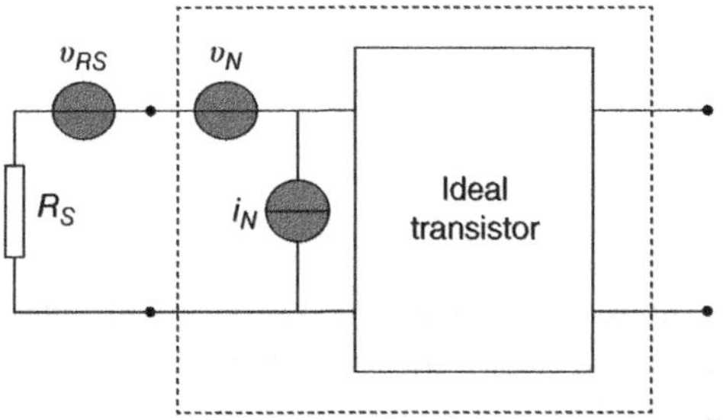

Figure 9.32 The uncorrelated noise sources v_N and i_N at the input of the two-port.

The formulas for R_N, F_{MIN}, and Y_{OPT} are also very simple. R_N characterizes the voltage source

$$R_N = \frac{C_{11}^A}{4kT} \tag{9.138}$$

Y_{OPT} is given by the ratio of current and voltage source

$$Y_{OPT} = \sqrt{\frac{C_{22}^A}{C_{11}^A}} = \frac{1}{R_{OPT}} \tag{9.139}$$

The formula for F_{MIN} is simplified to

$$F_{MIN} = 1 + \frac{C_{11}^A Y_{OPT}}{2kT} \tag{9.140}$$

And for F applies:

$$F = F_{MIN} + R_N R_S \left(\frac{1}{R_S} - \frac{1}{R_{OPT}} \right)^2 \tag{9.141}$$

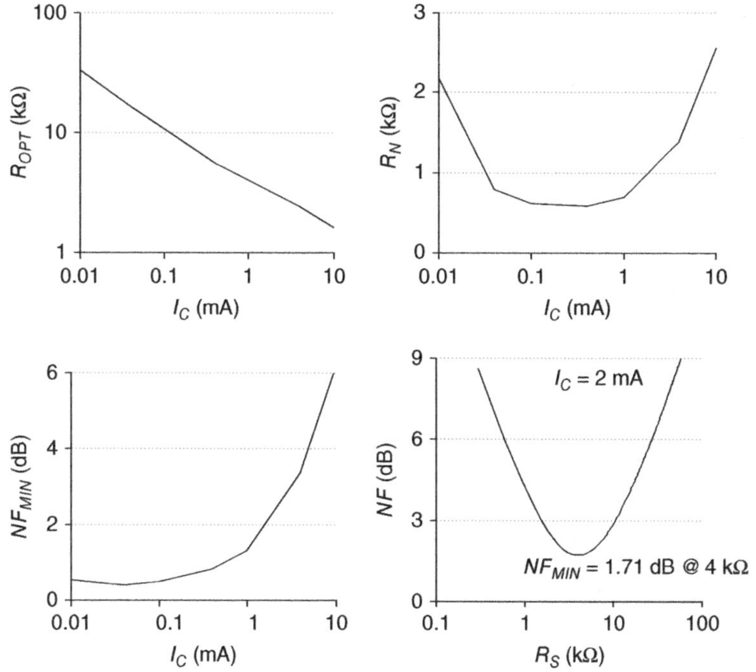

Figure 9.33 Results for 2N3392 @ 1 kHz calculated from the sources in Figure 9.31.

The contour diagram clearly summarizes the individual results from Figure 9.33. The right-hand scale gives an information of the input resistances of the transistor in the three configurations. This shows that the common-emitter circuit is best suited as an amplifier for simultaneous power matching ($R_S = R_{IN}$) and noise matching ($R_S = R_{OPT}$). One can see at a glance, with which combination of R_S and I_C a certain noise figure can be expected. But we can also see that the influence of the source resistance is very large. Good noise figures ($NF < 3$ dB) are only achieved in a narrow range $R_S = 2 \ldots 10$ kΩ. For the 200 Ω that are common in audio engineering the noise figure is $NF > 9$ dB. The good noise characteristics of the transistor itself do not bring any advantage here. Circuit elements, which are necessary to adjust the operating point, contribute to further deterioration (Figure 9.34).

Transistor in the Circuit

We can see from the contour diagram that the elements of the circuit in which the transistor is embedded have a decisive influence on the noise behavior of the

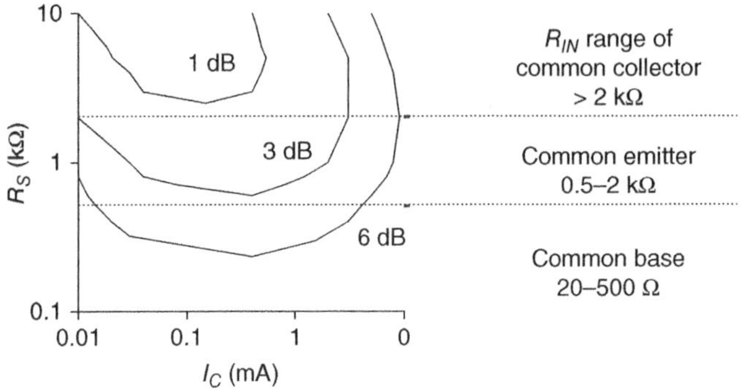

Figure 9.34 Contour diagram of the 2N3392 @1 kHz.

Table 9.3 H-parameters and noise sources from the BC546 data sheet.

h_{11}	h_{12}	h_{21}	h_{22}	e_n	i_n
4.5 kΩ	2×10^{-4}	330	30 μS	$3\ \mathrm{nV}/\sqrt{\mathrm{Hz}}$	$1\ \mathrm{pA}/\sqrt{\mathrm{Hz}}$

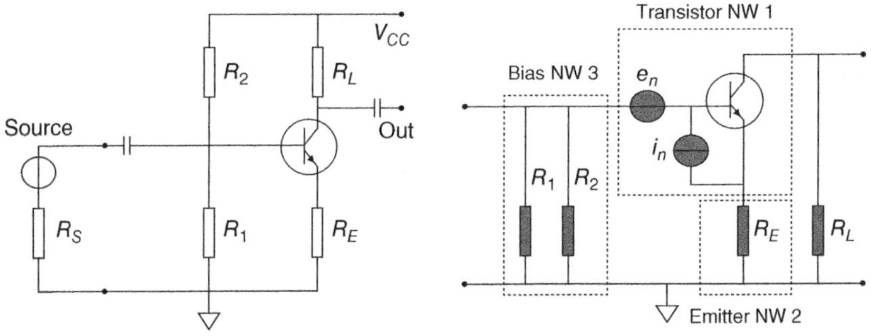

Figure 9.35 BJT BC546 with bias network and input circuit.

entire circuit. Especially the source resistance R_S is of importance. In the following example we analyze a BJT BC546 [17] with bias network and input circuit. Table 9.3 is taken from the data sheet.

Figure 9.35 on the right shows the noise sources in common-emitter circuit. We do not consider R_L because its noise is not amplified. The $\boldsymbol{H}$-matrix of the transistor is converted to the $\boldsymbol{Z}$-matrix

$$\boldsymbol{H}_T = \begin{pmatrix} h_{11} & h_{12} \\ h_{21} & h_{22} \end{pmatrix}; \quad \boldsymbol{Z}_T = \frac{1}{h_{22}} \begin{pmatrix} \Delta H_T & h_{12} \\ -h_{21} & 1 \end{pmatrix}; \quad \Delta H_T = h_{11}h_{22} - h_{12}h_{21} \tag{9.142}$$

The noise correlation matrix in chain form of the network 1 (transistor) is

$$\boldsymbol{C}_T^A = \begin{pmatrix} e_n^2 & 0 \\ 0 & i_n^2 \end{pmatrix} \tag{9.143}$$

This must be converted to the Z-form so that the emitter resistance R_E can be added. Network 1 and 2 are connected in series, i.e. the impedances are added.

$\boldsymbol{C}_T^A$ will be converted into $\boldsymbol{C}_T^Z$.

The transformation matrix of $\boldsymbol{C}_T^A$ to $\boldsymbol{C}_T^Z$ is $\boldsymbol{T}_1$

$$\boldsymbol{T}_1 = \begin{pmatrix} 1 & -Z_{T11} \\ 0 & -Z_{T21} \end{pmatrix}; \quad \boldsymbol{T}_1^H = \begin{pmatrix} 1 & 0 \\ -Z_{T11} & -Z_{T21} \end{pmatrix}; \quad \boldsymbol{C}_T^Z = \boldsymbol{T}_1 \boldsymbol{C}_T^A \boldsymbol{T}_1^H \tag{9.144}$$

The noise correlation matrix of network 2 (emitter resistance) in Z-form is simply derived from the impedance matrix

$$\boldsymbol{Z}_2 = R_E \begin{pmatrix} 1 & 1 \\ 1 & 1 \end{pmatrix}; \quad \boldsymbol{C}_2^Z = 4kTR_E \begin{pmatrix} 1 & 1 \\ 1 & 1 \end{pmatrix}; \quad \boldsymbol{C}_{12}^Z = \boldsymbol{C}_T^Z + \boldsymbol{C}_2^Z \tag{9.145}$$

This matrix $\boldsymbol{C}_{12}^Z$, which represents transistor T and emitter resistance R_E, must be converted back to the A-form so that it can be cascaded with bias network 3.

For this we need the chain matrix of this combination. First we calculate the $\boldsymbol{Z}$-matrix by addition and transform it into the $\boldsymbol{A}$-matrix $\boldsymbol{AT}_2$.

$$\boldsymbol{ZT}_2 = \boldsymbol{Z}_T + \boldsymbol{Z}_2$$

$$\boldsymbol{AT}_2 = \frac{1}{Z2_{21}} \begin{pmatrix} ZT2_{11} & \Delta ZT2 \\ 1 & ZT2_{22} \end{pmatrix} \tag{9.146}$$

We need the elements of $\boldsymbol{AT}_2$ to form the transformation matrix $\boldsymbol{T}_2$.

$$\boldsymbol{T}_2 = \begin{pmatrix} 1 & -AT2_{11} \\ 0 & -AT2_{21} \end{pmatrix}; \quad \boldsymbol{T}_2^H = \begin{pmatrix} 1 & 0 \\ -AT2_{11} & -AT2_{21} \end{pmatrix} \tag{9.147}$$

with $\boldsymbol{C}_{T2}^A = \boldsymbol{T}_2 \boldsymbol{C}_{12}^Z \boldsymbol{T}_2^H$.

This correlation matrix $\boldsymbol{C}_{T2}^A$ in chain form is cascaded with the correlation matrix of the bias network 3. For this we need the chain matrix $\boldsymbol{A}_3$ of the parallel connection of R_1 and R_2. They act in parallel with regard to their noise behavior, since R_2 is grounded at the top with respect to RF-currents.

$$G_P = \frac{1}{R_P} = \frac{1}{R_1} + \frac{1}{R_2}; \quad \boldsymbol{A}_3 = \begin{pmatrix} 1 & 0 \\ G_P & 1 \end{pmatrix} \tag{9.148}$$

We also need the noise correlation matrix of G_P. To do this, we start with the correlation matrix in Z-form analogous to the procedure for the emitter resistor R_E and transform it into the A-form

$$\boldsymbol{C}_3^Z = 4kTR_P \begin{pmatrix} 1 & 1 \\ 1 & 1 \end{pmatrix} \tag{9.149}$$

The transformation matrices $\boldsymbol{T}_3$ are

$$\boldsymbol{T}_3 = \begin{pmatrix} 1 & -1 \\ 0 & -G_P \end{pmatrix}; \qquad \boldsymbol{T}_3^H = \begin{pmatrix} 1 & 0 \\ -1 & -G_P \end{pmatrix} \tag{9.150}$$

This leads us to the correlation matrix of G_P in chain form.

$$\boldsymbol{C}_3^A = \boldsymbol{T}_3 \boldsymbol{C}_3^Z \boldsymbol{T}_3^H = 4kTG_P \begin{pmatrix} 0 & 0 \\ 0 & 1 \end{pmatrix} \tag{9.151}$$

Now we can calculate the correlation matrix of the complete circuit $\boldsymbol{C}^A$.

$$\boldsymbol{C}^A = \boldsymbol{A}_3 \boldsymbol{C}_{T2}^A \boldsymbol{A}_3^H + \boldsymbol{C}_3^A \tag{9.152}$$

This gives us the noise parameters of interest in the circuit, especially F_{MIN} and R_{OPT}. Since there are no reactive elements, Y_{OPT} is real.

$$R_N = \frac{C_{11}^A}{4kT}; \quad Y_{OPT} = \sqrt{\frac{C_{22}^A}{C_{11}^A}}; \quad F_{MIN} = 1 + \frac{C_{12}^A + C_{11}^A Y_{OPT}}{2kT} \tag{9.153}$$

The results (Figure 9.36) show at first glance that the use of a low-noise transistor does not automatically lead to a low-noise circuit, even if it contains only the most necessary elements. Obviously the resistance that the transistor "sees" at its input is crucial. In our example (BC546, Table 9.3) the transistor itself has $NF_{MIN} = 1.34$ dB at $R_{OPT} = 3$ kΩ. With a bias network and emitter resistance this would be achieved in the limit of a very small R_E and a very high-impedance base network (Figure 9.37 left). But this is not realistic, especially since the source resistance is still connected in parallel to the bias network. Our example would be a favorable solution if we want to amplify the signal of a high-impedance microphone with about $R_S = 2$ kΩ. Then, with $R_1 = 10$ kΩ and $R_E = 100$ Ω, we would get about $NF = 2.6$ dB. This is not a good solution for the $R_S = 200$ Ω that are common

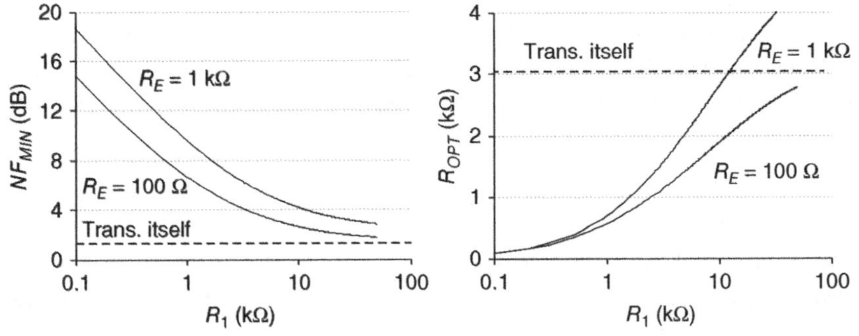

Figure 9.36 Noise parameters of the BJT BC546 in circuit Figure 9.35 as a function of the bias network. $R_2 = 10 \times R_1$.

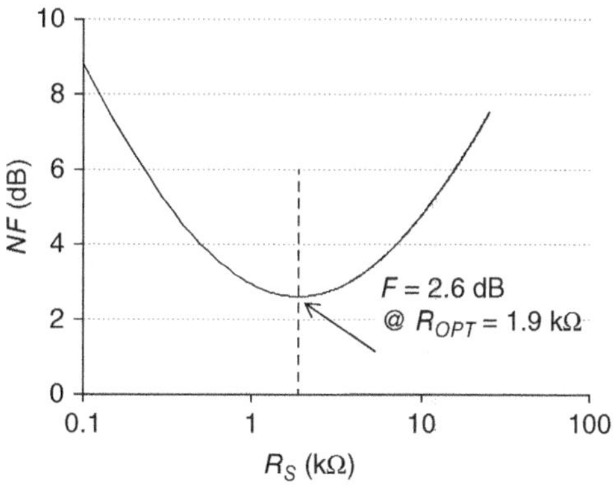

Figure 9.37 Noise figure of the circuit Figure 9.35 versus the source resistance R_S.

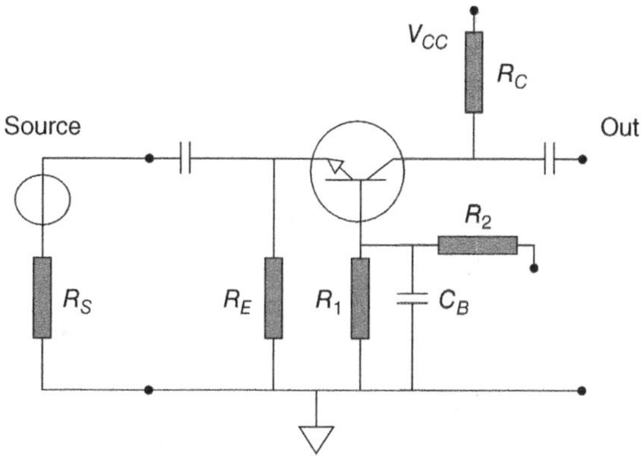

Figure 9.38 Noise model of the BJT BC546 in common-base circuit.

in audio engineering. In this case we would get $NF = 6.4$ dB. The dependence on R_S can be seen in Figure 9.37. Too high values of R_S are also unfavorable. But here a remedy can be a voltage divider at the source.

The analysis of the same type of transistor in common-base Figure 9.38 is carried out according to the same procedure. The circuit parameters are: $V_{CC} = 9$ V, $R_E = 400\,\Omega$, $R_C = 2$ kΩ; $R_2 = 100$ kΩ, $R_1 = 20$ kΩ. R_C is not taken into account. The noise of $R_1||R_2$ is ineffective because it is short circuited by C_B.

The noise figure in the comparison of the two circuits is shown in Figure 9.39. One can see that the common-base circuit is advantageous for small source resistances. This can also be seen from the contour diagram.

Using the Contour Diagram

In most data sheets one will find the contour diagram from which the current dependence of the noise sources e_n and i_n can be obtained. For this purpose we

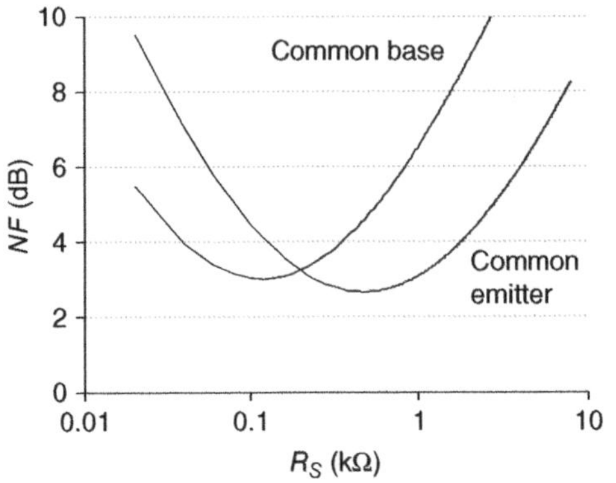

Figure 9.39 Comparison of the noise figure of the common-emitter and common-base circuit of the BC546 versus R_S. Common-emitter circuit with $R_E = 100\,\Omega$; $R_1 = 4\,\text{k}\Omega$.

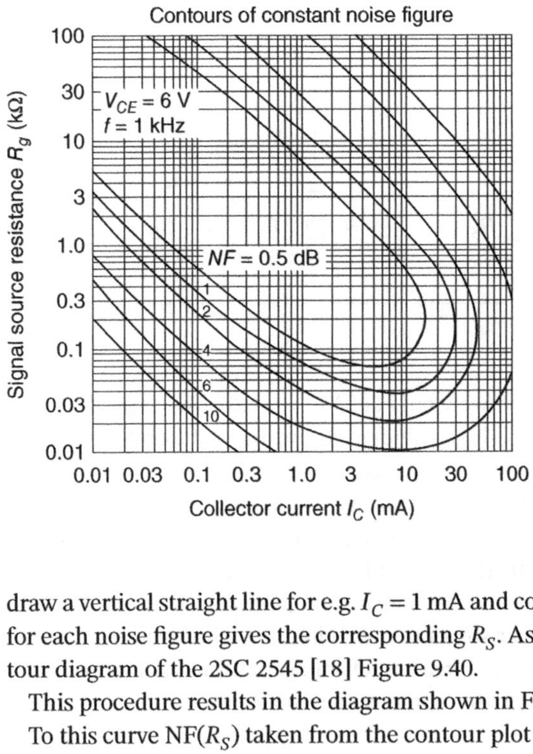

Figure 9.40 Contour diagram from the Hitachi 2SC 2545 data sheet.

draw a vertical straight line for e.g. $I_C = 1$ mA and compile a table of values, which for each noise figure gives the corresponding R_S. As an example, we take the contour diagram of the 2SC 2545 [18] Figure 9.40.

This procedure results in the diagram shown in Figure 9.41.

To this curve NF(R_S) taken from the contour plot we fit the formula:

$$F = F_{MIN} + R_N R_S \left(\frac{1}{R_S} - \frac{1}{R_{OPT}} \right)^2 \tag{9.154}$$

As starting values one extracts $R_{OPT} \approx 1$ kΩ and $F_{MIN} \approx 1.05$. This allows us to easily fit (9.154) to the values from the contour diagram. For Figure 9.41 one

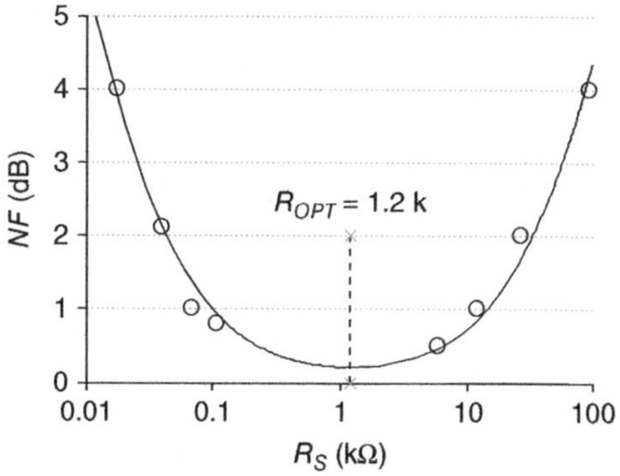

Figure 9.41 Noise figure of the 2SC 2545 taken from the contour diagram. $I_C = 1$ mA. Solid line: Eq. (9.154) fitted to the symbols.

obtains $R_N = 25\,\Omega$; $R_{OPT} = 1.2\,\text{k}\Omega$ and $F_{MIN} = 1.05$ (0.21 dB). With R_N we have the voltage source

$$e_N = \sqrt{4kTR_N} = 0.64\,\text{nV}/\sqrt{\text{Hz}}$$

With the knowledge of R_{OPT}, the current source

$$i_N = \frac{e_n}{R_{OPT}} = 0.54\,\text{pA}/\sqrt{\text{Hz}}$$

The same procedure for other currents I_C then provides the complete current dependence of the two sources.

References

1 Müller, R. (1990). *Rauschen*. Springer Verlag.
2 Ebers, J. and Moll, J.L. (1954). The large signal behaviour of junction transistors. *Proc. IRE* 42: 1761–1772.
3 Nielsen, E.G. (July 1957). Behaviour of noise figure in junction transistors. *Proc. IRE*: 957–963.
4 Hawkins, R.J. (1977). Limitations of Nielsen's and related noise equations applied to microwave bipolar transistors and a new expression for the frequency and current dependent noise figure. *Solid State Electron.* 20: 191–196.
5 Vendelin, G.D., Pavio, A.M., and Rohde, U.L. (2005). *Microwave Circuit Design Using Linear and Nonlinear Techniques*. Wiley Interscience.
6 Agilent Technologies AT-41400: 6 GHz Low Noise Silicon BJT Chip, Datasheet.
7 Infineon Technologies (2018). BFP 760 SiGe:C NPN RF Bipolar Transistor, Datasheet.
8 Schiek, B., Rolfes, I., and Siweris, H.-J. (2006). *Noise in High-Frequency Circuits and Oscillators*. Wiley Interscience.

9 Hillbrandt, H. and Russer, P. (1976). Efficient method for computer aided analysis of linear amplifier networks. *IEEE Trans. Circuits Syst.* 23: 235–238.

10 van der Ziel, A. (1986). *Noise in Solid State Devices and Circuits*. New York: Wiley.

11 Rudolph, M. (2006). *Introduction to Modeling HBTs*. Artech House.

12 Fukui, H. (Mar. 1966). The noise performance of microwave transistors. *IEEE Trans. Electron Devices* 13: 329–341.

13 Rudolph, M. and Heymann, P. (2007). Comparative Study of Shot Noise Models for HBTs. In: *Proc. European Microwave Integrated Circ. Conf.*, Munich, 191–194.

14 Rudolph, M. and Heymann, P. (2007). On compact HBT RF noise modeling. In: *IEEE MTT-S Dig.*, Honolulu, 1783–1786.

15 Rudolph, M., Korndorfer, F., Heymann, P., and Heinrich, W. (2008). Compact large-signal shot noise models for HBTs. *IEEE Trans. Microwave Theory Tech.* 56: 7–14.

16 Infineon Technologies (2011). BCW 60 NPN Silicon AF Transistor, Datasheet.

17 ON Semiconductor (2012). BC 546 NPN Silicon Amplifier Transistor, Datasheet.

18 Hitachi (1999). 2SC2545 Silicon NPN Epitaxial Transistor, Datasheet.

19 Fairchild Semiconductor (2001). 2N 3392 NPN General Purpose Amplifier, Datasheet.

10

Operational Amplifier

Operational Amplifier as Circuit Element

Operational amplifiers are widely used integrated circuits with very high gain. Different transfer characteristics can be easily realized by a few passive components before the input or in the feedback circuit. They can operate in the frequency range from 0 Hz (direct current coupling) to some 10 MHz. Important applications in both analogue and digital technology are consumer electronics (10 Hz–20 kHz), sensor technology, and signal processing in measurement technology (0 Hz–some MHz). Their noise behavior is particularly interesting for the optimization of the application in the audio frequency range [1].

Figure 10.1 shows an ideal operational amplifier with feedback. The operating voltage supply is omitted. The input voltage V_S is amplified to the output voltage V_O. The feedback network consisting of R_1 and R_2 determines the gain of the circuit.

The ideal operational amplifier has an infinitely large gain A_{OL} (open loop), an infinitely large input resistance, and an unlimited frequency range. It also has a power output stage with zero output resistance. This is called a virtual short circuit between the two input terminals + and −. This can be understood in this way: The output voltage V_O at a load resistor cannot become infinitely large. But if the gain A_{OL} is infinite, the input voltage must become zero. This corresponds to a short circuit. Although the real values are of course finite, the characteristics of the operational amplifier in the circuit are almost exclusively determined by the external elements. The noise properties are no exception. As shown in the example below, these are also essentially determined by the external resistors.

With the concept of the virtual short circuit, we can derive the gain of the real circuit. The source voltage V_S rests also against the resistor R_1. Since no current flows into the inverting input, the following must apply

$$\frac{V_S}{R_1} + \frac{V_S - V_O}{R_2} = 0 \tag{10.1}$$

A Guide to Noise in Microwave Circuits: Devices, Circuits, and Measurement, First Edition.
Peter Heymann and Matthias Rudolph.

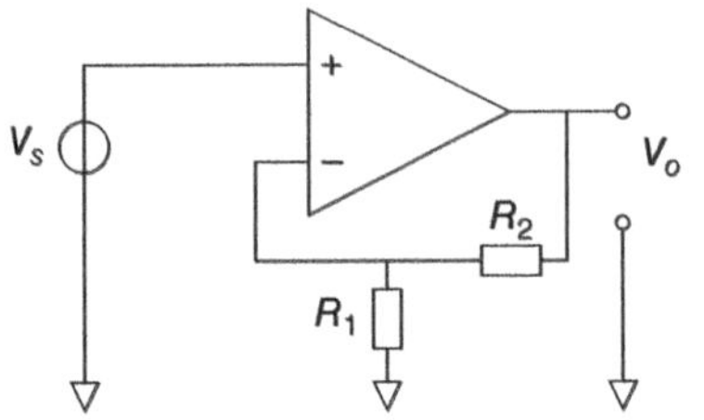

Figure 10.1 Circuit scheme of a non-inverting operational amplifier.

and thus for the gain A_{NI} of the (non-inverting) amplifier with a feedback loop according to Figure 10.1

$$A_+ = \frac{V_O}{V_S} = 1 + \frac{R_2}{R_1} \tag{10.2}$$

If the signal is applied to the inverting input, the same consideration results in a gain

$$A_- = \frac{R_2}{R_1} \tag{10.3}$$

Noise Sources of the Operational Amplifier

The noise characteristics are described by a noise current source i_N and a noise voltage source v_N in front of the amplifier, which is thought to be noise-free. This corresponds to the two-port representation in the form of the chain matrix. The quantities v_N and i_N are specified by the manufacturer. Figure 10.2 depicts the root of the spectral noise power density for two typical examples versus frequency. On the left the spectral noise voltage density v_N right the spectral noise current density i_N. These diagrams can be found in the data sheets [2].

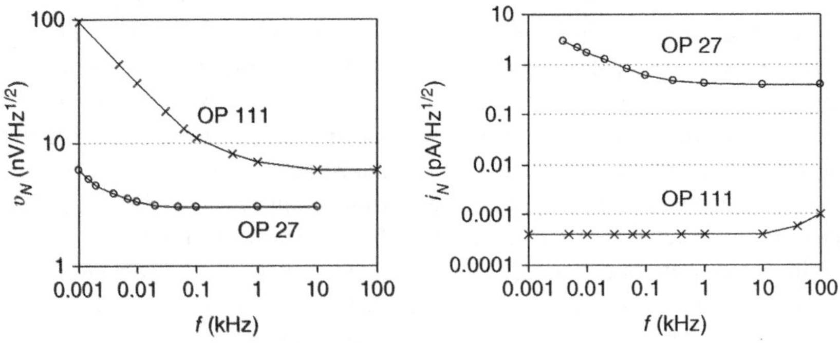

Figure 10.2 Noise voltage and current sources at the input of BJT OP 27 and FET OP 111 according to manufacturer's specifications.

In accordance with the data sheets, we use here v_N for the spectral density of the noise voltage with the unit $\mathrm{nV}/\sqrt{\mathrm{Hz}}$ and i_N for the spectral density of the noise current with the unit $\mathrm{pA}/\sqrt{\mathrm{Hz}}$. It can be seen that the bipolar technology (OP27) allows low voltage but higher currents. With operational amplifiers consisting of field-effect transistors (OP111), it is the other way around. Depending on the internal resistance of the source, this allows noise to be minimized by selecting the particularly suitable OPA, as we will see in the example, but within certain limits. In contrast to the general case of noise sources of two-port devices, the two sources are assumed to be uncorrelated. Noisy two-ports of microwave technology, e.g. amplifiers or single transistors, have complex input resistances within the vicinity of 50 or 75 Ω, i.e. the characteristic impedance of lines. In contrast, the input resistance of operational amplifiers operating in the MHz range is in the order of MΩ. In (7.2) we had seen that the equivalent sources v_{1A} and i_{1A} at the input of the two-port both depend on i_2. They are correlated via the output current. We repeat here:

$$v_{1A} = -\frac{i_2}{Y_{21}}; \quad i_{1A} = i_1 - i_2\frac{Y_{11}}{Y_{21}} \tag{10.4}$$

In the case of the operational amplifier we have $Y_{11} \cong 0$ so that $i_{1A} \cong i_1$ and therefore there is no correlation between current and voltage source, since the common part with i_2 is ineffective. A complete noise characterization with the determination of F_{MIN}, R_N and Γ_{OPT} with the intention of noise matching is therefore not useful for the operational amplifier. Especially as low-loss transformations with resistances are not possible in the MΩ-range. As already mentioned, however, noise can be minimized by suitable selection of the type of operational amplifier and the external circuitry.

Let us first look at the noise behavior without external circuit [3].

We see in Figure 10.2 that v_N and i_N are frequency dependent due to $1/f$ noise. In wide ranges, however, one can assume a constant value of the spectral noise power density and thus calculate the noise figure as a function of the source resistance R_S. For this purpose we use the noise equivalent circuit Figure 10.3. It is the well-known scheme: the amplifier is noise-free, and its noise characteristics in the sources v_N and i_N are summarized. A signal source (e.g. microphone) is connected to the input with the internal resistance R_S, which is noisy with T_0. This thermal noise adds to v_N.

The current source i_N generates a voltage drop at R_S, which also adds to v_N. This results in the spectral noise voltage density of the equivalent noise source at the input v'_N

$$v'_N = \sqrt{v_N^2 + \left(i_N R_S\right)^2 + 4kT_0R_S} \tag{10.5}$$

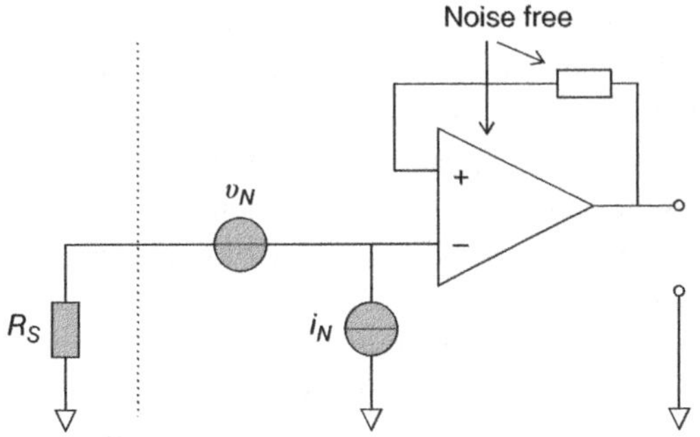

Figure 10.3 Noise equivalent circuit of an operational amplifier. R_S: noisy source resistance; intrinsic sources of the OPA: v_N and i_N.

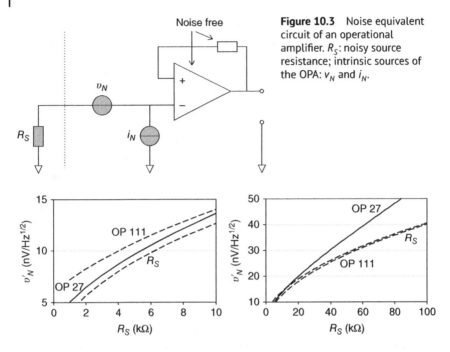

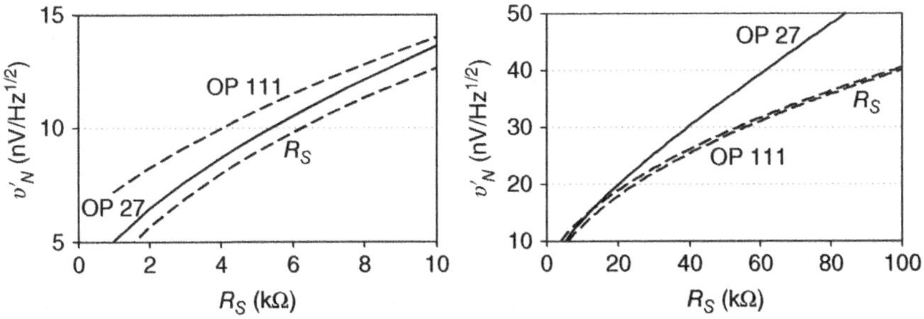

Figure 10.4 Equivalent noise voltage density v'_N at the input versus the source resistance. At high R_S the OP111 reaches the thermal limit.

The dependence (10.5) on the source resistance R_S is well suited for comparing different OPA types. Figure 10.4 shows OP27 (BJT) [4] and OP111 (FET) [5]. The calculation was made using the white noise values ($f = 10\,\text{kHz}$) from Table 10.1. For $R_S < 10\,\text{k}\Omega$ the OP27 is superior (Figure 10.4 right). For $R_S > 10\,\text{k}\Omega$ the OP111 almost reaches the theoretical limit of the thermal noise of the source resistance (Figure 10.4 left).

From the interaction of these three sources, one can also obtain the noise factor F of the device. However, this is theoretical and will not be achieved in a practical circuit, as we will see below. We first calculate the noise factor according to the definition (6.3) as the total noise power normalized to the noise power of the

Table 10.1 Noise data as specified by the manufacturer.

Type	Data sheet	Technology	f (kHz)	v_N (nV/Hz$^{1/2}$)	i_N (fA/Hz$^{1/2}$)
OP 27	Analog dev.	BJT	10	3	400
OP 111	Burr-brown	FET	10	6	0.5

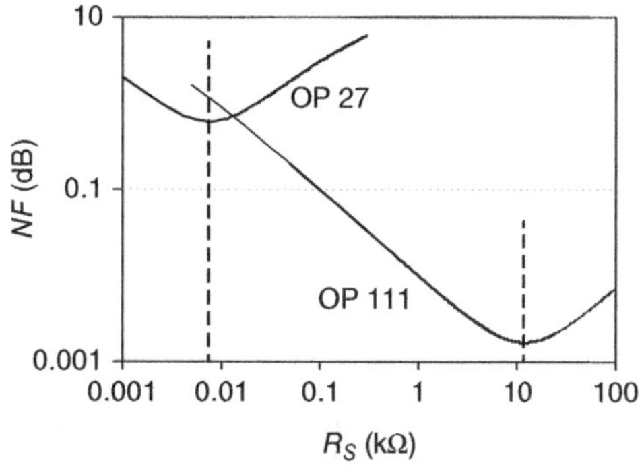

Figure 10.5 Noise figure @ 10 kHz for BJT (OP27) and FET (OP111) technology versus R_S (10.6). Dotted lines indicate R_{OPT}.

generator at the input. With the sources in Figure 10.3 we get

$$F = \frac{\overline{|v_N|^2} + \overline{|i_N|^2} R_S^2 + 4kT_0R_S}{4kT_0R_S} \tag{10.6}$$

From Figure 10.2 we can take the following values for $f = 10\,\text{kHz}$ (outside the $1/f$ range).

As mentioned earlier the unit of v_N and i_N does not provide pure voltage or current values. They are related to the root of the bandwidth $(\text{V}/\sqrt{\text{Hz}})$. This results in the spectral power density (V^2/Hz). The information can be found in the unit $(\text{nV}/\sqrt{\text{Hz}})$ in the data sheets. They are also depicted in Figure 10.2. In (10.6) this is considered, also the term $4kT_0R_S$ has the unit (V^2/Hz).

The noise figure NF as a function of the generator resistance R_S is shown in Figure 10.5.

Here the different dependence on R_S can be seen even more clearly. The strong influence can be seen in the second term of the numerator of (10.6). The minimum noise figure occurs at the resistor R_{OPT}

$$NF_{MIN} \quad @ \quad R_{OPT} = \frac{v_N}{i_N} \tag{10.7}$$

At this value the plots in Figure 10.5 have their minimum. In the range $R_S < R_{OPT}$ the voltage source is dominant, and in the range $R_S > R_{OPT}$ the current source. One can also clearly see the preferred areas of application of the various operational amplifiers. For low impedance sources the BJT technology (e.g. OP 27) is more suitable, and for high impedance sources the FET technology (e.g. OP 111). Figure 10.5 shows theoretical limit values. In a practical circuit the value of OP 111 of $F_{MIN} \cong 10^{-3}$ dB, which is at the thermal limit, cannot be reached, especially since a $R_S \cong 10\,\text{M}\Omega$ is also unrealistic, but one can see the tendency.

A realistic example is shown in Figure 10.6. In practice the noise voltage at the output of the OPA, i.e. v_{TOT}, is more important than the theoretical noise

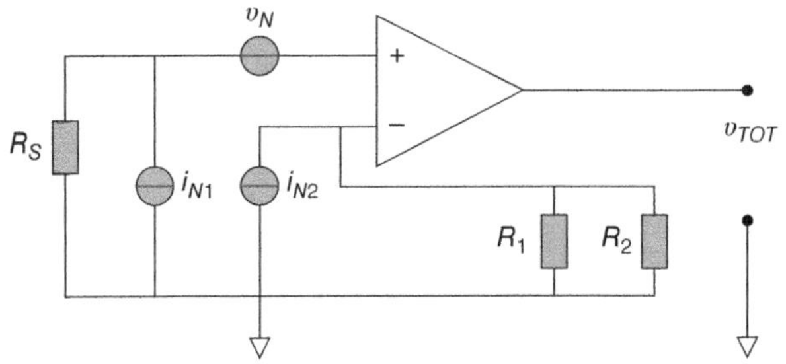

Figure 10.6 Noise equivalent circuit of the non-inverting OP27 with intrinsic sources and thermal noise of the external resistors.

factor (10.6) of the blank amplifier. We first consider the contributions of the individual sources to v_{TOT}. The values v_N and i_N we take again from the data sheet at $f = 10\,\text{kHz}$. The feedback resistors R_1 and R_2 generate thermal noise, as does the resistance of the source with $v_R = 0.13\sqrt{BR}$ (nV). As the bandwidth we choose the acoustic range $B = 20\,\text{kHz}$. With respect to the noise characteristics we assume a short-circuited output, therefore R_1 and R_2 are in parallel (different to Figure 10.1).

The individual contributions are summarized in Table 10.2. In line 1 is the source resistance, which we assume with $R_S = 10\,\text{k}\Omega$. Its noise voltage 1.82 µV

Table 10.2 Compilation of the values for the noise analysis of OP27 in the circuit Figure 10.6. $B = 20\,\text{kHz}$.

Circuit element	Value	Formula	Voltage at input (µV) (B = 20 kHz)	Ampl.	Voltage at output (µV)
R_S	10 kΩ	$\sqrt{4kTR_SB}$	1.82	A_+	20
$R_P = R_1 \| R_2$	9.1 kΩ	$\sqrt{4kTR_PB}$	1.74	A_-	17
v_N	$3\,\text{nV}/\sqrt{\text{Hz}}$	$v_N\sqrt{B}$	0.42	A_+	4.6
i_{N1}	$0.4\,\text{pA}/\sqrt{\text{Hz}}$	$i_{N1}R_S\sqrt{B}$	0.57	A_+	6.3
i_{N2}	$0.4\,\text{pA}/\sqrt{\text{Hz}}$	$i_{N2}R_P\sqrt{B}$	0.51	A_-	5.1
Total noise voltage at output	$v_{TOT} = \sqrt{\left[20^2 + 17^2 + (4.6)^2 + (6.3)^2 + (5.1)^2\right] \times 10^{-12}} = 28\,\mu\text{V}$				
Noise figure (NF)	$NF = 20\,log\left(\frac{v_{TOT}}{\sqrt{4kTR_SBA_{NI}}}\right) = 2.82\,\text{dB}$				
NF OPA alone	$NF = 20\,log\frac{\sqrt{v_N^2 + (i_NR_S)^2 + 4kTR_S}}{\sqrt{4kTR_S}} = 0.61\,\text{dB}$				

rests directly against the non-inverting input of the OPA. The feedback resistors $R_1 = 10\,\text{k}\Omega$ and $R_2 = 100\,\text{k}\Omega$ determine the gain to $A_+ = 11$ (10.2). The resistors R_1 and R_2 are in parallel to ground. They form $R_P = R_1 || R_2 = 9.1\,\text{k}\Omega$ (line 2). The voltage source of the OPA is v_N (line 3). It adds to the thermal noise of R_S and is amplified by A_+. In both inputs there is a noise current i_N. At the non-inverting input the current source i_{N1} drives a noise current through R_S. The voltage drop $i_{N1}R_S$ is amplified by A_+ (line 4). The current i_{N2} of the inverting input flows through R_P. The voltage drop is amplified by $A_- = 10$ (line 5).

We have thus covered all contributions.

Since all contributions are assumed to be uncorrelated, the total output voltage v_{TOT} is the root of the sum of the squares of the individual voltages (line 6). If one looks at the contributions of the individual sources, one sees that the resistors determine the noise behavior. The noise figure of the circuit is $NF = 2.8\,\text{dB}$ (line 7), while the operational amplifier itself contributes only about 5% to the noise voltage at the output of the circuit. It alone has the theoretical value $NF = 0.6\,\text{dB}$ (line 8). This is also approximately the value NF_{MIN} (Figure 10.5).

Consideration of 1/*f* Noise

In this example, we have replaced the integration of the noise voltage density $v_N(f)$ or noise current density $i_N(f)$ by a multiplication with $B = 20\,\text{kHz}$. Instead of the curves in Figure 10.2, we used a constant value outside the $1/f$-range. This is of course only an approximation and does not take the $1/f$ noise into account. Correctly one must consider both the frequency response of the gain and the dependency of the plots according to Figure 10.2, if these low frequencies can play a role in the application.

The approximation of the frequency dependence of the noise is based on the data sheet (Figure 10.2). It can be seen that a $1/f$-dependence in this display format is not a straight line but a hyperbola. The approximation used in the integration refers to the spectral power densities plotted on a frequency axis in logarithmic scaling. In Figure 10.2 the spectral voltage and current densities are plotted. We therefore take the spectral power densities S_V and S_I. The result is a straight line in double logarithmic scale. We describe the experimental curves in the $1/f$ range with

$$S_V^f = K_V\frac{1}{f}; \quad S_I^f = K_I\frac{1}{f} \tag{10.8}$$

This approximation is shown in Figure 10.7 as dashed lines. $K_{V,I}$ results from the level of the white noise S^W and the corner frequency f_C.

$$K_V = f_{CV}S_V^W; \quad K_I = f_{CI}S_I^W \tag{10.9}$$

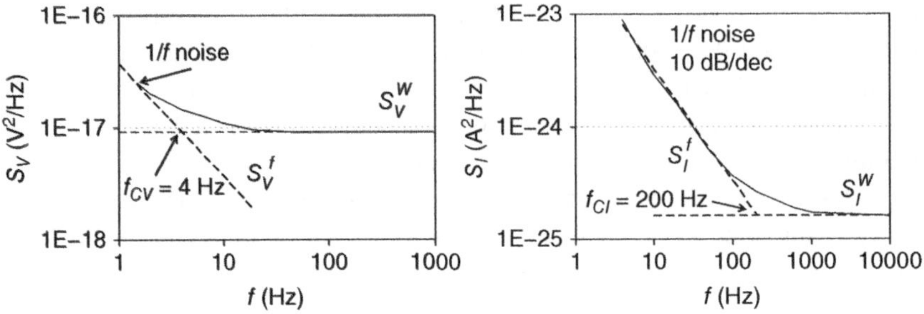

Figure 10.7 Modeling the LF noise of OP27 with (10.8).

In our example of OP 27 we have

$$K_V = 9 \times 10^{-18}\ {}^{V^2}\!/_{Hz} \times 4\,\text{Hz} = 3.6 \times 10^{-17}\ \text{V}^2$$

$$K_I = 1.6 \times 10^{-25}\ {}^{A^2}\!/_{Hz} \times 200\,\text{Hz} = 3.2 \times 10^{-23}\ \text{A}^2$$

Only the currents i_{N1} and i_{N2} generate a noticeable contribution to $1/f$ noise at the output. The resistors have only thermal noise when carefully selected. For example, if one uses metal film resistors. The voltage source v_N of the OP27 has the very low corner frequency ($f_C \cong 4\,\text{Hz}$) and is therefore not relevant for acoustic applications. Integration over the frequency range provides the noise voltage, the square of which we designate with $\overline{v^2}$. It has the correct unit $\left[\overline{v^2}\right] = \text{Volt}^2$. As a reminder: The information in data sheets v_N, i_N that we have also used here are spectral voltage or current densities, related to the square root of the bandwidth with the unit $\text{V}/\sqrt{\text{Hz}}$. The contribution of $1/f$ noise is obtained by integrating from the lowest frequency f_L to the corner frequency f_C.

$$\overline{v_1^2} = \int_{f_L}^{f_C} K\frac{1}{f}df = K\,ln\left(\frac{f_C}{f_L}\right) \tag{10.10}$$

The contribution of white noise is obtained by integration from the corner frequency f_C to the upper frequency limit f_U.

$$\overline{v_2^2} = S_V^W\,(f_U - f_C) \tag{10.11}$$

These two contributions add up to the total voltage $\overline{v_{TOT}}$. Averaging here means integration over the frequency range.

$$\overline{v_{TOT}} = \sqrt{\overline{v_1^2} + \overline{v_2^2}} = \sqrt{S_V^W\left[(f_U - f_C) + f_C\,ln\left(\frac{f_C}{f_L}\right)\right]} \tag{10.12}$$

The contribution of $1/f$ noise is usually low in the acoustic range. Figure 10.8 shows by the example of OP27 the contribution of the $1/f$ noise, i.e. $\sqrt{\overline{v_1^2}}$ to the voltage $\overline{v_{TOT}}$ at the output. The relative part of the $1/f$ noise current source is higher

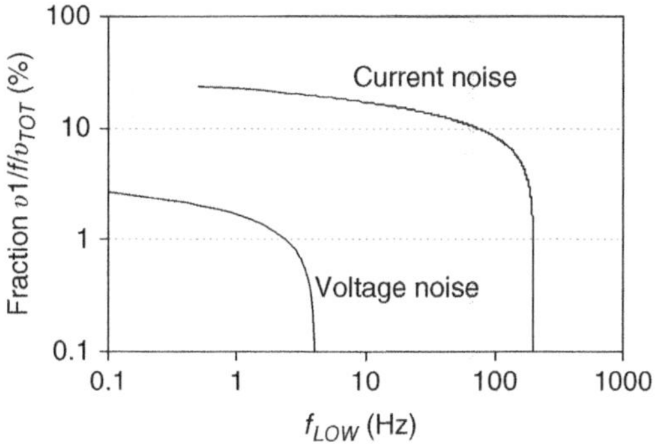

Figure 10.8 Contribution of the $1/f$ range to the total noise voltage at the output. The passband is from f_{LOW} to $f_U = 20\,\text{kHz}$.

than that of the voltage. This is due to the higher corner frequency. However, the contribution of the current to the total noise voltage at the output is usually very small. The upper frequency limit is $f_U = 20\,\text{kHz}$. Plotted is $\sqrt{\overline{v_1^2}/\overline{v_1^2} + \overline{v_2^2}}$ in percent above the lower frequency limit f_L. Above the corner frequency $f_C = 4\,\text{Hz}$ for voltage and $f_C = 200\,\text{Hz}$ for current, the contribution disappears. It can be seen that at the lower frequency limit of the acoustic range $f_L \cong 20\,\text{Hz}$, the contribution of $1/f$ noise is low. Even for the current it is less than 10%. In addition, the human ear is not very sensitive to noise below ≅100 Hz. For the application of the operational amplifier in measurement and sensor technology, however, the frequencies in the $1/f$ range are definitely important and must be estimated for the specific case.

When defining the frequency limits of the transmission range, the difference between 3 dB bandwidth and noise bandwidth must be taken into account. If the bandwidth is limited by a first-order low-pass filter, the noise bandwidth is larger than the 3 dB bandwidth by a factor of $\pi/2$.

The noise of a circuit with operational amplifiers can be minimized by the following measures, which are easy to understand from the considerations we made earlier.

1. Selection of the type according to the technology. BJT for low resistance sources. FET for high impedance sources.
2. Select feedback resistors with an impedance as low as possible. The limit is given by the power of the output stage.
3. Use metal film resistors.
4. The upper frequency limit f_U should be only as high as absolutely necessary.

Operational Amplifier as an Active Low-Pass Filter

We have seen that the knowledge of output noise voltage generated by the amplifier itself, the circuit elements, and the input resistance is more important than the

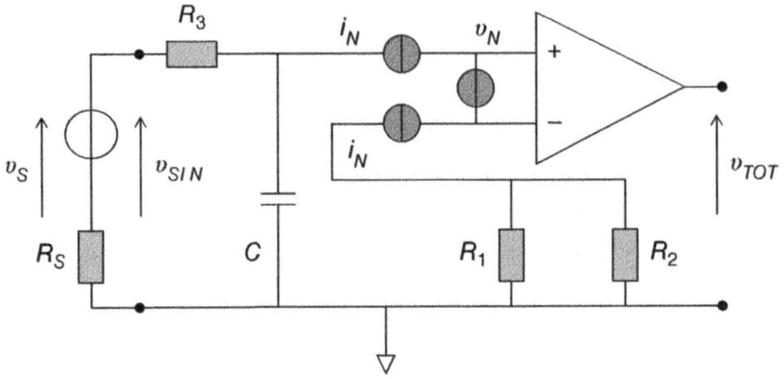

Figure 10.9 Noise equivalent circuit of the AD743 as active low-pass filter.

four IEEE noise parameters. Here we want to analyze an active filter consisting of an RC low-pass filter and the very low-noise operational amplifier AD 743 [6] (Figure 10.9).

The data sheet shows that the two-port parameters are almost ideal: The input resistance is very high (MΩ), the output resistance is very low (a few Ω). The gain is very high and the feedback is negligible. The noise data are:

Spectral noise voltage density: $v_N = 2.9\, n\text{V}/\sqrt{\text{Hz}}$

Spectral noise current density: $i_N = 7\,\text{fA}/\sqrt{\text{Hz}}$

The $1/f$ noise of the voltage source starts below $f_C = 10\,\text{Hz}$, that of the current source below $f_C = 100\,\text{Hz}$. It is neglected here.

As we have seen earlier, the feedback resistors $R_1 = 1\,\text{k}\Omega$ and $R_2 = 10\,\text{k}\Omega$ are connected in parallel with respect to their noise characteristics. $R_3 = 800\,\Omega$ and $C = 0.1\,\mu\text{F}$ form the low pass, R_S is the internal resistance of the source. To calculate the noise voltage at the output, we need the bandwidth of the filter. The cut-off frequency is given by

$$f_g = \frac{1}{2\pi\left(R_S + R_3\right)C} = 1.6\,\text{kHz} \tag{10.13}$$

The voltage gain of the non-inverting input results in

$$A_+ = \frac{{}^{R_2}/_{R_1} + 1}{1 + j\omega C\left(R_S + R_3\right)} \tag{10.14}$$

For the noise bandwidth we need the square of the magnitude

$$|A_+|^2 = \frac{{}^{R_2}/_{R_1} + 1}{1 + \left(\omega C\left[R_S + R_3\right]\right)^2} \tag{10.15}$$

Table 10.3 Contributions of the components from Figure 10.9 to the total noise.

Circuit element	Value	Formula	Voltage at input ($B = 2.5$ kHz)	Ampl.	Voltage at output (nV)
R_S	200 Ω	$\sqrt{4kTR_SB}$	0.091 μV	A_+	1000
$R_P = R_1 \| R_2$	0.91 kΩ	$\sqrt{4kTR_PB}$	0.19 μV	A_-	1900
R_3	0.8 kΩ	$\sqrt{4kTR_1B}$	0.18 μV	A_+	2000
v_N	$2.9\ \text{nV}/\sqrt{\text{Hz}}$	$v_N\sqrt{B}$	0.145 μV	A_+	1600
i_{N1} (R_3)	$7\ \text{fA}/\sqrt{\text{Hz}}$	$i_{N1}R_3\sqrt{B}$	0.28 nV	A_+	3.1
i_{N1} (R_S)	$7\ \text{fA}/\sqrt{\text{Hz}}$	$i_{N1}R_S\sqrt{B}$	0.07 nV	A_+	0.77
i_{N2} (R_P)	$7\ \text{fA}/\sqrt{\text{Hz}}$	$i_{N2}R_P\sqrt{B}$	0.32 nV	A_-	3.2

Total noise voltage at output $v_{TOT} = \sqrt{1000^2 + 1900^2 + 2000^2 + 1600^2 + 3.1^2 + 0.77^2 + 3.2^2} \times 10^{-9} = 3.3\ \mu\text{V}$

Noise figure (NF) $NF = 10\log\frac{(v_{TOT})^2}{\left(\sqrt{4kTR_SB} \times A_{NI}\right)^2} = 10.5\ \text{dB}$

NF OP alone $NF_{OA} = 10\log\frac{\left(v_N\sqrt{B}\right)^2 + 4kTR_SB + \left(i_{N1}R_S\sqrt{B}\right)^2}{4kTR_SB} = 5.5\ \text{dB}$

The noise bandwidth results from the integration of the transmission curve

$$\int_0^f \frac{df}{1 + \left(2\pi C\left[R_S + R_3\right]\right)^2 f^2} = \frac{1}{2\pi C\left(R_S + R_3\right)} arctan\left(2\pi fC\left[R_S + R_3\right]\right) \tag{10.16}$$

for $f \to \infty$ the noise bandwidth is obtained

$$B = \frac{1}{2\pi C\left(R_S + R_3\right)} \times \frac{\pi}{2} = \frac{\pi}{2} f_g \tag{10.17}$$

In our case $B = 2.5$ kHz.

As an example, we calculate the noise voltage at the output for a source resistance of $R_S = 200\ \Omega$. An increase of R_S would reduce the noise figure (Figure 7.12), but it would also change the filter curve due to the influence on f_g (10.13).

As shown earlier, we can calculate the output noise voltage v_{TOT} and the noise figure NF and estimate the effect of the circuit elements (Table 10.3).

We see in the example that the very good noise properties of the AD 743 cannot be utilized here. This is not primarily due to the influence of the additional resistors, but to the too low source resistance R_S. Thus the operational amplifier works far away from its noise minimum. It is unrealistic to operate the amplifier in the

Table 10.4 Comparison of noise characteristics of operational amplifiers.

Type	Technology	SNR (dB) ($V_S = 1\,mV$)	NF (dB)	$R_{OPT} = v_N/i_N$
TL 071	FET	60	20	1.8 MΩ
TLC 2272	CMOS	64	15	15 MΩ
LT 1028	BJT	70	10	850 Ω
LT 1167	BJT	66	14	60 kΩ
SSM 2017	BJT	69	11	480 Ω
AD 743	BJT	69	11	400 kΩ

noise minimum according to its optimum source resistance, because this is much too high. We see this in our example, if we derive the formula for the noise factor F with respect to R_S and set it to zero.

$$F = 1 + \frac{1}{4kTR_S}\left\{v_N^2 + i_N^2\left(R_3^2 + R_S^2 + R_P^2\right) + 4kT\left(R_3 + R_P\right)\right\} \tag{10.18}$$

$$\frac{\partial F}{\partial R_S} = -\frac{1}{4kTR_S^2}\left\{v_N^2 + i_N^2\left(R_3^2 + R_S^2 + R_P^2\right) + 4kT\left(R_3 + R_P\right)\right\} = 0 \tag{10.19}$$

For R_{SOPT}, this results in

$$R_{SOPT} = \sqrt{\frac{v_N^2}{i_N^2} + R_3^2 + R_P^2 + \frac{4kT\left(R_3 + R_P\right)}{i_N^2}} \tag{10.20}$$

In our numerical example we obtain $R_{SOPT} = 900\,k\Omega$, a completely unrealistic value, which is even higher than the value for the operational amplifier without all circuit elements, i.e. $v_N/i_N = 400\,k\Omega$. The thermal noise voltage of R_{SOPT} itself, at approx. 70 µV, is already in the order of magnitude of the signal voltage. Nevertheless bipolar operational amplifiers are still better for this purpose than those in FET technology. In Table 10.4 there are some examples for $R_S = 200\,\Omega$ and the OPA alone without all circuit elements.

Further useful details can be found on the CCInfo website.

References

1 Bergtold, F. (1973). *Schaltungen mit Operationsverstärkern*. Oldenbourg Verlag.

2 Landstorfer, F. and Graf, H. (1981). *Rauschprobleme der Nachrichtentechnik*. Oldenbourg Verlag.

3 Karki, J. (2005). Calculating Noise Figure in OP-Amps. *Texas Instr. Analog Appl. J.* 40: 31–38.

4 Analog Devices (2015). OP27 Low Noise Precision Operational Amplifier. Datasheet.

5 Burr-Brown (1995). OPA111 Low Noise Precision Difet Operational Amplifier. Datasheet.

6 Analog Devices (2003). AD743 Ultralow Noise BiFET Operational Amplifier. Datasheet.

11

Field Effect Transistors

JFET

It is often found in the literature that the channel of the JFET generates thermal noise [1, 2]. This drain noise source is usually described without comment with

$$\overline{i_D^2} = \frac{2}{3} 4kTg_m B \tag{11.1}$$

The factor $^2\!/_3$ and the occurrence of the transconductance g_m instead of a "real" conductance is based on a complicated approach with a relatively simple solution from van der Ziel. We want to present it here in simplified form. In doing so, we will abstain from an exact mathematical analysis in favor of understanding. But first we will start with the current–voltage characteristic of the FET.

Mode of Operation of the FET

The drain current in the channel is driven by the drain-source voltage V_D and controlled by the gate-source voltage V_G. Along the channel, below the gate, the voltage $V(x)$ depends on the position x, since it increases from 0 (directly at the source contact) to V_D (at the drain contact). The space charge layer under the gate is shaped accordingly. This drain current includes a fluctuation component $\overline{i_D^2}$.

We derive the V/I-characteristic for the junction field effect transistor with pn-junction. The same applies analogously to the other designs, e.g. MESFET or HEMT with Schottky barrier. We consider the open channel without gate as a piece of material with the conductivity σ, the cross-section $A = a \times w$ and the length L. w is the width of the cuboid, a its height, and L its length. Due to the n-doping the channel has an electron density n. According to the material these electrons have a mobility μ. With this approach, Ohm's law applies, which in its general form is

$$J = \sigma E \tag{11.2}$$

J is the current density, E the electric field strength.

A Guide to Noise in Microwave Circuits: Devices, Circuits, and Measurement, First Edition.
Peter Heymann and Matthias Rudolph.

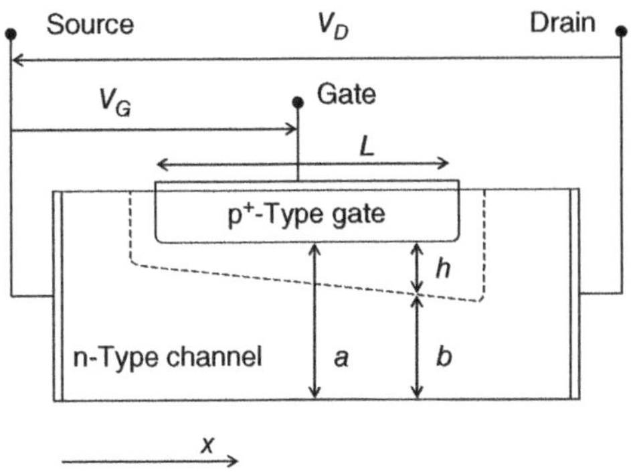

Figure 11.1 Schematic of a field effect transistor with a pn-diode as gate (JFET).

For the current in the conductive cuboid with the above-mentioned dimensions and electrodes on the end faces, the following results are obtained with $\sigma = qn\mu$ and $E = dV/dx$:

$$I = J \times A = qn\mu aw\frac{dV}{dx} \tag{11.3}$$

In contrast to the bipolar transistor, in which the diffusion current is the main component, this is a pure drift current. In the field effect transistor (Figure 11.1), a gate electrode is applied to this cuboid under which a space charge layer is formed. With its variable thickness h, the current in the channel can be controlled. Depending on the technology and material, there are variants: JFET with pn-junction under the gate (Figure 11.1), MOSFET with an oxide layer between gate metal and channel on silicon, and MESFET with the gate as Schottky diode, mostly on GaAs, InP, or GaN. The cross-section remaining for the current conduction has the height b. The thickness of the space charge layer h depends on the local potential difference between gate and channel and is therefore a function of x, since gate and local drain voltage are superimposed. h is largest at the drain end and can completely block the channel here. It holds:

$$h(x) = \sqrt{\frac{2\varepsilon\left(V(x) + V_G + \phi\right)}{qn}} \tag{11.4}$$

$V(x)$ is the voltage generated by the drain voltage V_D relative to source at location x, ε is the dielectric constant $\varepsilon_0\varepsilon_r$, and ϕ is the built-in voltage, also called diffusion voltage. The dotted line in Figure 11.1 shows the dependence $h(x)$. A negative gate voltage leads to a depletion of the channel and increases the thickness of the space-charge region. At the pinch-off voltage V_P the channel is totally blocked. So we have the following current–voltage characteristic: at constant gate voltage V_G (e.g. -1 V against source) we increase the drain voltage. At low values ($V_D < 1$ V) the transistor behaves like an ohmic resistor ($I \propto V_D$). By increasing h, it equals the channel thickness a at the drain end. After that, the current no longer increases, but continues to flow. We are in the saturation range. Further increase of V_D leads to a shift of the pinch-off point toward the source. The pinch-off voltage results from (11.4) with $h = a$

$$V_P = \frac{a^2 qn}{2\varepsilon} - \phi \tag{11.5}$$

The drain current I_D is determined by the open part of the channel. This is

$$b(x) = a - h(x) \tag{11.6}$$

If we insert (11.4) and (11.6) in (11.3), we obtain

$$I_D = qn\mu aw \left(1 - \sqrt{\frac{V(x) + V_G + \phi}{V_P + \phi}}\right)\frac{dV}{dx} \tag{11.7}$$

The integration over the length of the channel results in the current–voltage characteristic of the transistor. It is sufficient to integrate over the gate length L, i.e. $0 \leq x \leq L$.

$$I_D = qn\mu \frac{aw}{L}\left\{V_D - \frac{2}{3}(V_P + \phi)\left[\left(\frac{V_D + V_G + \phi}{V_P + \phi}\right)^{\frac{3}{2}} - \left(\frac{V_D + V_G}{V_P + \phi}\right)^{\frac{3}{2}}\right]\right\} \tag{11.8}$$

This relation only applies to the range of low drain voltage $V_D < V_P - V_G = V_{DS}$ in which the transistor behaves like an ohmic resistor controlled by the gate voltage. As the drain voltage V_D increases, the field strength at the drain end of the gate becomes higher and higher due to the constriction of the channel; in the limit it becomes infinite. This makes no sense. In addition, the conductivity formula (11.3) no longer applies, as it is based on the constant mobility μ. At high field strength, however, the velocity of the electrons in both silicon and GaAs remains constant. In GaAs, the formation of Gunn domains can also occur. A simple, practical solution is based on the assumption that for $V_D > V_{DS}$ the drain current remains constant. For this range of the IV-characteristic, we simply replace V_D by V_{DS} in (11.8). The following approximation is also often used:

$$I_D = I_{DS}\left(1 - \frac{V_G + \phi}{V_P}\right)^2 \tag{11.9}$$

with

$$I_{DS} = \frac{q^2 n^2 \mu}{6\varepsilon}\frac{a^3 w}{L} \tag{11.10}$$

An example for the IV-characteristic of a GaAs MESFET (without p^+-layer) according to (11.8) is shown in Figure 11.2. The parameters are $n = 1 \times 10^{17}\ \text{cm}^{-3}$; $\mu = 4.5 \times 10^3\ \text{cm}^2/\text{V s}$; $a = 0.2\ \mu\text{m}$; $L = 1\ \mu\text{m}$; $w = 1\ \text{mm}$; $V_P = -2.8\ \text{V}$. Also plotted is the transition to the saturation range at $V_D = V_{DS}$.

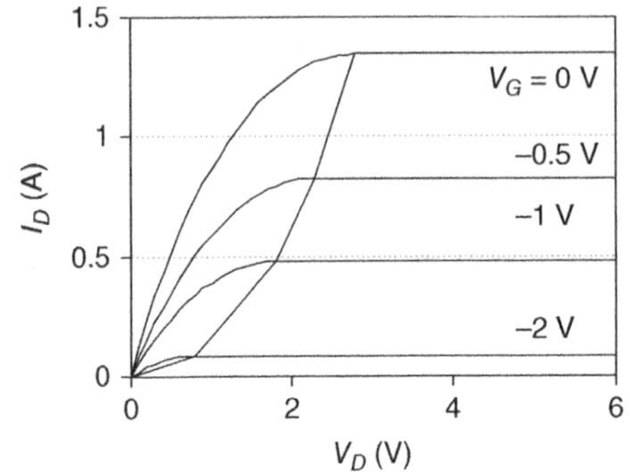

Figure 11.2 *IV*-characteristics of a GaAs FET according to (11.8) and (11.10) with the limiting curve to the saturation range.

The transconductance $g_m = {}^{\partial I_D}/_{\partial V_G}$ is an important quantity in the equivalent circuit, which, as we have seen earlier (11.1), determines the noise characteristics. We determine g_m by derivation of (11.9)

$$g_m = \frac{\partial I_D}{\partial V_G} = \frac{2I_{DS}}{V_P}\left(\frac{V_G + \phi}{V_P} - 1\right) \tag{11.11}$$

The Channel Noise

We compare conductance G (11.12) of the cuboid of semiconductor material, representing the channel of the FET (Figure 11.3)

$$G = \frac{qn\mu aw}{L} \tag{11.12}$$

and the transconductance, that both are identical except for the gate voltage dependence of g_m and a constant factor. If we insert in (11.11) the values for I_{DS} (11.10) and V_P (11.5) (without ϕ), we obtain

$$\frac{2I_{DS}}{V_P} = \frac{2}{3}\frac{qn\mu aw}{L} = \frac{2}{3}G \tag{11.13}$$

We have thus understood the approach (11.1), in which not the channel conductance G but the transconductance g_m is used for the thermal noise of the channel.

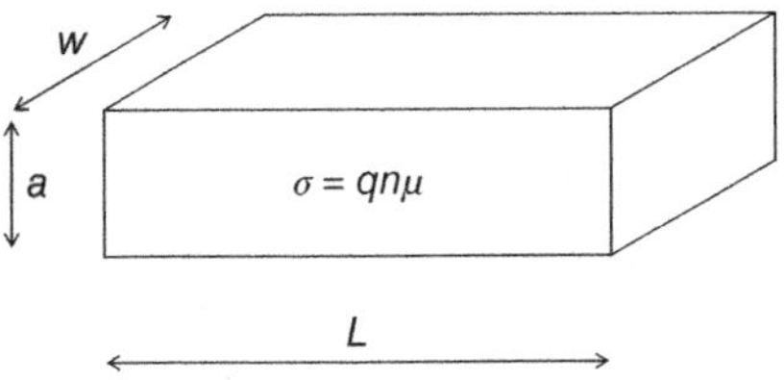

Figure 11.3 Cuboid of conductive material for understanding the transconductance.

The result of the integration over the channel, which is inhomogeneous in the longitudinal direction, is the noise current source at the drain:

$$\overline{i_D^2} = 4kTBg_m \times P \quad (11.14)$$

The factor P is a function of the voltages V_D, V_G which are applied to the transistor and of the voltage V_{DS}, which is determined by the technology. Its value is between 2/3 for saturation and 1 for $V_D = 0$, i.e. in thermal equilibrium. In this case, the channel is noisy like an ohmic resistor with conductance g_m. In active mode

$$\overline{i_D^2} = \frac{2}{3}4kTBg_m \quad (11.15)$$

As already written in (11.1).

Noise Sources at the Gate

Since the gate is operated with a bias voltage against source, a direct current can be existent [3]. In the JFET with its pn-diode, this is the well-known reverse current. In the MESFET and HFET it does not normally occur, occasionally with imperfect process technology. In any case, it is very small compared to the drain current, but it is a shot noise source which is at the input and therefore appears more strongly at the output [4, 5]. It is also possible that in non-linear operation of the transistor, e.g. in an oscillator circuit, the gate is driven up in the forward direction and then a stronger gate current flows. This source is described by the well-known formula for shot noise.

$$\overline{i_G^2} = 2qI_GB \quad (11.16)$$

I_G is the direct current in the gate-source circuit.

As the frequency increases, another noise source appears: induced gate noise. The fluctuating current in the channel is accompanied by potential fluctuations under the gate. The metal layer of the gate and the conductive channel, together with the space charge layer as dielectric, form the gate-source capacitance C_{GS}. The input circuit is capacitively coupled to the channel via this capacitance. A noise voltage v_{CH} in the channel generates a noise current i_{GI} in the input circuit.

$$i_{GI} = j\omega C_{GS}v_{CH} \quad (11.17)$$

We insert the channel noise (11.18) and square

$$v_{CH} = \sqrt{4kTB\frac{1}{g_m}} \quad (11.18)$$

This elementary consideration explains the essential physical effect, but is not a quantitative solution to the relatively complicated calculation. Therefore, similar to the factor P above one introduces a function R.

$$\overline{i_{GI}^2} = 4kTB\frac{\omega^2 C_{GS}^2}{g_m} \times R \tag{11.19}$$

At saturation, this function has the value $R = 0.12$. It follows

$$\overline{i_{GI}^2} \approx \frac{1}{2}kTB\frac{\omega^2 C_{GS}^2}{g_m} \tag{11.20}$$

This is the noise current square of the induced gate noise.

The Correlation

Since the induced gate current is derived directly from the channel noise, it is clear that both sources must be strongly correlated [6]. To calculate the correlation coefficient we need the equations for the currents. From (11.15) we obtain

$$i_D = \sqrt{4kTBg_m P} \tag{11.21}$$

in (11.17) we have the current directly. The unnormalized correlation coefficient (covariance) is

$$i_D i_{GI}^* = i_D\left(-j\omega C_{GS} v_{CH}\right) = -j\omega C_{GS} 4kTB\sqrt{PR} \times C \tag{11.22}$$

C is also a fitting factor. For the JFET is $C = 0.4$.

This gives the correlation coefficient to

$$\rho_Y = -j\frac{\omega C_{GS} 4kT\sqrt{PR}C}{\sqrt{\left(4kTB\frac{\omega^2 C_{GS}^2}{g_m}R\right)\left(4kTBg_m P\right)}} = -jC = -j0.4 \tag{11.23}$$

The influence of the induced gate current increases with ω^2. It therefore only plays a role at high frequencies near the cut-off frequency $\omega_G = g_m/C_{GS}$ of the transistor. The correlation is purely imaginary. Except in the range of highest frequencies this correlation is usually neglected.

Transformation to the Input

The noise sources (11.15) and (11.20) refer to the Y representation of the FET (Figure 11.4a). We saw in Chapter 9 that the ***A***-matrix representation (Figure 11.4b) is more suitable for further calculations and measurement

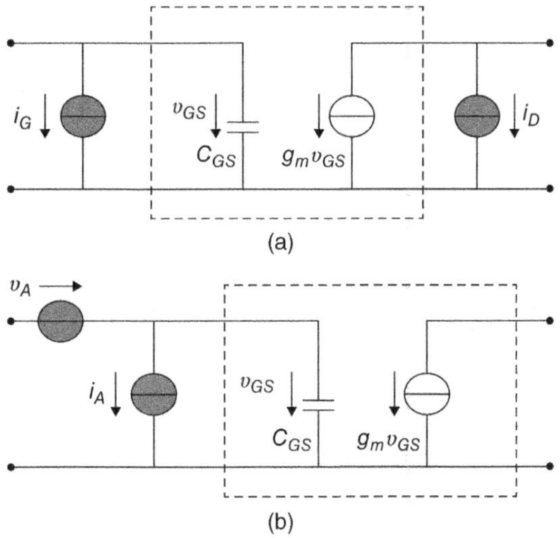

Figure 11.4 Noise model of the intrinsic FET in Y-form (a) and in A-form (b).

techniques. This corresponds also to the information, which – even if not always – can be found in data sheets.

The conversion can be done with the formula (7.2) or with the formalism of the correlation matrices. We write here once again:

$$v_A = -\frac{i_D}{Y_{21}}; \quad 1ei_A = i_{GI} - i_D\frac{Y_{11}}{Y_{21}} \tag{11.24}$$

The $\mathbf{Y}$-matrix of the simplest FET equivalent circuit is

$$Y = \begin{pmatrix} j\omega C_{GS} & 0 \\ g_m & 0 \end{pmatrix} \tag{11.25}$$

With this we obtain

$$v_A = -\frac{1}{g_m}\sqrt{4kTBg_mP}; \quad \overline{v_A^2} = \frac{4kTBP}{g_m} \tag{11.26}$$

and for i_A

$$i_A = j\omega C_{GS}\sqrt{\frac{4kTBR}{g_m}} - j\frac{\omega C_{GS}}{g_m}\sqrt{4kTBg_mP} \tag{11.27}$$

When squaring (11.24), we must take into account that i_{GI} and i_D are correlated. So it applies:

$$\overline{i_A^2} = \overline{i_{GI}^2} - 2\overline{i_{GI}i_D}\frac{Y_{11}}{Y_{21}} + \overline{i_D^2}\left(\frac{Y_{11}}{Y_{21}}\right)^2 \tag{11.28}$$

With (11.22) we receive

$$\overline{i_A^2} = \frac{4kT}{g_m}\omega^2 C_{GS}^2 R - j2\omega C_{GS} 4kT\sqrt{PR}C\frac{j\omega C_{GS}}{g_m} + 4kTg_m P\frac{\omega^2 C_{GS}^2}{g_m^2}$$
$$= \frac{4kT}{g_m}\omega^2 C_{GS}^2 (P + R + 2\sqrt{PR}C) \tag{11.29}$$

this is exactly C_{22}^A of the noise correlation matrix. In this way we have directly performed the transition from the Y-form to the A-form. This is still possible and clearly arranged with this simple structure, but is not recommended for more complicated equivalent circuits. There it is better to apply the formal transformation algorithm of the correlation matrixes and let the computer do the calculation. In our example, however, one can easily see how the correlation of the sources affects squaring and averaging.

The A-correlation matrix is

$$C^A = \frac{4kT}{g_m}\begin{pmatrix} P & -j\omega C_{GS}(P + C\sqrt{PR}) \\ j\omega C_{GS}(P + C\sqrt{PR}) & \omega^2 C_{GS}^2 (P + R + 2C\sqrt{PR}) \end{pmatrix} \tag{11.30}$$

We have thus calculated the two noise sources at the input of the FET, which is now assumed to be noise-free. The numerical factors for the JFET: $P = 0.67$; $R = 0.12$, and $C = 0.4$ are not strictly derived and may have different values in the literature, depending on the accuracy and complexity of the approach. The correlation of the two sources is usually not considered at frequencies below the microwave range. We can calculate it from the elements of the correlation matrix.

$$\rho_A = \frac{C_{12}^A}{\sqrt{C_{11}^A C_{22}^A}} = -j\frac{P + C\sqrt{PR}}{\sqrt{P^2 + PR + 2PC\sqrt{PR}}} = -j0.95 \tag{11.31}$$

It can be seen that the correlation coefficients of the sources in Y representation ($\rho_Y = -j0.4$) and in A representation ($\rho_A = -j0.95$) are naturally different.

We will now describe the direct analysis of the noisy network consisting of the input circuit with the generator resistance R_G. The source resistance R_S is added to the transistor (Figure 11.4). In the Chapters 3 and 6 we had denoted the generator resistance R_S. In calculations with FETs, however, it is usual to use R_S to denote the source path resistance.

The noise voltages or the noise current of the channel in Figure 11.5 are

$$v_G = \sqrt{4kTR_G B}; \quad v_R = \sqrt{4kTR_S B} \tag{11.32}$$

$$v_{IG} = i_{GI} R_G = j\sqrt{\frac{4kTBR}{g_m}}\omega C_{GS} R_G; \quad i_D = \sqrt{4kTBPg_m} \tag{11.33}$$

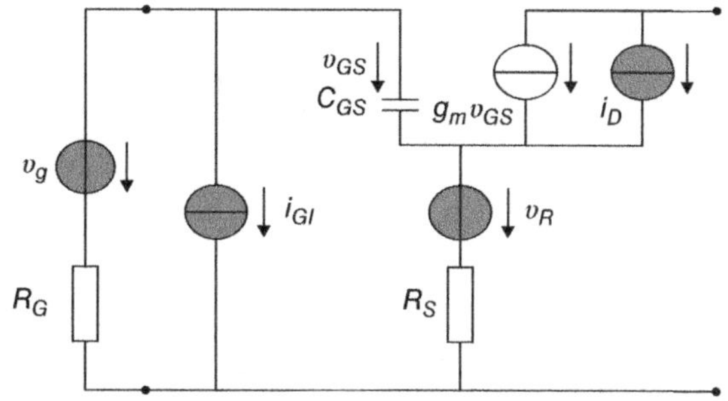

Figure 11.5 Model of intrinsic FET extended by the source resistance and generator at the input.

The short-circuit current in the output, generated by the noisy transistor and the generator resistance R_G, is

$$i_{OT} = g_m v_{GS} + i_D \tag{11.34}$$

The noise voltage v_{GS} at the gate results in

$$v_{GS} = \frac{v_G - v_{IG} - v_R}{1 + j\omega C_{GS}\,(R_G + R_S)} \tag{11.35}$$

We now form the square of i_{OT} over real and imaginary part of (11.34)

$$\overline{|i_{OT}|^2} = \overline{Re(g_m v_{GS} + i_D)^2} + \overline{Im(g_m v_{GS} + i_D)^2} \tag{11.36}$$

as a result of averaging, products of non-correlated variables are eliminated. For example,

$$\overline{v_G v_R} = 0$$

Thus the noise current square in the output circuit is

$$\overline{|i_{OT}|^2} = g_m^2 \frac{v_G^2 + v_{IG}^2 + v_R^2}{1 + \omega^2 C_{GS}^2 (R_G + R_S)^2} + i_D^2 - 2g_m \frac{\overline{v_{IG} i_D}}{1 + \omega^2 C_{GS}^2 (R_G + R_S)^2} \tag{11.37}$$

to obtain the noise factor F, we divide by the contribution $\overline{|i_O|^2}$ which is generated by the generator resistance R_G alone. We get it from (11.37) by setting the sources of the transistor v_{IG}, v_R, and $i_D = 0$

$$\overline{|i_O|^2} = \frac{g_m^2 v_G^2}{1 + \omega^2 C_{GS}^2 (R_G + R_S)^2} \tag{11.38}$$

This results in

$$F = \frac{\overline{|i_{OT}|^2}}{\overline{|i_O|^2}} = 1 + \frac{1}{v_G^2}\left\{ v_R^2 + v_{IG}^2 + \frac{i_D^2}{g_m^2}\left(1 + \omega^2 C_{GS}^2 [R_G + R_S]^2\right) - \frac{2\overline{v_{IG} i_D}}{g_m} \right\} \tag{11.39}$$

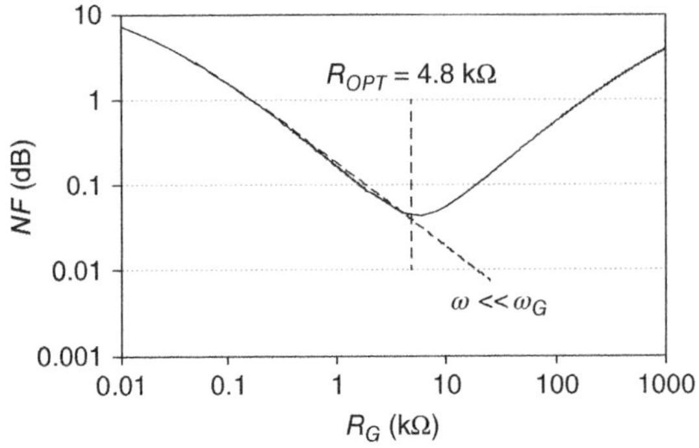

Figure 11.6 Noise figure of the circuit in Figure 11.5 versus the generator resistance R_G. f = 1 MHz. Calculated with (11.40).

Using the numerical values for P and R and ω_G, the final formula is

$$F = 1 + \frac{R_S}{R_G} + \frac{g_m R_G}{8}\left(\frac{\omega}{\omega_G}\right)^2 + \frac{2}{3 g_m R_G}\left\{1 + g_m^2 (R_G + R_S)^2 \left(\frac{\omega}{\omega_G}\right)^2\right\} - 0.58\frac{\omega}{\omega_G} \tag{11.40}$$

To calculate the optimum generator resistance we derive $\partial F/\partial R_G = 0$.

$$\frac{\partial F}{\partial R_G} = \frac{\omega^2 C_{GS}^2}{8 g_m} - \frac{2}{3 g_m R_{OPT}^2} + \frac{2\omega^2 C_{GS}^2}{3 g_m} = 0 \tag{11.41}$$

From this follows:

$$R_{OPT} \approx 0.9\frac{1}{\omega C_{GS}} \tag{11.42}$$

We apply (11.40) to the low-noise JFET 2SK 170 [7]. From the data sheet we take: g_m = 20 mS; C_{GS} = 30 pF. Thus the cut-off frequency $\omega_G/2\pi \approx 100$ MHz. We set R_S = 10 Ω. The result for f = 1 MHz is shown in Figure 11.6. At high generator resistance R_G, above R_{OPT} the induced gate noise is effective.

On the other hand, we can use the manufacturer's data for the noise voltage source (usually e_N (nV/$\sqrt{\text{Hz}}$)) at the input and the gate current and calculate the noise figure. These are v_A = 0.8 nV/$\sqrt{\text{Hz}}$ and i_G = 1 nA. The gate current generates a shot noise i_A = 20 fA/$\sqrt{\text{Hz}}$. With these two sources we calculate the noise factor

$$F = 1 + \frac{v_A^2 + i_A^2 R_G^2}{4kTR_G} \tag{11.43}$$

and $R_{OPT} = v_A/i_A$ = 40 kΩ. As expected, we have good agreement for $R_G < R_{OPT}$, but not for higher values of R_G, because the induced gate noise is neglected (Figure 11.7).

Figure 11.8 shows the frequency dependence of the noise figure resulting from (11.40). Parameter is the generator resistance.

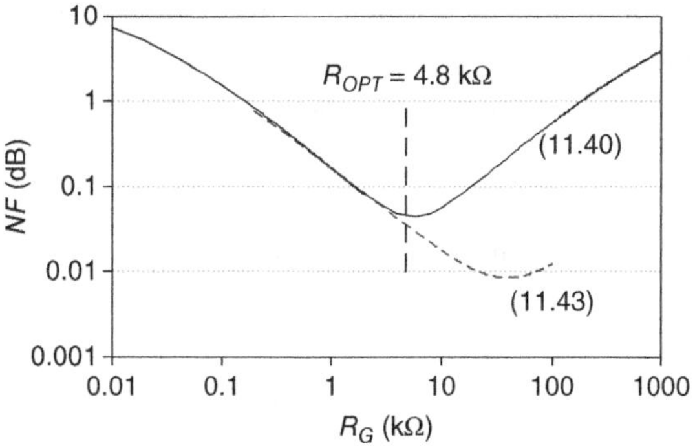

Figure 11.7 Noise figure of the circuit in Figure 11.5 versus the generator resistance R_G. $f = 1$ MHz. Calculated with (11.40) and (11.44).

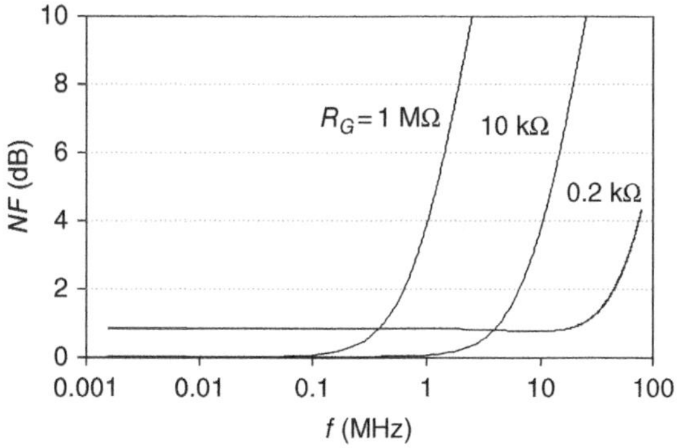

Figure 11.8 Frequency dependence of the noise figure at different R_G (11.40)

Simple Approximations

For low frequencies $\omega \ll \omega_G$, e.g. the acoustic range, a simple approximation of (11.40) is

$$F = 1 + \frac{1}{R_G}\left(R_S + \frac{2}{3g_m}\right) \tag{11.44}$$

Because the transconductance of the JFET is often $g_m < 10$ mS and $R_S < 20\ \Omega$ we can also set

$$F = 1 + \frac{2}{3g_m R_G} \tag{11.45}$$

In this limit case it is also possible to work with the equivalent noise resistor R_{EQ}. It is assumed that all noise of the component is produced by this resistor, which is located in the input circuit between generator and gate. This approach comes from vacuum tube technology and is justified by the fact that the noise current source $\overline{i_A^2}$ is negligible. This source is fed by the shot noise of the gate current, which is usually neglected in Si-JFET and MOSFETs. This also applies to the grid current in vacuum tubes. The other source is the induced gate noise, which also does not play a role at low to medium frequencies. R_{EQ} is derived from the noise voltage source, which is usually described in data sheets as $e_n(\mathrm{nV}/\sqrt{\mathrm{Hz}})$.

$$R_{EQ} = \frac{e_n^2}{4kT} \tag{11.46}$$

in our notation

$$R_{EQ} = \frac{v_A^2}{4kTB} \tag{11.47}$$

One sometimes finds statements that the product of R_{EQ} and g_m for field effect transistors has a constant value $\lambda \approx 0.2 \ldots 1$

$$\lambda = R_{EQ} g_m \tag{11.48}$$

This comes from vacuum tube technology, where for triodes $\lambda \approx 3 \ldots 4$ applies. As Table 11.1 (last column) shows, this is not true for FETs.

From (11.48), however, follows the useful information that the noise figure decreases with increasing g_m, which on the other hand increases with the square root of the drain current. This is shown in Figure 11.9 for the 2N4416 [8]. In contrast to the BJT, it is therefore favorable to select the operating point at high current.

In the lower frequency range we have to consider the $1/f$ noise more than with the bipolar transistor, because it is significantly stronger. We take Figure 11.10 from the data sheet of the 2N 4416.

The approximation with (10.8) and $K = 5.9 \times 10^3\ \mathrm{nV}^2$ and $f_C = 450$ Hz is not very accurate. However, it is justified here because of the small influence on the overall noise. It can be seen that generation recombination processes obviously also take place in the channel with a frequency near 1 kHz.

Table 11.1 Manufacturer's specifications for different JFETs and the R_{EQ} and λ calculated from them.

Type	Technology	$e_n(\mathrm{nV}/\sqrt{\mathrm{Hz}})$	g_m(mS)	$1/g_m(\Omega)$	$R_{EQ}(\Omega)$ (11.47)	λ (11.48)
2N4416	n-Channel JFET	6	5	200	2200	11
2N5460	p-Channel JFET	60	2	500	217k	435
2SK170	n-Channel JFET	0.9	22	45	50	1.1
2SK2394	n-Channel JFET	2	30	33	240	7.1
LSK489	n-Channel JFET	1.8	1	1000	200	0.2
SST201	n-Channel JFET	4	0.6	1700	970	0.5
ECC88	Triode	—	12.5	80	200	3.7

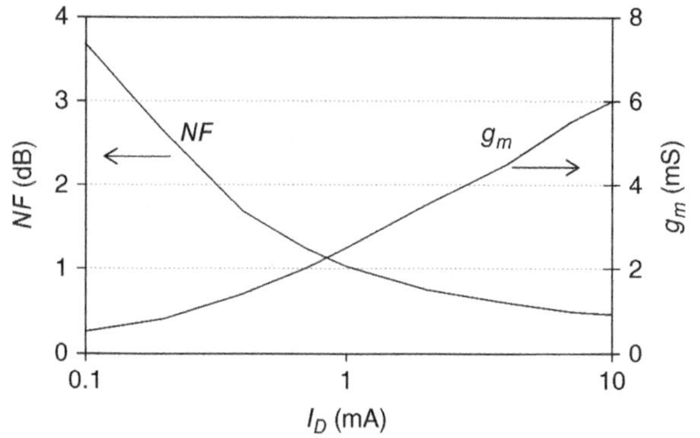

Figure 11.9 Noise figure and transconductance of JFET 2N4416 versus the drain current. Source: Based on Vishay Siliconix [8].

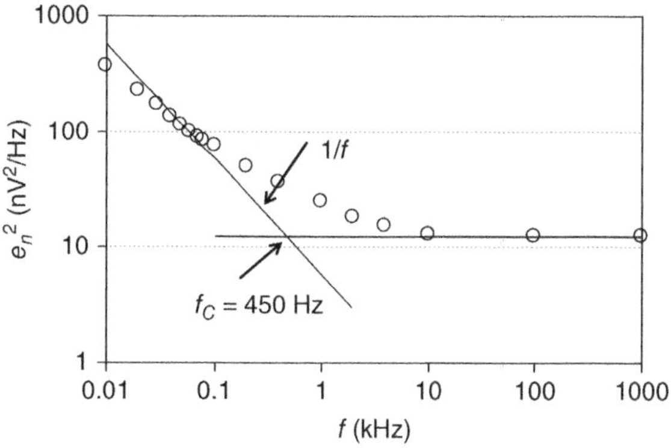

Figure 11.10 Noise voltage square of JFET 2N4416 in the $1/f$ range according to manufacturer's specifications.

Figure 11.11 shows the calculated noise figure for $R_G = 1\,\text{k}\Omega$ in the whole frequency range from the $1/f$-range ($f < 1\,\text{kHz}$) over the range of thermal noise ($1\,\text{kHz} < f < 100\,\text{MHz}$) to RF-frequencies ($f > 100\,\text{MHz}$). Here the theoretical result (lower curve) is compared with the experimentally supported result (upper curve). The theoretical curve results from the assumption of the purely thermal channel noise and the induced gate noise above $f = 1\,\text{MHz}$ (11.40). For the range $f < 1\,\text{MHz}$ we use the thermal channel noise augmented with a term for the $1/f$ noise.

$$\overline{v_n^2} = \frac{2}{3} 4kT \frac{1}{g_m} \left(1 + \frac{f_C}{f}\right) \tag{11.49}$$

As the only information from manufacturer's data we use $f_C = 450\,\text{Hz}$ from Figure 11.10. Then we have with $\overline{i_{GI}^2}$ from (11.20)

$$F = 1 + \frac{\overline{v_n^2} + \overline{i_{GI}^2} R_G^2}{4kTR_G} \tag{11.50}$$

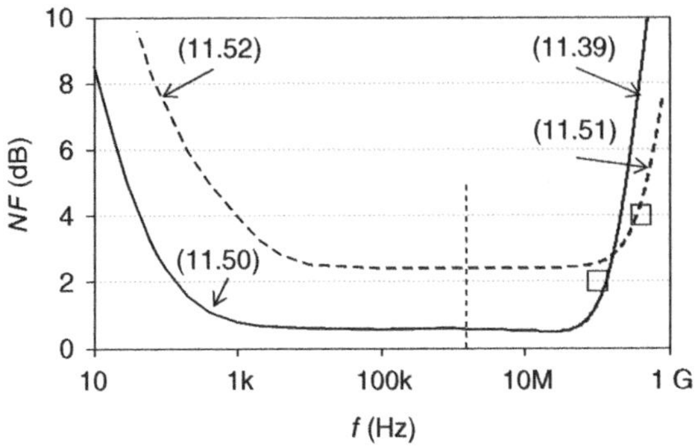

Figure 11.11 Noise figure of the JFET2N4416. Full line: Calculated according to the FET theory. Dotted line: Calculated according to manufacturer's specifications. Symbol: Measurements according to data sheet. $R_G = 1\,\text{k}\Omega$.

In the upper curve based on the data sheet. The spectral noise voltage density outside the $1/f$ range is $\overline{e_n} = 3.5\,\text{nV}/\sqrt{\text{Hz}}$ and the gate current is $I_G = 100\,\text{pA}$. However, the latter does not play a role with $R_G = 1\,\text{k}\Omega$, since it only generates a spectral noise current density due to shot effect of $\overline{i_g} = \sqrt{2qI_G} = 5.7\,\text{fA}/\sqrt{\text{Hz}}$. We use in the RF range

$$F = 1 + \frac{\left(3.5 \times 10^{-9}\right)^2 + \overline{i_{GI}^2} R_G^2}{4kTR_G} \tag{11.51}$$

and for $f < 1\,\text{MHz}$

$$F = 1 + \frac{\overline{e_n^2} + \overline{i_G^2} R_G^2}{4kTR_G} \tag{11.52}$$

with $\overline{e_n(f)}$ from the data sheet (Figure 11.10). The approximation is only needed for an integration over the frequency range to calculate the total noise voltage at the output.

We can see that the purely theoretical approach obviously does not take into account all noise contributions in the thermal range and therefore yields too low noise figures, whereas the approach based on the manufacturer's data yields good results. In the rising range $f > 100\,\text{MHz}$, both calculations agree well with the manufacturer's specifications.

Field Effect Transistors for the Microwave Range (MESFET, HFET)

The operating principle of the field effect transistor was discussed earlier using the example of the JFET [9]. There the gate control electrode is a pn-junction biased in reverse direction. Depending on the gate voltage, the depletion layer expands into the channel and thus controls the drain current. The MESFET is a barrier field effect transistor with a Schottky contact as gate electrode. The material is usually Gallium Arsenide (GaAs), for special applications also Gallium Nitride (GaN) or Indium Phosphide (InP). While the characteristic properties and the mode of operation are very similar to the JFET, there are some special features in the noise properties. The dimensions of the device chips are very small and the typical channel lengths are a few micrometers. The gate lengths $L < 0.5\ \mu\text{m}$. Therefore very high field strengths occur in the channel in longitudinal direction. The simple approach to the thermal behavior of the channel with constant mobility μ, as we did for the JFET (11.3), is no longer applicable here. The electrons move in the range of the saturation drift velocity, or in the range of the negative differential mobility (above 5 kV/cm) in the case of GaAs. Thus they are no longer in thermal equilibrium with the bulk material, and they have an increased electron temperature. However, the term electron temperature is not correct, because it implies a Maxwellian velocity distribution, which is also not present here. In addition, the small spatial dimensions are in the range of the free path, so that physical concepts like mobility and temperature are not applicable. The contribution of these hot electrons to the noise of the channel is summarized under the general term high-field diffusion noise. Its quantitative description is the subject of complicated theories. In practice, fitting parameters are widely used. In the microwave range, in addition to this overthermal noise, the thermal noise of the parasitic resistors also plays an increasing role.

The Pucel Model

Pucel's theory takes into account these particularities of the microwave FET and provides formulae for the four noise parameters, which include three coefficients (K_G, K_C, and K_R) [10]. These can also be used as fitting parameters for modeling certain types. The equivalent circuit (Figure 11.12) of the intrinsic FET includes the drain noise current i_d, the induced gate noise i_{gi} and the thermal noise resistances R_g (gate metal resistance), R_s (source resistance), and R_i (gate–source coupling resistance). In the formalism already discussed in Chapter 7, the following formulae result for the noise sources transformed to the input (Figure 7.3)

The voltage source

$$\overline{v_N^2} = 4kT\left(R_s + R_g + \frac{1 + (\omega C_{gs} R_i)^2}{g_m}\right) \times K_R \tag{11.53}$$

The uncorrelated current source

$$\overline{i_N^2} = 4kT\frac{(\omega C_{gs})^2}{g_m} \times K_g; \quad G_N = \frac{\overline{i_N^2}}{4kT} \tag{11.54}$$

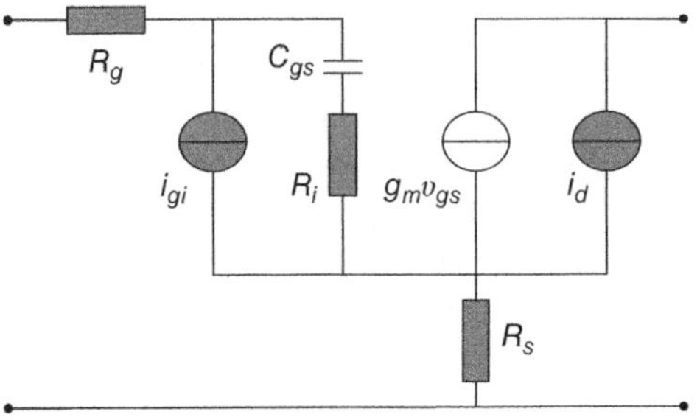

Figure 11.12 Equivalent circuit of a FET according to the Pucel model.

The correlation resistance or conductance

$$Z_{COR} = R_s + R_g + \frac{K_C}{Y_{11}}; \quad Y_{COR} = G_{COR} + jB_{COR} = \frac{1}{Z_{COR}} \tag{11.55}$$

with

$$Y_{11} = \frac{1}{R_s + R_g + R_i + \dfrac{1}{j\omega C_{gs}}} \tag{11.56}$$

This gives us for the four noise parameters

$$R_N = \frac{\overline{v_N^2}}{4kT}$$

$$F_{MIN} = 1 + 2R_N G_{COR} + 2\sqrt{R_N G_N + R_N^2 G_{COR}^2} \tag{11.57}$$

$$G_{OPT} = \sqrt{\frac{G_N}{R_N} + G_{COR}^2}; \quad B_{OPT} = -B_{COR}$$

The coefficients of the Pucel theory for MESFET and HEMT show a current dependence (except K_c), which is not present for P, R, C in the JFET. This is caused by the dominant influence of the high field diffusion noise in the channel.

The numerical values in Figure 11.13 can be seen as initial values for a fitting to a special transistor. For example, we compare the model calculation with the manufacturer's data of the PHEMT Avago VMMK 1225 [11] at the bias point $V_D = 3$ V, $I_D = 20$ mA. The elements of the equivalent circuit according to the data sheet are presented in Table 11.2.

With this data we calculate the noise parameters of the PHEMT. From Figure 11.13 we take the coefficients: $K_R = 0.05$, $K_G = 0.05$, $K_C = 3.0$ at the bias point mentioned above. The result is shown in Figure 11.14.

When comparing the model and measurement in Figure 11.14, it must be considered that the manufacturer's specifications refer to a packaged transistor, whereas the model (Figure 11.12) describes only the intrinsic transistor, i.e. approximately the chip. Nevertheless, with only minor fitting of the coefficients,

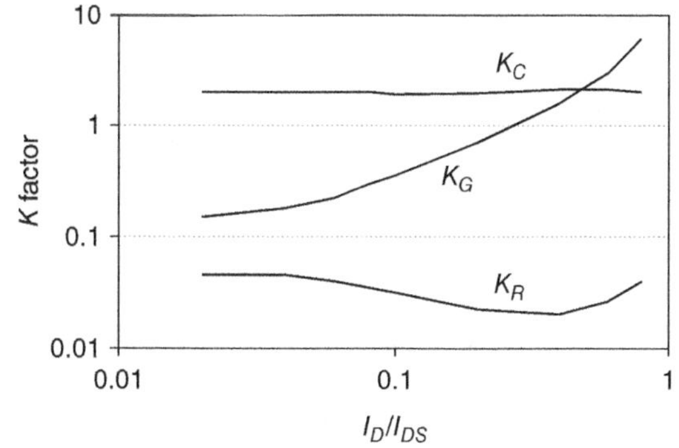

Figure 11.13 Coefficients of the Pucel model versus the drain current normalized to I_{DS}.

Table 11.2 Elements of the PHEMT Avago VMMK 1225 equivalent circuit according to the data sheet.

g_m	τ	C_{GS}	R_I	C_{DG}	C_{DS}	R_{DS}	R_G	R_S
0.13 S	2 ps	0.56 pF	2.2 Ω	0.022 pF	0.11 pF	480 Ω	4.5 Ω	1.7 Ω

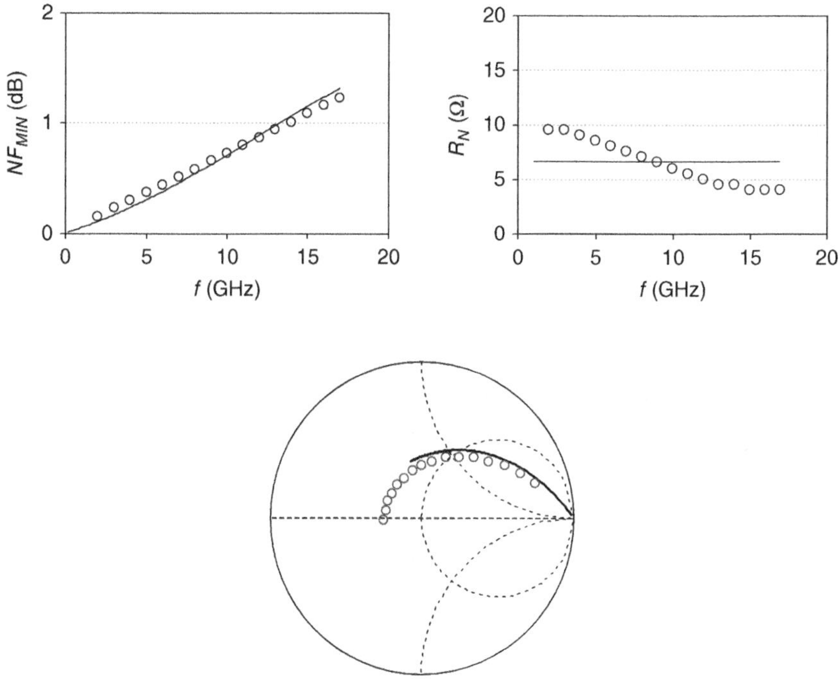

Figure 11.14 Lines: Calculated noise parameters according to the Pucel model for the PHEMT Avago VMMK 1225. Symbols: manufacturer's data. $V_D = 3$ V; $I_D = 20$ mA.

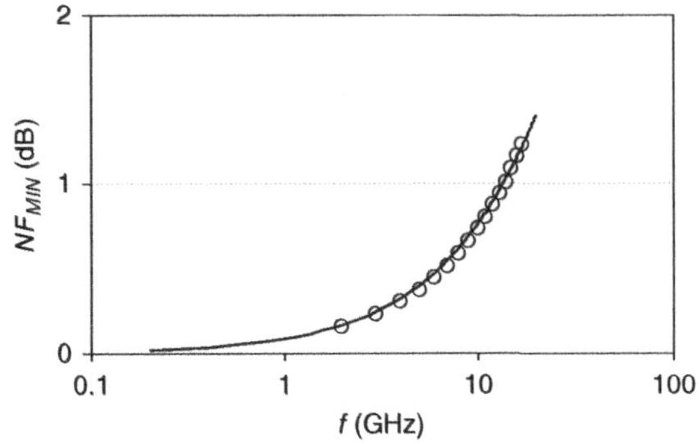

Figure 11.15 Fukui (11.58) for the VMMK 1225.

the modeling of NF_{MIN} is very good. With increasing frequency, the deviations in R_N and Γ_{OPT} are due to the influence of the reactive elements of the package.

The empirical formula of Fukui [12] provides a very useful interrelation with the data of transistor technology.

$$F_{MIN} = 1 + K_L L f \sqrt{g_m (R_I + R_S)} \tag{11.58}$$

Here K_L is a constant, L the gate length in μm, and f the frequency in GHz. Our example transistor VMMK 1225 has $L = 0.25$ μm (Figure 11.15).

The Pospieszalski model

The Pospieszalski noise model of the field effect transistor is very well suited for the practice of modeling MESFET and HEMT [13]. It requires only two additional parameters besides the knowledge of the small-signal equivalent circuit. These are the equivalent temperatures T_g of the internal gate resistance R_i and T_d of the drain resistance R_{ds}. Both can be reliably extracted from the measurement of the four noise parameters in the frequency domain (Figure 11.16).

The noise voltage of the internal gate resistor R_i is thermal with the equivalent temperature T_g. We have in $B = 1$ Hz bandwidth

$$v_{Ri} = \sqrt{4kT_g R_i} \tag{11.59}$$

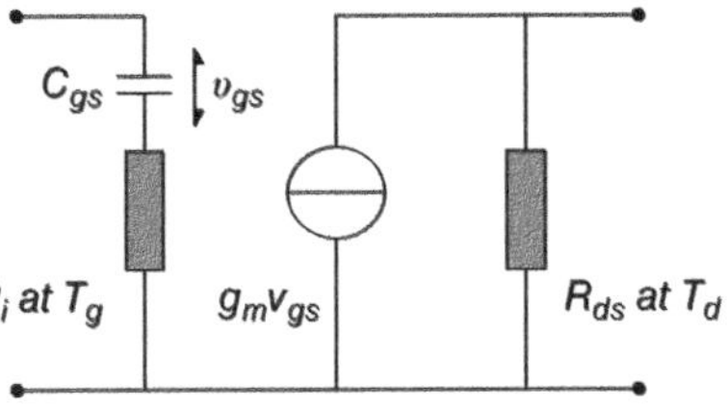

Figure 11.16 The intrinsic transistor in the Pospieszalski model.

This drives the noise current in the short-circuited input circuit i_1, which we write separately for real and imaginary part:

$$i_1 = \frac{v_{Ri}}{R_i + \frac{1}{j\omega C_{gs}}} = \frac{\omega C_{gs} v_{Ri}}{1 + \omega^2 C_{gs}^2 R_i^2} \left(\omega C_{gs} R_i + j1\right) \tag{11.60}$$

its square is

$$\overline{i_1^2} = \frac{\left(\omega^2 C_{gs}^2 v_{Ri} R_i\right)^2}{\left(1 + \omega^2 C_{gs}^2 R_i^2\right)^2} + \frac{\left(\omega C_{gs} v_{Ri}\right)^2}{\left(1 + \omega^2 C_{gs}^2 R_i^2\right)^2} \tag{11.61}$$

with the temperature T_g from (11.59)

$$\overline{i_1^2} = \frac{4kT_g R_i \omega^2 C_{gs}^2}{1 + \omega^2 C_{gs}^2 R_i^2} \tag{11.62}$$

The current i_1 generates at C_{gs} the voltage drop v_{gs}. This voltage in turn drives a current in the short-circuited output which is determined by g_m. In addition, there is the overthermal noise current of the drain–source resistor R_{ds} which is based on the equivalent drain temperature T_d. By definition, gate and drain noise sources are not correlated.

$$i_2 = v_{gs} g_m + i_d \tag{11.63}$$

$$v_{gs} = \frac{i_1}{j\omega C_{gs}}; \quad i_d = \sqrt{4kT_d \frac{1}{R_{ds}}} \tag{11.64}$$

v_{gs} results in

$$v_{gs} = \frac{v_{Ri}}{1 + j\omega C_{gs} R_i} = \frac{v_{Ri}}{1 + \omega^2 C_{gs}^2 R_i^2} \left(1 - j\omega C_{gs} R_i\right) \tag{11.65}$$

and thus for i_2

$$i_2 = \left(\frac{g_m v_{Ri}}{1 + \omega^2 C_{gs}^2 R_i^2} + i_D\right) - j\left(\frac{g_m \omega C_{gs} R_i v_{Ri}}{1 + \omega^2 C_{gs}^2 R_i^2}\right) \tag{11.66}$$

and for the square of i_2

$$\overline{i_2^2} = \frac{4kT_g R_i g_m^2}{1 + \omega^2 C_{gs}^2 R_i^2} + \frac{4kT_d}{R_{ds}} \tag{11.67}$$

Although the noise of the two resistors is uncorrelated, the two currents i_1 and i_2 are partially correlated, since i_2 contains contributions of R_i and R_{ds}. We write the product

$$\begin{aligned} \overline{i_1^* i_2} = &\left\{\frac{\omega C_{GS} v_{RI}}{1 + \omega^2 C_{GS}^2 R_I^2} \left(\omega C_{GS} R_I - j1\right)\right\} \\ &\times \left\{\left(\frac{g_m v_{RI}}{1 + \omega^2 C_{GS}^2 R_I^2} + i_D\right) - j\left(\frac{g_m \omega C_{GS} R_I v_{RI}}{1 + \omega^2 C_{GS}^2 R_I^2}\right)\right\} \end{aligned} \tag{11.68}$$

Mixed products, which comprise of v_{RI} and i_D are zero, because they are uncorrelated.

$$\overline{i_1^* i_2} = -j\frac{1}{\left(1+\omega^2 C_{GS}^2 R_I^2\right)^2}\left(g_m \omega C_{GS} v_{RI}^2 + g_m \omega^3 C_{GS}^3 v_{RI}^2 R_I^2\right) \tag{11.69}$$

The correlation coefficient is purely imaginary

$$\overline{i_1^* i_2} = -j\frac{4kT_G g_m \omega C_{GS} R_I}{1+\omega^2 C_{GS}^2 R_I^2} \tag{11.70}$$

With the formulae (11.62), (11.67), and (11.70), we have at the same time the elements of the Y noise correlation matrix $\boldsymbol{C}^Y$ of the intrinsic transistor according to Figure 11.15.

We now want to apply the Pospieszalski model to a microwave field effect transistor. For comparison with the model we rely on manufacturer's data. As above we take the E-PHEMT VMMK-1225, for which a very detailed data sheet is available. It is a low noise pseudomorphic HEMT in enhancement technology, i.e. the gate voltage has the same polarity as the drain voltage. The gate length is 0.25 μm. The chip is located in a special housing without lead wires. The prerequisite for a noise model is the best possible adaptation of the small-signal equivalent circuit to the measured S-parameters. The equivalent circuit (Figure 11.17) corresponds to the ADS model given in the data sheet. Only the source inductance L_S and the external feedback capacity C_{pgd} are neglected. In extension of the model shown above in the Pucel model we consider here all elements of the packaged transistor specified by the manufacturer. First we make a slight correction of the elements of the equivalent circuit, based on the comparison of measured and calculated S-parameters.

With the values in Table 11.3 one obtains adequate agreement with the S-parameters from the data sheet. It is not a question here of exact agreement in the entire frequency range, as is usually only achieved with one's own measurements, but of demonstrating the calculation process for the noise

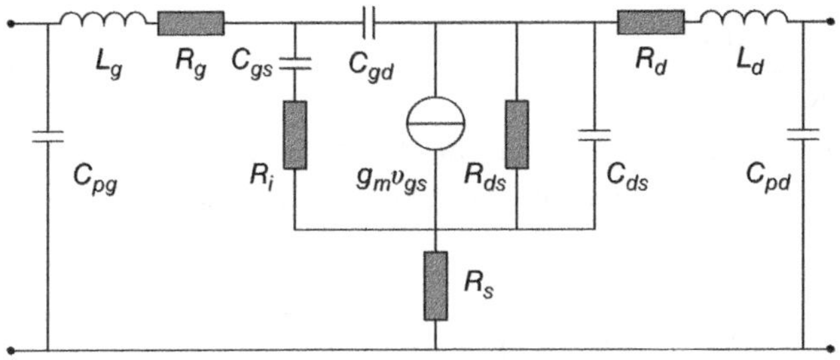

Figure 11.17 Extended equivalent circuit of VMMK1225 for noise analysis according to the Pospieszalski model.

Table 11.3 Values of the elements in Figure 11.17 optimized for $f = 2\text{–}26$ GHz.

Intrinsic

Element	g_m	R_i	R_{ds}	C_{gs}	C_{ds}	C_{gd}	τ
Value	130 mS	2 Ω	480 Ω	0.5 pF	0.05 pF	0.04 pF	2 ps

Parasitic

Element	R_g	R_s	R_d	L_g	L_d	C_{pg}	C_{pd}
Value	4 Ω	1 Ω	1.9 Ω	0.35 nH	0.2 nH	0.05 pF	0.05 pF

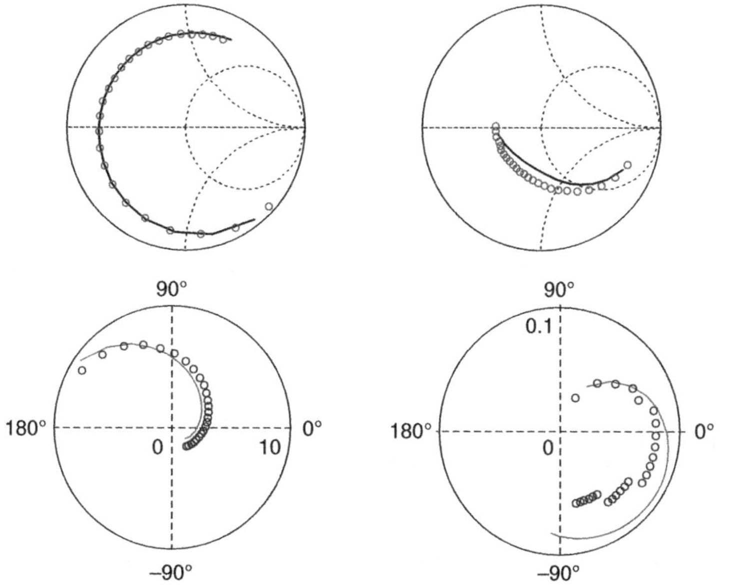

Figure 11.18 S-parameters of VMMK1225 $f = 2\text{–}26$ GHz. Full lines: Calculated with Table 11.3. Symbols: manufacturer's data. Arrangement like the matrix elements.

model. For this purpose, the manufacturer's specifications were slightly modified (Figure 11.18).

The step-by-step procedure for noise modeling is shown schematically in Figure 11.19. We start with the intrinsic transistor (network *I*), whose Y-correlation matrix we have with the formulae (11.62), (11.67), and (11.70).

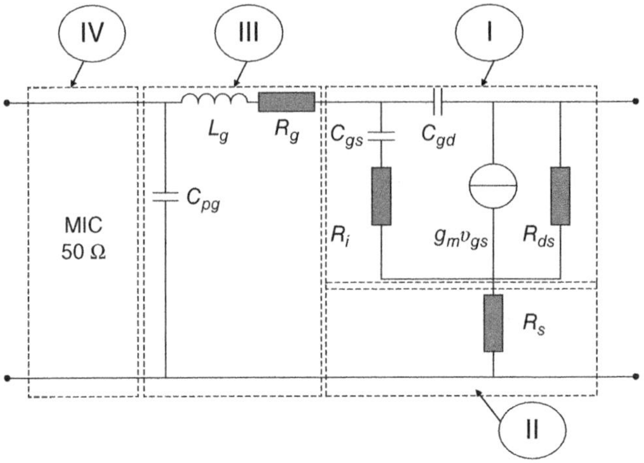

Figure 11.19 Splitting the equivalent circuit Figure 11.17 into the networks I–IV to apply the correlation matrices,

We start with the $\boldsymbol{C}_\mathrm{I}^\mathrm{Y}$-matrix of the intrinsic transistor

$$\boldsymbol{C}_I^Y = \frac{4k}{1+(\omega C_{gs}R_i)^2}\begin{pmatrix} T_g R_i (\omega C_{gs})^2 & jT_g R_i g_m \omega C_{gs} \\ -jT_g R_i g_m \omega C_{gs} & T_g R_i g_m^2 + T_d g_{ds}\left[1+(\omega C_{gs}R_i)^2\right] \end{pmatrix} \tag{11.71}$$

The gate–drain feedback capacitance C_{gd} and the drain–source capacitance C_{ds} are not included, because they do not generate noise. However, they are taken into account in the Y network matrix. This is as follows:

$$\boldsymbol{Y_I} = \begin{pmatrix} j\omega\left[\dfrac{C_{gs}}{1+j\omega C_{gs}R_i} + Cgd\right] & -j\omega C_{gd} \\ \dfrac{g_m exp(-j\omega\tau)}{1+j\omega C_{gs}R_i} & j\omega\left(C_{gd}+C_{ds}\right)+g_{ds} \end{pmatrix} \tag{11.72}$$

Here the drain conductance $g_{ds} = {}^1/_{R_{ds}}$ is used.

The first step in completing the noise model is the addition of the source network II, which here consists only of the source resistor R_s. Since it is a series connection, both the Z-network matrices and the Z-correlation matrices are added. The Z-matrices of network II are

$$\boldsymbol{Z_{II}} = R_s\begin{pmatrix} 1 & 1 \\ 1 & 1 \end{pmatrix}; \quad \boldsymbol{C_{II}^Z} = 4kTR_s\begin{pmatrix} 1 & 1 \\ 1 & 1 \end{pmatrix} \tag{11.73}$$

The $\boldsymbol{Z}$-matrix of the intrinsic network I results from the transformation of (11.72)

$$\boldsymbol{Z_I} = \frac{1}{DY_I}\begin{pmatrix} Y_{I22} & -Y_{I12} \\ -Y_{I21} & Y_{I11} \end{pmatrix}; \quad DY_I = Y_{I11}Y_{I22} - Y_{I12}Y_{I21} \tag{11.74}$$

For the transformation of the Y-correlation matrix of the intrinsic transistor (11.71) into the Z-correlation matrix the matrix equation is used

$$\boldsymbol{C}_I^Z = \boldsymbol{T}_I \boldsymbol{C}_I^Y \boldsymbol{T}_I^H \tag{11.75}$$

The transformation matrix is: $\mathbf{T}_I = \mathbf{Z}_I$. There is $\mathbf{T}_I^H$ again the Hermitic Matrix to $\mathbf{T}_I$ (conjugate complex and rows and columns interchanged). We obtain the Z correlation matrix of the two networks I and II by adding (11.73) and (11.75)

$$\boldsymbol{C}_{I,II}^Z = \boldsymbol{C}_I^Z + \boldsymbol{C}_{II}^Z \tag{11.76}$$

The corresponding Z-network matrix is also obtained simply by adding

$$\boldsymbol{Z}_{I,II} = \boldsymbol{Z}_I + \boldsymbol{Z}_{II} \tag{11.77}$$

The next step is to add the Gate network III, which is cascaded with the network I + II that we just obtained. We now have to multiply the two chain matrices with each other to get the new network matrix. For the correlation matrices the rule (11.84) applies. First the $\boldsymbol{Z}$-matrix (11.77) is transformed into the $\boldsymbol{A}$-matrix.

$$\boldsymbol{A}_{I,II} = \frac{1}{Z_{I,II;21}} \begin{pmatrix} Z_{I,II;11} & DZ_{I,II} \\ 1 & Z_{I,II;22} \end{pmatrix}; \quad DZ_{I,II} = Z_{I,II;11} Z_{I,II;22} - Z_{I,II;12} Z_{I,II;21} \tag{11.78}$$

To convert the correlation matrix from the Z-form (11.76) to the A-form we need the transformation matrix

$$\boldsymbol{T}_{I,II} = \begin{pmatrix} 1 & -A_{I,II;11} \\ 0 & -A_{I,II;21} \end{pmatrix} \tag{11.79}$$

This leads us to the A-correlation matrix of networks I and II, which is needed for this step. The matrix equation is

$$\boldsymbol{C}_{I,II}^A = \boldsymbol{T}_{I,II} \boldsymbol{C}_{I,II}^Z \boldsymbol{T}_{I,II}^H \tag{11.80}$$

Now we have to calculate the A-matrices of gate network III. The network matrix is

$$\boldsymbol{A}_{III} = \begin{pmatrix} 1 & R_g + j\omega L_g \\ j\omega C_{pg} & (1 - \omega^2 L_g C_{pg}) + j\omega L_g C_{pg} \end{pmatrix} \tag{11.81}$$

We have to calculate the correlation matrix of III via the intermediate step of the $\boldsymbol{Y}$-matrix. As a reminder: Although $\mathbf{C}^Y = 4kT\,Re(Y)$ and $\mathbf{C}^Z = 4kT\,Re(Z)$ but this does not apply to the $\boldsymbol{A}$-matrix.

$$\boldsymbol{C}_{III}^Y = 4kT\,Re\left(\boldsymbol{Y}_{III}\right) = \frac{4kTR_g}{R_g^2 + \left(\omega L_g\right)^2} \begin{pmatrix} 1 & -1 \\ -1 & 1 \end{pmatrix} \tag{11.82}$$

The transformation $\mathbf{C}_{III}^Y \to \mathbf{C}_{III}^A$ is performed again according to the scheme $\mathbf{C}_{III}^A = \mathbf{T}_{III} \mathbf{C}_{III}^Y \mathbf{T}_{III}^H$ with the transformation matrix

$$\boldsymbol{T}_{III} = \begin{pmatrix} 0 & A_{III;12} \\ 1 & A_{III;22} \end{pmatrix} \tag{11.83}$$

We now have all matrices together and can calculate the A-correlation matrix of the combination of networks I, II, and III according to Figure 11.19. This combination should describe the noise behavior of the transistor. We now use the index T to avoid confusion with the increasing number of indices.

$$\mathbf{C}_T^A = \mathbf{A}_{III}\mathbf{C}_{I,II}^A\mathbf{A}_{III}^H + \mathbf{C}_{III}^A \tag{11.84}$$

To model the noise behavior we have used the two temperatures T_g and T_d as free parameters. We had adapted the elements of the small signal equivalent circuit (Table 11.3) to the S-parameters from the data sheet of VMMK-1225. For comparison with the noise parameters given there we calculate them in the known way from the elements of the matrix $\mathbf{C}_T^A$ (Figure 11.20).

$$Y_{OPT} = \frac{1}{C_{T;11}^A}\sqrt{C_{T;11}^A C_{T;22}^A - \left[Im\left(C_{T;12}^A\right)\right]^2} + jIm\left(\frac{C_{T;12}^A}{C_{T;11}^A}\right) \tag{11.85}$$

$$R_N = \frac{C_{T;11}^A}{4kT}; \quad F_{MIN} = 1 + \frac{Re\left(C_{T;12}^A\right) + Re\left(Y_{OPT}\right)C_{T;11}^A}{2kT} \tag{11.86}$$

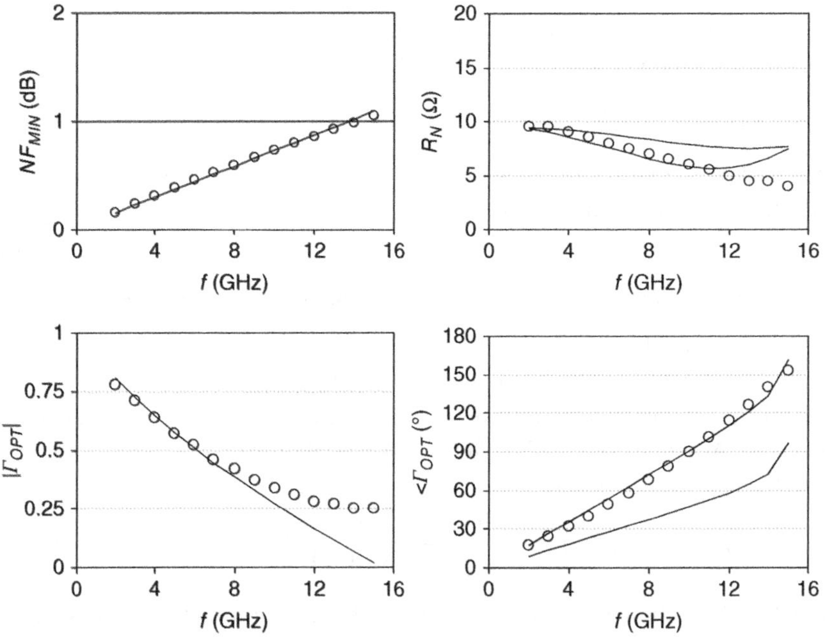

Figure 11.20 Pospieszalski model after Figure 11.19. Full Lines: Calculated $T_g = 500$ K; $T_d = 2500$ K. Symbols: Manufacturer's data. Upper curve in the $\angle\Gamma_{OPT}$-diagram after correction with MIC.

We get these results with $T_g = 500$ K; $T_d = 2500$ K. The angle of $\angle\Gamma_{OPT}$ is not well modeled. We apply a correction length at the input, i.e. a very short section of a loss-free 50 Ω microstrip line on an alumina substrate. Its parameters are: substrate thickness $h = 0.625$ mm; conductor width $w = 0.6$ mm; $\varepsilon_{eff} = 6.7$. According to the well-known scheme, the $\boldsymbol{A}$-matrices are added, because the line section and the transistor are cascaded.

The $\boldsymbol{A}$-matrix of the transmission line is

$$\boldsymbol{A}_{IV} = \begin{pmatrix} cos(\beta L) & j50sin(\beta L) \\ \frac{j}{50}sin(\beta L) & cos(\beta L) \end{pmatrix} \tag{11.87}$$

With $\beta = {}^{\omega\sqrt{\varepsilon_{eff}}}/_{c}$, $c = 3 \times 10^{10}$ cm/s, and L is the physical length of the line, here by adapting to $\angle\Gamma_{OPT}$ we set $L = 0.7$ mm. A very short line section, which may occur as an uncertainty of the reference plane during noise measurement. Since the line is assumed to be loss-free, it does not provide any noise contribution. The new correlation matrix thus results in

$$\boldsymbol{C}^A = \boldsymbol{A}_{IV}\boldsymbol{C}_T^A\boldsymbol{A}_{IV}^H \tag{11.88}$$

Discussion of the Results

We have shown the successive extension of the noise equivalent circuit from the intrinsic transistor to the parasitic resistors on the chip to the outer reactive elements of the package. Calculating with correlation matrices is the method of choice, as it provides a very good overview of the individual steps. It cannot be the goal here to achieve complete agreement with published data without own measurements. However, one can see that the adaptation of the model to measurement results should be carried out in plausible agreement with the physical processes in the transistor. This is more likely to be the case with the Pospieszalski model than with the Pucel model with its three relatively freely available parameters [10].

There is a plurality of other models, which, depending on the application and frequency range, are quite simple. For example, the Gupta model [14, 15] only requires an equivalent drain temperature, which is often sufficient. A comprehensive description can be found, for example, in Vasilescu [16] and in [17–19] and [20].

Criteria for Noise Data

From the basic model (Chapter 7) of the noise of a two-port device discussed earlier, criteria can be derived which measured noise parameters, e.g. of a transistor,

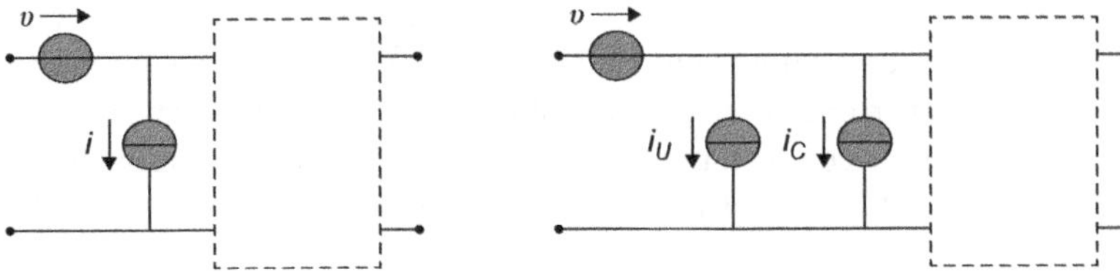

Figure 11.21 Noise sources at the input of the two-port to derive the criteria for the model parameters.

must fulfill. The correlation of voltage and current source is the decisive quantity. We write here once again the equations according to Figure 11.21 for the bandwidth $B = 1$ Hz.

$$\overline{v^2} = 4kTR_n; \quad \overline{i^2} = 4kTG_n; \quad \overline{i_U^2} = 4kTG_U; \quad i_C = Y_{COR}v \tag{11.89}$$

$$Y_{COR} = G_{COR} + jB_{COR} = \rho\sqrt{\frac{G_n}{R_n}}; \quad G_U = G_n\left(1 - |\rho|^2\right) \tag{11.90}$$

In the literature [21, 22] one can find criteria that must be fulfilled by noise data, simply because they refer always to two-ports.

We start from the known formulas:

$$F_{MIN} = 1 + 2R_nG_{COR} + 2\sqrt{R_nG_U + R_n^2G_{COR}^2} \tag{11.91}$$

$$G_{OPT} = \sqrt{\frac{G_U}{R_n} + G_{COR}^2} \tag{11.92}$$

We look at the range $0 \leq \rho \leq 1$ of the possible correlation of voltage and current source with real ρ. With (11.90) inserted in (11.91) and (11.92)

$$F_{MIN} = 1 + 2R_n\rho\sqrt{\frac{G_n}{R_n}} + 2\sqrt{R_nG_n\left(1 + |\rho|^2\right) + \left(R_n\rho\sqrt{\frac{G_n}{R_n}}\right)^2} \tag{11.93}$$

$$G_{OPT} = \sqrt{\frac{G_n\left(1 - |\rho|^2\right)}{R_n} + \left(\rho\sqrt{\frac{G_n}{R_n}}\right)^2} \tag{11.94}$$

F_{MIN}, R_n, and G_{OPT} are measured values which in any case must meet the condition in (11.93) and (11.94)

$$1 \leq \frac{F_{MIN} - 1}{2R_nG_{OPT}} \leq 2 \tag{11.95}$$

We denote the value of the fraction $\kappa 1$.

The equal sign applies to $\rho = 0$, which can be easily seen by inserting it.

$$F_{MIN} = 1 + 2\sqrt{R_n G_n} = 1 + 2R_n G_{OPT} \tag{11.96}$$

for $\rho = 1$ results analogously

$$F_{MIN} = 1 + 4R_n G_{OPT} \tag{11.97}$$

Pospieszalski [21] specifies a temperature criterion using the dimensionless quantity N (no noise power)

$$N = G_n R_{OPT} \tag{11.98}$$

As is well known, the noise temperature T_R and the noise factor F have the relationship

$$T_R = (F - 1)T_0; \quad T_0 = 290\,\text{K} \tag{11.99}$$

With the quantity N, the noise temperature as a function of the source impedance Z_g is

$$T_R = T_{MIN} + NT_0 \frac{\left|Z_g - Z_{OPT}\right|^2}{Re\left(Z_g\right) R_{OPT}} \tag{11.100}$$

For measured values N, T_{MIN}, and Z_{OPT} the following must apply

$$1 \leq \frac{4NT_0}{T_{MIN}} \leq 2 \tag{11.101}$$

We denote the value of this fraction $\kappa 2$.

Figure 11.22 shows the plot of (11.95) and (11.101) versus $|\rho|$.

$\kappa 1$ has the advantage that the measured values are directly included. G_{OPT} results from Γ_{OPT}. If we compare the manufacturer's specifications in Figure 11.23, we see that the chip TGF 2018 [23] better fulfills the criteria

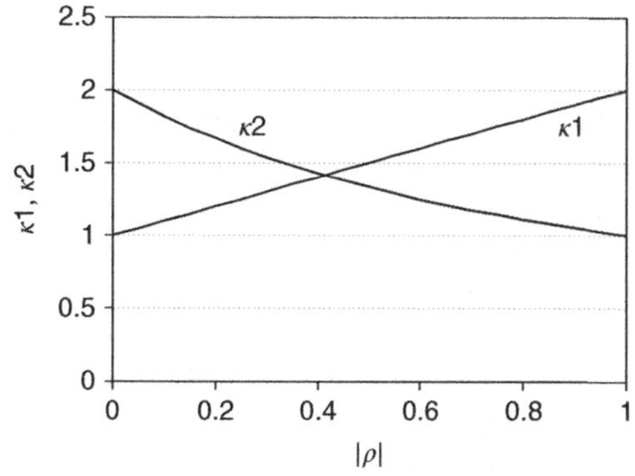

Figure 11.22 The criteria $\kappa 1$ and $\kappa 2$ versus the correlation coefficient $|\rho|$.

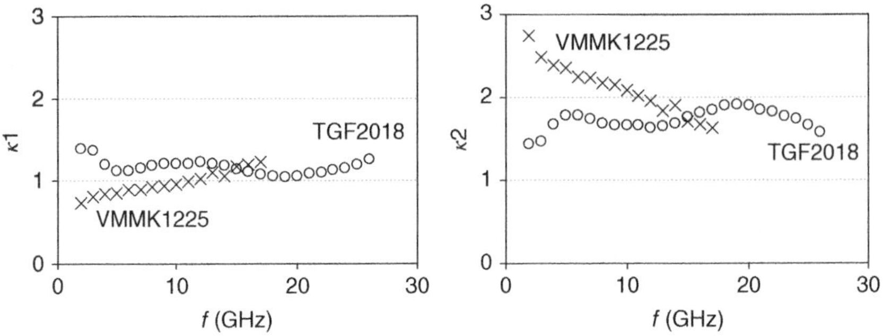

Figure 11.23 Application of criteria (11.95) and (11.101) to manufacturer's data of TGF2018 and VMMK1225.

than the packaged transistor VMMK 1225. The reasons are probably the many uncertainties associated with noise measurements on the highly mismatched objects. We will come back to this in Chapter 14. One indication was already the shift of the reference plane in the model (11.87).

The quantity N was introduced by Lange [24], there in the form $N = R_n G_{OPT}$. Using this the fundamental equation (7.14) can be written:

$$F = F_{MIN} + N \frac{\left| Y_g - Y_{OPT} \right|^2}{G_g G_{OPT}} \tag{11.102}$$

Exactly the quantity N describes the increase of the noise factor, if a circuit is not in noise matching ($Y_g = Y_{OPT}$) state. Furthermore, one has the advantage that N and F_{MIN} are fundamental characteristic values of the intrinsic transistor and are not changed by networks of the package.

A further criterion for measured noise data that are self-consistency is given by W. Wiatr [25]:

$$R_n \geq \frac{F_{MIN} - 1}{4G_{OPT}}; \quad F_{MIN} - 1 \leq 4N; \quad T_{MIN} \leq 4NT_{MIN} \tag{11.103}$$

An example of the R_n criterion is shown in Figure 11.24. The manufacturer's data (MwT LN240, 2.5 V; 20 mA) fulfill the inequality (11.103).

In summary, it should be emphasized that the noise modeling examples in this chapter were not intended to achieve the closest possible agreement. Rather, the aim was to show how the models are used and how the calculations are performed in detail. It becomes apparent very quickly that during the simulation for a circuit design, own measurements of the components are very advantageous, since one can support these with model calculations. In this way, a coherent model of the component can best be developed. A good example can be found in [26], where a comprehensive transistor model with all network elements and internal

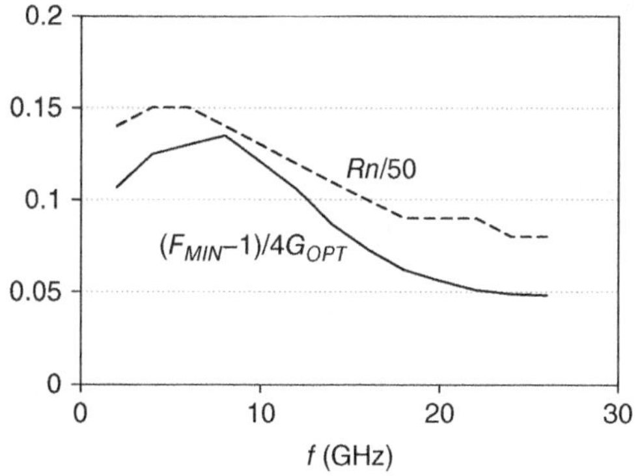

Figure 11.24 Wiatr's criterion R_n for the PHEMT LN240 (data sheet) is fulfilled for all frequencies.

noise sources is created based on measurements of the S-parameters and the noise parameters. In the lower gigahertz range a complete noise characterization is usually not useful, because simplified models are sufficient here. Especially since the usual measurement technique is susceptible to errors due to the strong mismatch of the FET [27, 28].

References

1 Müller, R. (1990). *Rauschen*. Springer Verlag.

2 Beneking, H. (1971). *Praxis des elektronischen Rauschens*. Bibliographisches Institut Mannheim.

3 van der Ziel, A. (1986). *Noise in Solid State Devices and Circuits*. New York: Wiley.

4 Heymann, P., Kantelberg, G., and Prinzler, H. (1989). Gate–drain breakdown and microwave noise of GaAs-MESFETs. *Frequenz* 43: 112–115.

5 Prinzler, H. and Heymann, P. (1992). Intrinsic noise sources of GaAs field effect transistors: theory and experiments in the 10 MHz–12 GHz frequency range. *Frequenz* 46: 53–59.

6 Schiek, B., Rolfes, I., and Siweris, H.-J. (2006). *Noise in High-Frequency Circuits and Oscillators*. Wiley Interscience.

7 Toshiba (1997). 2SK170 Silicon N-Channel Junction Type. Datasheet 1997.

8 Vishay Siliconix (2001). 2N4416 N-Channel J-FET. Datasheet 2001.

9 Kellner, W. and Kniepkamp, H. (1985). *GaAs-Feldeffekttransistoren*. Springer Verlag.

10 Pucel, R.A., Haus, H.A., and Statz, H. (1975). Signal and noise properties of gallium arsenide microwave field effect transistors. In: *Advances in Electronics and Electron Physics*, vol. 38 (ed. Elsevier Inc.), 195–265. New York: Academic Press.

11 Avago Technologies (2014). VMMK-1225 0.5 to 26GHz Low Noise E-PHEMT. Datasheet 2014.

12 Fukui, H. (1979). Optimal noise figure of microwave GaAs MESFETs. *IEEE Trans. Electron Devices* 25: 1032–1037.

13 Heymann, P., Rudolph, M., Prinzler, H. et al. (1999). Experimental evaluation of microwave field effect transistor models. *IEEE Trans. Microwave Theory Tech.* 47: 156–163.

14 Gupta, M.S., Pitzalis, O., Rosenbaum, S.F., and Greiling, P.T. (1987). Microwave noise characterization of GaAS-MESFETs: evaluation by on-wafer, low frequency output noise current measurements. *IEEE Trans. Microwave Theory Tech.* 35: 1208–1217.

15 Gupta, M.S. and Greiling, P.T. (1988). Microwave noise characterization of GaAs-MESFETs: determination of extrinsic noise parameters. *IEEE Trans. Microwave Theory Tech.* 36: 745–751.

16 Vasilescu, G. (2005). *Electronic Noise and Interfering Signals*. Springer Verlag.

17 Danneville, F., Happy, H., Dambrine, G. et al. (1994). Microscopic noise modeling and macroscopic noise modeling: how good a connection? *IEEE Trans. Electron Devices* 41: 779–786.

18 Danneville, F. (2010). Microwave noise and FET devices. *IEEE Microwave Mag.* 11: 53–60.

19 Pospieszalski, M.W. (2010). Interpreting transistor noise. *IEEE Microwave Mag.* 11: 61–69.

20 Heymann, P., Rudolph, M., Prinzler, H., Doerner, R., Klaproth, L. and Böck, G. (1999). Experimental evaluation of microwave field effecttransistor models. *IEEE Trans. Microwave Theory Tech.* 47: 156–163.

21 Pospieszalski, M.W. (1989). Modeling of noise parameters of MESFETs and MODFETs and their frequency and temperature dependence. *IEEE Trans. Microwave Theory Tech.* 37: 1340–1350.

22 Maas, S.A. (2005). *Noise in Linear and Nonlinear Circuits*. Artech House.

23 TriQuint (2013). TGF2018 180μm Discrete GaAs pHEMT. Datasheet 2013.

24 Lange, J. (1967). Noise characterization of linear two-ports in terms of invariant parameters. *IEEE J. Solid-State Circuits* 2: 37–40.

25 Wiatr, W. (1994). Accuracy verification of a technique for noise and gain characterization of two-ports. *Proc. MIKON* 1: 525–529.

26 Lee, S., Webb, K.J., Tilak, V., and Eastman, L.F. (2003). Intrinsic noise equivalent-circuit parameters for AlGaN/GaN HEMTs. *IEEE Trans. Microwave Theory Tech.* 51: 1567–1577.

27 Heymann, P., Doerner, R., and Prinzler, H. (1997). Improved measurement procedure for extremely low noise figures of FETs in the frequency range below 3 GHz. In: *49th ARFTG Conf. Dig.* (ed. IEEE), 161–170.

28 Heymann, P. and Prinzler, H. (1992). Improved noise model for MESFETs and HEMTs in lower GHz-frequency range. *Electron. Lett.* 28: 611–612.

12

Theory of Noise Measurement

Measurements of Two-Ports

First of all, let's look again at the definitions of the noise factor F, which are most important for the different measurement methods [1].

1 The noise factor is the quotient of the signal to noise ratio at the input and the signal to noise ratio at the output.

$$F = \frac{S_{IN}/N_{IN}}{S_{OUT}/N_{OUT}} \geq 1 \quad (12.1)$$

2 The noise power N_{OUT} available at the output of the two-port in units of kT_0 is normalized to the equivalent noise bandwidth B and the power gain G.

$$F = \frac{N_{OUT}}{kT_0} \frac{1}{BG} \quad (12.2)$$

3 The available signal power of a source at the input, in units of kT_0 and normalized to the bandwidth, produces a signal to noise ratio of 1 at the output.

$$F = \frac{S_{IN}}{kT_0} \frac{1}{B} \quad (12.3)$$

4 The noise factor is the ratio of the noise power present at the output of the two-port to that which would be present there if the two-port were noise-free and had only gain (or attenuation).

Simple conversion shows that (12.1)–(12.3) all represent the same context in a slightly different form. Assuming (12.2), we need an accurate measurement of the total noise power at the output N_{OUT}, the power gain G, and the equivalent noise bandwidth B. The power gain requires a separate measurement and makes the procedure more difficult.

A Guide to Noise in Microwave Circuits: Devices, Circuits, and Measurement, First Edition.
Peter Heymann and Matthias Rudolph.

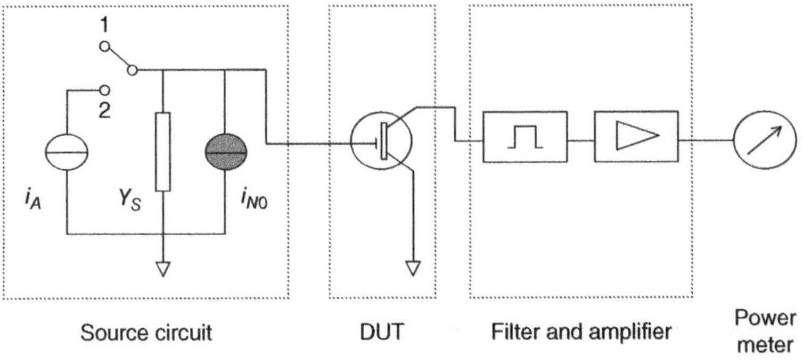

Figure 12.1 Basic configuration for noise measurement of a two-port, here a transistor. The current source i_A provides a sinusodial or a noise signal.

In practice, noise measurement is based on Figure 12.1 [2].

For reasons of clarity, we do not denote pure noise power with N here, but use P for mixed power. At the output we measure P_1, if the additional power $P_{IN} = 0$ (switch position 1)

$$P_1 = GkT_0B + GF_ZkT_0B = GFkT_0B \tag{12.4}$$

It is in any case a pure noise power, no matter whether the additional source is a signal or a noise generator.

If we switch the additional power to $P_{IN} > 0$ (switch position 2), we get at the output

$$P_2 = G\left(kT_0B + P_{IN}\right) + GF_ZkT_0B = GFkT_0B + GP_{IN} \tag{12.5}$$

When using the signal generator, a mixed signal is produced at the output. It consists of the monochromatic signal and the broadband noise. The display unit should add both powers linearly regardless of their nature.

We can solve (12.4) for the gain

$$G = \frac{P_1}{kT_0BF} \tag{12.6}$$

Inserted in (12.5)

$$P_2 = P_1 + \frac{P_1P_{IN}}{kT_0BF} \tag{12.7}$$

When solved for F, the result is

$$F = \frac{P_{IN}}{kT_0B}\frac{1}{{}^{P_2}\!/_{P_1} - 1} \tag{12.8}$$

This formula is the basis for the various measuring methods. Depending on the quality standard of the measurement and depending on the equipment available, two methods can be used in principle.

1 One keeps the input power ratio constant, i.e. P_{IN} is constant and measures the ratio of the output power P_2/P_1.
2 One keeps the output power ratio constant, e.g. 3 dB and varies P_{IN} until this value is reached.

The additional power of the source P_{IN} required for measurement can be a sinusoidal signal from a signal generator or the noise signal from a noise generator. In the first case it is called the signal generator method; in the second case it is called the noise generator method.

This source is represented in Figure 12.1 by the current source i_A. From a signal generator we have the available additional power

$$P_{IN} = \frac{\overline{|i_A|^2}}{4Re(Y)} \tag{12.9}$$

If this source is a noise generator with the equivalent noise temperature T_E, we have

$$P_{IN} = kT_E B \tag{12.10}$$

The admittance Y of the source should be variable if a complete noise characterization of the DUT is to be achieved. This can be done by adding *RLC* networks to the source, in the microwave range by inserting an impedance tuner.

The thermal noise power of the real part of Y is always present in both switch positions:

$$P_{N0} = kT_0 B = 4\,Re(Y)\overline{|i_{N0}|^2}B = 4 \times 10^{-21} B(\text{W s}) \tag{12.11}$$

The noise generator method has long been the method of choice. This had several reasons.

The source has constant spectral power density within the relevant bandwidth and a normal distribution of amplitudes.

A separate determination of the bandwidth is not necessary, because it is omitted. That means no matter how is the shape of the passband curve, noise generator and DUT fill it out equally.

The noise power of the source can be varied by means of attenuators.

With a signal generator, its level must usually be reduced by very strong attenuation to make it comparable with the noise level. This is an additional uncertainty.

In the frequency range 10 MHz $< f <$ 50 GHz almost exclusively commercial noise generators are used, which have a semiconductor diode as source operated in avalanche breakdown. Their noise temperature is about $T_E = 10{,}000$ K. The gas discharge noise sources formerly widely used in the microwave range are now only used in the millimeter wave range or for special applications. Depending on the type of gas, they have $T_E = 12{,}500$ K for argon and $T_E = 23{,}000$ K for neon.

These values result from the electron temperature of the positive column. They have the advantage of easy inserting in a waveguide, which ensures good matching and are characterized by high stability over many years.

These sources provide a constant ratio of input power when switched between on and off states. So one measures the ratio of the output powers in the two generator states and calculates F from (12.8). It is also possible to use a variable attenuator at the output, with which the level $P_2 = 2 \times P_1$ is set. If this ratio is denoted $m < 1$, then

$$F = m\frac{T_E}{T_0} \tag{12.12}$$

The Equivalent Noise Resistance

The simplest way to describe the noise of a two-port is to introduce an equivalent noise resistor R_{EQ} at the input. This comes from vacuum-tube technology and is in many cases sufficient for field effect transistors in the MHz-range. The basic idea is that of the two noise sources at the input of a two-port only the voltage source is effective. The current source is neglected. This voltage source has the value $\overline{v_N^2} = 4kT_0BR_{EQ}$. The resistor is ineffective in terms of circuitry, and it does not influence the transmission behavior. A configuration as shown in Figure 12.2 is used for its measurement. There, bias supply of the practical circuit is not taken into account, e.g. the short circuit in position 1 must be realized by a capacitance. If a voltmeter is used, an amplifier with high voltage gain G_V and low inherent noise is required. The filter is rather uncritical [3].

If the input (switch 1) is short-circuited, we measure the output voltage

$$v_S = \sqrt{4kTR_{EQ}BG_V^2} \tag{12.13}$$

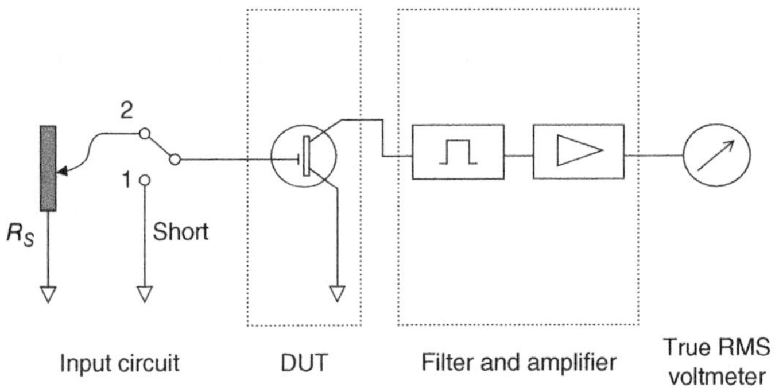

Figure 12.2 Scheme for measuring the equivalent noise resistance.

In switch position 2 we have the resistor R_S, and the output voltage v_R

$$v_R = \sqrt{4kT\left(R_{EQ} + R_S\right) BG_V^2} \tag{12.14}$$

By combining both equations we can obtain R_{EQ}

$$R_{EQ} = \frac{R_S}{\left(\frac{v_R}{v_S}\right)^2 - 1} \tag{12.15}$$

If we tune R_S to a value where the output is $v_R = \sqrt{2} \times v_S$, i.e. double the power, we immediately have $R_{EQ} = R_S$

Voltage and Current Source

The next step in describing the noise behavior is to know the two sources at the input, which we initially assume to be uncorrelated. This is a good approach for operational amplifiers, as we have seen in Chapter 10. In Figure 12.3 we have the scheme of a measurement circuit for the frequency range 0.1 Hz to 100 kHz. The digital spectrum analyzer Keysight 35670A [4] is especially suitable for the $1/f$ range due to its possibility of low bandwidth [5]. The voltage amplification of the DUT is given by

$$A_{DUT} = 1 + {}^{R_1}/_{R_2}; \quad R_P = \frac{R_1 + R_2}{R_1 R_2} \tag{12.16}$$

The equivalent circuit with noise sources shows Figure 12.4.

The DUT model comprises the noise sources of the intrinsic OPA e_N and i_N in the usual representation in data sheets in front of the input, as well as the thermal

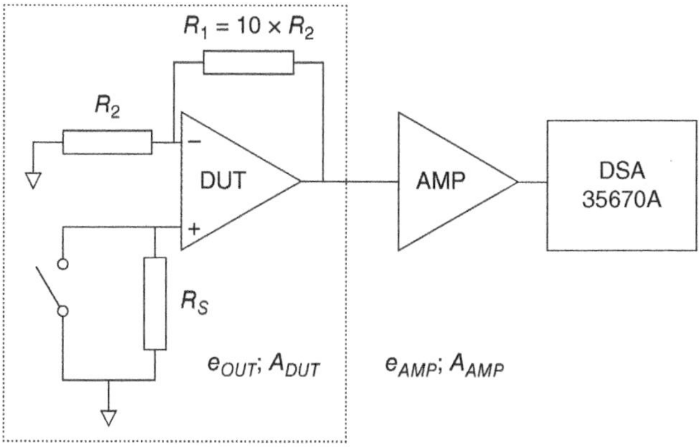

Figure 12.3 Scheme for measuring the noise properties of an OPA in the $1/f$ range.

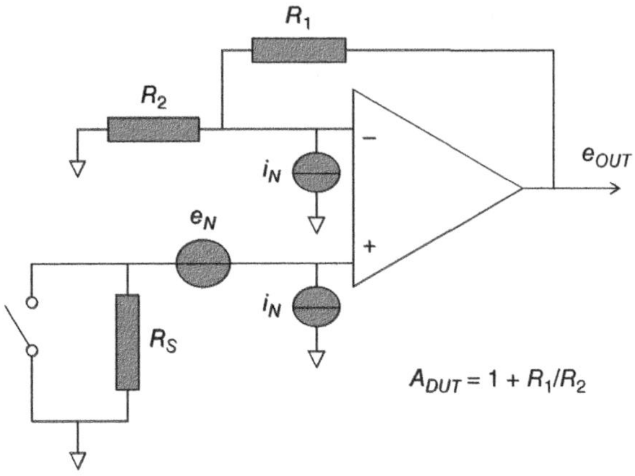

Figure 12.4 Wired OPA with the uncorrelated noise sources e_N and i_N.

noise resistors of the circuit R_1, R_2, and R_S. The noise sources express the spectral voltage or current densities with the unit nV/$\sqrt{\text{Hz}}$, respectively, pA/$\sqrt{\text{Hz}}$.

All sources sum up at the DUT output when the switch is set to R_S

$$e_{OUT;1} = \sqrt{e_N^2 + (R_S i_N)^2 + (R_P i_N)^2 + 4kT(R_S + R_P)} \times A_{DUT} \tag{12.17}$$

If the input is short-circuited, we have

$$e_{OUT;2} = \sqrt{e_N^2 + (R_P i_N)^2 + 4kT(R_S + R_P)} \times A_{DUT} \tag{12.18}$$

This noise voltage reaches the input of the DSA via the measuring amplifier. On the screen there appears the frequency spectrum of the power at the input resistance of the DSA.

Either

$$e_{RS}^2(f) = \overline{|e_{OUT;1}|^2} A_{DUT}^2 + \overline{|e_{AMP}|^2} A_{AMP}^2$$

or with short-circuited input

$$e_0^2(f) = \overline{|e_{OUT;2}|^2} A_{DUT}^2 + \overline{|e_{AMP}|^2} A_{AMP}^2 \tag{12.19}$$

Let us calculate the difference between these two curves $e_{RS}^2(f)$ and $e_0^2(f)$ so we get

$$e_{RS}^2 - e_0^2 = \left\{ (R_S i_N(f))^2 + 4kTR_S \right\} \times A_{DUT}^2 A_{AMP}^2 \tag{12.20}$$

We can solve this for $i_N(f)$

$$i_N^2(f) = \frac{e_{RS}^2(f) - e_0^2(f)}{(A_{DUT} A_{AMP} R_S)^2} - \frac{4kT}{R_S} \tag{12.21}$$

With $R_S = 0$ (12.18) and the knowledge of i_N we can calculate e_N. It has to be considered that, according to (12.19), the inherent noise of the amplifier stage e_{AMP}

Table 12.1 Noise sources at the input of operational amplifiers.

Technology	Type	e_N (nV/$\sqrt{\text{Hz}}$)	i_N (pA/$\sqrt{\text{Hz}}$)
BJT	ISL28290	1	3.5
MOS	ISL28148	25	0.01

Data sheet AN1560 $f = 1$ kHz.

is added. This must be determined separately. With the i_N determination according to (12.21) it is omitted.

$$e_N^2(f) = \frac{e_0^2(f) - e_{AMP}^2 A_{AMP}^2}{\left(A_{DUT} A_{AMP}\right)^2} - \left(R_P i_N(f)\right)^2 - 4kTR_P \qquad (12.22)$$

Now we have determined the two uncorrelated sources.

In this way, very low noise source values could be measured on Intersil OPA [5] at $f = 1$ kHz (Table 12.1).

Voltage and Current Source with Correlation

The consideration of the two sources in Section "Voltage and Current Source" is not sufficient for a full description of noisy two-ports, since there is no correlation. This simplification is particularly incorrect in the microwave range. As we have seen in Chapter 7, we need four quantities. The determination by measurements is called complete noise characterization [6].

With the designations introduced in Chapter 7 these are the following quantities:

1 $\overline{|i_U|^2} = 4kTG_U B$: The uncorrelated part of the noise current source. In Section "Voltage and Current Source", we had used the spectral noise current density i_N in accordance with data sheets.
2 $\overline{|v_N|^2} = 4kTR_N B$: The voltage source, above we had used the spectral noise voltage density e_N.

Above we had neglected.

3 $Re(Y_{COR}) = G_{COR}$
4 $Im(Y_{COR}) = B_{COR}$

The following quantities are accessible for a measurement:

1 $F(Y_S)$: The noise factor as a function of the input termination with the source admittance $Y_S = G_S + jB_S$.

2 $F_{MIN} = 1 + 2R_N(G_{COR} + G_{OPT})$ (7.19): The minimum noise factor at $Y_S = Y_{OPT}$.
3 $Y_{OPT} = \sqrt{\frac{G_U}{R_N} + G_{COR}^2} - jB_{COR}$ (7.21)

Here we write once again the general dependence of the noise factor on the input admittance. In (7.14)

$$F = 1 + \frac{G_U + R_N |Y_S + Y_{COR}|^2}{G_S} \tag{12.23}$$

This form is not particularly favorable for a measurement because it contains Y_{COR} explicitly and the fraction exceeding F_{MIN} does not appear separated. We therefore combine (7.19) with (12.23), such as Beneking [7], to create a form

$$F = F_{MIN} + F_+$$

in which F_+ is the additional contribution compared to the minimum possible value.

$$F_+ = \frac{G_U + R_N \left\{ (G_{COR} + G_S)^2 + (B_{COR} + B_S)^2 \right\}}{G_S} - 2R_N (G_{OPT} + G_{COR}) \tag{12.24}$$

G_{COR} and B_{COR} can be eliminated by using (7.21). Inserted in (12.24):

$$F_+ = \frac{G_U}{G_S} + \frac{R_N \left(G_{OPT}^2 - \frac{G_U}{R_N} \right)}{G_S} + R_N G_S + \frac{R_N B_{OPT}^2}{G_S} - \frac{2R_N B_{OPT} B_S}{G_S} + \frac{R_N B_S^2}{G_S} - 2R_N G_{OPT} \tag{12.25}$$

This bulky expression can be summarized:

$$F_+ = \frac{R_N}{G_S} |Y_S - Y_{OPT}|^2 \tag{12.26}$$

Thus we have for F the clear expression, which one finds common in literature:

$$F = F_{MIN} + \frac{R_N}{G_S} |Y_S - Y_{OPT}|^2 \tag{12.27}$$

We show how to determine the four quantities F_{MIN}, R_N, and $Y_{OPT} = G_{OPT} + jB_{OPT}$ step by step. For this purpose we write (12.23) in the form

$$F = 1 + \frac{G_U}{G_S} + \frac{R_N \left\{ (G_{COT} + G_S)^2 (B_{COR} + B_S)^2 \right\}}{G_S} \tag{12.28}$$

First we keep the real part of the source admittance constant (G_S = const.) and vary the imaginary part, e.g. a capacitance at the input. We obtain a quadratic dependence of the measured noise factor $F\left(B_S^2\right)$ from B_S. The derivation

$$\partial F / \partial B_S = \frac{2R_N}{G_S} (B_{COR} + B_S) = 0 \tag{12.29}$$

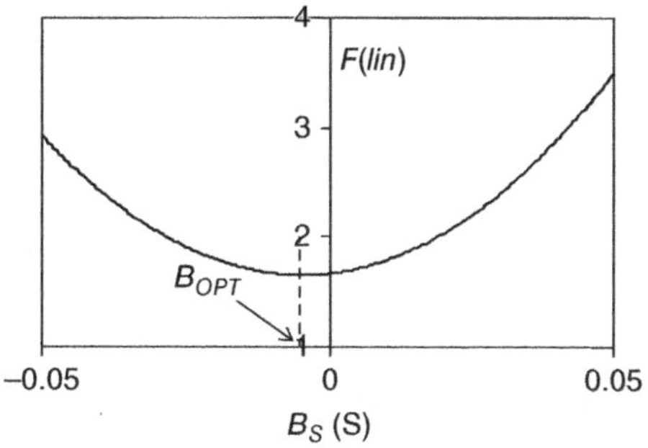

Figure 12.5 Noise factor with variation of the imaginary part of the source admittance B_S at constant G_S.

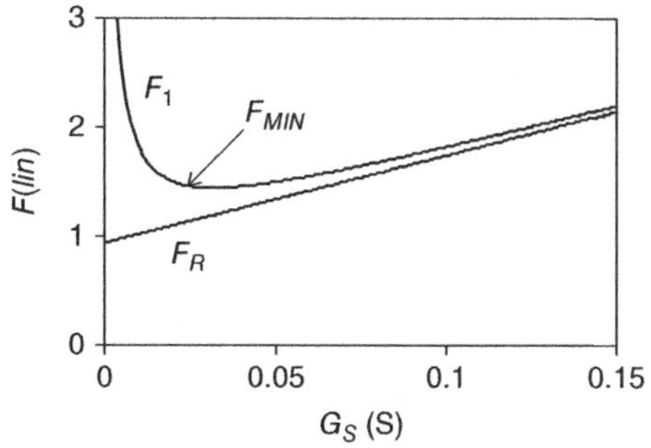

Figure 12.6 Noise minimum of the AT41400 with variation of G_S. $F_{MIN} = 1.45$ (1.6 dB) with $Y_{OPT} = 31$ mS $- j5$ mS. $R_N = 8\,\Omega$. $f = 2$ GHz.

shows that the minimum occurs at $B_{COR} = -B_S$ (Figure 12.5). If we keep this tuning constant and now vary the real part G_S, we get the curve $F_1(G_S)$ (Figure 12.6).

$$F_1 = 1 + \frac{G_U + R_N\left(G_{COR} + G_S\right)^2}{G_S} = 1 + 2R_N\left(G_{COR} + G_S\right) + \frac{G_U + R_N G_{COR}^2}{G_S} \tag{12.30}$$

From the position of the minimum we can take G_{OPT}

$$\partial F_1/\partial G_S = 2R_N - \frac{2\left(G_U + R_N G_{COR}^2\right)}{G_S^2} = 0 \tag{12.31}$$

$$G_{S;MIN} = G_{OPT} = \sqrt{G_{COR}^2 + \frac{G_U}{R_N}} \tag{12.32}$$

It is clear that with this input circuit $Y_S = Y_{OPT}$ the value of the noise factor must be $F_1 = F_{MIN}$.

Let us take the example of measuring the RF-BJT AT41400 [8] at $f = 2$ GHz. We first connect the input with a complex admittance, whose real part we keep constant with $G_S = 0.013$ S (corresponding to 75 Ω). The imaginary part B_S is varied until we have reached a minimum of the noise factor. The process is shown in Figure 12.5.

The minimum occurs at $B_S = -5$ mS. With this we have $B_{COR} = -B_{OPT}$ determined. We now keep this value constant and vary the real part G_S. We obtain the plot shown in Figure 12.6 $F_1(G_S)$. The minimum is $G_S = G_{OPT}$ and has the value F_{MIN}.

As we can see from (12.31), the increase of $F_1(G_S)$ for large G_S is given by the noise resistance R_N. In Figure 12.6 this straight line is denoted F_R.

$$F_R = 1 + 2R_N G_{COR} + R_N G_S \tag{12.33}$$

From the zero crossing ($G_S = 0$) we can take G_{COR}.

$$G_{COR} = \frac{F_R(0) - 0}{2R_N} \tag{12.34}$$

The uncorrelated noise current source can now also be calculated (12.32).

$$G_U = R_N \left(G_{OPT}^2 - G_{COR}^2\right) \tag{12.35}$$

From the measurement curves (Figures 12.5 and 12.6) we have extracted the four noise parameters F_{MIN}, R_N, and $Y_{OPT} = G_{OPT} + jB_{OPT}$. These can be converted into the elements of the noise equivalent circuit (e.g. Figure 7.4).

$$\overline{e_N^2} = 4kTR_N; \quad \overline{i_N^2} = 4kTG_U$$

$$G_{COR} = \frac{F_R(0) - 0}{2R_N}; \quad B_{COR} = -B_{OPT} \tag{12.36}$$

In our example the sources are: $e_N = 0.36\ \mathrm{nV}/\sqrt{\mathrm{Hz}}$ and $i_N = 11\ \mathrm{pA}/\sqrt{\mathrm{Hz}}$.

Since in the microwave range one works preferably with reflection coefficients instead of admittances, we calculate the reflection coefficient of the source for noise matching. The characteristic impedance is $Z_0 = 50\ \Omega$.

$$\Gamma_{OPT} = \frac{1 - Y_{OPT} Z_0}{1 + Y_{OPT} Z_0} \tag{12.37}$$

In our example at $f = 2$ GHz, $\Gamma_{OPT} = 0.24/163°$.

The most commonly used representation of noise parameters is in wave variables (S-parameters, reflection coefficients), as these are the appropriate parameters for signal analysis at higher frequencies anyway. We therefore want to convert (12.27) to reflection coefficients by replacing Y by Γ according to (12.38).

$$Y = \frac{1 - \Gamma}{Z_0(1 + \Gamma)} \tag{12.38}$$

This conversion is somewhat laborious. At first we arrive at

$$F = F_{MIN} + \frac{4R_N}{G_S Z_0^2} \frac{\left|\Gamma_{OPT} - \Gamma_S\right|^2}{\left|(1 + \Gamma_S)\,(1 + \Gamma_{OPT})\right|^2} \tag{12.39}$$

This is only an intermediate result, as it still contains the quantity G_S which is not suitable here.

Mostly the following formula is used, which contains only wave quantities:

$$F = F_{MIN} + \frac{4R_N}{Z_0} \frac{|\Gamma_{OPT} - \Gamma_S|^2}{|1 + \Gamma_{OPT}|^2 \left(1 - |\Gamma_S|^2\right)} \tag{12.40}$$

3 dB and Y-Method

The full noise characterization block diagram is shown in Figure 12.7. It is not fundamentally different from the configurations discussed above. The 3 dB method uses a noise generator with variable power. This is usually a noise source with an attenuator. More rarely in the MHz range a saturated vacuum diode, whose power can be controlled easily by varying the current and does not require calibration, but is no longer available today. If the gain G_{AV} of the DUT is sufficiently high, the inherent noise of the receiver (filter, amplifier, attenuator) need not be taken into account. First the attenuator is set to 0 dB and the noise generator is switched off ($T_C = T_0$) [1]. We then have at the output of the DUT the power

$$N_0 = kT_0BG_{AV} + N_A \tag{12.41}$$

Increasing the temperature of the noise generator to $T_1 = m \times T_0$ doubles the output power. The attenuator makes one independent of the characteristic of the detector, because by an attenuation of 3 dB the displayed value is the same as that produced by T_0.

We now have the output power

$$N_1 = kT_1BG_{AV} + N_A \tag{12.42}$$

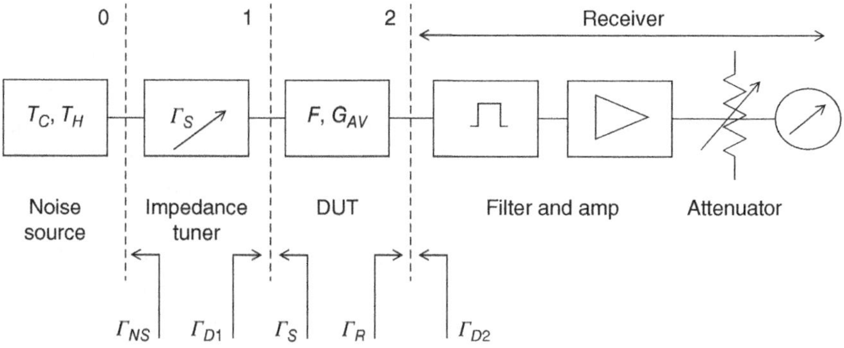

Figure 12.7 Setup for complete noise characterization of a two-port with noise generator and impedance tuner. Source: Schiek et al. [9] / John Wiley & Sons.

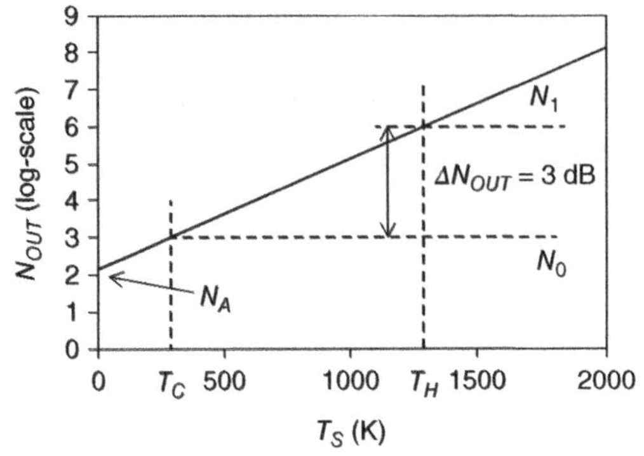

Figure 12.8 Output power for the 3 dB method. Variation of the temperature T_S of the noise generator.

because of the doubling $N_1 = 2N_0$, which corresponds to 3 dB, we can write (Figure 12.8)

$$kT_1BG_{AV} + N_A = 2\left(kT_0BG_{AV} + N_A\right) \tag{12.43}$$

Solved for N_A

$$N_A = kBG_{AV}\left(T_1 - 2T_0\right) \tag{12.44}$$

The noise factor F can be obtained from the definition

$$N_A = kT_0BG_{AV}(F - 1) \tag{12.45}$$

inserted in (12.44) results in

$$F - 1 = \frac{T_1}{T_0} - 2 \tag{12.46}$$

using the factor $m = T_1/T_0$, which can be read on some commercial noise measuring devices, we receive

$$F = m - 1 \tag{12.47}$$

The 3 dB method is based on a fixed ratio of output power $N_1/N_2 = 2$ and the input power tuned to it. This is a disadvantage for the automatic measurement of many values, at different frequencies, which is overcome by the Y-method. Furthermore, in the microwave range continuously adjustable noise generators are not common. Moreover the vacuum diode formerly used in the MHz range is no longer available.

Radio and audio amateurs occasionally use readily available semiconductor components as noise sources. As with the vacuum diode, the shot effect is the source, which can be changed by the direct current. These are, e.g. photodiodes, which are controlled by the illumination or the base–emitter path of bipolar transistors. The currents are in the mA range and produce only a low level of overthermal noise. There are also digital noise generators. These generate a spectrum consisting of signals with discrete frequencies. Superimposed they show the typical characteristics of noise: Gaussian distribution of the amplitudes and a white spectrum in *a*, mostly low frequency range.

Noise generators in the microwave range are usually sources that are switched between two temperatures $T_C \cong T_0$ and T_H. These are semiconductor diodes which are operated in avalanche breakdown, more rarely gas discharge tubes in waveguide technology in the millimeter wave range. In contrast to the 3 dB method, here we have the method of constant input power. The two associated output powers are measured with the detector, for which the highest requirements on linearity must be ensured.

$$Y = \frac{N_1}{N_0} = \frac{kBG_{AV}\left(T_H + T_E\right)}{kBG_{AV}\left(T_C + T_E\right)} \tag{12.48}$$

Solved for T_E

$$T_E = \frac{T_H - YT_C}{Y - 1} \tag{12.49}$$

with $F = {}^{T_E}/_{T_0} + 1$ we get the noise factor of the DUT

$$F = \frac{T_H - T_0 - Y\left(T_C - T_0\right)}{T_0(Y - 1)} \tag{12.50}$$

depending on the accuracy requirements, it is often possible to set $T_C = T_0$ and obtain

$$F = \frac{T_H - T_0}{T_0(Y - 1)} \tag{12.51}$$

In the chapters 14–16, we will discuss the components from Figure 12.7 in detail.

References

1 Hewlett Packard (2004). Fundamentals of RF- and Microwave Noise Figure Measurements. Application Note 57-1.

2 Paul, R. (1966). *Transistor Messtechnik*. Berlin: Verlag Technik.

3 Bittel, H. and Storm, L. (1971). *Rauschen*. Springer Verlag.

4 Keysight Technologies (2017). 35670A Dynamic Signal Analyzer. Datasheet.

5 LaFontaine, D. (2011). Making Accurate Voltage Noise and Current Noise Measurements on Operational Amplifiers Down to 0.1Hz. Intersil Application Note 1560.

6 Müller, R. (1990). *Rauschen*. Springer Verlag.

7 Beneking, H. (1971). *Praxis des elektronischen Rauschens*. Bibliographisches Institut Mannheim.

8 Agilent Technologies. AT-41400: 6GHz Low Noise Silicon BJT Chip. Datasheet.

9 Schiek, B., Rolfes, I., and Siweris, H.-J. (2006). *Noise in High-Frequency Circuits and Oscillators*. Wiley Interscience.

13

Basics of Measuring Technique

Principles of the RF-Receiver

When measuring the power provided by an RF source, the choice of the indicator depends heavily on the level. Power sensors are available for signals in the milliwatt range [1]. This range can be extended by series attenuators for use in the watt range. For very low levels, especially in the noise limit range, special measuring receivers based on the superheterodyne principle are required [2]. There are three operating principles for the usual power sensors [3]: (i) Heating of a thermocouple. (ii) Heating of thermistors. (iii) Rectification of the RF voltage on the square-law characteristic of a semiconductor diode. The smallest measurable power here is about 10^{-10} W. An improvement by orders of magnitude, up to about 10^{-14} W, is achieved by superheterodyne reception. Instead of the detectors 1–3, mixer stages of nonlinear semiconductor components are used. Combination frequencies in the intermediate frequency range are created, whose amplitude and phase correspond to the RF signal and can be analyzed.

The Detection Limit

According to the definition of the noise factor in Chapter 6 and (12.1), the total power N_{OUT} available at the output within the frequency band B and the available power gain G_{AV} are required to determine the noise factor of a two-port.

A noise measurement is therefore always a power measurement of a level that is usually very small. For every power measurement in the RF or microwave range, a number of sources of error must be taken into account. These are, e.g. mismatch of source and power sensor, driving the diode into nonlinearity, harmonic content of the source, calibration to sinusoidal signal and not "True RMS" [4, 5]. Since noise is the signal to be measured, the influence of the receiver's own noise further complicates the measurement. We therefore use the term: noise signal. Usually this noise

A Guide to Noise in Microwave Circuits: Devices, Circuits, and Measurement, First Edition.
Peter Heymann and Matthias Rudolph.

signal passes through tuning elements, e.g. filters, is amplified and displayed by a detector. With a well-defined filter, the display corresponds to the spectral power density of the noise at the center frequency of the filter. In addition to special noise measurement receivers (noise figure analyzer) for the highest requirements, simpler components of the measurement technology may be sufficient from case to case and depending on the frequency range.

To estimate the levels involved, we first consider a high-quality power sensor as a possible test receiver, e.g. the R&S NRV Z4 [5]. It works with Schottky diodes up to 6 GHz and is therefore particularly sensitive. Its detection limit is specified to 100 pW. It is advisable to limit the bandwidth by a filter $B < 5\,\text{MHz}$, depending on the measurement frequency. Let's take an amplifier with $NF = 2.5\,\text{dB}$ and $G_A = 20\,\text{dB}$ at $f = 2\,\text{GHz}$ as a measurement object, whose noise figure we want to verify with a bandwidth of $B = 4\,\text{MHz}$ via an interconnected filter with the just mentioned detector head. Using the formula for the output power

$$N_{OUT} = FkT_0BG_A = 3\,\text{pW} \tag{13.1}$$

Obviously, without an additional amplifier of $G > 20\,\text{dB}$ we will not see any effect.

Another example is a VHF-antenna at $f = 100\,\text{MHz}$, which has a low directivity. The galactic noise produces an antenna temperature $T_A = 1500\,\text{K}$. We calculate the voltage that appears at its output resistance $Z_F = 240\,\Omega$.

$$v = \sqrt{4kT_ABZ_F} \approx 10\,\mu V \tag{13.2}$$

In comparison, the lowest voltage measurable with the highly sensitive probe head R&S URV-Z7 is 200 μV. This also does not work, and one is an order of magnitude below the detection limit.

Historically, radio astronomy has been the great challenge, in addition to the development of sensitive receivers for communication technology and radar. It started in the 1940s on the basis of a barely developed high-frequency measurement technology. The receivers produced much more noise than today, so there were higher noise levels that had to be reduced. On the other hand, radio astronomy was interested in extremely weak noise signals from space. In communications engineering, noise is a disturbance, in radio astronomy or thermometry it is the desired signal. In noise measurement technology, on the other hand both become signal noise, whose measurement is disturbed by the inherent noise of the measuring device. Since then, great progress has been made in the field of noise measurement. Thanks to semiconductor technology, noise figures far below 1 dB up to 100 GHz can be achieved today with cooled transistor amplifiers. In the early days, one was dependent on diode mixers as the first stage of the receiver with not less than $NF = 2$ dB even with extreme cooling. Figure 13.1 shows an overview of noise figures for the various techniques.

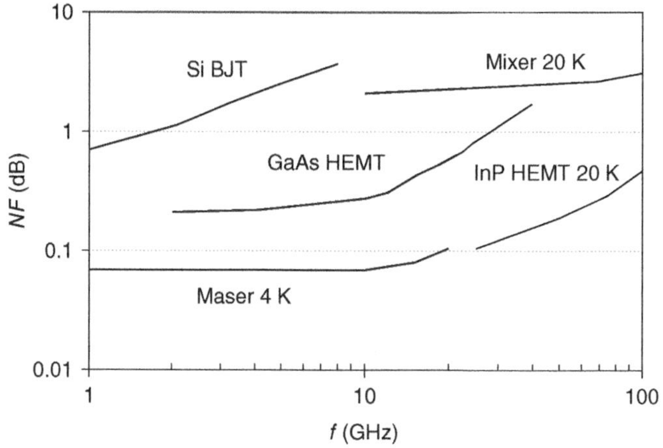

Figure 13.1 Noise figures of input stages and transistor technologies.

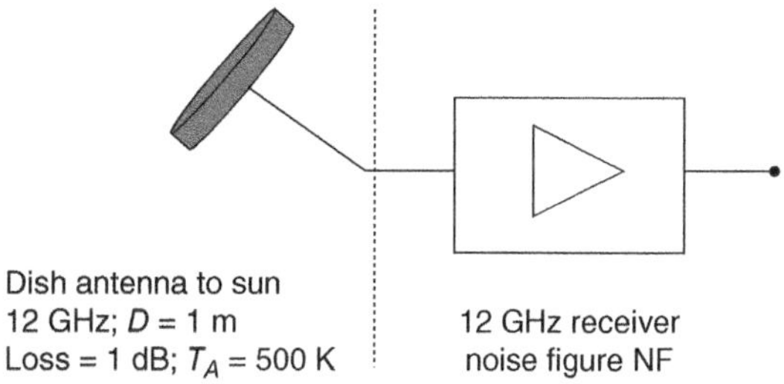

Figure 13.2 12GHz receiving system directed toward the sun.

In the early days of radio astronomy, the situation was roughly as follows: As shown in Figure 13.2, an X-band parabolic antenna was directed to the sun. The sun is the strongest radio source with 15.000 K photosphere temperature at $f = 12\,\text{GHz}$ and produces an antenna temperature $T_A = 500\,\text{K}$. For details see Chapter 6.

We calculate the power of the antenna signal N_{SUN} compared with the inherent noise of the receiver N_{REC}

$$N_{SUN} = kGB\left\{(F-1)\,T_0 + T_A + T_L\right\}$$
$$N_{REC} = kGB\left\{(F-1)\,T_0 + T_L\right\} \tag{13.3}$$

$T_A = 500\,\text{K}$; $T_L = 60\,\text{K}$ corresponding to 1 dB antenna loss; $B = 1\,\text{MHz}$, $G = 10\,\text{dB}$ cancel out in the comparison.

The ratio $^{N_{SUN}}/_{N_{REC}}$ corresponds to the commonly used signal-to-noise ratio $^{S}/_{N}$. For the example in Figure 13.2 this is plotted in Figure 13.3 versus the noise figure of the receiver. Even with the strong radio source of the sun, there was no measurable effect in the 1950s with $NF = 8 \ldots 10\,\text{dB}$ receiver noise figure ($^{S}/_{N} \leq 1\,\text{dB}$

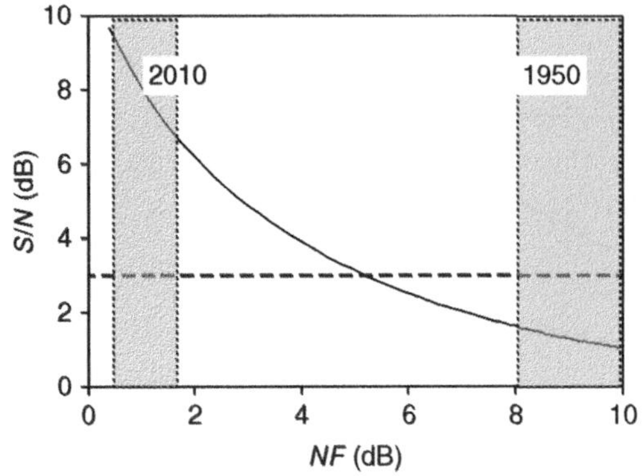

Figure 13.3 Ratio of sun signal to receiver noise. Measurement is possible for $NF < 5$ dB.

Table 13.1 Noise voltages of different sources at 50 Ω in $B = 1$ MHz.

Noise source	50 Ω resistance	Amplifier[a)]	VHF aerial (13.2)	Parabolic to Sun (13.3)
Noise voltage (μV) @ $Z_0 = 50\,\Omega$, $B = 1$ MHz	0.9	6	10	1

a) $NF = 2.5$ dB; $G = 20$ dB.

Table 13.2 Detection limit of typical receivers.

Receiver	RMS voltmeter[a)]	Power meter	Oscilloscope	Spectrum analyzer	NFA	VNA
Sensitivity Limit (μV)	50	70	90	10	1	6

a) High impedance input.

on the right side of the picture). With today's transistor technology, cooling of the system yields $NF < 1$ dB. There we have a clear effect ($S/N > 8$ dB on the left side of the picture).

Typical noise power levels encountered in measurement technology are summarized in Tables 13.1 and 13.2.

In the LF and MHz range, mainly the sources with noise voltages and currents are characterized, in the microwave range by the four noise parameters. To measure a quantity reliably, it should be at least 10 times greater than the effect produced by the measuring instrument. It can be seen from the aforementioned diagram and table that this is not the case with these examples for the noise of a one-port, with the exception of NFA. This limit can be overcome by special radiometer technology.

Let us first analyze the RF detection at the diode characteristic and the sensitivity that can be derived from this, since diode measuring heads are also relevant for noise measurement, at least if they are equipped with a pre-amplifier.

Diode as RF Receiver (Video Detector)

The simplest way to measure small RF powers is direct rectification at the non-linear characteristic of a Schottky diode. A system of matching element, diode, and following amplifier (Figure 13.4) is called a video detector. The term is derived from the early days of radar technology, when signals were observed visually. The sensitivity is given by the efficiency of the conversion of RF power into DC power. It is limited by the inherent noise of the diode (Chapter 9) and the subsequent amplifier. The sensitivity is about 30–40 dB worse than in the case of superheterodyne reception. Nevertheless there are some advantages in application. The setup can be designed broadband, only one diode is needed and no local oscillator. For analysis we use the equivalent circuit with load resistance in Figure 13.5 [6, 7].

The complex input resistance of the Schottky diode is

$$Z = R_B + \frac{R_j\left(1 - j\omega C_j R_j\right)}{1 + \omega^2 C_j^2 R_j^2} \tag{13.4}$$

In a detector head, a more or less broadband matching to this resistance is achieved by a suitable transformation circuit. In this way the input power P is completely absorbed in Z.

$$P = |I|^2 Re\,(Z) \tag{13.5}$$

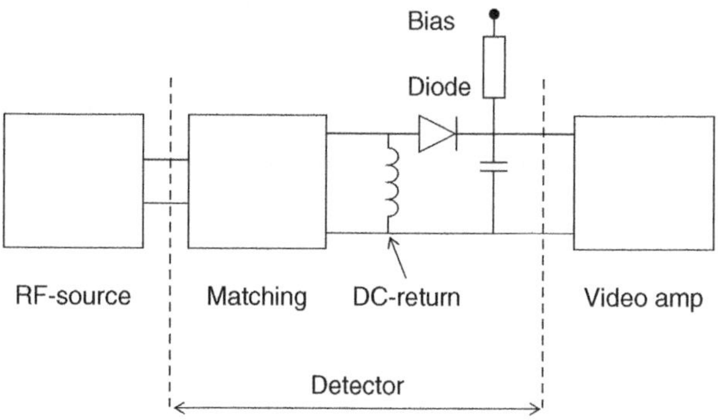

Figure 13.4 Video detector consisting of diode and matching element between source and amplifier.

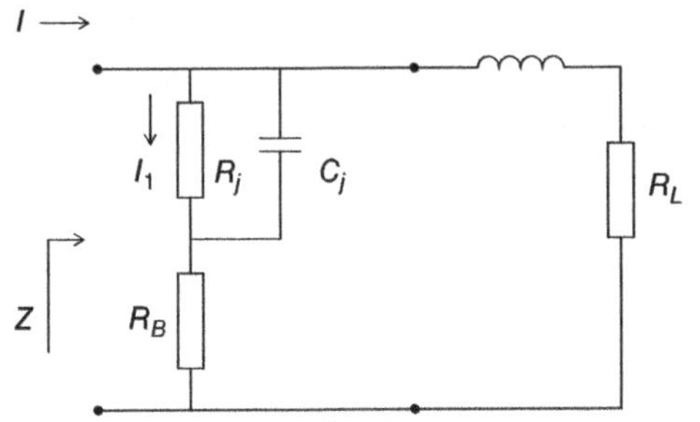

Figure 13.5 Equivalent circuit of a diode with load resistor R_L. Notation: R_B = parasitic path resistance, R_j = non-linear junction resistance, C_j = junction capacitance. Source: Based on Skyworks Solutions [7].

However, only the part absorbed in the junction resistor *Rj* is converted into the DC-power P_{Rj}.

$$P_{Rj} = |I_1|^2 R_j \tag{13.6}$$

The ratio of the two active powers is determined by the current distribution

$$\frac{P_{Rj}}{P} = \frac{1}{1 + \frac{R_B}{R_j}\left(1 + \omega^2 R_j^2 C_j^2\right)} \tag{13.7}$$

The principle of the video detector is shown in Figure 12.5. The matching element is designed according to the sensitivity and broadband requirements.

Maximum sensitivity is achieved by matching to the highest possible value *Rj*. This would be the case with "zero bias" operation with *Rj* of some 100 kΩ. Of course, strong transformation only works in narrow bands and is also limited by the path resistance R_B and losses in the impedance transformer. Extreme sensitivities of = 30 mV/μW and $TSS = -70$ dBm are thus possible. A broadband detector works with moderate transformation, so that R_j from some 100 Ω can be matched. This operation requires a bias current of about 100 μA. The sensitivities are = 3 mV/μW and $TSS = -55$ dBm. Small reflection coefficient in a wide frequency band is achieved by forced matching with a 50 Ω resistor in parallel to the diode, instead of the impedance transformer.

The following consideration holds for diodes with pn-junction as well as for those with the Schottky contact. As already mentioned, the nonlinear I/V characteristic is essential (see also Chapter 9)

$$I = I_S\left\{exp\left(\frac{qV_j}{mkT}\right) - 1\right\} \tag{13.8}$$

V_j is the junction voltage at the p-n junction. $V_j = V - IR_B$. When an RF-voltage of the frequency ω is applied, the current consists of the bias current I_0 and an RF component i:

$$I = I_0 + i\cos(\omega t) \tag{13.9}$$

We solve (13.8) for V_j and receive (13.10)

$$V_j = \frac{mkT}{q} ln\left\{\frac{I_S + I_0 + i\ \cos(\omega t)}{I_S}\right\} \tag{13.10}$$

After converting the ln

$$V_j = \frac{mkT}{q} ln\left\{\frac{I_S + I_0}{I_S}\left(1 + \frac{i\cos(\omega t)}{I_S + I_0}\right)\right\} \tag{13.11}$$

We can divide the product under the logarithm into a sum

$$V_j = \frac{mkT}{q} ln\left\{\frac{I_S + I_0}{I_S}\right\} + \frac{mkT}{q} ln\left\{1 + \frac{i\cos(\omega t)}{I_S + I_0}\right\} \tag{13.12}$$

For the RF-term we use the series expansion

$$ln(1 + x) = x - \frac{1}{2}x^2 + \frac{1}{3}x^3 - \cdots \tag{13.13}$$

Assuming a low level, we neglect higher terms than x^2. This leads to

$$V_j = \frac{mkT}{q} ln\left\{\frac{I_S + I_0}{I_0}\right\} + \frac{mkT}{q}\left\{\frac{i\cos(\omega t)}{I_S + I_0} - \frac{i^2\cos^2(\omega t)}{2(I_S + I_0)^2}\right\} \tag{13.14}$$

We have a DC voltage V_{DC}, determined by the operating point of the diode (1st term) and a rectified part of the RF voltage (2nd term). Only the quadratic term gives a contribution to V_j, because the mean value $\overline{\cos(\omega t)} = 0$, while $\overline{\cos^2(\omega t)} = 1/2$. If we eliminate the RF current i by the voltage drop at R_j, we obtain

$$V_j = iR_j = i\frac{mkT}{q(I_0 + I_S)};\ i^2 = V_j^2\left(\frac{q}{mkT}\right)^2 (I_0 + I_S)^2 \tag{13.15}$$

And

$$V_{DC} = \frac{mkT}{q} ln\left\{\frac{I_S + I_0}{I_S}\right\} - \frac{qV_j^2}{4mkT} = V_0 - 9 \times V_j^2 \tag{13.16}$$

What is the relationship between the power P_{RF} of the source and the voltage at the output of the diode in this case? The sinusoidal signal of a source to be measured with the detector diode produces the voltage drop V_j at the diode resistance R_j. We consider the broadband case with a $Z_0 = 50\,\Omega$ resistance instead of the transformer. The total voltage range V, from the positive peak to the negative peak is

$$V = 2 \times \sqrt{2P_{RF}Z_0};\ V^2 = 8P_{RF}Z_0 \tag{13.17}$$

Due to the forced matching, Z_0 and R_j are connected in parallel.

$$V_j^2 = \frac{8P_{RF}Z_0}{\left(1 + \frac{Z_0}{R_j}\right)^2} \tag{13.18}$$

R_B and C_j are assumed to be sufficiently small. Thus we obtain the voltage sensitivity

$$\beta = \frac{V_0 - V_{DC}}{P} = 9 \times \frac{8\, Z_0}{\left(1 + \frac{Z_0}{R_j}\right)^2} \tag{13.19}$$

The sensitivity increases if the junction resistor is matched to a narrow band, i.e. $Z_0 = R_j$. Then $\beta_m = 18 \times R_j$.

When comparing with manufacturer's specifications of diode parameters based on measurements, the approximate nature of the above considerations must be taken into account. Especially the broadband matching transformation considered here is only imperfect. Therefore no complete agreement can be expected. However, the principle and the dependence on the bias current (indirectly R_j) are explained.

Avago HSCH 53 Beam Lead Schottky Diode ($B = 2\,\text{MHz}$)

Another measure of the sensitivity of a diode is the Tangential Sensitivity (TSS) [8] in the unit Watt or dBm, as already mentioned earlier. It describes the situation when a pulsed signal is detected by the diode with an oscilloscope. The tangential sensitivity is defined by the following situation. The lower edge of the wanted signal emerging from the noise just coincides with the upper edge of the noise (Figure 13.6).

The definition of this subjective value is that the signal to noise ratio at the output is 8 dB. For the input this means a ratio of 2.5 corresponding to 4 dB. The diode operates in the quadratic range, i.e. input: $10\,log\,(2.5/1) = 4$ dB output: $10\,log(2.5/1)^2 = 8$ dB.

When calculating the tangential signal sensitivity (TSS), we have to relate the inherent noise of the detector unit consisting of diode and video amplifier to the signal sensitivity. The noise voltage of the diode is (see also Chapter 9)

$$v_{ND} = \sqrt{4kTBR_S + 2kTBR_j\left(1 + \frac{I_S}{I_0 + I_S}\right)} \tag{13.20}$$

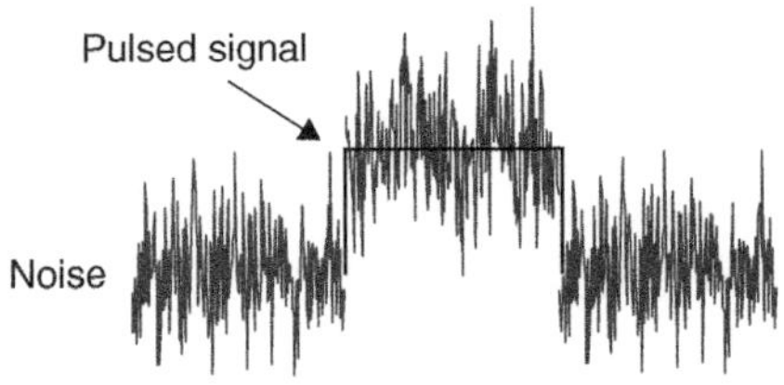

Figure 13.6 Screen display on an oscilloscope for tangential sensitivity. The signal is visible above the noise.

The noise voltage of the amplifier is described by its equivalent noise resistance $R_A \approx 1200\,\Omega$.

$$v_{NA} = \sqrt{4kTBR_A} \tag{13.21}$$

Together

$$v_N = v_{ND} + v_{NA} = \sqrt{2kTB\left\{R_j\left(1 + \frac{I_S}{I_0 + I_S}\right) + 2R_S + 2R_A\right\}} \tag{13.22}$$

The TSS definition states that the lower edge of the noise band with signal is at the level of the upper edge of the noise band without signal. If we set the peak value of the noise voltage to 1.4 of the RMS value, we have as a criterion

$$V_{DC} + 1.4v_N = V_O - 1.4\,v_N;\, V_O - V_{DC} = 2.8\,v_N \tag{13.23}$$

If we use the voltage sensitivity (13.19) we get for $Z_0 = 50\,\Omega$:

$$\frac{V_0 - V_{DC}}{P_{TSS}} = \frac{3600}{\left(1 + \frac{50}{R_j}\right)^2} \tag{13.24}$$

And in comparison with the noise voltage (13.22)

$$\frac{3600P_{TSS}}{\left(1 + \frac{50}{R_j}\right)^2} = 2.8v_N \tag{13.25}$$

$$TSS\,(\text{dBm}) = 10log\left(0.78\left(1 + \frac{50}{R_j}\right)^2\sqrt{2kTB\left\{R_j\left(1 + \frac{I_S}{I_0 + I_S}\right) + 2R_S + 2R_A\right\}}\right) \tag{13.26}$$

The derivation of this formula is plausible, but it does not include the junction capacitance C_j and is only valid far below the cut-off frequency of the diode. However, it gives a good value in Table 13.3 in comparison with the data sheet.

Table 13.3 Comparison of the sensitivity of (13.19) with the factory specifications of a Schottky diode (Avago HSCH53) [9].

	I (μA)	*Rj* (k)	(mV/μW)	*TSS* (dBm)
Data sheet	20	1.4	6.6	−54
Calculation	20	1.4	3.6	−53

Source: Based on Avago Technologies [9].

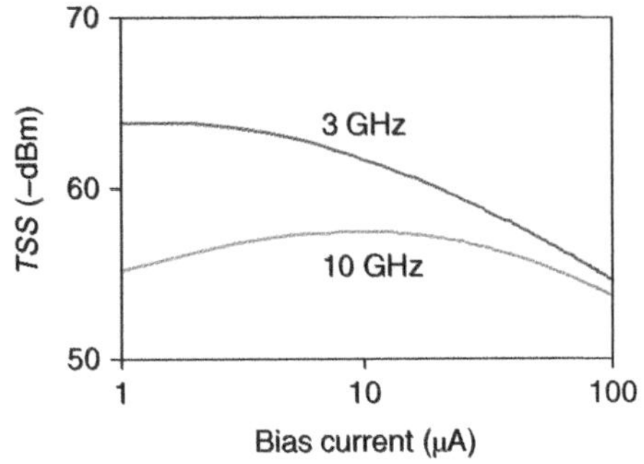

Figure 13.7 TSS of a Schottky diode for two frequencies versus the bias current according to (13.27).

More details, especially the frequency dependence, are given in the following formula from Agilent AN 923 [8], which we reproduce here without derivation

$$TSS\,(\text{dBm}) = -107 + 5\ \log(B) + 10\ \log(I_0) + 5log\left\{R_A + \frac{28}{I_0}\left(1 + \frac{f_N}{B}\,ln\frac{B}{f_L}\right)\right\} + 10log\left(1 + \frac{R_B C_j^2 f^2}{I_0}\right) \qquad (13.27)$$

The parameters are to be entered in the following units:

B: Video bandwidth (Hz)
I_0: Diode bias current (μA)
f_N: Corner frequency of $1/f$ noise (Hz)
f_L: Lower cut-off frequency of the video amplifier (Hz)
R_B: Series resistance of the diode (Ω)
R_A: Equivalent noise resistance of the video amplifier (kΩ)
f: Operating frequency (GHz)

$$C_j = \frac{C_j(0)}{\sqrt{1 - 0.1\log\left(1300 I_0\right)}}\ (\text{pF})$$

An example of the above diode with $C_j(0) = 0.1$ pF; $R_B = 13\,\Omega$; $f_N = 3$ kHz; $f_L = 100$ Hz; $B = 2$ MHz is shown in Figure 13.7. It is clear that the sensitivity decreases with increasing frequency. On the other hand, there is an optimal diode current I_0.

RF and Microwave Range Receiver

The receiver (Figure 13.8) must have the following characteristics: Low noise figure, high gain, and stability of this gain, well-defined bandwidth, and possible tunability in the frequency range. Before low noise figures could be realized, it was

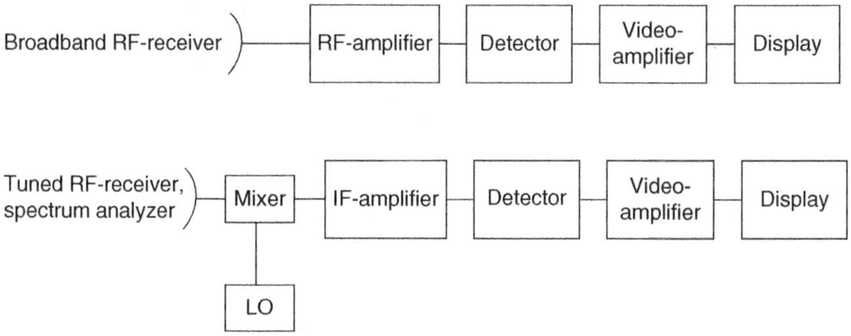

Figure 13.8 Block diagrams of test receivers in the RF and microwave range.

possible to circumvent this problem by the invention of the Dicke radiometer (Figure 13.12).

The detection limit of a receiver is given by the fluctuations of the output voltage at the indicator when there is no signal at all. These fluctuations have two main causes: fluctuations, because they are noise, and spontaneous, short-term fluctuations in the system's gain. Both can in principle be compensated by integration times of any length at the indicator. However, this does not make sense for most applications because the system then has a long response time and changes in the incident radiation can no longer be detected without distortion. A drift of the gain cannot be eliminated either. Nevertheless, in the early days of radio astronomy, so-called compensation systems were used. Because of the high noise figure of the receivers, there was a very strong background noise and a very small useful signal. The background noise was compensated at the output by a DC voltage to zero and the small signal was made visible. Of course, the demands on the stability of the system were extreme.

Here we show once again the relationship between the noise signal and the inherent noise at the input and the levels available for measurement at the output, as already discussed in Chapter 6. This interaction forms the basis for the calibration of the noise measurement receiver.

In the example Figure 13.9 the receiver has a noise figure $NF = 3$ dB (linear $F = 2$, $T_E = 290$ K). We recall the definition of F and its relation to the effective noise temperature T_E. This is the temperature of a resistor with which the noise-free receiver is terminated at the input so that the same noise power is produced at the output as in the real receiver.

$$F = \frac{N_A + kGBT_0}{kGBT_0} = \frac{T_E}{T_0} + 1 \tag{13.28}$$

G is the available power gain of the receiver, B is the noise bandwidth, T_S is the equivalent noise temperature of the one-port to be measured.

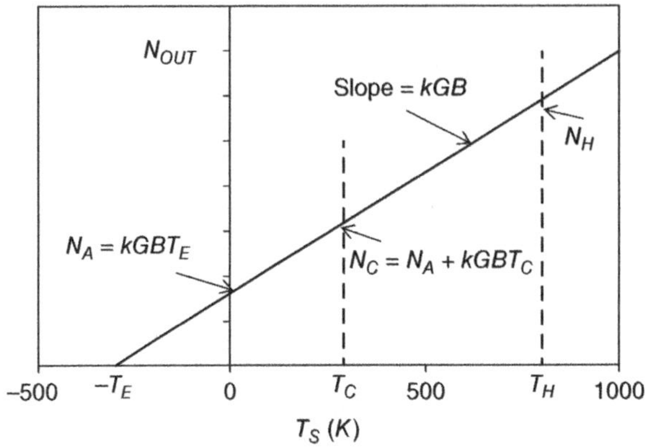

Figure 13.9 Noise power N_{OUT} at receiver output versus the temperature of the source T_S.

When calibrating the system, the *kGB* factor must be determined. The straight line

$$N_{OUT}\left(T_S\right) = N_A + kGBT_S \tag{13.29}$$

goes through zero at $T_S = -T_E$. This results in

$$N_A = kGBT_E = kGBT_0\left(F - 1\right) \tag{13.30}$$

The calibration procedure is based on these relationships. We need a noise generator with two different temperatures and the same impedance as the test object. The temperatures are T_C and T_H with $\Delta T = T_H - T_C$.

The two display levels are

$$N_C = N_A + kGBT_C; \qquad N_H = N_A + kGB\left(T_C + \Delta T\right) \tag{13.31}$$

With this we have calibrated the receiver

$$kGB = \frac{N_H - N_C}{\Delta T} \tag{13.32}$$

If the filter curve is very well defined, e.g. with digital signal processing in the system, we do not need a noise generator, but can work with a high-quality signal generator (details see below).

As mentioned above, the display of the noise level at the output is superimposed by the "noise of the noise." It can be minimized by long integration times. However, this must be optimized to achieve short measuring times with sufficient accuracy.

The exact analysis of the detection limit requires complicated considerations regarding the non-linear change in the amplitude distribution of the noise after rectification [2, 10]. The result is

$$\frac{\Delta T_S}{T_S} \approx 0.3 \frac{1}{\sqrt{B\tau}} \tag{13.33}$$

τ is the time constant of the averaging element of the display.

We need a large RF bandwidth and a small video bandwidth to improve sensitivity. We had already seen this in Chapter 5 when analyzing the spectrum analyzer and we will come back to it in the following text.

Let us look again at the signal in the time domain as it passes through the receiver chain.

The RF passband is a rectangular filter curve of bandwidth B at frequency f_0. At the input of this filter we have broadband white noise.

Behind the input filter we see in the time domain a fluctuating RF voltage with the accentuated frequency f_0, whose envelope is governed by the fluctuations. In the further course of the signal processing we will deal with these fluctuations. Their frequency spectrum is determined by the RF bandwidth B, since the system does not react to voltage changes that are faster than the settling time $1/B$. This voltage (Figure 13.10 left hand side) is applied to the rectifier. Behind the rectifier we have the time response (Figure 13.10 right). The lower half of the RF oscillations is truncated, only the envelope of the upper half remains. We have a direct current component (mean value) and a superimposed noise spectrum $\overline{\delta^2}$. The quantitative analysis is not trivial, as we no longer have a normal distribution of amplitudes and white noise. The spectral noise power density behind the rectifier is shown in Figure 13.11. We had already made these considerations in Chapter 5.

We have a triangular LF-spectrum that extends to the bandwidth B and a RF-component at twice the receiving frequency. The latter is of no interest

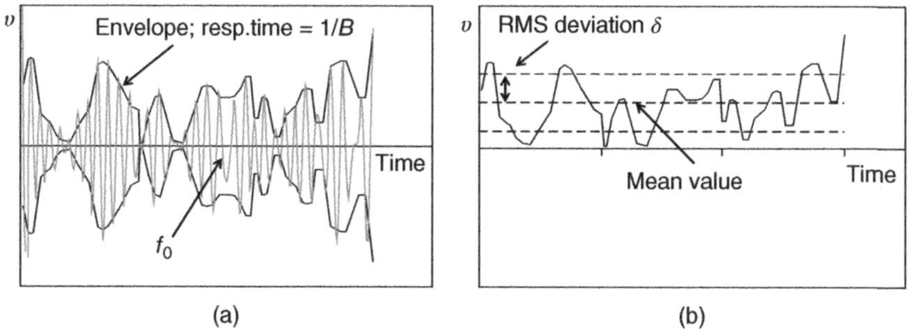

Figure 13.10 Time response of the noise voltage behind a rectangular bandpass of the center frequency f_0. (a): before rectifier. (b): behind rectifier.

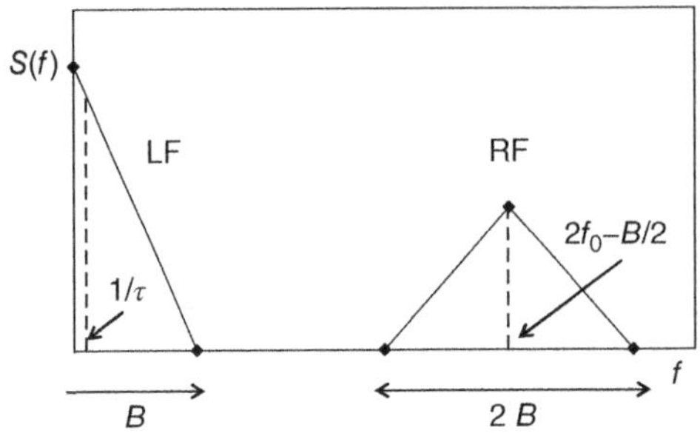

Figure 13.11 Spectral power density of the time domain curve Figure 13.10 behind the rectifier.

because it is filtered out. The first triangle corresponds to the difference frequencies, the second to the sum frequencies. The area of the first triangle is the fluctuation $\overline{\delta^2}$. The voltage fluctuations are about 50% of the direct current component. The RC filter with the time constant τ behind the rectifier now provides the actual effect of reducing the "noise of the noise." From the triangular area which contains all the noise, the filter cuts out a small part: the frequency range $0 < f < {}^1/_\tau$. In Figure 13.11 given by the dotted line. The figure is "not to scale," because ${}^1/_\tau << B$. The new noise voltage is

$$\sqrt{\overline{\nu^2}} = \sqrt{\overline{\delta^2}}\sqrt{\frac{2}{\tau B}} \tag{13.34}$$

We have thus obtained a plausibility explanation for the sensitivity formula (13.33).

Dicke Radiometer

With the modulation method introduced by R. H. Dicke, it was possible to overcome the extreme discrepancy between the radio astronomical antenna signal and the inherent noise of the apparatus as it was given in the 1940s. A scheme is shown in Figure 13.12. The modulation is realized by a high-frequency switch which alternately connects the antenna and a reference standard of known temperature to the receiver input. There is a periodic fluctuation of the input noise, corresponding to the temperature difference between the antenna and the standard. The clock generator controls not only the RF switch, but also a synchronously running LF switch, which converts the modulation AC voltage into a DC voltage. This is also known as a phase-sensitive detector (PSD). It is the LOCK-IN principle, which is widely used today. However, half of the useful signal is lost due to the switchover [11].

In front of the rectifier we have a square-wave modulated RF signal (Figure 13.13). The constant inherent noise of the apparatus is always present.

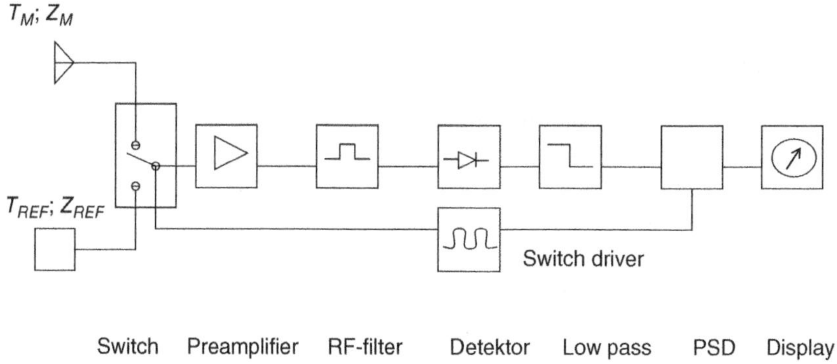

Figure 13.12 Block diagram of Dicke's radiometer for measuring microwave radiation.

Figure 13.13 Time curve of the noise power behind the switch.

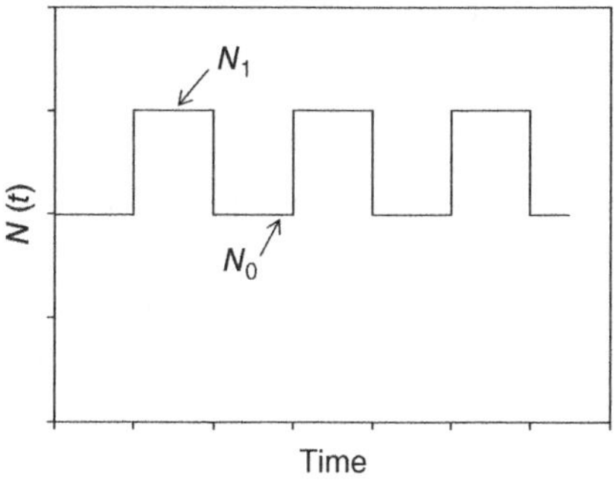

The lower value N_0 corresponds to this intrinsic noise increased by the contribution of the reference source, the upper value N_1 adds the measured object instead of the reference source. We can adjust the difference to zero and then get $T_M = T_{REF}$.

Van der Ziel [10] has analyzed the sensitivity in case $T_{REF} = T_0$. The first radiometers using waveguide technology had a switch in which a dielectric disc rotated in axial direction in the waveguide. Half of this disk was coated with absorbing material. If this part was immersed in the waveguide, it acted like a matched absorber that generated noise with T_0. If the other part was immersed, the antenna signal was passed through.

In this case

$$N_0 = kGBFT_0 \tag{13.35}$$

the gain constant of the apparatus is

$$kGB = \frac{N_0}{FT_0} \tag{13.36}$$

It follows for the amplitude of the square wave signal

$$\Delta N = N_1 - N_0 = kGB\Delta T \tag{13.37}$$

Thus the antenna temperature can be determined from the amplitude of the square wave signal, since N_0 and F are constants of the apparatus

$$\Delta T = \frac{\Delta N}{N_0} F T_0 \tag{13.38}$$

According to the above considerations, the fluctuations behind the PSD with integration are

$$\sqrt{\overline{\Delta T^2}} = 0.28 \frac{FT}{\sqrt{B\tau}} \tag{13.39}$$

In our example with $T = 500$ K antenna temperature and a Dicke-radiometer as receiver ($NF = 3$ dB; $B = 1$ MHz; and $\tau = 1$ s) we would have an uncertainty of the temperature measurement

$$\frac{\Delta T}{T}(\%) = 0.28 \frac{3}{\sqrt{10^6 \times 1}} \times 100 \approx 0.06\%$$

This is a very good value. However, it cannot be improved arbitrarily, since with a higher RF bandwidth B the matching of the input becomes more difficult to realize and with an increase in the time constant τ the reaction time increases unacceptably.

With every power measurement there is the problem of matching the object to the input impedance of the measuring device. This can never be realized exactly for a broadband frequency range. In noise measurements, there is another source of error in addition to the reduced power transmission: the input stage emits a noise wave in the direction of the object under test. This is partially reflected there and runs back and forth between the test object and the input. These re-reflections falsify the measurement, since the value of the correlation coefficient between the outgoing and incoming wave depends on the length of the cable.

In radiometer applications, this can be overcome by a compensation circuit [12]. In addition to the actual receiver, it contains a switch and a circulator (Figure 13.14)

In switch position I, the noise power N_I is applied to the amplifier

$$N_I = kT_{REF}B|\Gamma|^2 \times \frac{1}{2} + kT_M B\left(1 - |\Gamma|^2\right) \tag{13.40}$$

It is composed of the part from the reference source that is reflected by the object (1st term) and the part of the measured object that is reduced due to mismatch (2nd term).

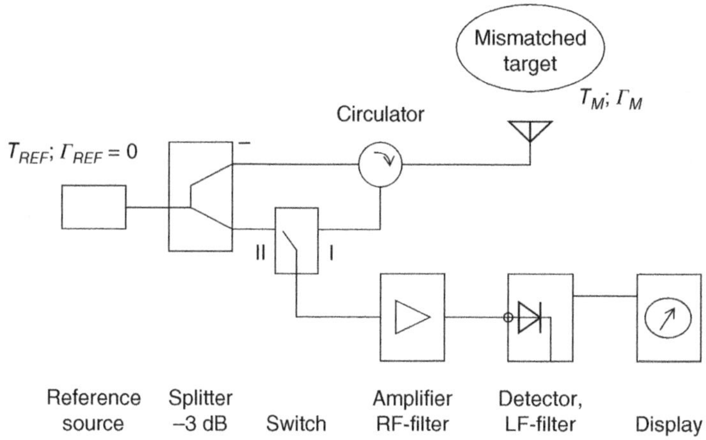

Figure 13.14 Radiometer for mismatched sources. Source: Schiek et al. [12] / John Wiley & Sons.

In switch position *II* the noise power N_{II} reaches the amplifier.

$$N_{II} = kT_{REF}B \times \frac{1}{2} \tag{13.41}$$

The factor is ½ due to the 3 dB splitting. By varying the reference temperature T_{REF}, the difference between the two displays can be tuned to zero. Then the reflection coefficient Γ drops out and we get

$$T_M = \frac{1}{2}T_{REF} \tag{13.42}$$

Correlation Radiometer in the Microwave Range

Figure 13.15 shows the scheme of a correlation radiometer.

In the hybrid coupler at the input, the measurement and reference signal are combined. This generates the sum signal at one output and the difference signal at the other output. Both signals are amplified, band-limited by filters and fed to a correlator. In the microwave range, this is usually an analogue circuit with diodes. This is followed by an averaging stage and the display [12].

At the input of the correlation radiometer there is no switch as in the Dicke-radiometer, but a 3 dB hybrid coupler. A signal at one input port is divided equally between the two output ports. There is a phase shift of 90° or 180° at the two output ports, depending on the design of the hybrid [13]. Figure 13.16 shows a 180° hybrid, as it is usually realized in microstrip technology.

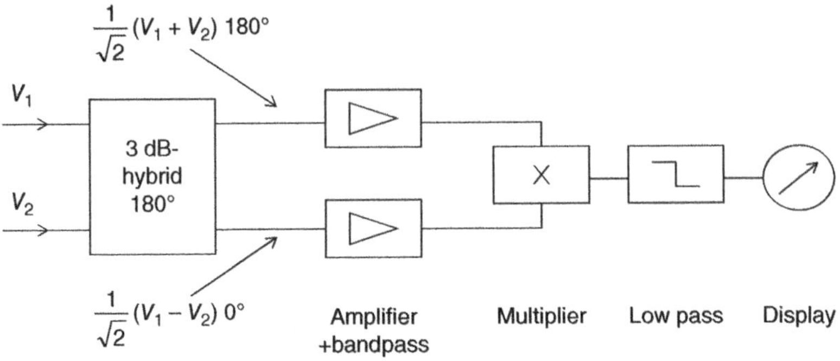

Figure 13.15 Block diagram of a correlation radiometer. At the input there are the measurement signal V_1 and reference signal V_2.

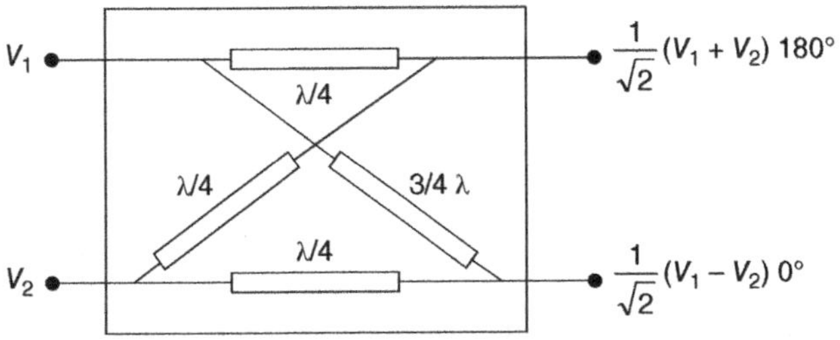

Figure 13.16 Schema of a 3 dB-hybrid 180°.

The multiplier attains the input noise amplified by G and the inherent noise of the amplifier.

$$\frac{G}{\sqrt{2}}(V_1 + V_2) + V_{NA1}; \frac{G}{\sqrt{2}}(V_1 - V_2) + V_{NA2} \tag{13.43}$$

All quantities, the average noise voltages of DUT V_1 and the reference standard V_2, as well as the inherent noise of amplifiers VNA_1 and VNA_2, are not correlated, because they stem from different sources. The two branches are connected to the multiplier, at whose output the following noise power appears

$$N = \frac{1}{2}AG^2\left(V_1^2 - V_2^2\right) \tag{13.44}$$

A is a proportionality factor. The intrinsic noise of the amplifiers drops out because of the lack of correlation in mixed products. If we change over to temperatures, we have to take into account that $V^2 \propto kTB$

$$N = \frac{1}{2}AG^2kB\left(T_1 - T_2\right) \tag{13.45}$$

By adjusting the reference standard to the temperature of the measured object, the output will be zero.

Network Analyzer as a Noise Measurement Device

Modern network analyzers have highly sensitive input stages with well-defined bandwidth. These are comparable to spectrum analyzers and are also suitable as noise measurement receivers. Figure 13.17 shows a diagram of a VNA with digital signal processing in the configuration for a noise measurement [14–16].

This measuring method is made possible by the highly developed digital signal processing in modern measuring instruments, here the VNA. At the measuring frequency f_0 a signal is fed into the DUT. At the output, the amplified input signal is superimposed to the noise contribution of the DUT. Figure 13.18 shows schematically the voltage curve in the time domain at port 2 of the VNA.

The data processing is done in the following way: The RF signal of frequency f_0 is mixed down into the IF range and digitized. This digital signal is analyzed by two detectors. One determines the effective value (RMS), the other the average value (AVG). Both have highest linearity, which is achieved in the lower range

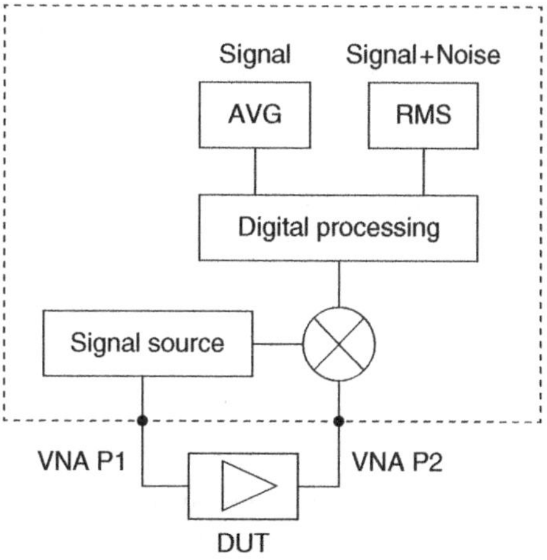

Figure 13.17 Schema of a noise measurement with signal generator and VNA.

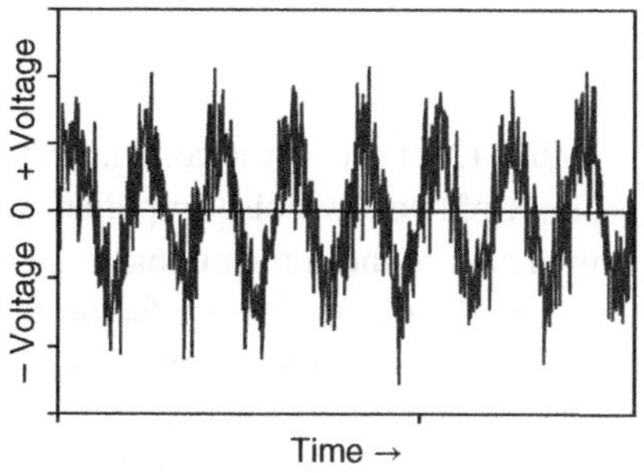

Figure 13.18 Signal after passing through the DUT. A noise voltage is superimposed on the sine wave.

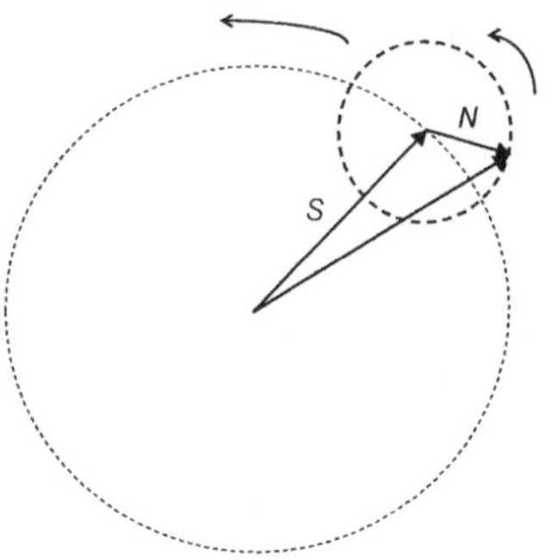

Figure 13.19 Phasor of the signal with superimposed noise Figure 13.18.

by dithering. From the difference of both values the noise power at the output of the DUT can be obtained. The operation can be understood by the vector diagram Figure 13.19.

The vector of the signal voltage S rotates with $\omega_0 = 2\pi f_0$ in the large circle. Its magnitude fluctuates due to the superimposed noise. This forms the vector N, which rotates with approximately the same frequency within the measurement bandwidth in the small circle. The RMS detector averages over a large number of digitally measured values.

$$v_{RMS}^2 = \overline{|S+N|^2} = |S|^2 + \overline{|N|^2} \tag{13.46}$$

The AVG detector first calculates the average value and then squares it. The noise falls out, because the average value is zero.

$$v_{AVG}^2 = \left|\overline{S+N}\right|^2 = |S|^2 \tag{13.47}$$

The noise signal remains as the difference

$$|N|^2 = v_{RMS}^2 - v_{AVG}^2 \tag{13.48}$$

In the evaluation it should be noted that $|N|^2$ is the contribution from the upper and lower sidebands (DSB noise measurement), since no image band filter is used. The noise power is therefore

$$P_N = \frac{|N|^2}{2Z_0} \tag{13.49}$$

Figure 13.20 shows the comparison of the three measuring methods using the example of two low-noise amplifiers at $f = 6\,\text{GHz}$ [17]. On the left of the diagram is the result of the conventional Y-method with noise generator and NFA. In the center of the diagram is a Y-factor measurement with noise generator, but with port 2 of a VNA ZVAB from R&S as receiver. The measurement points on the right were obtained with the VNA without noise generator using the signal generator method. Apart from the VNA, which is necessary anyway, no other device

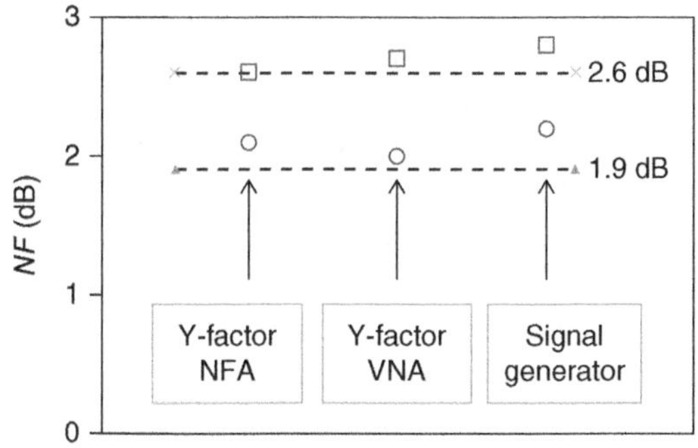

Figure 13.20 Comparison of the methods on a 6 GHz amplifier. Left: Standard measurement with noise figure analyzer. Center: VNA and noise generator. Right: VNA and signal generator.

is required. As the amplifiers are provided with SMA connectors, only $Z_S = 50\,\Omega$ and no multi-impedance measurements were carried out. Manufacturer's specifications for $f = 6\,\text{GHz}$ are: LNA1: $NF = 1.9\,\text{dB}$; $G = 24\,\text{dB}$; LNA2: $NF = 2.6\,\text{dB}$; $G = 25\,\text{dB}$. The performance of the signal generator method is comparable with the Y method, with less equipment required. However, it must be noted that the VNA inputs have a much higher noise figure (10 times) than the NFA. For objects with low inherent gain, a low-noise preamplifier must be used. An S-parameter measurement is then not possible in this configuration.

It has been shown that these devices, which are widespread laboratory equipment, are well suited for a variety of noise measurements. A calibrated noise generator is not required for this.

References

1 Agilent Technologies (2000). Fundamentals of RF- and Microwave Power Measurements. Application Note 64-1B.
2 Schiek, B. (1984). *Messsysteme der Hochfrequenztechnik*. Hüthig Verlag.
3 Kummer, M. (1989). *Grundlagen der Mikrowellentechnik*. Berlin: Verlag Technik.
4 Groll, H. (1969). *Mikrowellen Messtechnik*. Braunschweig, Vieweg & Sohn.
5 Rohde & Schwarz (2010). Spannungs- und Leistungsmesstechnik. Application Note.
6 Agilent Technologies (1999). Schottky Barrier Diode Video Detectors. Application Note 923.
7 Skyworks Solutions (2008). Mixer and Detector Diodes. Application Note.
8 Agilent Technologies (1990). The Criterion for the Tangential Sensitivity Measurement. Application Note 956-1.
9 Avago Technologies (2006). HSCH53 Beam Lead Schottky Diodes for Mixers and Detectors. Datenblatt.

10 van der Ziel, A. (1955). *Noise*. London: Chapman and Hall.
11 Dicke, R.H. (1946). The measurement of thermal radiation at microwave frequencies. *Rev. Sci. Instrum.* 17: 268–275.
12 Schiek, B., Rolfes, I., and Siweris, H.-J. (2006). *Noise in High-Frequency Circuits and Oscillators*. Wiley Interscience.
13 Detlefsen, J. and Siart, U. (2012). *Grundlagen der Hochfrequenztechnik*. Oldenbourg Verlag.
14 Boss, H., Schmidt, K., and Freidhof, N. (2007). Verfahren und Vorrichtung zum Bestimmen einer Rauschgröße eines elektronischen Meßobjektes. Patent, DE10302362A1.
15 Rohde & Schwarz (2009). ZVAB-K30: Noise Figure Measurement Option. R&S App. Note, Munich, Germany.
16 Maury Microwave Corp (2012). 50GHz Noise Parameter Measurements Using Agilent N5245-Series PNA-X with Noise Option 029. Application Note 5c-088.
17 Rudolph, M., Heymann, P., and Boss, H. (Oct. 2010). Impact of receiver bandwidth and nonlinearity on noise measurement methods. *IEEE Microwave Mag.* 11: 110–121.

14

Equipment and Measurement Methods

Noise Measurement Receiver

The requirements vary depending on the application and frequency range. The receiver must be selected or modified specifically for each DUT. A low-noise amplifier, for example, has a low noise figure with high gain. A mixer has a high noise figure with negative gain. The power emitted by the DUT is usually so low that it must be amplified considerably in order to measure it reliably at the indicator. This amplification is usually done in several steps with frequency conversion to avoid oscillations. In Figure 13.8 the essential components of the noise measurement receiver are outlined: amplifier and indicator. In addition, there is usually a filter at the input, which determines the frequency range. If its bandwidth is small with respect to the center frequency, the spot noise figure is measured at this center frequency. Within the bandwidth, both the noise power of the DUT and the noise figure of the test receiver must be constant (white noise). Tunable filters in the microwave range usually consist of coupled resonant circuits. Their passband curve is well defined. Table 14.1 shows the relationship between noise bandwidth and −3 dB bandwidth. Modern spectrum analyzers have a highly developed filter and receiver technology and are therefore very well suited as noise measurement receivers. In the frequency range $f < 100\,\text{kHz}$, digital FFT analyzers are optimal choice for noise measurements, since analogue filters with the required narrow bandwidth cannot be implemented here.

Amplifier and indicator should have a very good linearity. Where this is not assured, one can use the 3 dB method or generally an attenuator to set the indicator to constant deflection. This avoids nonlinearities in the receiver.

The indicator shows the average value of the noise power within the selected bandwidth over a sufficiently long period. This time averaging is necessary because a statistically fluctuating quantity is measured. The value v to be measured fluctuates continuously with the value Δv. As we already analyzed in Chapter 13, the larger the *RF* bandwidth B and the greater the integration time, the smaller these

A Guide to Noise in Microwave Circuits: Devices, Circuits, and Measurement, First Edition.
Peter Heymann and Matthias Rudolph.

Table 14.1 Noise bandwidth of typical filters.

Circuits	Application	NBW/−3 dB BW
1	Simple set-up	1.57
4	Spectrum analyzers, NFA	1.13
Digital	FFT spectrum analyzer	1.06

Source: Based on Agilent Technologies [4].

fluctuations are. This is certainly to be taken into account for fast sweeps in the frequency range. At $B = 1$ MHz and $\tau = 5$ ms, the fluctuation is about 1% of the measured value. The displayed value is obtained by averaging higher and lower instantaneous values. How large these instantaneous values are depends on the Crest factor (ratio of peak voltage to RMS value) of the noise signal. With the Gaussian distribution, this value is theoretically infinite. In practice, the correct recording of amplitude values of three times the average measured value is sufficient. Higher values occur only in $^1/_{1000}$ of all cases. Therefore a linearity of amplifier and indicator up to three times the average value is sufficient. This should be checked for high accuracy measurements.

The indicator is in a broader sense an alternating voltage voltmeter. With simple equipment and low demands on measurement accuracy, the noise level can be measured on an oscilloscope. This corresponds approximately to a measurement of the peak voltage. With an analogue oscilloscope we have about $v_{eff} \cong 0.2v_{pp}$. For a digital storage oscilloscope (DSO) $v_{eff} \cong 0.17v_{pp}$ because even large deflections are stored and visible here.

The indicator is usually an AC voltmeter with a square-law characteristic. If such a device is used in an experimental setup, it should be noted that the linear mean value of the rectified signal is measured here. The scale is calibrated in RMS values. However, this only applies to sinusoidal voltages. It is better to use a True-RMS voltmeter. The relationship between RMS value v_{eff} and rectified value v_g for full-wave rectification is $^{v_{eff}}/_{v_g} = 1.11$ (*Sine*); 1.25 (Gauss). However, it should be noted that a Gaussian distribution of the amplitudes is only present at the indicator if the lower frequency limit is sufficiently low in a broadband noise measurement. When measuring the band-limited noise with filters whose bandwidth is small with respect to the measuring frequency, which is usually the case, we have a Rayleigh distribution at the indicator. In practice, this is not important when working with a noise generator using the Y-method. If one has a modern spectrum analyzer as test receiver, this is already taken into account in the noise mode for the marker display. In Chapter 2 we had mentioned that this correction of 2.5 dB is achieved when using the logarithmic scale. A detailed explanation is given in the following text.

This must be taken into account for other configurations. In addition, the linearity requirement is somewhat higher than for Gaussian distribution, as the Rayleigh distribution comprises higher components.

Depending on the facilities in the laboratory and accuracy requirements, the following configurations are suitable:

(1) Self-made combination of measuring amplifier and indicator, with or without filter or noise generator.
(2) Spectrum analyzer as test receiver, usually with low-noise preamplifier.
(3) For highest requirements noise figure analyzer (NFA), e.g. Keysight N8973A … N8975A [1], with the corresponding noise generators.
(4) Vector network analyzers of the latest design, e.g. Rohde & Schwarz ZVA and ZVT [2, 3] can also be used for noise measurements (Chapter 13).

Spectrum Analyzer

Let us consider a measurement with a simple spectrum analyzer (SPA), a device that is available in most RF laboratories and workshops [4, 5]. This has several advantages. In the tuning range of the SPA we can measure the spot noise figure with selectable resolution (RBW). One can monitor the interfering radiation in the environment and see if there is a possible error due to external interference. Even if the actual noise measurement is carried out with very good equipment, a simple procedure for function check and overview is useful [6]. Spectrum analyzers are broad-band tunable receivers whose input usually is directly connected to a mixer and which therefore cannot be designed for low noise. For a good instrument, e.g. R&S FSV, a noise floor (DANL = Displayed Average Noise Level) of −152 dBm at RBW = 1 Hz is specified. This is a theoretical value because RBW = 1 Hz is too low for a measurement. Let us first calculate the noise factor of the spectrum analyzer used as a test receiver. According to the definition of F

$$F = \frac{N_{OUT}}{kT_0B} + 1 \tag{14.1}$$

The DANL is measured on a logarithmic scale and is expressed in dBm. The following applies

$$N_{OUT}(\text{mW}) = 10^{\frac{DANL}{10}} \tag{14.2}$$

In logarithmic notation, the noise figure of the spectrum analyzer is given by

$$NF(\text{dB}) = DANL + 174(\text{dBm}) + 2.5\ \text{dB} \tag{14.3}$$

In our example: $NF = 24.5$ dB. The 2.5 dB correction comes from underestimating the noise with the average detector and logarithmic averaging [7].

The level displayed by the spectrum analyzer represents an average value over the noise power present in the resolution bandwidth RBW. The calculation of this average value leads to a problem of signal statistics. As we have seen in Chapter 2, the amplitude distribution for band-limited noise is no longer the Gaussian distribution of white noise, but the Rayleigh distribution. Measuring the amplitudes on a logarithmic scale complicates the interpretation of the average value.

Modern spectrum analyzers have different types of detectors, of which the sample detector is the most suitable for noise measurements. It takes values from the envelope of the noise voltage at given times and thus records the entire distribution of amplitudes over a certain time range. The root mean square of N measured voltage values v_i $(i = 1 \ldots N)$ is

$$\overline{v^2} = \frac{1}{N}\sum_{i=1}^{N} v_i^2 \tag{14.4}$$

Correspondingly for the power

$$\overline{P} = \frac{1}{N}\sum_{i=1}^{N} P_i \tag{14.5}$$

The measured voltages are fluctuations with the statistical distribution $W_v(v)$. In general holds for the transition of probability densities, in this case from voltage v to power P

$$W_P(P) = W_v(v)\frac{dv}{dP} \tag{14.6}$$

The power will be normalized to the root mean square value of v.

$$P - \overline{P} = 10log\left(\frac{v^2}{\overline{v^2}}\right) \quad \text{(dB)} \tag{14.7}$$

For further consideration, the curve shape of the probability density distribution W is decisive. In our case, W is a logarithmic Rayleigh distribution, since the measured values are in the dB scale. The Rayleigh distribution of the voltages is

$$W_v(v) = \frac{2v}{\overline{v^2}}\, exp\left(-\frac{v^2}{\overline{v^2}}\right) \tag{14.8}$$

Since the measured power P is in dB, the following conversion is necessary. We remain with the 1 Ω-reference, since the resistance is not important for the consideration, but instead of the decadic logarithm (log) we go to the natural (ln). The well-known equation $P = 20\,log\,(v)$ is then

$$P = 20 \times 0.434\, ln(v) \tag{14.9}$$

The transition from the distribution (14.8) to the logarithmic Rayleigh distribution of power is achieved by transformation (14.6). For this we need $^{dv}/_{dP}$. From

(14.9) results $v = exp(aP)$; $a = 0.115$. And the derivation

$$\frac{dv}{dP} = a \times exp(aP) \tag{14.10}$$

The logarithmic distribution is

$$W(P) = \frac{2a}{\overline{v^2}} exp(2aP)\, exp\left\{ -\frac{exp(2aP)}{\overline{v^2}} \right\} \tag{14.11}$$

Instead of the root mean square of the voltage (14.4) we introduce the mean value of the power corresponding to (14.9)

$$\overline{P} = 4.34 \times ln\left(\overline{v^2}\right); \quad \overline{v^2} = exp(2a\overline{P}) \tag{14.12}$$

So we can write (14.11):

$$W(P) = 2aexp\{2a(P - \overline{P}) - exp[2a(P - \overline{P}]\} \tag{14.13}$$

It is to be expected that the averaging of the logarithmic measured values P will result in a different value than the linear average $\overline{P}$ according to (14.5). So we need the expected value of the log-Rayleigh distribution. The Rayleigh distribution is a special case of the Weibull distribution. Its density distribution has the general form

$$W(x) = \frac{a}{\eta} x^{a-1} exp\left\{ -\left(\frac{x}{\eta}\right)^{a} \right\} \tag{14.14}$$

We refer to a as the form parameter and η the scale parameter. In our case $a = 2$ and $\eta > 0$. By the transition $x \rightarrow log(x)$ we obtain the double exponential distribution, which corresponds to the form derived above (14.13). We set in (14.13) $y = ln(x)$; $\mu = ln(\eta)$; $z = 2(y - \mu)$ and receive the form of distribution

$$WE(y, \mu) = 2exp\{z - exp(z)\} \tag{14.15}$$

With regard to y, it has the expected value and the variance

$$E(y) = \mu - \frac{C}{2}; \quad Var(y) = \frac{\pi^2}{24} \tag{14.16}$$

$C = 0.577$ is Euler's constant.

The relation to the measured values of the noise power and its density distribution (14.13) is determined by the location parameter μ. It is $\eta = \sqrt{\overline{v^2}}$.

For comparison with the average power (14.5), we have to convert $E(y)$ to the dB scale from ln to log: $20 \times 0.434 \times E(y)$. This gives the correction value of 2.5 dB which is always used for noise measurements with the spectrum analyzer. By averaging the logarithmic distribution, the band-limited noise is underestimated by 2.5 dB compared with linear averaging

Figure 14.1 shows an example of the probability density distribution (14.13). The difference between the average value (14.5) and the expected value of the distribution is illustrated here. This is exactly 2.51 dB less than $\overline{P}$.

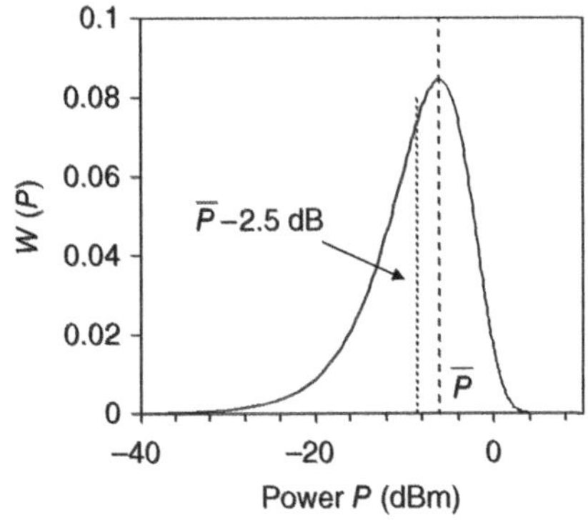

Figure 14.1 Logarithmic Rayleigh distribution with linear average $\overline{P}$ and the expected value of the distribution $\overline{P} - 2.5$ dB.

The DANL value results in a noise figure of spectrum analyzers of $NF > 20$ dB. However, this is too much for a noise measurement receiver. The following should apply for a measurement:

$$NF_{REC}(\text{dB}) < NF_{DUT}(\text{dB}) + G_{DUT}(\text{dB}) - 5\text{ dB} \tag{14.17}$$

For example, if we want to measure an amplifier with $NF = 4$ dB and $G = 13$ dB, the receiver should have $NF_{REC} < 12$ dB. So one has to insert a low-noise preamplifier. How does the noise figure of the test receiver, now consisting of preamplifier and spectrum analyzer, decrease? This is derived from the Friis formula for cascaded two-ports.

$$F_{REC} = F_{AMP} + \frac{F_{SPA} - 1}{G_{AMP}} \tag{14.18}$$

For example, we take ZX60-33LN [8], a low-cost amplifier in coaxial technology from Mini-Circuits. It has the following characteristics in the frequency range 0.5–8 GHz: $NF_{AMP} = 1.3$ (1.2 dB) and $G_{AMP} = 32$ (15 dB). Used in (14.18) the combined noise measurement receiver has $NF_{REC} = 3$ dB. A very good value! Of course, the additional gain must be taken into account in the evaluation. Since it is not constant over the whole frequency range, it should be measured with a signal generator. In linear scale, when N_{SPA} is measured in watts, the result is

$$F_{DUT} = \frac{N_{SPA}}{kT_0BG_{DUT}G_{AMP}} - \frac{F_{AMP} - 1}{G_{DUT}} \tag{14.19}$$

Usually the second term is sufficiently small, so that the noise figure NF_{DUT} results in a logarithmic scale:

$$FN_{DUT} = N_{SPA}(\text{dBm}) - \left(G_{DUT} + G_{AMP}\right) - 10\,log(B) + 174 + 2.5 \tag{14.20}$$

N_{SPA} is the noise level measurement reading at the spectrum analyzer in the dBm scale. If the instrument has a noise marker, the 2.5 dB correction and the correction from 3 dB bandwidth to noise bandwidth are performed automatically. Figure 14.2

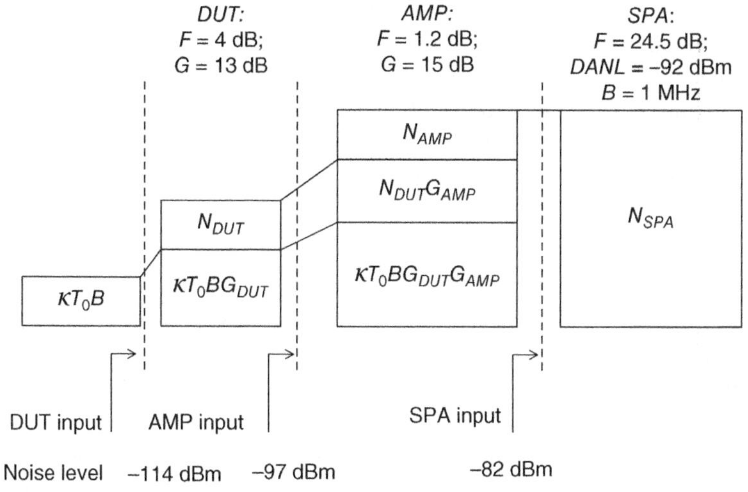

Figure 14.2 Example of the increasing noise level in the chain: DUT – preamplifier – spectrum analyzer.

shows the formation of the noise level at the input of the spectrum analyzer when measuring an amplifier (DUT) using a low-noise preamplifier (AMP).

The Y-Method

The measurement with the Y-method requires the use of a noise generator. If an NFA is used, it is periodically switched between the two states "Hot" and "Cold." It can also be operated in a stationary mode, e.g. if a spectrum analyzer is used as a receiver [9]. Let us consider the two levels in our example Figure 14.2. The lower level in the "Cold" state is unchanged. In the "Hot" state the temperature is no longer T_0 but T_H. From the calibration table of the noise generator we take the ENR value at the selected frequency. This is e.g. at $f = 6\,\text{GHz}$, $ENR = 13.52\,\text{dB}$ (Table 15.2). This means $T_H = 6800\,\text{K}$. The part of the total level coming from the source is now $kT_HBG_{DUT}G_{AMP}$ (Table 14.2).

Table 14.2 Level with the Y method.

Noise source	**Input (dBm)**	N_{SPA} **(dBm)**	***Y* (lin)**	***NF* (dB)**
Cold	$kT_0B = -114$	−82	10	4
Hot	$kT_HB = -100$	−72		

Noise figure NF according to (12.51).

The big advantage of the Y-method with noise generator is that the noise figure can be read directly and the bandwidth is omitted because it is the same for both operating states.

In general, however, the noise contribution of the receiver is not negligible. One must then consider the circuit as a cascade of noisy two-ports. Its first stage is the DUT, but its second stage also makes a contribution. The Friis formula is used to separate the contributions. It reads for this case:

$$F_{MEAS} = F_{DUT} + \frac{F_{REC} - 1}{G_{DUT}} \tag{14.21}$$

To obtain F_{DUT} from the measured value F_{MEAS}, one needs the noise factor of the receiver F_{REC} and the available power gain of the DUT G_{DUT}. F_{REC} is obtained by calibrating the receiver with directly connected noise generator. This step also allows the determination of G_{DUT} (Figure 14.3).

The gain of the DUT is automatically determined in an NFA after calibration, so that after the measurement the quantities F_{MEAS}, F_{REC}, and G_{DUT} are known and F_{DUT} can be calculated from (14.21). Figure 14.4 shows how the Y-method obtains G_{DUT} in addition to F_{REC} and F_{MEAS}.

During calibration, we have the two noise levels N_{CR} and N_{HR} on the indicator. Their ratio $Y = {}^{N_{HR}}/_{N_{CR}}$ corresponding to the noise temperatures of source T_C and T_H gives the value F_{REC}. Similarly, N_{CRD} and N_{HRD} give the value F_{MEAS} with DUT

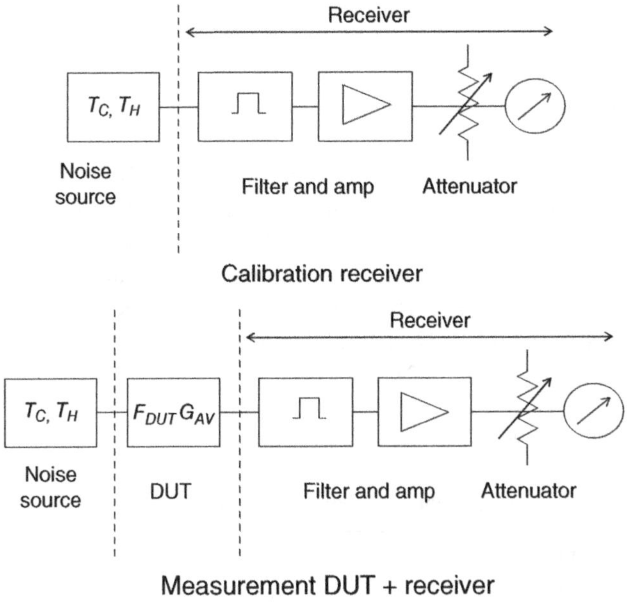

Figure 14.3 Calibration and measurement to determine F_{REC}, G_{DUT}, and F_{DUT} according to (14.21).

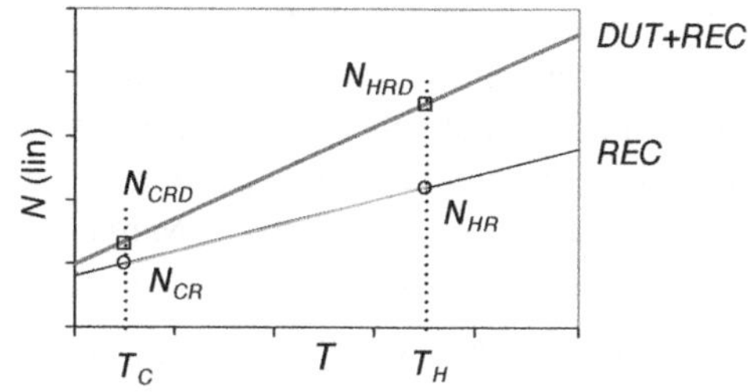

Figure 14.4 Level at the receiver versus the temperature of the noise generator. Lower line: calibration without DUT. Upper line: With DUT.

inserted. Now the total gain is higher, because the increase of the straight lines is given by the product kGB. The receiver straight line has the increase $kG_{REC}B$, and the straight line with DUT has the increase $kG_{REC}G_{DUT}B$. This results in the calculation of G_{DUT} to

$$G_{DUT} = \frac{kG_{REC}G_{DUT}B}{kG_{REC}B} = \frac{N_{HRD} - N_{CRD}}{N_{HR} - N_{CR}} \tag{14.22}$$

In the modern NFA these evaluations are carried out automatically. F_{DUT} and G_{DUT} are displayed in their frequency dependence, because the ENR table of the noise generator can be stored.

Measurements in the Microwave Range

With increasing frequency, there are increasing requirements of equipment. The simplest case is when the frequency range of the amplifier agrees with the frequency range of the NFA.

Here are some examples:

For a long time, this device was unsurpassed [10]. It is still widely used today because of its reliability and ease of use.

Hewlett–Packard noise figure meter: HP 8970 B
Frequency range: 10 MHz–1.6 GHz; $B = 4$ MHz
Measuring range: $NF = 0$–30 dB; $G = -20$ to +40 dB

The range of products is nowadays considerably extended, so that the choice is determined not only by the task but also to a large extent by the budget. Further developments are the devices of the Keysight NFA series [1]

N 8973 A	10 MHz–3 GHz	NF < 11 dB
N 8974 A	10 MHz–6.7 GHz	NF < 11 dB
N 8975 A	10 MHz–26.5 GHz	NF < 13 dB (57 T$)

The bandwidth is adjustable on all models $B = 100$ kHz–4 MHz.

The Rohde & Schwarz FSW series [11] is an example of a spectrum analyzer as a noise measurement receiver. It covers frequencies up to 85 GHz. Of course, the noise figure is relatively high in these devices, which are not specifically designed for noise measurements. Therefore, it is usually necessary to use a preamplifier. For example, the combination of a Narda–Miteq AMF amplifier [12] with an FSW makes a very good receiver for $f = 40\,\text{GHz}$.

Noise figure of the FSW: $DANL = -144\,\text{dBm}$ @ $B = 1\,\text{Hz}$; $NF = 32\,\text{dB}$
Noise figure of the AMF: $NF = 4\,\text{dB}$ at $f = 36–40\,\text{GHz}$; gain $G = 39\,\text{dB}$.

The cascading results in $NF = 4.3\,\text{dB}$. A very good value!

If the desired measuring frequencies are higher than the range of the available device, one has to downconvert into its frequency range. Therefore, a suitable mixer and local oscillator are also required.

The principle of frequency conversion (heterodyne reception) is used in almost all communications systems. By non-linear superposition with an auxiliary frequency (LO), an intermediate frequency (IF) is formed from the incoming frequency, which is then further processed. This has several advantages:

1. The IF is located in a frequency range in which it can be further processed with little effort. For microwave noise measurements, e.g. mixing into the frequency range of the HP 8970B.
2. When changing the incoming frequency, the LO can be tuned and the IF filters remain unchanged. This is the case, for example, with the spectrum analyzer.
3. If the total gain is very high, partial amplifications can be shifted to different frequencies, thus reducing the risk of oscillation.

One problem is the ambiguity of frequency conversion. The IF f_{IF} at the output of the mixer can result from two different frequencies of the high-frequency input signal. These are separated by twice the IF and are located above and below the frequency of the local oscillator f_{LO}. They are called lower $(f_{LO} - f_{IF})$ and upper sideband $(f_{LO} + f_{IF})$. The frequency plan is shown in Figure 14.5. The frequencies f_M measured, if the RF noise is broadband, are therefore

$$f_M = f_{LO} \pm f_{IF} \tag{14.23}$$

The equipment must be extended with a suitable mixer and the local oscillator corresponding to the incoming frequency (Figure 14.6). If only one sideband is to be received, the mirror band must be suppressed by a single sideband (SSB) filter at the input (image reject).

Both RF sidebands therefore produce the same IF. No image reject filter is required for broadband DUTs, since calibration with the noise generator is also

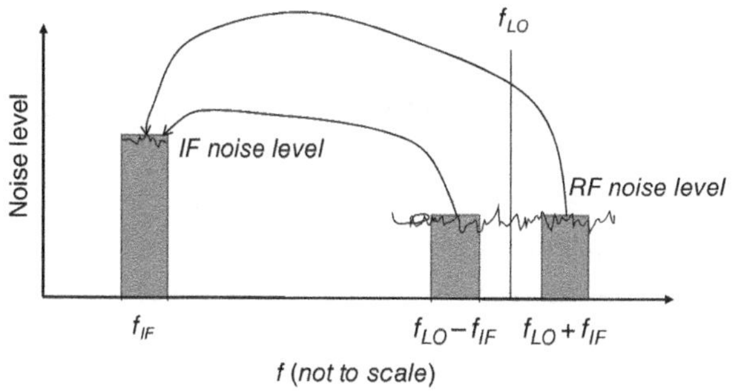

Figure 14.5 Formation of the intermediate frequency f_{IF} from the two sidebands in the RF range.

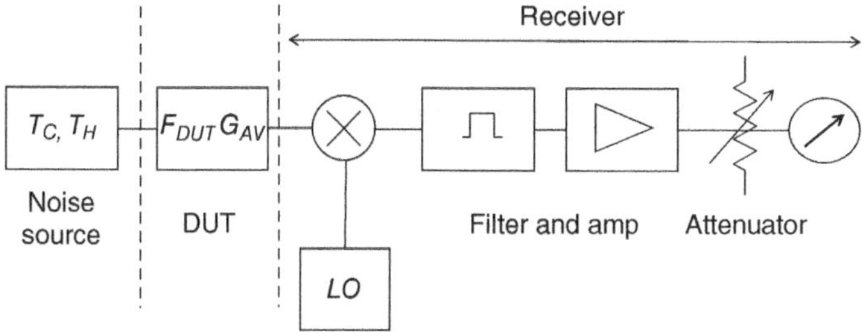

Figure 14.6 Heterodyne receiver with mixer and local oscillator.

broadband. For narrow-band DUTs, e.g. transistors with impedance tuners, this must be taken into account very carefully. It is possible that the impedances are already significantly different within the frequency range of $2 \times f_{IF}$ and the measurement may therefore be incorrect.

If a heterodyne system is characterized, it must be clarified whether SSB or double sideband (DSB) reception is present and whether the wanted signal is received in the upper or lower band. This depends on the design of the mixer and its application in the system. If the mixer has mirror rejection, a noise figure meter (HP 8970A) measures SSB. If both bands are received, it measures DSB. Confusion can now arise when DSB results are used for a system where the wanted signal occurs in only one sideband. The gain displayed by the Noise Figure Meter is too high by 3 dB because the calibration was only made in the 4 MHz IF band of the Noise Figure Meter, but the noise comes from two 4 MHz RF bands. If we apply the noise figure obtained in this way to an SSB communication system, the values are 3 dB too good. The effective noise figure is therefore 3 dB higher.

Selection Criteria of the Mixer

The mixer must be selected in the frequency ranges of RF and IF according to the task.

Other important criteria are:

Noise figure: As low as possible, approximately equal to the conversion loss.
LO isolation: To what extent the LO passes through to the RF input and thus reaches the DUT.
Input matching
Image frequency suppression: There are specially developed mixers with image frequency rejection. A distinction is made between "Image Rejection Mixer" with approx. 20 dB mirror suppression and "Enhanced Image Rejection Mixer" with approx. 30 dB.

There is a wide range of commercial mixers suitable for noise measurements. The selection is determined by the frequency range, the characteristics of the LO and by the budget. There are also downconverters where mixer and LO are combined. A well-known example is the low noise blocks (LNBs), which are located directly at the antenna for satellite reception. Here are some examples of mixers (Figure 14.7).

Suitable mixers for noise measurements are the example

Frequency range: 2–18 GHz
Model: MITEQ DB0218LAa
Local oscillator: $P = 7$ dBm; $IF = 0$–750 MHz
Conversion loss: 7 dB
Input matching: $VSWR = 1.5:1$; $|\Gamma| = 0.2$
Insulation RF – LO: 30 dB
Frequency range: 75–110 GHz
Harmonic mixer for spectrum analyzer: Quinstar W 75–110 GHz [13]
Local oscillator: $P = 7$–14 dBm
Intermediate frequency: <1 GHz
Detection limit at $B = 1$ kHz–85 dBm

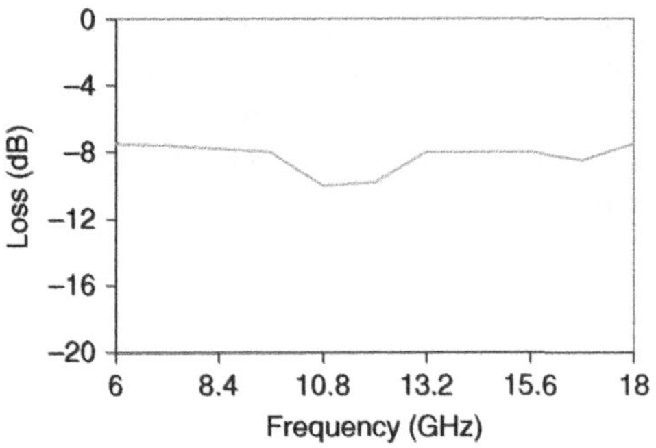

Figure 14.7 Conversion loss diagram of a commercial mixer in microwave range. Source: L3Harris Narda-MITEQ.

Image Rejection

For an SSB measurement it is necessary to suppress the image band. This can be done by a filter at the input. A receiver thus completed is shown in Figure 14.8.

If the IF is very small compared with the measurement frequency, the filter must have a high quality to clearly separate the receive and image band. This limits the applicability. Frequency variation can also be a problem, because tunable YIG filters always have a certain transmission loss ($\approx$2 dB), which increases the noise figure of the receiver. They are also not well matched ($VSWR \approx 2:1$). Despite these disadvantages, this is the method of choice for spectrum analyzers and NFA. Especially since these drawbacks can be avoided by connecting a broadband LNA at the input.

An advisable method for suppressing the image band is implemented in the Image Reject Mixer [14]. A circuit in which the image band is deleted by phase-correct superposition of the mixed products and only the desired signal remains from the RF band. This task is performed, for example, by the Hartley I/Q mixer (Figure 14.9)

It consists of one mixer in an A-branch and one in a B-branch. The local oscillator is split via a 90° hybrid. The $\cos(\omega_{LO}t)$ part is applied to the A-branch, thus creating the I-signal (in phase) at the output of the mixer. In the B-branch the Q-signal (quadrature phase) is generated, due to the modulation by the $sin(\omega_{LO}t)$ part of LO. The position of the phasors in the reference plane 11′ is shown in Figure 14.10. Higher mixing products are of no importance for further considerations and are omitted.

The phasor or complex amplitude combines the amplitude a (here ½) and the phase angle ωt in a complex quantity ($ae^{j\omega t}$).

An important role plays negative frequencies. At first glance they seem unreal to the practitioner, but they result clearly from the Euler formulas of the complex notation of the trigonometric functions. In signal theory their use is standard.

$$cos(\omega t) = \frac{1}{2}\left\{e^{j\omega t} + e^{-j\omega t}\right\}; \quad sin(\omega t) = -j\frac{1}{2}\left\{e^{j\omega t} - e^{-j\omega t}\right\} \tag{14.24}$$

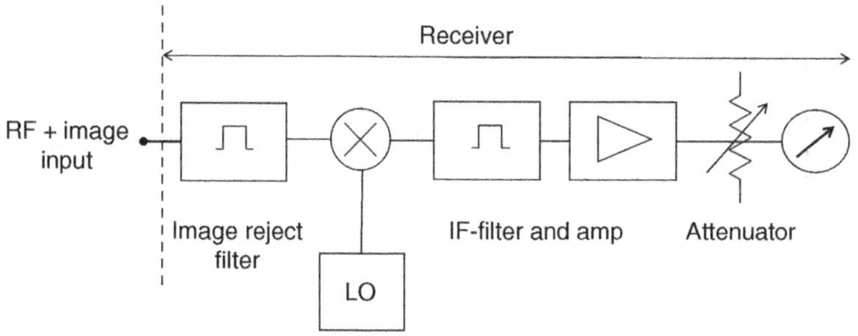

Figure 14.8 Heterodyne receiver with single sideband input filter.

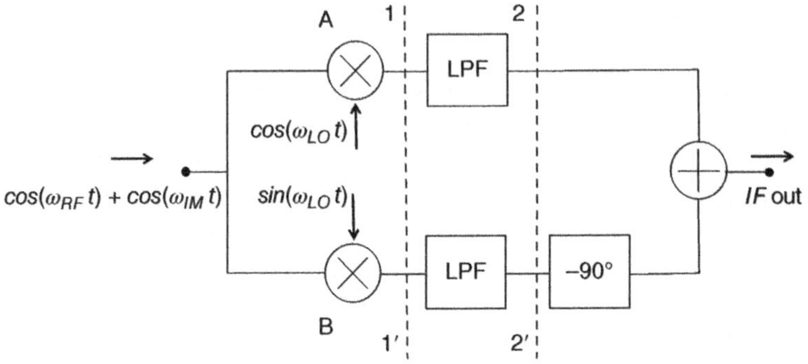

Figure 14.9 Diagram of a mixer with internal suppression of the image band. Source: Archer, Batchelor [22].

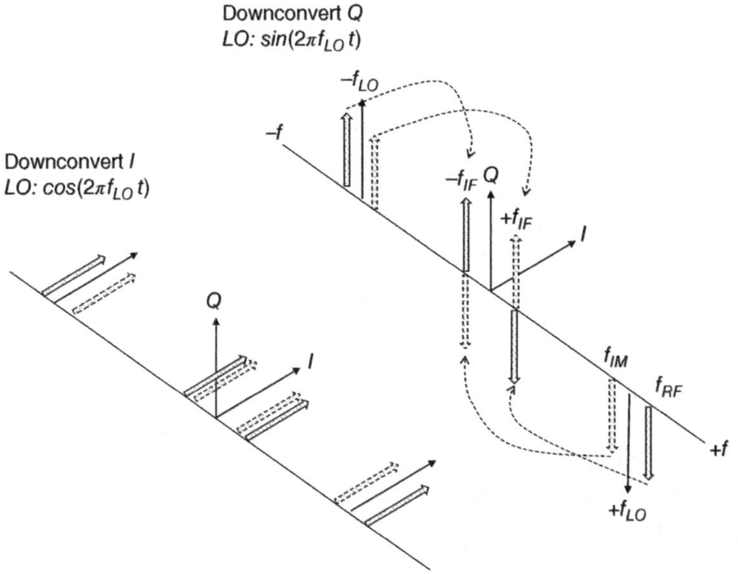

Figure 14.10 Orientation of the phasors in the two branches. In the A branch: *I*; in the B branch: *Q*. Dashed arrows: Mirror band. Solid arrows: Reception band.

In the brackets there are the two phasors rotating in opposite directions $e^{j\omega t}$ and $e^{-j\omega t}$the latter with negative frequency. Their rotation in the complex plane with the angular velocity ω or $-\omega$ is shown in Figure 14.11. Their addition always gives a real value. The *j*-operator rotates a complex number by 90°. $j = e^{j\pi/2}$. These considerations result in the phase position of the mixed products before the low-pass filter shown in Figure 14.10. Behind the low-pass filter, the high

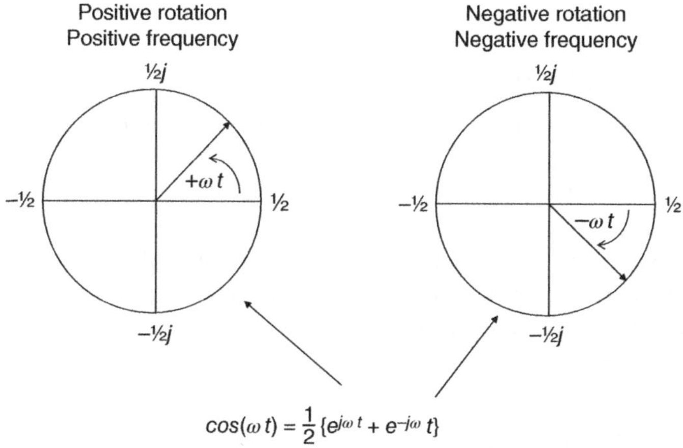

Figure 14.11 Rotation of the phasors representing positive and negative frequencies.

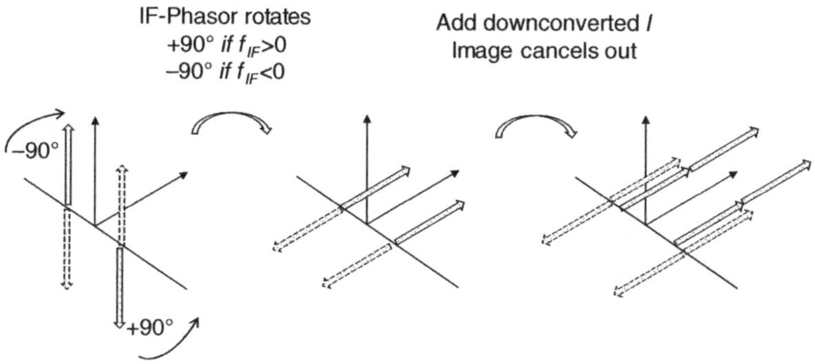

Figure 14.12 Cancelation of the mirror band (dashed) by the effect of the phase shifter in the B branch and subsequent superposition with the A branch.

frequencies near the LO are eliminated. We now look at the IF phasors in the Q branch B in the reference plane 22′, before the phase shifter. These are shown in Figure 14.12 on the left. They are the IF components from Figure 14.10 top right.

The 90° phase shifter is now the decisive element, as it turns positive and negative frequencies in different directions. This can be understood analytically in the following way. Well known is the trigonometric conversion

$$sin\,(x - \pi/2) = -cos(x)$$

In the complex notation we have:

$$sin\,(x - \pi/2) = -\frac{j}{2}\left\{e^{jx}e^{-j\pi/2} - e^{-jx}e^{j\pi/2}\right\} \tag{14.25}$$

With $e^{j\pi/2} = j$ we can write:

$$sin\,(x - \pi/2) = -\frac{j}{2}\left\{e^{jx}(-j) - e^{-jx}(j)\right\} \tag{14.26}$$

One can see that positive frequencies are multiplied by $-j$ (first term in brackets) and negative frequencies by $+j$ (second term in brackets). Multiplication by j, however, means phase rotation by 90°, $+j$ means +90° and $-j$ means −90°.

This is the basis for the effect of the 90° phase shifter, which leads to the cancelation of the image band. The rotation in branch B (Q-component) after the phase shifter is shown in Figure 14.12 center. If the signal from branch A (I-component) is now added, the phasors of the image band interfere to zero and those of the receiving band add up (Figure 14.12 right).

Complete Noise Characterization

We know that the noise factor F is not an independent quantity of the network, but depends on the input circuitry, but not on the output circuitry [15, 16]. The variation of the source reflection factor Γ_S enables the possibility to determine the four noise parameters F_{MIN}, R_N, $|\Gamma_{OPT}|$, $\angle\Gamma_{OPT}$. The variation of Γ_S is performed by means of an impedance tuner according to Figure 12.7. The direct search for the noise minimum by adjusting the tuner and subsequent measurement of Γ_S is not recommended for accurate measurements. Different settings cause a change in the gain of the DUT and also an alteration of the attenuation of the tuner. In this way one cannot obtain R_N either. Since we want to determine four unknowns, it is advisable to measure the noise factor F_i with at least four tuner settings. We then have four equations

$$F_i = F_{MIN} + 4\frac{R_N}{Z_0}\frac{|\Gamma_{Si} - \Gamma_{OPT}|^2}{\left(1 - |\Gamma_{Si}|^2\right)|1 + \Gamma_{OPT}|^2} \qquad i = 1 \ldots 4 \tag{14.27}$$

If one has more measurements, which is useful, the overdetermined system of equations can be solved by the method of least squares. If one applies the Hot/Cold method with noise generator, one must first determine the s-parameters of the tuner with a network analyzer to calculate the selected $\Gamma_{S,i}$. This is necessary to know the attenuation of the tuner in the individual positions. Then the noise generator is connected directly to the receiver, in order to determine the system gain kGB and its noise factor F_{REC}. One must be aware that there are a number of sources of error. The input termination of the receiver is the output of the DUT. If this is poorly matched, or changes with the tuner setting, which is the case with transistors, the calibration may be incorrect. The attenuation of the tuner must be known exactly in order to correct T_{HOT} for each setting. It is also assumed that the matching of the noise generator does not change in the Hot/Cold states. The elimination of these sources of error requires a precise analysis of every detail of the measurement setup.

Analysis of Multi-impedance Measurements

A clearly arranged approach of the overdetermined system of equations obtained from more than four measurements can be achieved by rewriting and linearizing (14.27) [17]. For this we start from the admittance form

$$F = F_{MIN} + \frac{R_N}{G_S}\left|Y_S - Y_{OPT}\right|^2 \tag{14.28}$$

with $Y_{OPT} = G_S + jB_S$. We introduce four new parameters A, B, C, and D. With them (14.28) becomes a linear equation. The relation with the terms of (14.28) is

$$F_{MIN} = A + BG_S + \frac{C + BB_S^2 + DB_S}{G_S}; \quad R_N = B$$

$$Y_{OPT} = \sqrt{\frac{C}{B} - \frac{D^2}{4B^2}} - j\frac{D}{2B} \tag{14.29}$$

The equation for F is now

$$F = A + BG_S + \frac{C + BB_S^2 + DB_S}{G_S} \tag{14.30}$$

The fitting to the n measuring points is carried out according to the method of least squares. For this purpose we define the error criterion:

$$\varepsilon = \frac{1}{2}\sum_{i=1}^{n}\left\{A + B\left(G_i + \frac{B_i^2}{G_i}\right) + \frac{C}{G_i} + \frac{DB_i}{G_i} - F_i\right\}^2 \tag{14.31}$$

It means: F_i measured noise factors, $G_i + jB_i$ source admittance for the ith measuring point.

To minimize the deviation, we form a linear system of equations from the partial derivatives.

$$\frac{\partial \varepsilon}{\partial A} = \sum_{i=1}^{n} P = 0$$

$$\frac{\partial \varepsilon}{\partial B} = \sum_{i=1}^{n}\left(G_i + \frac{B_i^2}{G_i}\right)P = 0$$

$$\frac{\partial \varepsilon}{\partial C} = \sum_{i=1}^{n}\frac{1}{G_i}P = 0$$

$$\frac{\partial \varepsilon}{\partial D} = \sum_{i=1}^{n}\frac{B_i}{G_i}P = 0 \tag{14.32}$$

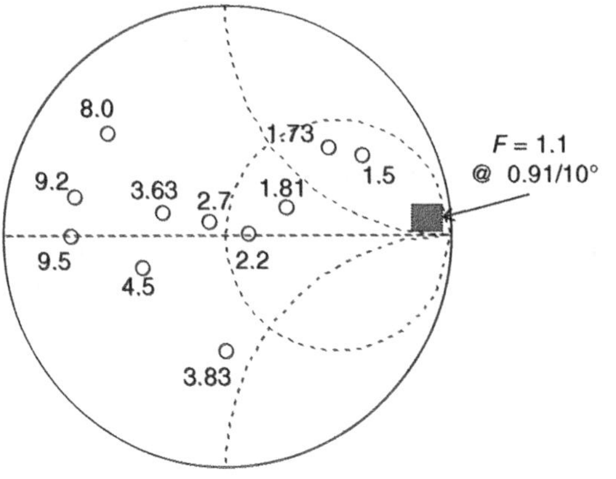

Figure 14.13 Source reflection coefficients Γ_S in the Smith chart (circles). Corresponding noise factors F from measurements. From (14.32) results Γ_{OPT} (square).

It means

$$P = A + B\left(G_i + \frac{B_i^2}{G_i}\right) + \frac{C}{G_i} + \frac{DB_i}{G_i} - F_i$$

The solution of the system of equations provides the optimum values for the four unknowns, from which the four noise parameters can be calculated using (14.29).

As an example, we present the complete noise characterization of a GaAs-FET chip at $f = 2.5\,\text{GHz}$. There were 11 measurements. The reflection coefficients at the output of the tuner are shown in the Smith chart. For each tuner position there is a noise factor, which is also shown in the diagram. In addition, a measurement at $50\,\Omega$ ($\Gamma_S = 0$) $F_{50} = 2.44$. The solution of the linear system of equations with 12 unknowns (14.32) results in $F_{MIN} = 1.1$ (0.41 dB), $R_N = 75\,\Omega$ and $\Gamma_{OPT} = 0.91/6°$ (Figure 14.13).

A good test for the quality of a measurement is the plot of F according to (14.27) as a linear equation. The abscissa is

$$x_i = \frac{|\Gamma_{Si} - \Gamma_{OPT}|^2}{|1 + \Gamma_{OPT}|^2\left(1 - |\Gamma_{Si}|^2\right)} \tag{14.33}$$

If all measuring points are ideal, they all lie on a straight line with the slope R_N/Z_0 that intersects the ordinate axis at F_{MIN}. This is shown for our example in Figure 14.14.

One can see that it is not necessary to set the Γ_{OPT} value directly with the impedance tuner. This is also not possible with on-wafer measurements, because due to the unavoidable losses between tuner and DUT input. In particular at low frequencies the range $|\Gamma_S| > 0.8$ can hardly be reached. Rather, it is important to select the Γ_S values in such a way that the largest possible x_i range is covered. This may require an iterative procedure. It is also important to avoid areas where the DUT can oscillate. Stability check with a spectrum analyzer is recommended in any case.

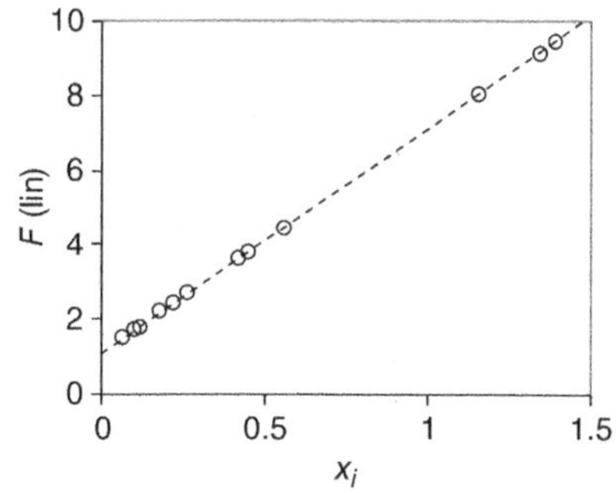

Figure 14.14 Measurement points versus x_i (14.33) ideally form a straight line.

Cold Source Method

When using the Y-method, the power of the noise generator is fed into the DUT via the impedance tuner [18, 19]. The tuner has a different attenuation in each setting. This must be determined very carefully by characterizing the tuner for each position and for each frequency. This is done by measuring the s-parameters with a network analyzer. Additional coupling elements, e.g. cables, bias-tees, and wafer probes, are also added. At high frequency and strong transformation, the attenuation can quickly rise above 5 dB, which cannot be easily corrected. This is where the cold source method creates remedy. Let us look again at the contributions of the components to the noise power displayed by the receiver. They are shown in Figure 14.15 for the simple configuration consisting of noise generator, DUT, and receiver. Here N_R also contains the noise contribution of the receiver, i.e. it is the power which the detector indicates.

The individual contributions are amplified by the transducer gain G_T of the receiver, so that we have the measured power N_R, from which F_{DUT} should be extracted.

$$N_R = kG_{T\,REC}B\left\{T_{NS} + T_0\left(F_{DUT} - 1\right)G_{A\,DUT} + T_0\left(F_{REC} - 1\right)\right\} \tag{14.34}$$

If the DUT is a transistor, and the source reflection factor is varied with an impedance tuner, then in (14.34) we have a number of equations from which the

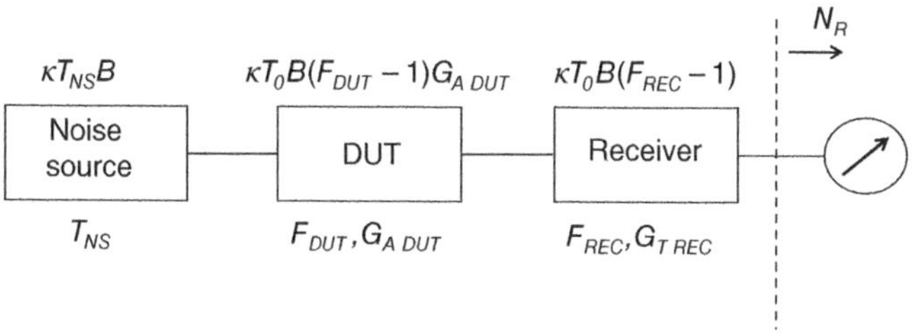

Figure 14.15 Noise contributions of the components for the display N_R on the detector.

final extraction of F_{DUT} is possible, but at least intransparent. T_{NS} varies by the states "Hot" and "Cold," and also at the output of the tuner due to different attenuation, depending on the setting. It varies $G_{A\,DUT}$, the available gain of the DUT, as it depends on the source impedance. F_{REC} and $G_{T\,REC}$ also vary because they depend on the output impedance of the DUT, which in turn depends on the source impedance. The evaluation is done by recording all these quantities and is also used commercially.

The situation is made somewhat clearer by the following analysis of the cold source method. We do not directly measure the noise factor as in the Y-method but the output power N e.g. with a spectrum analyzer. N is the noise power at the output of the DUT, it does not include the receiver component. According to the scheme in Figure 14.16, the noise sources generate the voltage v_{IN} at the input of the DUT. The noise power N_{IN} is fed into the input of the DUT.

$$N_{IN} = \overline{|v_{IN}|^2} Re\left(Y_{IN}\right) = \frac{\overline{i_{tot}^2}}{|Y_S + Y_{IN}|^2} Re\left(Y_{IN}\right) \tag{14.35}$$

At the output this appears amplified as $N = GN_{IN}$. If we insert for i_{tot}, we get

$$N = G \times Re\left(Y_{IN}\right) \frac{\overline{i_S^2} + \overline{i_U^2} + \overline{v_N^2}|Y_S + Y_{COR}|^2}{|Y_S + Y_{IN}|^2} \tag{14.36}$$

$\overline{i_S^2}$ and Y_S are characteristics of the input termination, i.e. of the noise generator and tuner. In the Y method, the noise generator is periodically switched so that $\overline{i_S^2}$ varies between two values. Each tuner setting provides a Y_S and thus an equation (of at least four) for determining the four noise parameters. As (14.36) shows, one can also obtain conditional equations if $\overline{i_S^2}$ is constant and only Y_S varies. The noise generator can be switched off and then generates noise at room temperature, hence the name Cold Source Method. For the evaluation, however, the unknowns G and Y_{IN} must still be determined. The input admittance Y_{IN} of the DUT can be

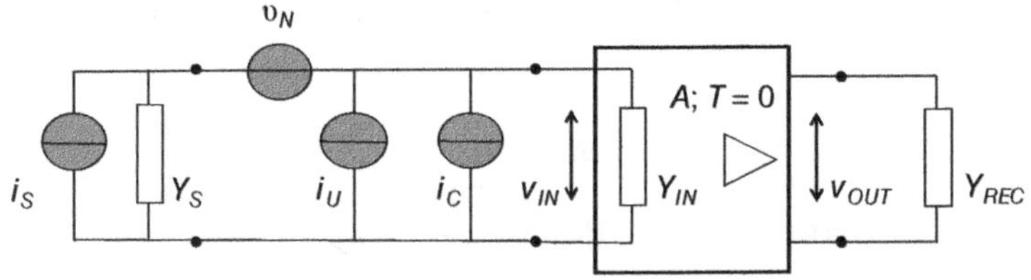

Figure 14.16 Two-port noise measurement between source and receiver.

measured with a network analyzer. With the 7-State method described in the following, it results automatically. The gain G is determined appropriately by a noise measurement with the noise generator set to T_{HOT}. Therefore, it is not possible to operate without a calibrated noise generator.

With constant operating point of DUT and tuner we measure the levels N_H at T_{HOT} and N_C at T_{COLD}. This results in the amplification straight line

$$kGB = \frac{\left(N_H - N_C\right) |Y_S + Y_{IN}|^2}{4\left(T_{HOT} - T_{COLD}\right) Re\left(Y_{IN}\right)} \tag{14.37}$$

With this we have calibrated the equipment.

The 7-State Method

This procedure extends the information from the cold source method [20]. With seven individual measurements in an unmodified structure, all desired data can be obtained.

Details of the situation of a noisy two-port in a measurement circuit can be seen in Figure 14.16. It is the well-known approach where the noise sources are placed in front of the two-port. The input is terminated with the admittance Y_S, the output with the admittance of the receiver Y_{REC}.

The sources have the following values:

$$\overline{i_S^2} = 4kT_S G_S B; \quad \overline{v_N^2} = 4kT_0 R_N B; \quad i_C = v_N Y_{COR}; \quad \overline{i_U^2} = 4kT_0 G_N B \tag{14.38}$$

For the admittances applies:

$$Y_S = G_S + jB_S; \quad Y_{IN} = G_{IN} + jB_{IN} \tag{14.39}$$

This results in the input voltage at the two-port:

$$v_{IN} = \frac{1}{Y_{IN} + Y_S} \left\{ i_S + i_U + v_N \left(Y_S + Y_{COR}\right) \right\} \tag{14.40}$$

In the transition to power we must square to $\overline{|v_{IN}|^2}$.

$$\overline{|v_{IN}|^2} = \overline{|i_S|^2} + \overline{|i_U|^2} + \overline{|v_N|^2}|Y_{COR}|^2 + \overline{|v_N|^2}|Y_S|^2 + 2\overline{|v_N|^2} Y_{COR} Y_S^* \tag{14.41}$$

The mixed terms of uncorrelated quantities are omitted, since the average is zero.

The input power amplified with A is displayed on the receiver.

$$N = A\overline{|v_{IN}|^2} Re\left(Y_{REC}\right) \tag{14.42}$$

If Y_S is varied, A also changes, so that it is one of the unknowns for each measurement of a multi-impedance series. It holds with $g = ARe(Y_{REC})$

$$N|Y_{IN} + Y_S|^2 = g\left\{\overline{|i_S|^2} + \overline{|i_U|^2} + \overline{|i_C|^2} + \overline{|v_N|^2}|Y_S|^2 + 2\overline{|v_N|^2 Y_{COR} Y_S^*}\right\} \tag{14.43}$$

Ideally, one needs seven tuner positions (variations of Y_S) to determine the seven unknowns. These are:

$G_{IN} = Re(Y_{IN})$; $B_{IN} = Im(Y_{IN})$; g; i_U; v_N; $G_{COR} = Re(Y_{COR})$; and $B_{COR} = Im(Y_{COR})$.

Increase of accuracy by additional measurements is recommended in any case. One then obtains an over-determined system of equations, the solution of which can be carried out using the method of least squares. For this purpose we calculate the difference

$$D = N|Y_{IN} + Y_S|^2 - g\left\{\overline{|i_S|^2} + \overline{|i_U|^2} + \overline{|i_C|^2} + \overline{|v_N|^2}|Y_S|^2 + 2\,\overline{|v_N|^2 Y_{COR} Y_S^*}\right\} \tag{14.44}$$

This means

$$D = N|Y_S|^2 + 2NG_{IN}G_S + 2NB_{IN}B_S + N|Y_{IN}|^2 - g\overline{|i_S|^2} - g\overline{|i_U|^2} - g\overline{|i_C|^2} \\ - g\overline{|v_N|^2}|Y_S|^2 + g2\overline{|v_N|^2 Y_{COR}} Y_S^* \tag{14.45}$$

The minimization of the deviations is achieved by partial derivation from (14.45) to the unknowns. By zeroing the individual derivatives, one obtains a system of equations for the seven unknowns.

$$\frac{\partial D}{\partial G_{IN}} = 0; \quad \frac{\partial D}{\partial B_{IN}} = 0; \quad \frac{\partial D}{\partial i_U} = 0; \quad \frac{\partial D}{\partial v_N} = 0; \quad \frac{\partial D}{\partial g} = 0; \\ \frac{\partial D}{\partial G_{COR}} = 0; \quad \frac{\partial D}{\partial B_{COR}} = 0 \tag{14.46}$$

In addition to the actual four two-port quantities v_N, i_U and $G_{COR} + jB_{COR}$, we still get the input admittance $G_{IN} + jB_{IN}$ and the characteristic value of the gain g.

On-Wafer Measurement of Cold Source

A calibration substrate is required for all on wafer measurements in the microwave range [21]. Calibration of the noise measurement apparatus is done with a Thru, as shown in Figure 14.17.

However, the input network consisting of wafer probe, cable, tuner, bias tea, and switch up to the noise generator must first be characterized. One needs a set

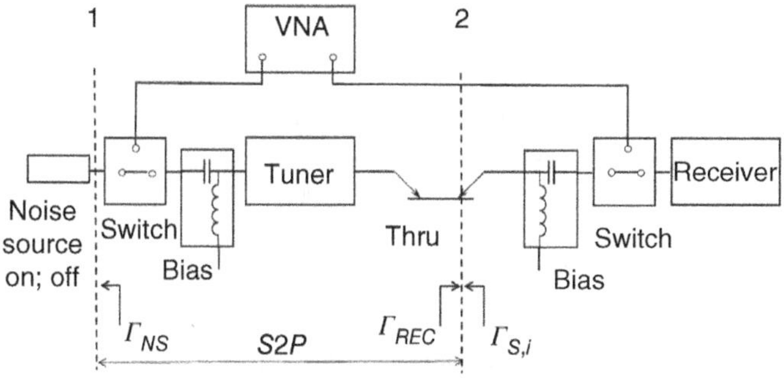

Figure 14.17 Noise measurement set-up for transistor measurement is calibrated with a Thru. The noise generator is operated in both states.

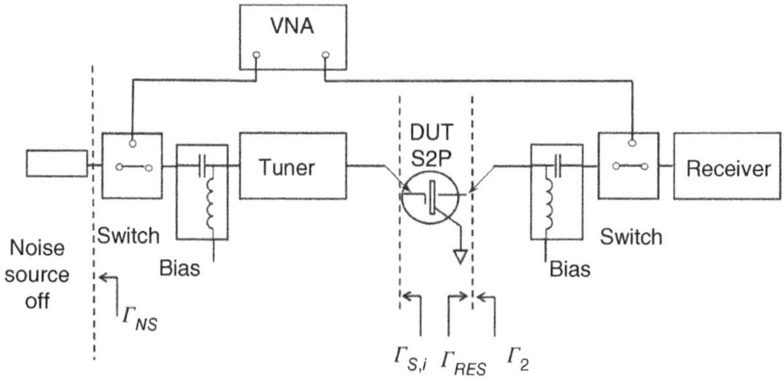

Figure 14.18 Arrangement for measuring a transistor with operating point setting via bias tees. With the cold source method the noise generator is not used.

of s-parameters for each frequency for the different tuner settings. For the actual measurement, the transistor is then used instead of the Thru (Figure 14.18).

The input network is preferably characterized as follows. A passive network can be determined from a single-port measurement with three known terminations. The VNA is calibrated in plane 2 (Figure 14.17). At high frequencies the reference plane must be shifted by the length of the Thru. We then measure with the VNA three sets of reflection coefficients Γ_L, Γ_S, and Γ_O versus the frequency. These three terminations are placed in plane 1 instead of the noise generator. In the simplest case they are Load (50 Ω load resistor), Short and Open. Since $S_{12} = S_{21}$ in the passive network, we only need these three equations.

The input reflection coefficient Γ_{IN} of a two-port terminated with Γ_L is

$$\Gamma_{IN} = S_{11} + \frac{S_{12}S_{21}\Gamma_L}{1 - S_{22}\Gamma_L} \tag{14.47}$$

Table 14.3 Standards for measuring the S-parameters of a passive two-port from single-port measurements.

Standard	Load	Short	Open
Value of standard Γ_L	0	−1	$exp(j\phi)$
Measured value	Γ_{50}	Γ_S	Γ_O

We have to take into account that the open standard in coaxial technology is defined with an offset to the short circuit. This can be given by a frequency-dependent angle, which includes the additional line length and the stray capacitance.

The standards are provided in Table 14.3.

According to (14.47) the measured values are

$$\Gamma_{50} = S_{22}; \quad \Gamma_S = S_{22} - \frac{S_{12}S_{21}}{1 + S_{11}}; \quad \Gamma_O = S_{22} + \frac{S_{12}S_{21}exp(-j\phi)}{1 - S_{11}exp(-j\phi)} \tag{14.48}$$

Directly S_{22} is obtained. S_{11} we receive from

$$S_{11} = \frac{(\Gamma_{50} - \Gamma_S)\, exp(-j\phi) + \Gamma_{50} - \Gamma_O}{(\Gamma_S - \Gamma_O)\, exp(-j\phi)} \tag{14.49}$$

The product $S_{12}S_{21}$ is derived from

$$S_{12}S_{21} = (\Gamma_{50} - \Gamma_S)\,(1 + S_{11})\,; \quad |S_{21}|^2 = |S_{12}S_{21}| \tag{14.50}$$

For the cold source measurement we need the s-parameters of the network only for the calibration with noise generator "On" and tuner in zero position. This allows us to calculate the noise temperature present in reference plane 2. The ENR of the source must be corrected for the attenuation between the planes 1 and 2. When measuring the DUT, the ongoing value $\Gamma_{S,i}$ is sufficient.

It is important to note at this point that the manufacturer's ENR calibration of the noise source is performed with a non-reflecting load. The ENR specification refers to an effective noise temperature T_{NE}

$$ENR = 10log\frac{T_{NE} - T_0}{T_0} \tag{14.51}$$

T_{NE} therefore does not directly correspond to the available power $N_{A,NS}$ of the source, but rather

$$|b_S|^2 = kT_{NE}B = N_{A,NS}\left(1 - |\Gamma_{NS}|^2\right) \tag{14.52}$$

We had already shown that for the correction the available power gain G_A (Available Gain) of the network between reference planes 1 and 2 should be applied. It is defined as the ratio of the available power at the output of the network to the available power of the source at the input. In the present case of the passive network with low attenuation, $G_A < 1$ and is also very low, so we neglect its inherent noise compared to the ENR of the source.

$$\alpha = G_A = |S_{21}|^2 \frac{1 - |\Gamma_{NS}|^2}{|1 - \Gamma_{NS} S_{11}|^2 \left(1 - |\Gamma_S|^2\right)} \tag{14.53}$$

The receiver sees the reflection coefficient Γ_S, according to Figure 14.17.

This means we have to reduce the available power of the source. At plane 2, the power N_2 is available.

$$N_2 = N_{A,NS} \times \alpha \tag{14.54}$$

However, only a part of this power is transferred to the receiver, since there is mismatch in plane 2 between the tuner output and the receiver input.

$$N_{REC} = N_2 \left(1 - |\Gamma_S|^2\right) \frac{1 - |\Gamma_{REC}|^2}{|1 - \Gamma_{REC} \Gamma_S|^2} \tag{14.55}$$

With (14.52)

$$N_{REC} = kT_{NE} B |S_{21}|^2 \frac{1 - |\Gamma_{REC}|^2}{|1 - \Gamma_{NS} S_{11}|^2 |1 - \Gamma_{REC} \Gamma_S|^2} \tag{14.56}$$

This corresponds to the transducer power gain. It is the ratio of the power transmitted by a network to any load to the available power of the source. N_{REC} is fed into the input of the receiver. At the output of the receiver there is the power meter, which displays a value amplified by G_{REC}. This value G_{REC} must be determined by calibration.

$$N_N = G_{REC} N_{REC} + kG_{REC} T_0 B \left(F_{REC} - 1\right) \tag{14.57}$$

The second term is the contribution of the internal noise of the receiver to the value displayed on the power meter. However, this is constant for the value of Γ_S present during calibration and is eliminated when the difference $N_{NH} - N_{NC}$ is calculated. The two powers N_{NH} with noise source on T_{NE} and N_{NC} with noise source on T_C, are measured. This results in the gain-bandwidth product.

$$kG_{REC} B = \frac{N_{NH} - N_{NC}}{T_{NE} - T_C} \frac{|1 - \Gamma_{NS} S_{11}|^2 \, |1 - \Gamma_{REC} \Gamma_S|^2}{|S_{21}|^2 \left(1 - |\Gamma_{REC}|^2\right)} \tag{14.58}$$

Furthermore, the calibration of the system includes the determination of the four noise parameters of the receiver. These can be determined by varying the tuner setting, i.e. different Γ_S. The configuration remains unchanged with the Thru in Figure 14.17. If the noise generator is very well matched, it is used as a terminating resistor, it remains switched off. Otherwise, it is replaced by a 50 Ω standard or its influence on Γ_S is calculated using the S parameters.

We assume that it is very well matched $\Gamma_{NS} \approx 0$ and that the system is at the temperature T_C, which in laboratories is about 5 K above T_0. Depending on the tuner setting, the available power N_{A2} is present at plane 2. It applies (14.56) as earlier, but with T_C instead of T_{NE}.

Due to mismatch the part

$$N_{REC,i} = kT_C B \frac{\left(1 - \left|\Gamma_{S,i}\right|^2\right)\left(1 - \left|\Gamma_{REC}\right|^2\right)}{\left|1 - \Gamma_{REC}\Gamma_{S,i}\right|^2} \tag{14.59}$$

is fed to the receiver. At the power meter we have this part plus the receiver's own noise, both are amplified with G_{REC}.

The measurement consists in setting about $i = 7 \ldots 14$ tuner positions, so that we measure i different powers $N_{N,i}$

$$N_{N,i} = kT_0 B G_{REC} \left(F_{REC,i} - 1 + \frac{T_C}{T_0}\right) \frac{\left(1 - \left|\Gamma_{S,i}\right|^2\right)\left(1 - \left|\Gamma_{REC}\right|^2\right)}{\left|1 - \Gamma_{REC}\Gamma_{S,i}\right|^2} \tag{14.60}$$

For the noise factor this results in

$$F_{REC,i} = \frac{N_{N,i}}{kT_0 B G_{REC}} \frac{\left|1 - \Gamma_{REC}\Gamma_{S,i}\right|^2}{\left(1 - \left|\Gamma_{REC}\right|^2\right)\left(1 - \left|\Gamma_{S,i}\right|^2\right)} + 1 - \frac{T_C}{T_0} \tag{14.61}$$

This results in an over-determined system of equations which can be solved, for example, with the above described method according to the four noise parameters.

During the actual measurement we replace the Thru with the DUT, e.g. a transistor, which we provide with the appropriate bias voltages. First, we measure the s-parameters S_D for each operating point. The frequencies should match those at which the noise parameters of the receiver were also measured. The displayed noise powers now contain the amplified contributions of the DUT, which together with the receiver represents the cascading of two noisy two-ports, which together have the noise factor F_{TOT}. For the DUT, the transducer power gain is effective because the available power at the input is transferred to a mismatched load. This results in

$$N_{N,i} = kT_0 B G_{REC} \left(F_{TOT,i} - 1 + \frac{T_C}{T_0}\right) \left|S_{D21}\right|^2 \frac{\left(1 - \left|\Gamma_{S,i}\right|^2\right)\left(1 - \left|\Gamma_{REC}\right|^2\right.}{\left|1 - S_{D11}\Gamma_{S,i}\right|^2 \left|1 - \Gamma_{REC}\Gamma_{2,i}\right|^2} \tag{14.62}$$

With the output reflection factor of the DUT, i.e. $\Gamma_{2,i}$ the receiver "sees."

$$\Gamma_{2,i} = S_{D22} + \frac{S_{D12}S_{D21}\Gamma_{S,i}}{1 - \Gamma_{S,i}S_{11}} \tag{14.63}$$

The value of F_{DUT} is derived from the Friis formula for cascaded noisy networks

$$F_{DUT,i} = F_{TOT,i} - \frac{F_{REC,i} - 1}{G_{A,DUT,i}} \tag{14.64}$$

Note that both F_{REC} and $G_{A,DUT}$ are different for each tuner setting, as $G_{A,DUT}$ depends on Γ_S and F_{REC} on Γ_2, which in turn also depends on Γ_S. From a series of 7 … 14 suitable tuner settings one can now calculate the four noise parameters of the DUT. A graphical representation according to the Lane Method is quite useful for this purpose, since one can view the position of the individual measuring points and estimate their usefulness.

These considerations are based on Meierer, Tsironis [21]. There the term $(1 - |\Gamma_{REC}|^2)$ does not appear as it is a constant value for the system and is already included in the G_{REC} calibration.

On-Wafer with Noise Generator According to the Y-method

The following analysis of the measurement procedure is based on Archer and Batchelor [22]. Here, temperatures and absorption coefficients are used. The scheme of the configuration is shown in Figures 14.19 and 14.20. Essential components are the NFA, the switch as impedance tuner and the feed of the noise generator via a directional coupler.

The NFA measures N_H if noise generator is "On" and N_C if noise generator is "Off." Feeding the noise generator via directional couplers corresponds exactly to the anechoic load, as when calibrating the noise source. We set $T_H = T_{NE}$. So the source switches between T_H and T_C.

The NFA determines the insertion gain G_M

$$kG_M B = \frac{N_H - N_C}{T_H - T_C} \tag{14.65}$$

The measured temperature T_M results from $Y = {}^{N_H}/_{N_C}$.

$$T_M = \frac{T_H - YT_C}{Y - 1} \tag{14.66}$$

This is of course not the temperature of the DUT, and T_M must be de-embedded to compensate for the losses at input and output.

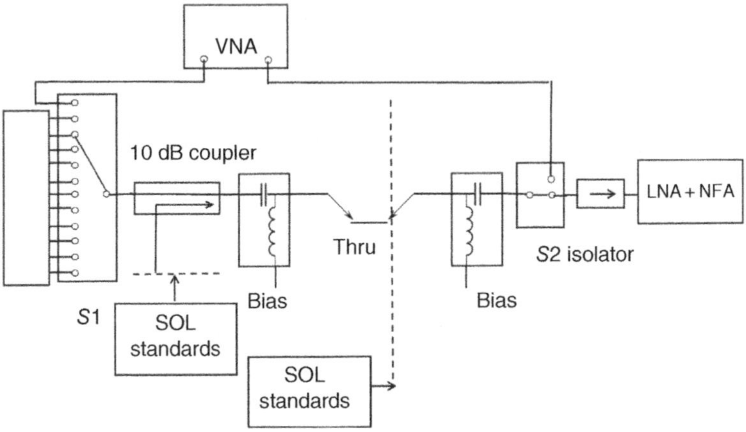

Figure 14.19 On-wafer set-up with NFA, switch and directional coupler in calibration mode. Source: Based on Narda-Miteq [12].

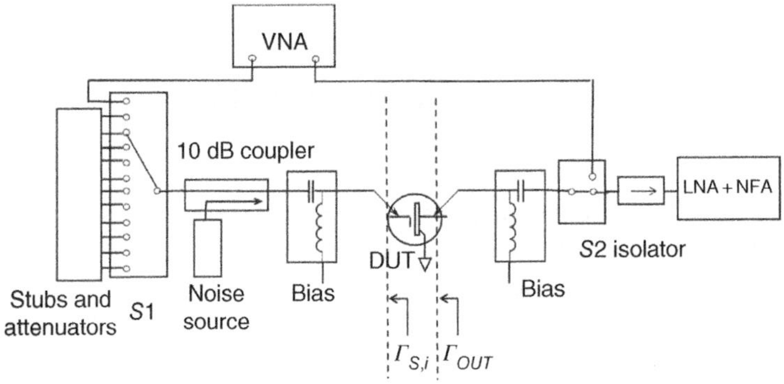

Figure 14.20 Set-up with transistor as DUT.

The authors calculate with temperatures and use the term L as notation of the "Loss." In Chapter 4 we have shown how a series attenuator changes the temperature of a noise source. There we have shown how the linear quantities

$$Re + Ab + Tr = 1 \tag{14.67}$$

are used. It means Re = reflection, Ab = absorption, Tr = transmission. In the absence of reflection holds $N_{OUT} = TrN_{IN}$ and $Tr = 1 - Ab$. We have seen that usually for the losses in coupling networks their available gain is used. This comprises both reflection and absorption. Since these are passive networks, the attenuation is calculated positively $1 \leq \alpha \leq \infty$. The relation to (14.67) is given by

$$\alpha = \frac{1}{1 - Ab} = \frac{1}{Tr} \tag{14.68}$$

The Loss L used by the authors is therefore

$$L = Tr = \frac{1}{\alpha} = 10^{\frac{-\alpha(\mathrm{dB})}{10}} \tag{14.69}$$

In the measurement setup we have a strong attenuation between noise generator and DUT input plane ($\alpha_1 > 10\,\mathrm{dB}$) and a low attenuation between DUT output plane and receiver ($\alpha_2 < 1\,\mathrm{dB}$).

The temperature measured by the NFA in the "On" state is

$$T_{MH} = Tr_2\left(G_{AV}\left[Tr_1\left(T_H - T_C\right) + T_C + T_{DUT}\right]\right)\left(1 - \left|\Gamma_{OUT}\right|^2\right) + Ab_2T_C \tag{14.70}$$

$Tr_1(T_H - T_C) + T_C$ is the part of the noise generator attenuated by the directional coupler and amplified by the available gain G_{AV} of the DUT. The T_{DUT} contribution is also amplified. This contribution is reduced by attenuation in the output network (here Tr_2) and the reflection at the output of the mismatched DUT (here Γ_{OUT}). The receiver is decoupled by using the isolator. Its remaining mismatch is constant and is determined during calibration. The last term records the contribution of the output network. When the noise generator is in the "Off" state, i.e. on T_C, the NFA measures the temperature

$$T_{MC} = Tr_2\left(G_{AV}\left[T_C + T_{DUT}\right]\right)\left(1 - \left|\Gamma_{OUT}\right|^2\right) + Ab_2T_C \tag{14.71}$$

From both values the gain G_M is determined, which contains the available gain G_{AV} of the DUT and the attenuation Tr_1 and Tr_2 at the input and output respectively. From (14.65) we get

$$G_M = \frac{T_{MH} - T_{MC}}{T_H - T_C} \tag{14.72}$$

This results in (14.70) and (14.71) for the measured gain

$$G_M = Tr_1Tr_2G_{AV}\left(1 - \left|\Gamma_{OUT}\right|^2\right) \tag{14.73}$$

From the same values the NFA calculates the temperature T_M according to the Y-method (14.66). With $Y = T_{MH}/T_{MC}$ using (14.72) applies

$$T_M = \frac{T_{MC}T_H - T_{MH}T_C}{G_M\left(T_H - T_C\right)} \tag{14.74}$$

Using this, this results in

$$T_M = \frac{T_{DUT}}{Tr_1} + Ab_1\frac{T_C}{Tr_1} + \frac{Ab_2T_C}{G_M} \tag{14.75}$$

Solved for T_{DUT}

$$T_{DUT} = Tr_1T_M - Ab_1T_C - \frac{Ab_2Tr_1T_C}{G_M} \tag{14.76}$$

The available gain of the DUT is calculated as follows (14.70)

$$G_{AV} = \frac{G_M}{Tr_1 Tr_2 \left(1 - |\Gamma_{OUT}|^2\right)} \tag{14.77}$$

It is now possible to perform a series of measurements with switch 1 (variation of Γ_S) and calculate the four noise parameters according to the method described above. If necessary the Γ_S values are to be optimized iteratively. This can be done by changing the cable lengths or the attenuators on the impedance switch. This can also be done by varying the frequency. A combination that is optimal for one frequency may be useless for another frequency. This must be left out or another combination is chosen. The lane diagram offers a good check of the usability of the measuring points.

Figure 14.19 shows the scheme for characterizing the system. One needs the source reflection coefficients $\Gamma_{S,i}$ for the different positions of switch S1, the attenuation between noise generator and DUT input plane (Ab_1) in the respective position and the losses between output plane of the DUT and receiver input (Ab_2). These quantities are usually very stable and only need to be rechecked occasionally.

To evaluate the measurement, the s-parameters of the DUT are needed to calculate Γ_{OUT}. Therefore the VNA is calibrated with the usual on-wafer standards.

References

1 Agilent Technologies (2007). N8973A – N8975A NFA Series Noise Figure Analyzers. Datasheet.

2 Rohde & Schwarz (2009). ZVAB-K30: Noise Figure Measurement Option. R&S App. Note, Munich, Germany.

3 Maury Microwave Corp. (2012). 50GHz Noise Parameter Measurements Using Agilent N5245-Series PNA-X with Noise Option 029. Application Note 5c-088.

4 Agilent Technologies (2003). Spectrum Analyzer Measurements and Noise. Appl. Note 1303.

5 Reeve, W.D. (2017). Noise Tutorial VI: Noise Measurements with a Spectrum Analyzer.

6 Heymann, P. and Wiatr, W. (1997). Measuring noise parameters of two-ports with spectrum analyzer FSM. *News from Rohde&Schwarz* 153: 20–23.

7 Härtler, G. (2016). *Statistik für Ausfalldaten*. Springer Spektrum.

8 Mini Circuits Coaxial Low Noise Amplifier 50–3000 MHz. Datasheet.

9 Hewlett Packard (2004) Fundamentals of RF- and Microwave Noise Figure Measurements. Application Note 57-1.

10 Hewlett Packard 8970B Noise Figure Meter 10–1600 MHz. Datasheet.

11 Rohde & Schwarz (2015). R&S FSW-K30, Noise Figure Measurements. User Manual.
12 Narda-Miteq (2005). AMF Amplifier Products. Catalogue.
13 Quinstar Technology (2020) Mixer Products QMA. Catalogue.
14 Banerjee, D. (2018). Image Reject Mixers Demystified. Planet Analog.
15 Müller, R. (1990). Rauschen. Springer Verlag.
16 TriQuint (2013). TGF2018 180μm Discrete GaAs pHEMT. Datasheet.
17 Lane, R.Q. (1969). The Determination of Device Noise Parameters. Proc. IEEE 57: 1461–1462.
18 Adamian, V. and Uhlir, A. (1984). Simplified noise evaluation of microwave receivers. IEEE Trans. Instrum. Meas. IM-33: 136–140.
19 Paech, A., Neidhardt, S., and Beer, M. (2010). Noise Figure Measurement without a Noise Source on a Vector Network Analyzer. Rohde & Schwarz Application Note 1EZ61-2E.
20 Schiek, B., Rolfes, I., and Siweris, H.-J. (2006). Noise in High-Frequency Circuits and Oscillators. Wiley Interscience.
21 Meierer, R. and Tsironis, C. (1995). An on-wafer noise parameter measurement technique with automatic receiver calibration. Microwave J. 38: 22–37.
22 Archer, J.W. and Batchelor, R.A. (1992). Fully automated on-wafer noise characterization of GaAs MESFETs and HEMTs. IEEE Trans. Microwave Theory Tech. 40: 209–216.

15

Noise Generators

Vacuum Diode

With the saturated vacuum diode (Figure 15.1), the noise power is directly given by the shot effect of the diode direct current I_D [1]. A calibration by comparison with a thermal standard is not necessary. It is the only noise source with continuously adjustable noise temperature. By varying the heating current of a directly heated cathode, the anode current I_D is continuously varied. For this purpose, the cathode must be a thin tungsten wire whose emission is given by the temperature. The anode voltage must be so high that the saturation current is reached. For example, with the diode GA560 one can achieve a maximum $T_E \approx 20,000$ K. Tubes with oxide cathode are not suitable, because they do not have a saturation current. This source is no longer important in high-frequency and microwave measurement technology. Possibly for special physical applications below $f = 1$ GHz. The parasitic circuit elements L and C already cause an error of 14% at $f = 300$ MHz.

Figure 15.2 shows a detailed photo of a GA560 noise diode. Heated filament as cathode and tubular anode can be seen.

The electron flow generates the noise current

$$\overline{i_S} = \sqrt{2qI_DB} \tag{15.1}$$

The resistance R generates thermal noise and supplies the current

$$\overline{i_R} = \sqrt{4kT_0B^1/_R} \tag{15.2}$$

Both are uncorrelated and can therefore be added quadratically to the noise current of the source

$$\overline{i_{NS}} = \sqrt{\overline{i_S^2} + \overline{i_R^2}} = \sqrt{2qI_DB + 4kT_0{}^1/_RB} \tag{15.3}$$

A Guide to Noise in Microwave Circuits: Devices, Circuits, and Measurement, First Edition.
Peter Heymann and Matthias Rudolph.

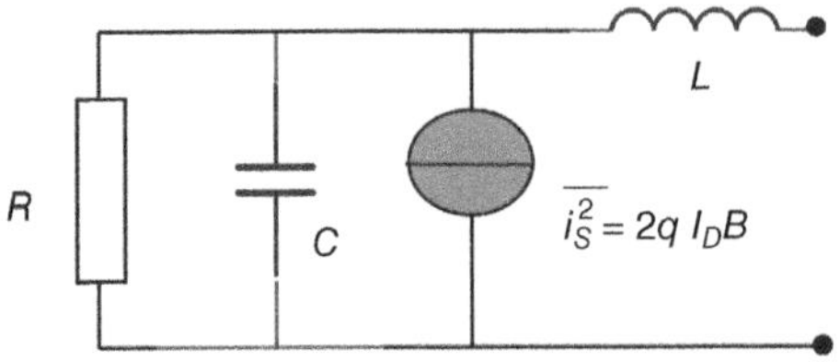

Figure 15.1 Equivalent circuit of a vacuum diode as noise source.

Figure 15.2 Vacuum diode GA560. Source: Radtke [2].

The available noise power is

$$N_V = \frac{\overline{i_{NS}^2} R}{4} = kT_0 B \left(1 + \frac{qI_D R}{2kT_0}\right) \tag{15.4}$$

Since in most cases $R = Z_0 \geq 50\ \Omega$ the thermal part is negligible.

Gas Discharge

Since this source was no longer suitable for higher frequencies, especially in waveguide technology, the plasma of gas discharge tubes was used as a noise source [3]. The positive column of the glow discharge, which can be seen, e.g. in the so-called neon glow discharge lamps, represents a very stable noise source of high temperature. The electron gas has a much higher temperature than the atoms of the gas, which is a few 10^4 K, depending on the type of gas and the

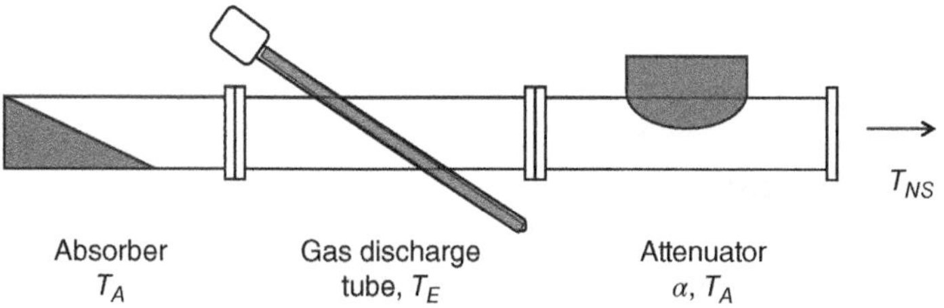

Figure 15.3 Waveguide noise source with absorber and attenuator. T_{NS} is adjustable.

diameter of the tube. A tube inbuilt at an angle in the waveguide is also a very well matched absorber [4]. This source is not tunable, it only has the two states $T_{OFF} = T_A \approx T_0$ or $T_{ON} = T_E$. T_A is the ambient temperature, T_E the electron temperature of the plasma. An attenuator is connected in series to adjust the noise power. We had seen in Chapter 4 how the noise power is influenced by a lossy component. There we had derived the relation between the attenuation value $\alpha(\text{dB})$ ($\alpha > 0$) and the linear absorption

$$\alpha(\text{dB}) = 10\, log\left(\frac{1}{1 - Ab}\right) \tag{15.5}$$

If we consider the setup in Figure 15.3 as an uniform noise source with the adjustable temperature T_{NS}, the following interaction of the components results: In the off state of the discharge and 0 dB attenuation we have the thermal noise of the absorber with the ambient temperature T_A. At higher attenuation nothing changes, because then the attenuator acts like the absorber. In the on-state of the discharge and 0 dB attenuation we have the temperature T_E. In intermediate positions of the attenuator, the temperature is lower than T_E, because the noise power is attenuated, at the same time the attenuator's contribution increases. Quantitatively, as already derived in Chapter 4, the following results are obtained

$$T_{NS} = (1 - Ab)(T_E - T_A) + T_A = \frac{T_E - T_A}{10^{\frac{\alpha}{10}}} + T_A \tag{15.6}$$

Gas-discharge noise generators were very common for a long time, as they were the only sources operating in the microwave range and are stable over a long period of time. Their temperature is derived from the plasma parameters, which can be determined by independent methods, e.g. Langmuir probe measurements [5]. They are still used above 100 GHz today. With the accelerated progress of the transistor in microwave technology, however, the disadvantages became clear. The tubes require a burning voltage of several 100 V, which is no longer common in electronics at this level. The change of on and off states, which is part of the Y-method, requires repeated ignition impulses of several kilovolts. These stray in any experimental setup and can easily destroy the sensitive FETs of an input stage. Furthermore, ionization waves in the kHz range can occur in the gas discharge, which are associated with temperature fluctuations and thus lead to measurement errors.

Semiconductor Diodes

Semiconductor diodes are suitable as noise sources, as the electron flow takes place across a barrier and therefore shows a shot effect [6]. In Chapter 9 (9.14) we had seen that the channel resistance of the currentless diode generates thermal noise with T_A and that in the case of forward direction the equivalent noise temperature is only $\frac{1}{2}T_A$. The temperature differences that can be achieved are small, but are of interest for special applications [7].

Technically important is the breakthrough range of the negatively biased Zener diode (Figure 15.4).

In Z-diodes with breakdown voltages $V_{BR} < 5\,\text{V}$, the reverse current flows due to the Zener effect. In these diodes, the space charge layer is very thin, since p and n regions are highly doped. The electrons pass through this thin layer as a result of the tunnel effect. The current is low, therefore, the shot effect is also small. Z-diodes with $V_{BR} > 15\,\text{V}$ are suitable as noise sources. Here the current flow is amplified by the avalanche effect. The electrons are accelerated in the electric field, resulting in impact ionization with an avalanche-like increase in the density of charge carriers. This is connected with a strong noise. It is caused by this multiplication effect and the formation of local microplasmas. These are narrow conductive channels that statistically form breakthrough paths and immediately disappear again. If the current is limited, this breakdown is reversible. Noise sources based on this effect are common today. They are characterized by a high noise temperature and very good long-term stability. This effect on the noise power is shown in Figure 15.5.

Figure 15.6 shows the circuit of noise diode and attenuator in a noise source [8]. The diode has a resistance that does not correspond to the characteristic

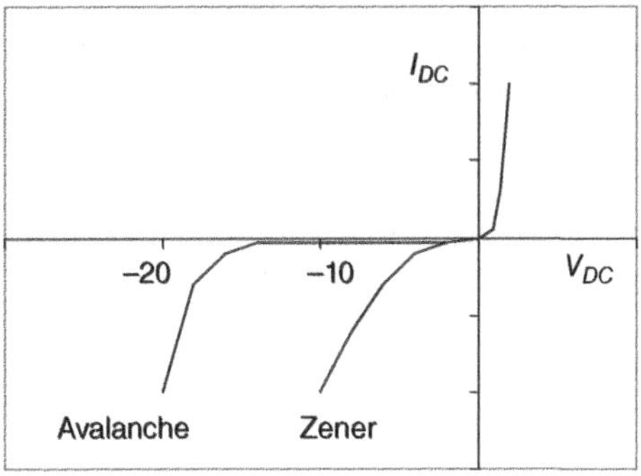

Figure 15.4 Breakdown characteristics of avalanche and Zener diode.

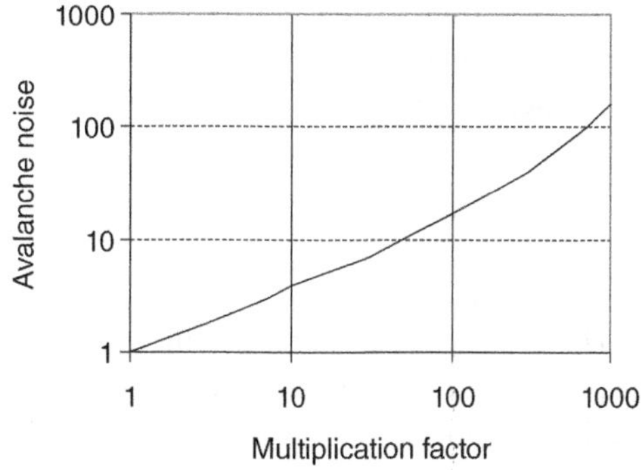

Figure 15.5 Avalanche diode noise amplification.

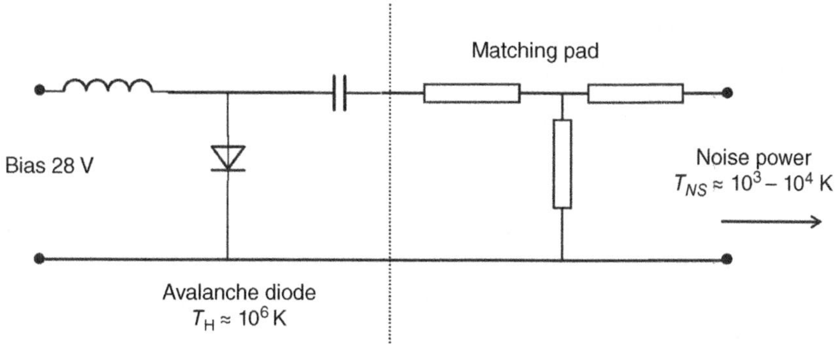

Figure 15.6 Noise generator with avalanche diode and attenuator for matching.

impedance Z_0 of the measuring system. Furthermore, it is different in the on and off state. The matching is achieved by an attenuator. In broadband applications, however, a certain mismatch and a difference between the two states always remains.

Unlike vacuum diode and gas discharge, the noise temperature of these sources is not given by a simple physical relationship. They must be calibrated by comparison with a standard source. The result is a table showing the "excess noise ratio" (ENR) in dB for a frequency range.

Excess Noise Ratio (ENR)

The ENR is a measure of the hot temperature of a noise source T_{NS} [9].

$$ENR(\mathrm{dB}) = 10\,log\left(\frac{T_{NS} - T_0}{T_0}\right) \tag{15.7}$$

It is important to note how this quantity is determined: It results from the noise power that the source delivers to a non-reflecting ($\Gamma_L = 0$), non-emitting load. This is a difference, although a very small one, compared with a conjugate complex matched load for maximum power transmission.

$$T_{NS} = \left\{10^{\frac{ENR(\mathrm{dB})}{10}} + 1\right\} \times 290\,\mathrm{K} \tag{15.8}$$

Here are some values in the Table 15.1.

The coaxial noise sources of the Keysight Series 346 [8] are designed for the application in combination with laboratory noise figure measuring devices (e.g. Agilent 8970B, Keysight N8975A). In case of a fixed installation in a system, e.g. for the continuous function control of radar systems, a higher noise figure is usually required, but only in a narrow frequency band [10]. This is done with a series insulator as protection against high RF pulse power.

Special components in SMD technology have been developed for permanent installation, which fit into the technology of printed circuit boards (PCBs). There are also modules with amplifiers, which can deliver a higher power in the frequency range up to 10 GHz. The continuous monitoring of the operational reliability of receivers makes high demands on the stability of the noise sources. Typical applications of this BIT (Built-in Test) are radar systems, radio astronomy, and magnetic resonance imaging (MRI) in medical technology.

Figure 15.7 shows a very common diode noise source in coaxial technology with high bandwidth.

Table 15.2 shows an example of a table with factory calibration (extract).

Table 15.1 ENR values and corresponding noise temperatures.

ENR (dB)	T_{NS} (K)
15	9460
0	580
−5.43	373 (100 °C)
$-\infty$	290 (T_0)

Figure 15.7 Microwave noise source from Keysight in coaxial technology. Source: Keysight Technologies.

Table 15.2 Frequency table of calibration data of an Keysight noise source (excerpt) [8].

Model Keysight 346 C		Serial number xxx				
		Γ_{ON}		Γ_{OFF}		
Frequency (MHz)	*ENR* (dB)	*MAG*	*ANG*	*MAG*	*ANG*	$\lvert\Gamma_{ON}-\Gamma_{OFF}\rvert$
10	13.01	0.04	−122.9	0.072	−17.8	0.091
100	13.14	0.026	172.6	0.067	−17.5	0.093
1,000	13.05	0.029	30.4	0.056	−155.1	0.085
2,000	13.04	0.043	−99.8	0.038	16.7	0.069
3,000	13.09	0.043	153.8	0.058	−139.3	0.057
4,000	13.20	0.015	19.1	0.047	102.7	0.048
5,000	13.33	0.028	−166.8	0.024	−57.8	0.043
6,000	13.52	0.051	82.6	0.046	128.4	0.038
Up to 26,500	…	…	…	…	…	…

Source: Based on Keysight Technologies [8].

The ENR is given in tabular form for a noise source. It is a measure for the "hot" temperature T_{NS}. In commercial noise sources, this temperature is significantly lower than the noise temperature of the actual source, the diode, due to the series attenuator. The attenuator ensures broadband matching for both the ON and OFF states. However, this does not succeed completely, as can be seen in the last column. There are also models with better matching. However, these have a lower ENR, since the matching is achieved by a higher attenuation. They are recommended for measuring extremely low noise figures.

Hot-Cold Sources

For special and highly accurate calibrations, so-called absolute calibrations, thermometric fixed points can be used in hot/cold loads [11, 12]. Suitable are the boiling points of liquid nitrogen (77 K) or water (373 K). Higher temperatures can be achieved with heated terminating resistors in coaxial or waveguide technology. The effort is considerable and only necessary to create absolute standards. A source of error here is the transmission line. It is not completely loss-free, possibly contains a switch and these losses are at different temperatures. Furthermore, a coaxial load resistor must have the value $Z_0 = 50\,\Omega$ at the ongoing temperature.

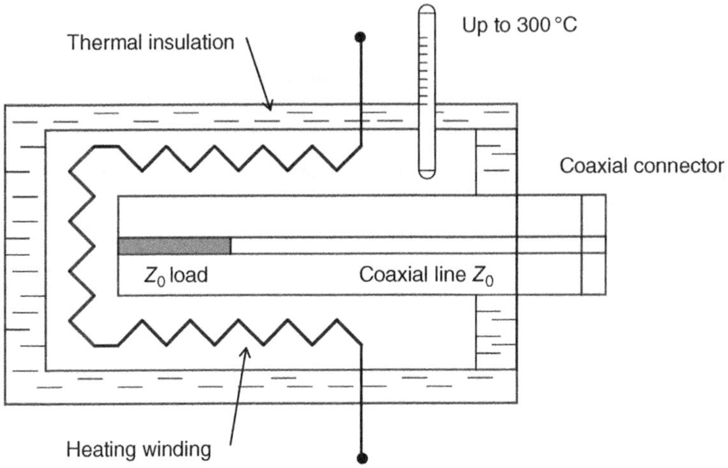

Figure 15.8 Hot load with a coaxial transmission line in a thermostat.

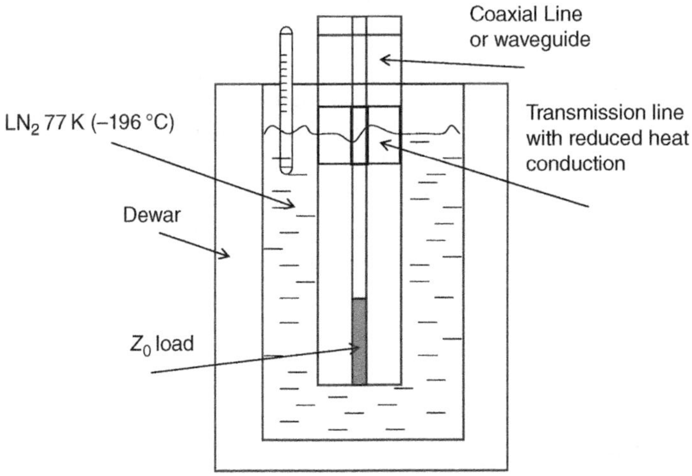

Figure 15.9 Cold load with a coaxial transmission line in a Dewar vessel with liquid nitrogen.

In Figures 15.8 and 15.9 we show the scheme of such systems.

The configurations can also be made using waveguide technology. Their application requires a lot of experience and does not belong to the routine methods. Especially the analysis of losses at different temperatures must be performed very accurately.

References

1 Meinke, H.H. and Gundlach, F.W. (1992). *Taschenbuch der Hochfrequenztechnik*. Springer Verlag.
2 Radtke, U. RFT GA560 Rauschdiode. Tubecollection Udo Radtke.
3 Mumford, W.W. (1949). A broadband microwave noise source. *Bell Syst. Tech. J.* 28: 608–618.
4 Philips (1960) Standard Noise Sources K50A, K51A. Datasheet.
5 Kaufmann, M. (2003). *Plasmaphysik und Fusionsforschung*. Vieweg und Teubner Verlag.
6 Sze, S.M. and Ng, K.K. (2007). *Physics of Semiconductor Devices*. Wiley Interscience.
7 Gaspard, I. (2013). Decreased noise figure measurement uncertainty in Y-factor method. *Adv. Radiosci.* 11: 1–5.
8 Keysight Technologies (2016). 346 A/B/C Noise Sources. Operating and Service Manual.
9 Hewlett Packard (2004) Fundamentals of RF- and Microwave Noise Figure Measurements. Application Note 57-1.
10 Mercury Systems (2020). Noise Sources. Catalogue.
11 Maury Microwave (2018). Noise Calibration Systems and Accessories. Datasheet 4N-062.
12 NoiseCom (2013) Cold Attenuator Noise Measurements on Cryogenic LNA. Application Note.

16

Impedance Tuners

Impedance Transformation with Simple Methods

A specific variation of the impedance of the noise source is not necessary for every measurement. If we only want to determine the noise figure of a matched DUT using the Y-method, it is sufficient to connect the source directly to the input. This is the case, for example, with an amplifier or a receiving system with a matched input. However, if one wants to perform a complete noise characterization, i.e. to determine the four noise parameters (NF_{MIN}, R_N, $|\Gamma_{OPT}|$, $\angle\Gamma_{OPT}$) of a two-port, one has to vary the impedance of the noise source.

The impedance transformation is a task that is very often solved in high frequency technology. The circuit design usually involves matching a mismatched component, e.g. a transistor, to the characteristic impedance of the system. Another case is the conjugate complex matching of an upstream stage to the input of the following stage. In the microwave range, this is done with transmission line sections and in the RF range with concentrated elements, as the line sections would become too long here. It requires some experience to make these transformation elements as broadband and low-loss as possible. As discussed in Chapter 14, the task of the impedance tuner in the noise measurement system is somewhat different. Here the task must be performed to transform the impedance of the noise generator ($Z_0 = 50\,\Omega$) at the input of the DUT as low-loss as possible into desired values of the complex plane.

We first consider the transformation by an LC two-port with variable capacitance, as it could be used in the RF range ($f < 500$ MHz). The calculation is done appropriately with the chain matrix

$$Z_{IN} = \frac{A_{11}R + A_{12}}{A_{21}R + A_{22}} \tag{16.1}$$

A Guide to Noise in Microwave Circuits: Devices, Circuits, and Measurement, First Edition.
Peter Heymann and Matthias Rudolph.

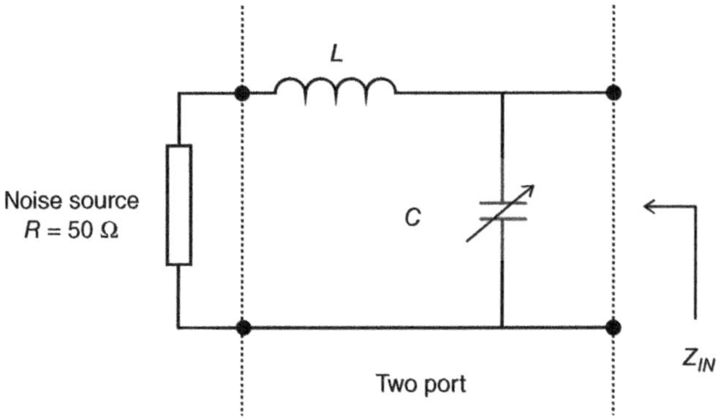

Figure 16.1 LC network for transformation from $R = 50\,\Omega$ to Z_{IN}.

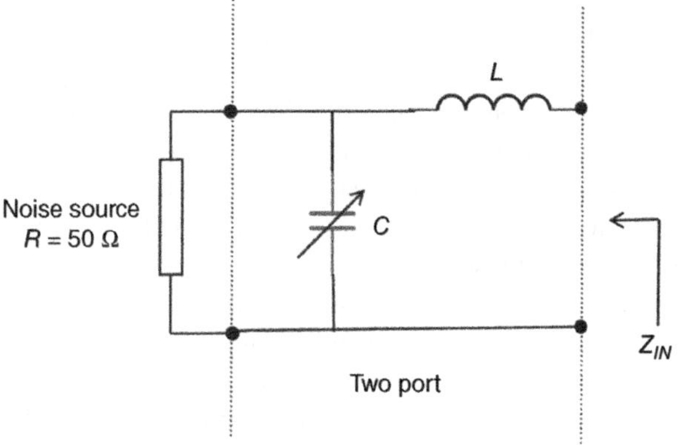

Figure 16.2 CL network for transformation from $R = 50\,\Omega$ to Z_{IN}.

For the network LC in Figure 16.1 the A-matrix is

$$\boldsymbol{A}_{LC} = \begin{pmatrix} 1 & j\omega L \\ j\omega C & 1 - \omega^2 LC \end{pmatrix} \tag{16.2}$$

Other transformation properties have the CL-circuit in Figure 16.2, whose matrix is

$$\boldsymbol{A}_{CL} = \begin{pmatrix} 1 - \omega^2 LC & j\omega L \\ j\omega C & 1 \end{pmatrix} \tag{16.3}$$

An estimation of the achievable values in the complex Z_{IN} plane is shown in Figure 16.3, assuming a varactor diode with capacitance variation $C = 4 \ldots 40$ pF. This is realistic for abrupt junction technology. The LC-circuit is much more effective. Nevertheless, this setup is only of limited use as an impedance tuner. The frequency range is limited to some 100 MHz and the achievable values are very

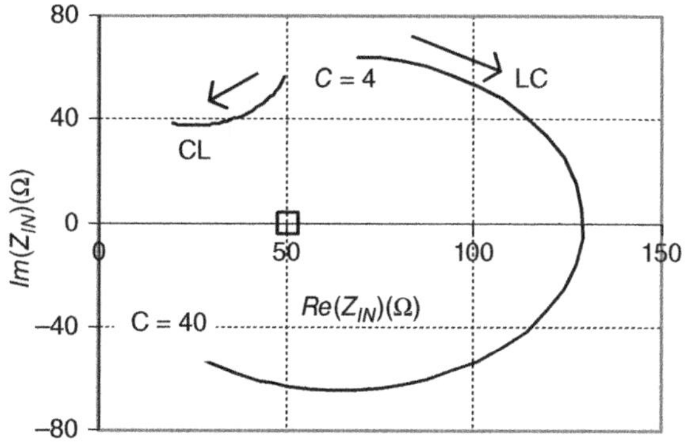

Figure 16.3 Complex impedances Z_{IN} achieved with LC and CL networks with capacity variation $C = 4–40$ pF; $L = 0.1$ μH; $f = 100$ MHz Starting point $R = 50\,\Omega$.

limited. Nevertheless it can be sufficient for a noise measurement, because theoretically only four impedance values are needed to determine the four noise parameters. In practice, however, it turns out that complete noise characterization below 1 GHz is usually not necessary, since powerful noise-models of transistors exist for which only one parameter is needed. This parameter can be obtained by a measurement with 50 Ω source impedance. The tuner problem therefore only arises in the microwave range.

Mechanical Components for the Microwave Range

The impedance transformation in the microwave range can be achieved by a combination of transmission lines of suitable length [1]. We first consider the mode of operation of the line transformation using a frequently occurring example: The matching of a mismatched load Z_L to the characteristic impedance of a line Z_0. This means that the input resistance of the mismatched load and the transformation element together is $Z_{IN} = Z_0$. However, this is valid only for the frequency to which the line lengths are designed.

The classical method is the matching with phase shifter and stub line. In noise measurement the task is exactly the other way round, the well adapted noise generator is to be mismatched in a defined way (Figure 16.4).

The mode of operation can be understood by the following calculation. The extendible line of length ℓ_1 has the normalized input conductance y_1.

$$y_1 = \frac{Z_0 + jZ_L \tan(bl_1)}{Z_L + jZ_0 \tan(bl_1)} \tag{16.4}$$

Here is λ the wavelength and $b_i = {}^{2\pi l_i}/_{\lambda}$; $i = 1;2$.

It is practical to calculate with conductance values, since the reactive conductance of the stub y_2 is connected in parallel at the point ℓ_1 to y_1. This is a line of

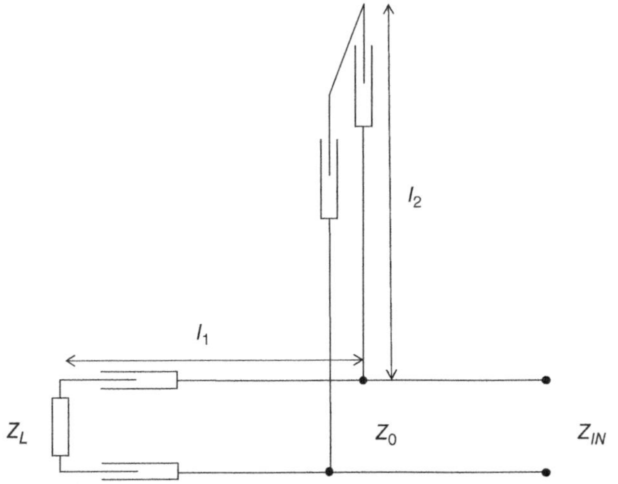

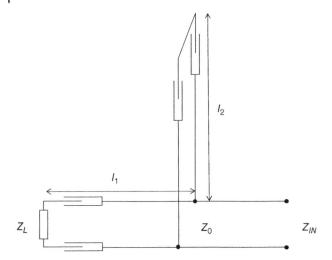

Figure 16.4 The load impedance Z_L is transformed into the impedance Z_{IN} with phase shifter (left) and stub (top).

tunable length ℓ_2 shorted circuited at the end and connected to the main line via a T-junction. Its normalized input conductance is

$$y_2 = -j\cot(b_2) \tag{16.5}$$

For the example of transforming the impedance $Z_L = (100 - j200)\ \Omega$ i.e. $\Gamma_L = 0.825\angle -22.8°$ into the matching point $Z_{IN} = Z_0 = 50\ \Omega$, we want to calculate the line lengths ℓ_1 and ℓ_2.

In Figure 16.5 the transformation path is shown in the admittance diagram. Starting point is the normalized value $y_L = {}^{Z_0}/_{Z_L} = 0.1 + j0.2$ which corresponds to the impedance Z_L. These are the two points for the normalized line length ${}^{l_1}/_{\lambda} = 0$. This length is increased until the real part of the input conductance reaches the value $y_1 = 1$ (symbol). In our case, at $\lambda = 10\,\text{cm}$ wavelength ${}^{l_1}/_{\lambda} = 0.171$ (dotted line). The capacitive reactive component present at this point $Im(y_1) = 2.95$ is compensated by the inductive conductance y_2 of the stub. For this purpose, the stub must have a length of ${}^{l_2}/_{\lambda} = 0.052$. This effect is indicated by the downward pointing arrow parallel to the dotted line. With these two tunings of the line lengths, a narrow-band matching is achieved at $\lambda = 10\,\text{cm}$ wavelength.

The reverse is true for noise measurement: a well matched noise generator must be mismatched in a defined way. The impedance tuner used for this purpose should be as low-loss as possible and cover a wide frequency band. In addition, another property is important. It must not present any impedances outside the frequency range under consideration that cause the DUT to oscillate. Therefore stubs, which are short circuits at low frequencies, are only of limited use.

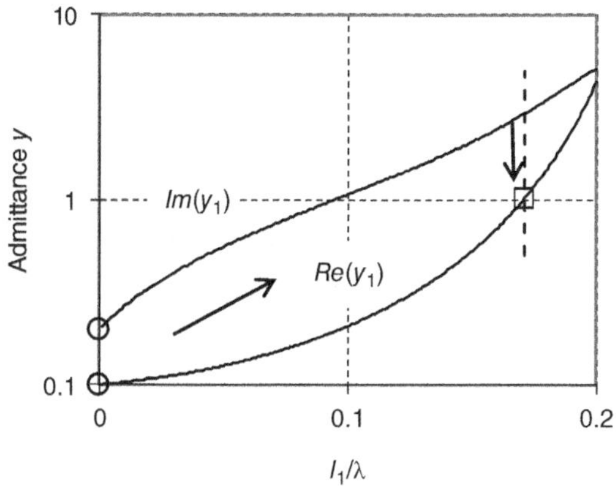

Figure 16.5 Transformation paths of $Re(y_1)$ and $Im(y_1)$ with variation of l_1. The stub compensates the imaginary part to $y = 1$ (symbol).

Capacitive tunings, on the other hand, are ineffective at low frequencies, so that the broadband DUT (e.g. a transistor) "sees" at input 50 Ω. Since modern transistors have a very high gain, especially at $f < 500$ MHz, a reflection-free termination at the input is necessary to avoid oscillations.

Electronic Components

In equipment used for the complete noise characterization of transistors, electronic or mechanical impedance tuners are used to vary the reflection coefficient of the noise source [2]. Electronic tuners are switches based on varactor diodes or MEMs capacitances, which produce a combination of line lengths. Mechanical tuners are usually variable capacitances that can be moved along a precision transmission line. The capacitance is a small plate capacitor whose plate distance is varied by dipping it against the inner conductor. This capacitance is mounted on a slide which can be moved along the line. The adjustment can be done manually or by stepper motors. These are devices of the highest mechanical precision, whereby low losses must be ensured by the construction, especially of the contacts.

An example of an impedance tuner based on switchable capacitors in MEMS technology (Microelectromechanical System) is given in [3]. The switchable capacitors integrated on coplanar lines are used to vary the electrical length of stubs and their relative distance. The entire circuit can be designed so small that it can be placed on a wafer probe for on-wafer measurements. In this case, however, a bias tea would also have to be integrated to prevent unwanted variations in capacitance. Figure 16.6 shows an example with 10 MEMS on a system with open stubs. Which of the 1024 possible settings are selected for a measurement depends on the object.

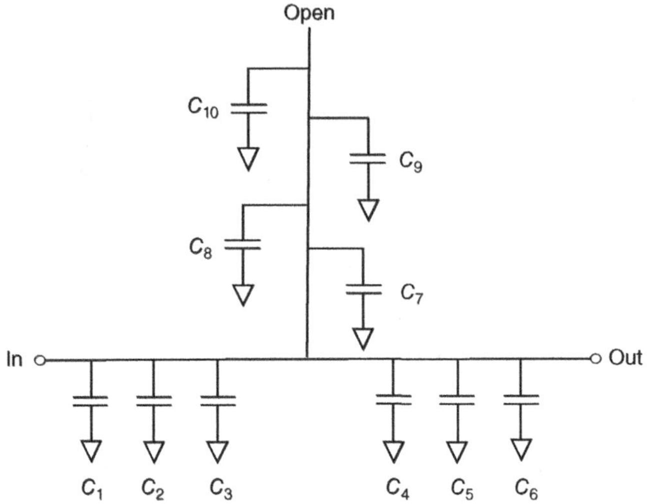

Figure 16.6 Integrated impedance tuner using MEMS technology. Source: [3].

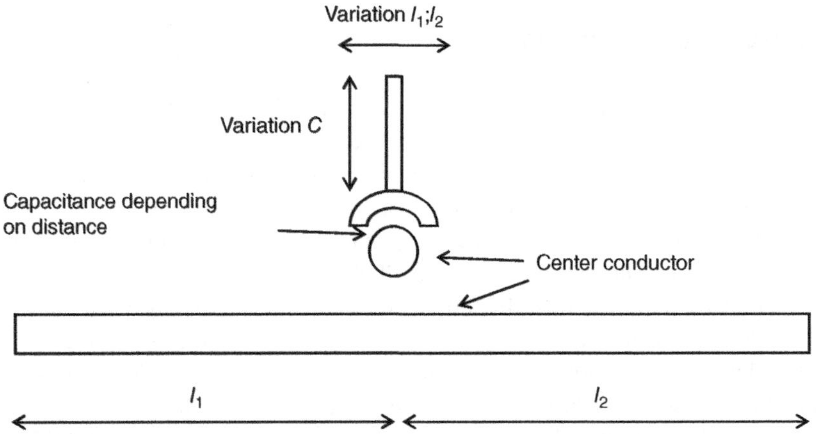

Figure 16.7 Structure of a mechanical tuner. The variable capacity is formed by an optimally shaped electrode and the inner conductor. This is moved along the line [4, 5].

This circuit allows a wide transformation range at 6–26 GHz. It was realized on glass substrate. The mechanical length between input and output is only about 5 mm. The capacity swing is about 80/400 fF.

Figure 16.7 shows the scheme of a mechanical tuner.

Precision Automatic Tuner

The measuring principle with automatic tuner and suitable receiver has been realized to highest performance in commercial systems and has been continuously improved [6, 7] (Figure 16.8).

To calculate its mode of operation, we consider the scheme in Figure 16.9. We divide the tuner into three networks, which are described by their chain matrices $\boldsymbol{A}_1$, $\boldsymbol{A}_C$ and $\boldsymbol{A}_2$. Here $\boldsymbol{A}_1$ and $\boldsymbol{A}_2$ represent loss-free lines. $\boldsymbol{A}_C$ represents the variable capacitance to ground, with the loss resistance R. Noise generator and tuner form a unit, whose output reflection factor Γ_T must be known. In addition, the transformation losses must be taken into account, since they determine the effective noise temperature in the output plane.

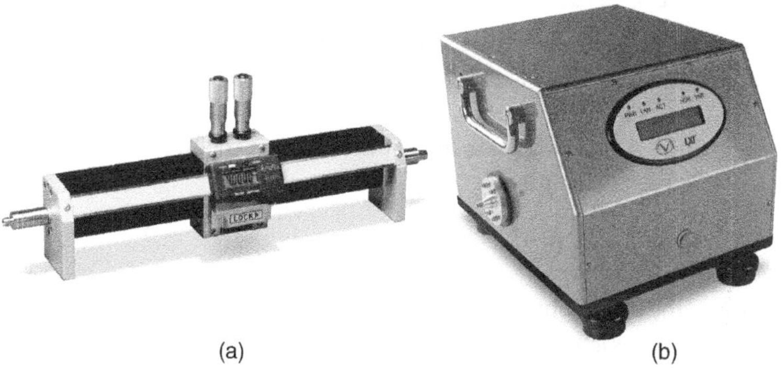

Figure 16.8 Impedance tuner for manual adjustment (a) and automatic tuner with stepper motors (b). Source: Maury Microwaves [2].

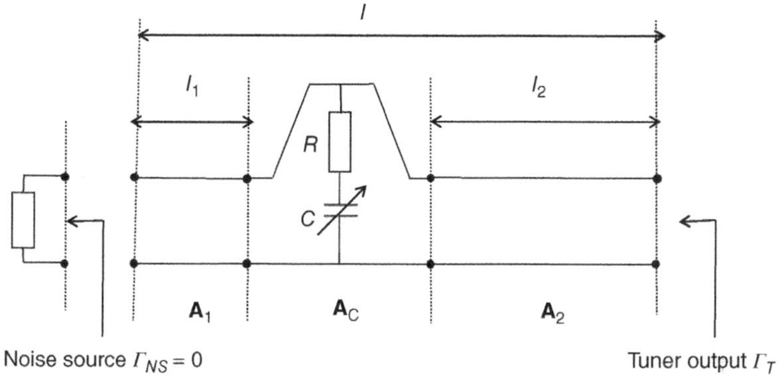

Figure 16.9 Electrical diagram of a mechanical tuner with capacitance, whose value and position on the line are variable.

The line matrices are

$$A_1 = \begin{pmatrix} cos(b_1) & jZ_0 sin(b_1) \\ j\dfrac{sin(b_1)}{Z_0} & cos(b_1) \end{pmatrix}; \quad A_2 = \begin{pmatrix} cos(b_2) & jZ_0 sin(b_2) \\ j\dfrac{sin(b_2)}{Z_0} & cos(b_2) \end{pmatrix} \tag{16.6}$$

The chain matrix of the capacitance contains a resistor R, which stands for the losses in the displacement mechanism.

$$A_C = \begin{pmatrix} 1 & 0 \\ \dfrac{j\omega C}{1 + j\omega CR} & 1 \end{pmatrix} \tag{16.7}$$

By multiplication we get the chain matrix of the tuner, which we convert into the s-matrix.

$$A_T = A_1 \times A_C \times A_2 \tag{16.8}$$

When converting to s-parameters, we must take into account that $\mathbf{A_1}$, $\mathbf{A_C}$, and $\mathbf{A_2}$ are not dimensionless. This is achieved by dividing A_{12} by Z_0 and multiplying $\mathbf{A}_{21}$ by Z_0.

After the transition $A \rightarrow S$ we can calculate the reflection coefficient of the noise source Γ_T at the output of the tuner.

$$\Gamma_T = S_{22} + \frac{S_{12}S_{21}}{1 - S_{11}\Gamma_{IN}}\Gamma_{IN} \tag{16.9}$$

We assume that the noise source is well matched, i.e. $\Gamma_{IN} = 0$, and calculate some transformation curves as a function of C. A displacement of the capacitance by $\lambda/2$ along the line is shown in Figure 16.10. By changing the capacitance in the range $0.1\,\text{pF} \leq C \leq 10\,\text{pF}$ moves $|\Gamma_T|$ on the Smith chart to the outside. In our example in the range $0.05 \leq |\Gamma_T| \leq 0.94$. The transformation range also depends on the losses

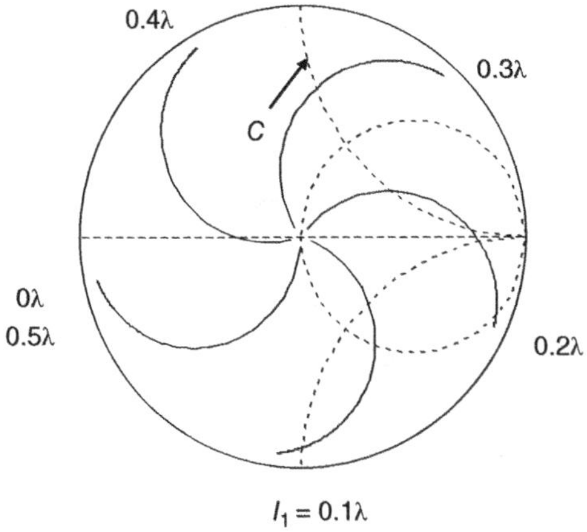

Figure 16.10 Tuner model with $R = 1\,\Omega, f = 3\,\text{GHz}$. At a certain position, increasing C in the Smith chart moves $|\Gamma_T|$ outwards on an arc of a circle.

within the tuner, which we considered in the model with $R = 1\ \Omega$. One can see that with this configuration practically every point in the Smith chart can be attained. It is the ideal construction for measurement purposes. For simple applications the manual tuner is sufficient, for complex tasks the automatic tuner is the appropriate choice. For on-wafer measurements of transistors the above mentioned electronic tuner is also used.

Attenuation of the Tuner

The increasing attenuation in the case of strong transformation is a problem. The noise power of the source must be corrected by this value, which significantly limits the usable range. There are a number of different definitions that describe gain or loss of a two-port. They are to be used according to the specific problem.

As already discussed in Chapter 14, the available gain is the appropriate parameter for noise considerations [8]. It is the ratio of the available power at the output to the available power of the source at the input. It is a function of the network parameters and the reflection coefficient of the source, but it is independent of the matching situation at the output. For the passive network $\alpha < 1$ and the noise factor is $= {}^{1}/_{\alpha}$.

$$\alpha = \frac{|S_{21}|^2 \left(1 - |\Gamma_{IN}|^2\right)}{\left(1 - |\Gamma_2|^2\right) |1 - S_{11}\Gamma_{IN}|^2} \tag{16.10}$$

With

$$\Gamma_2 = S_{22} + \frac{S_{21}S_{12}\Gamma_{\mathrm{IN}}}{1 - S_{11}\Gamma_{\mathrm{IN}}} \tag{16.11}$$

Since we assume that the noise source is well matched ($\Gamma_{IN} = 0$), (16.10) is simplified to

$$\alpha = \frac{|S_{21}|^2}{1 - |S_{22}|^2} \tag{16.12}$$

However, for accurate measurements of very low noise figures it may be necessary to take into account the always present, although small, mismatch of the noise source. In any case, a network analyzer is required to measure the s-parameters of the tuner in the appropriate setting.

An example of increasing attenuation with increasing transformation is shown in Figure 16.11. We assume $l_1 = 0.5\lambda$; $C = 0.1 \ldots 10$ pF. One can see that the attenuation increases strongly the further we move outwards in the Smith chart. The calibration table of the noise generator must be corrected by this value. This is an additional source of error. One should therefore avoid strong transformations. In the past, when VNA were not available, one could help oneself with the method

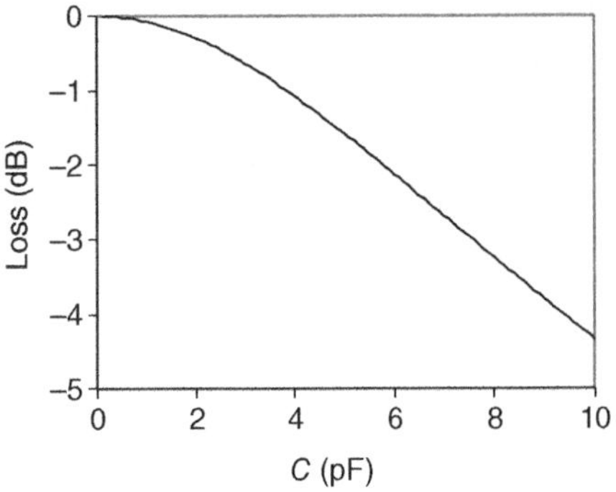

Figure 16.11 The attenuation of the tuner increases rapidly with increasing C (stronger transformation). Model Figure 16.9, $f = 3\,\text{GHz}$.

of tuner "back-out." A method which is prone to errors and is no longer recommended today. A second tuner of the same type is needed. The first tuner is set to the desired value. Then the second tuner is connected and adjusted so that there is again matching at its exit. At the interface between the two there is conjugate complex matching ($S_{22} = S_{11}^*$). One can now measure the attenuation of the series connection in the $50\,\Omega$-system and take 1/2 of it, since there is matching on both sides.

References

1 Groll, H. (1969). *Mikrowellen Messtechnik*. Braunschweig: Vieweg & Sohn.

2 Maury Microwaves Automated Tuners/Impedance Tuners. Catalogue 2018.

3 Vähä-Heikkilä, T., van Caekenberghe, K., Varis, J. et al. (2007). RF-MEMS impedance tuners for 6–24 GHz applications. *Int. J. RF Microwave Comput. Aided Eng.* 17: 265–278.

4 Focus Microwaves, Delta Series, Electromechanical Tuners. Catalogue 2018.

5 Tischer, F.J. (1958). *Mikrowellen Messtechnik*. Springer Verlag.

6 Maury Microwave Corp. (2009). Setting up Ultrafast Noise Parameters Using the Agilent PNA-X. Application Note 5c-084.

7 Simpson, G., Ballo, D., Dunsmore, J. et al. (2009). A New Noise Parameter Measurement Method Results in more than 100x Speed Improvement and Enhanced Measurement Accuracy. Maury Microwave Application Note, 5A-042.

8 Strid, E.W. (1981). Measurement of losses in noise matching networks. *IEEE Trans. Microwave Theory Tech.* 29: 247–252.

17

Examples of Measurement Problems

The most commonly used noise measurement setups are designed for packaged amplifiers and on-wafer systems. We have described them in the previous two chapters. Here are two additional problems from microwave technology.

Transistor in a Test Fixture

This setup is similar to that of a complete amplifier in a housing. Here we measure the noise parameters with respect to the reference planes of the coaxial connectors, e.g. PC7, but we are interested in the transistor as a mismatched two-port within the fixture. Therefore the four noise parameters have to be de-embedded from the coaxial connectors to the reference planes of the transistor.

The division of the configuration Figure 17.1 into networks is shown in Figure 17.2.

We have the following experimental situation when measuring a transistor in a test fixture using microstrip technology:

The mismatched DUT (packaged transistor) is embedded in two networks each at the input and output. These networks consist of the transmission lines, in the photo with transformation elements and the transitions from microstrip to coaxial line. The connection planes are coaxial, e.g. PC7 connectors.

(1) In the connection planes, the measurement of the s-parameters and the complete noise characterization is performed in a coaxial environment. The network analyzer and noise factor meter are calibrated with coaxial standards.
(2) These values are converted into chain parameters.
(3) The transformation properties of networks $\boldsymbol{A}_1$ and $\boldsymbol{A}_2$ must be determined by suitable calibration standards at the level of the DUT. This can be quite

A Guide to Noise in Microwave Circuits: Devices, Circuits, and Measurement, First Edition.
Peter Heymann and Matthias Rudolph.

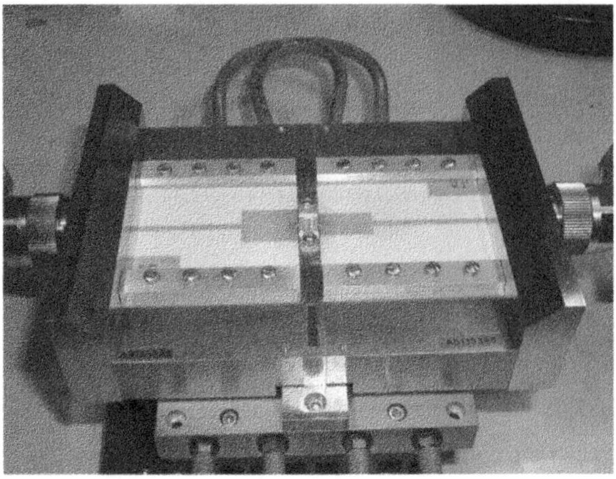

Figure 17.1 Photo of a transistor test fixture using microstrip technology. The transformation elements on the transistor are important for power components. They are not required for low-noise transistors.

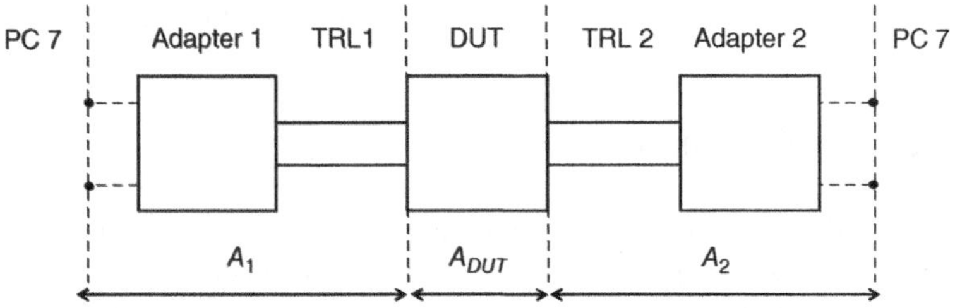

Figure 17.2 Block diagram of a transistor test fixture using microstrip technology. Coaxial connectors: PC7 connectors; adapters 1, 2: PC7 to microstrip line. TRL 1,2: Microstrip line with transformation elements, DUT: Transistor in a casing with reference planes on the edge of the housing.

problematic because there is no easily mountable terminating resistor in microstrip technology. Here, e.g. TRL calibration is recommended. These networks are also described in chain matrix form ($\boldsymbol{A}_1, \boldsymbol{A}_2$).

(4) Using the correlation matrices, the noise parameters of the DUT are then de-embedded. This means that the four noise parameters determined in the coaxial connection planes are converted to the reference plane of the DUT.

How is the "de-embedding" of a transistor done?

From measurements we know the following quantities:

(1) The four noise parameters of the entire network between the coaxial ports, with the DUT in the center. In the formalism of the correlation matrices we have the chain correlation matrix $\boldsymbol{C}_M^A$.
(2) The s-parameters of this network are converted into the chain matrix $\boldsymbol{A}_M$.

(3) The chain parameters of the "embedding" networks are A_1 and A_2. They each consist of a junction and a piece of the microstrip line.

The S-matrix of the entire test fixture measured with the network analyzer is converted into a chain form $S_M \rightarrow A_M$. It is important that the transistor is operated at the same operating point where the noise measurement was made.

$$A_M = A_1 \, A_{DUT} \, A_2 \tag{17.1}$$

The chain matrix of the DUT is obtained by "de-embedding" the transition networks:

$$A_{DUT} = A_1^{-1} \, A_M \, A_2^{-1} \tag{17.2}$$

We now need the correlation matrices of the transitions in chain form. As an intermediate step we first calculate these in Y-form. For this purpose we need the network matrices in Y-form, which up to now have been available in A-form.

So first the transformation $A_1 \rightarrow Y_1 \quad A_2 \rightarrow Y_2$.

Then the calculation of the correlation matrices of the left or right side.

$$C_1^Y = 2kT_0 Re\left(Y_1\right) \quad C_2^Y = 2kT_0 Re\left(Y_2\right) \tag{17.3}$$

Now the chain correlation matrix is calculated

$$C_1^A = T\, C_1^Y \, T^+ \quad C_2^A = T\, C_2^Y \, T^+ \tag{17.4}$$

According to Table 9.1, the transformation matrix is **T** for the transition from the Y-form to the A-form is:

$$T = \begin{pmatrix} 0 & A_{12} \\ 1 & A_{22} \end{pmatrix} \tag{17.5}$$

The two transitions must be de-embedded in two steps, as the matrices are directional.

From the four noise parameters of the entire fixture, which refer to the PC7 connection planes, we first calculate the correlation matrix $\mathbf{C}_M^A$ in chain form. The relation to the internal networks is given by

$$C_M^A = C_1^A + A_1 C_{DUT,2}^A A_1^+ \tag{17.6}$$

$C_{DUT,2}^A$ is the combined matrix of DUT and output network 2. Our goal is C_{DUT}^A. We must first $C_{DUT,2}^A$ calculate.

$$C_{DUT,2}^A = \left(A_1\right)^{-1} \left(C_M^A - C_1^A\right) \left(A_1^+\right)^{-1} \tag{17.7}$$

The next step is the formula

$$C_{DUT,2}^A = C_{DUT}^A + A_{DUT} \, C_2^A \, A_{DUT}^+ \tag{17.8}$$

which we solve for $\boldsymbol{C}^A_{DUT}$. Now we have reached our solution. In the following formulas, we only write the following for the sake of clarity $\boldsymbol{C}^A$ instead of $\boldsymbol{C}^A_{DUT}$. The four noise parameters of the transistor are after de-embedding:

$$F_{MIN} = 1 + 2\frac{Re\left(C^A_{12}\right) + G_{OPT}C^A_{11}}{2kT_0}$$
$$R_N = \frac{C^A_{11}}{2kT_0}$$
$$B_{OPT} = \frac{Im\left(C^A_{12}\right)}{C^A_{11}} \quad (17.9)$$
$$G_{OPT} = \frac{1}{C^A_{11}}\sqrt{C^A_{11}C^A_{22} - \left\{Im\left(C^A_{12}\right)\right\}^2}$$

It is clear that these calculations are usefully carried out in computer programs, as they contain a large degree of formal processing.

The Low Noise Block (LNB) of Satellite Television

In a system for satellite television, a low-noise signal converter is used which is located in the focus of a parabolic antenna [1]. It is called an LNB (Low Noise Block). It converts the satellite frequency from 10.7–11.75 GHz (FSS band) and 11.8–12.75 GHz (BSS band) to the 950–2150 MHz range. Its input stage is an extremely low-noise transistor amplifier with the best single transistors available. These are High Electron Mobility Transistors (HEMTs, HFET), for which mainly the layer sequence aluminum gallium arsenide/gallium arsenide is used. In gallium arsenide, a two-dimensional electron gas is formed at the interface, which has a very high mobility and acts as a channel of the field effect transistor.

A transistor that is preferably used in the LNB is, for example, the CE3514 M4 [2]. Its properties at $f = 12$ GHz are $NF = 0.42$ dB; $G_A = 12.2$ dB with noise matching.

The microwave side of the LNB consists of a Ku-band circular waveguide ($f = 10.7$–17.5 GHz) with a small horn antenna at the input and a transition to microstrip at the end of the waveguide. The HFET of the first amplifier stage is connected via a noise matching network. Then a second amplifier stage is added before the mixing stage follows via filters (Figure 17.3).

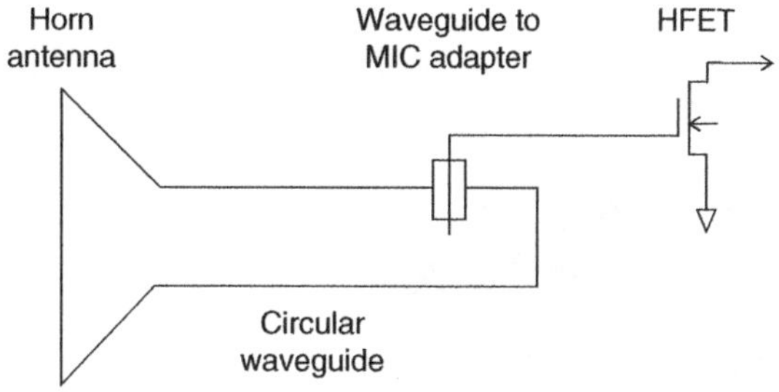

Figure 17.3 Scheme of LNB input.

The estimation of the signal power at the receiving horn gives an indication of the required noise figure of the LNB. We consider the transmission distance from the satellite at an altitude of 36,000 km to the waveguide input of the LNB.

The transmitter in the satellite has approximately the following parameters: Wavelength $\lambda = 2.5$ cm, transmitter power $P_S = 45$ dBm, antenna gain $G_S = 36$ dB. This results in an $EIRP = 45$ dBm + 36 dB = 81 dBm.

The following holds for the propagation in space: $L = 3.6 \times 10^8$ cm, attenuation

$$\alpha = 10 log(\lambda/_{4\pi L})^2 = -205 \text{ dB}$$

The receiver consists of a parabolic antenna and LNB: 60 cm parabolic aerial, $G_P = 38$ dB.

The signal power at the LNB is therefore: $EIRP + \alpha + G_P = 81$ dBm − 205 dB + 38 dB = −86 dBm.

For the LNB we assume $B = 33$ MHz, $NF = 1$ dB, $P_N = F \times kT_0B = -98$ dBm.

Signal to noise ratio: $S/N = 12$ dB, BER at 4PSK: 10^{-4}.

We can see that a noise figure of the LNB of $NF = 1$ dB ensures good reception. A slightly better noise figure does not produce a clear effect. It is also doubtful whether LNBs, some of which are advertised with $NF = 0.1$ dB, will achieve this value. As we saw above, the best transistors already have values $NF > 0.4$ dB. In addition, there is the second amplifier stage, as well as the losses in the transition to strip line and in the transformation element of the noise matching. So one has to assume $NF = 0.7 \ldots 1$ dB.

If the television viewer lives in an area that is not centrally illuminated, e.g. on the border of Europe, a larger antenna is required. Figure 17.4 shows a situation in Cyprus.

Measuring the noise figure of the LNB is not possible without considerable, particularly mechanical, effort. The problem is the feeding of the test noise

Figure 17.4 Typical antenna system in the Mediterranean region.

signal into the horn antenna. One possibility is to remove the horn and attach a waveguide flange. A waveguide-coaxial adapter can then be connected to this flange, to which in turn the noise generator is connected. The mechanical problems of the reconstruction should not be underestimated, however, as they must be carried out with the highest precision. In addition, losses are occur via the adapter, which must be taken into account. Another method, which can only be carried out by the manufacturer, is to make a model whose waveguide has a flange at the input and no horn (Figure 17.5) [3].

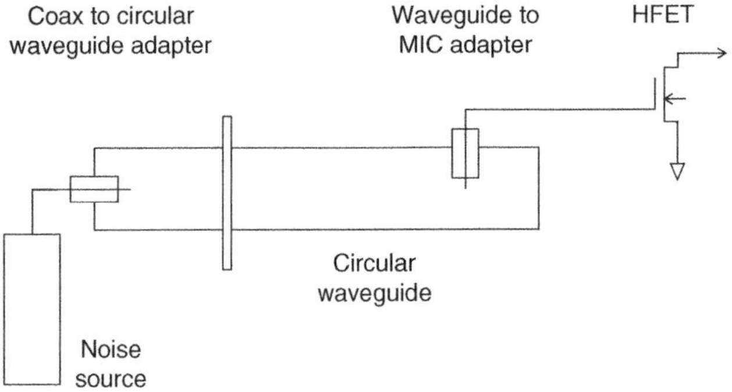

Figure 17.5 Noise measurement on the model of an LNB with the noise source connected directly to the waveguide. Source: Based on NXP Semiconductors [3].

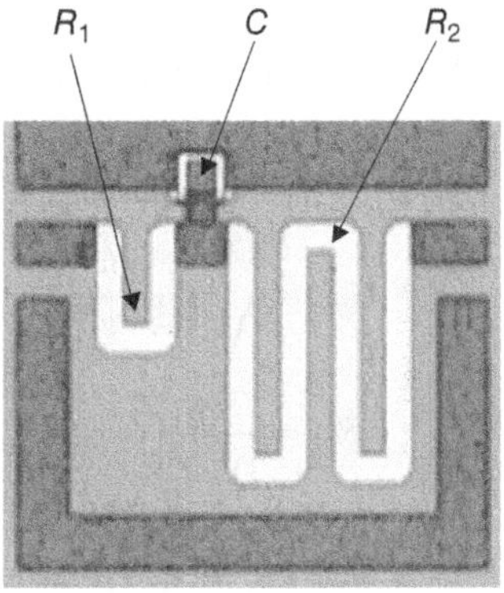

Figure 17.6 Photo of a RC noise test structure using coplanar technology. The light gold is a thin vapor deposition layer, the dark areas are galvanic layers.

A noise figure $NF = 1\,\text{dB}$ was measured on this model. For the first two amplifier stages the HFET CE3514 M4 was used. This shows that specifications of $NF < 0.7\,\text{dB}$ are obviously unrealistic for this inexpensive technology.

Verification of a Noise Measurement

A passive network can be used as a noise standard because its noise parameters are derived from the s-parameters using correlation matrices. The simplest case is an attenuator. A 3 dB attenuator has a noise figure of 3 dB. With this the possibilities in coaxial technology are mostly depleted. For on-wafer measurements passive structures can be designed, whose four noise parameters are similar to those of a FET. This allows test measurements of the apparatus, which are much more meaningful than those of an attenuator. Such an RC structure is shown in Figure 8.12.

The equivalent circuit is shown in Figure 17.6 (Figure 17.7).

There are two ways to obtain the four noise parameters of the test structure. (i) Determination of the equivalent circuit diagram and calculation of the four noise parameters. (ii) Direct calculation from the measured s-parameters. At low

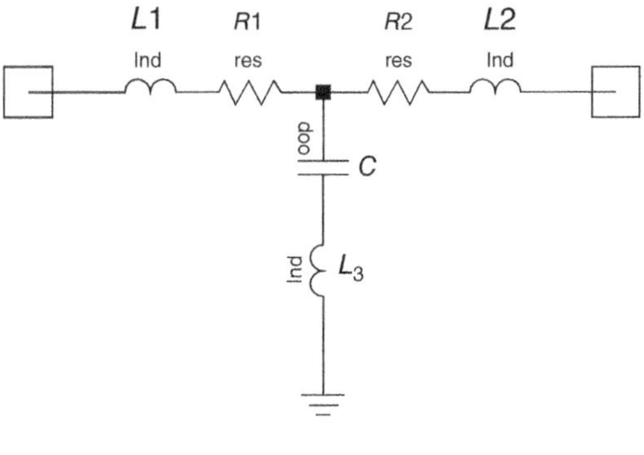

Figure 17.7 Simple circuit diagram of the test structure in Fig. 17.6. $R_1 = 6\,\Omega$, $R_2 = 17\,\Omega$, $C = 0.45\,\text{pF}$; $L_1 = 180\,\text{pH}$; $L_2 = 320\,\text{pH}$; $L_3 = 20\,\text{pH}$.

frequencies the first method is better, because VNA measurement inaccuracies below 1 GHz are compensated by using the equivalent circuit (Figure 17.8).

$$
\begin{aligned}
\mathbf{C}^Y &= Re(Y) \\
G_{opt} &= \sqrt{|Y_{21}|^2 \frac{C_{11}^Y}{C_{22}^Y} - |Y_{21}|^2 \frac{\left|C_{12}^Y\right|^2}{C_{22}^Y} + \left\{ Re\left(Y_{11} - Y_{21} \frac{C_{12}^Y}{C_{22}^Y} \right) \right\}^2} \\
R_N &= \frac{C_{22}^Y}{|Y_{21}|^2} \\
B_{opt} &= -Im\left\{ Y_{11} - Y_{21} \frac{C_{12}^Y}{C_{22}^Y} \right\} \\
F_{MIN} &= \frac{C_{22}^Y}{|Y_{21}|^2} \left\{ G_{opt} + Re\left(Y_{11} - Y_{21} \frac{C_{12}^Y}{C_{22}^Y} \right) \right\}
\end{aligned}
\tag{17.10}
$$

In this example the four noise parameters are calculated from the Y-parameters with the Y-correlation matrix $\mathbf{C}^Y$. The result is shown in Figure 17.9 together with noise measurements.

It must be noted, however, that possible linearity problems are not covered here because the test structure has no gain. When measuring high-gain transistors, however, a high dynamic range of the noise levels must always be expected. This circumstance, which can lead to measurement errors, is dealt with in Chapter 19.

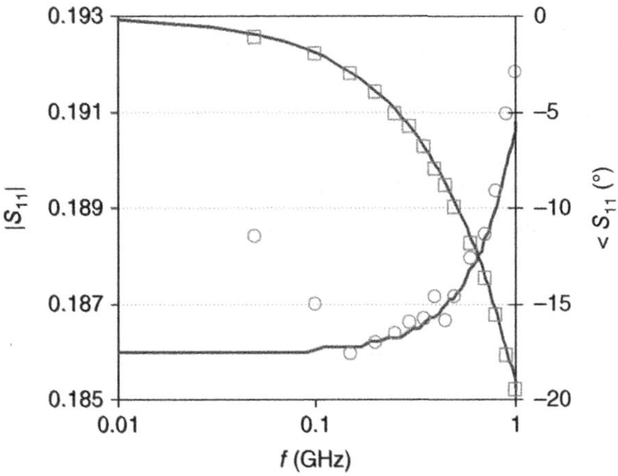

Figure 17.8 Comparison of S_{11} from direct measurement (symbols) and from the equivalent circuit (full line).

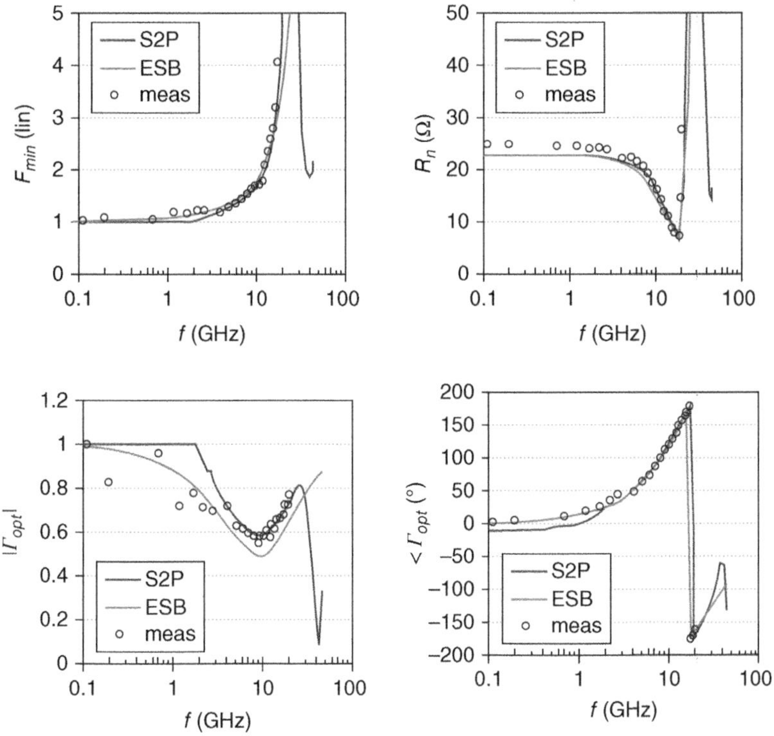

Figure 17.9 Verification of an on-wafer system with the passive test structure Figure 17.6.

References

1 Fischer, H.-M. (2006). *Europäische Nachrichtensatelliten*. Stedinger Verlag: Von Intelsat bis TV-Sat.

2 CEL (2016). RF-Low Noise FET CE3514M4. Datasheet.

3 NXP Semiconductors (2012). Universal Single LNB with TFF101xFIMOD IC. Application Note 11144.

References

18

Measurement and Modeling of Low-Frequency Noise

Correlation Radiometer for Low Frequencies ($f < 10$ MHz)

Noise measurements in the $1/f$-range are often carried out on semiconductor components. The voltages or currents to be measured are very small, as the bandwidths are only a few 10 Hz. With a BJT in active operation with 10 times amplification, one has about 50 nV/$\sqrt{\text{Hz}}$. The correlation spectrum analyzer (CSA) [1] shown in Figure 18.1 can be used to measure these very low noise levels. The example shows the measurement of the short-circuit current of a field effect transistor in the $1/f$ range. The current flows through the two inputs of two parallel transimpedance amplifiers whose input resistance is practically zero.

The transimpedance amplifier converts an input current into a proportional output voltage. It is well suited for measuring the LF noise of a component, as it directly provides the short circuit at the output of the DUT. The current flows into the inverting input, the + input is connected to ground, and it can be connected to the supply voltage (here V_{DS}). Due to the high gain and the feedback with R_F, the voltage drop between both inputs becomes practically zero, so that the non-inverting input is a virtual ground. The DUT sees a short circuit at its output.

The effect of the CSA is based, as with the correlation radiometer, on the fact that the inherent noise of the two channels is uncorrelated and results in the average value zero in the subsequent multiplication. The useful signal passes through both channels in phase and gives a finite value, when the cross-correlation is calculated. The remaining fluctuations can be user defined reduced by sufficiently long integration [2].

The detection limit of the real system in Figure 18.1 results from the contribution of the noise voltage sources $\overline{e_n^2}$ of the two operational amplifiers. These drive a current through the DUT, which, like the current to be measured, passes through both channels and cannot be separated from it because it is correlated. So here

A Guide to Noise in Microwave Circuits: Devices, Circuits, and Measurement, First Edition.
Peter Heymann and Matthias Rudolph.

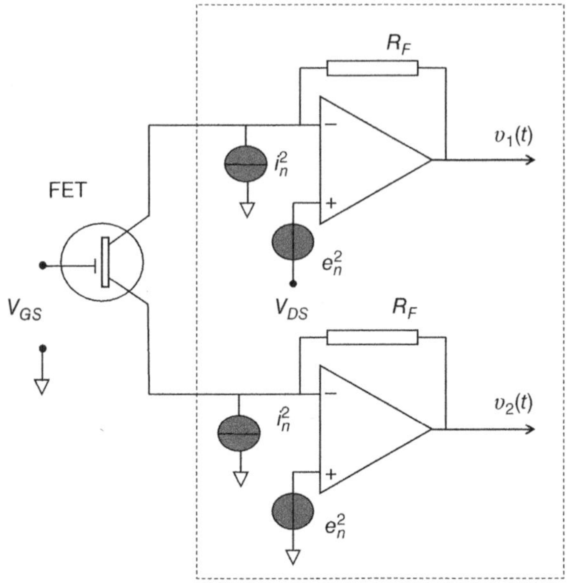

Figure 18.1 Input stage of the correlation spectrum analyzer for the 1/*f* range. Source: Based on Ferrari and Sampietro [1].

too, there is a limitation of the sensitivity due to the inherent noise of the receiver, although this is very small.

The current source $\overline{i_n^2}$ makes no contribution in this circuit because it is short-circuited in the channel by the low impedance input and therefore only contributes to uncorrelated noise. The same applies to the thermal noise of the feedback resistor R_F.

Figure 18.2 shows the circuits for voltage measurement (a) and for current measurement (b). The aim is to measure the noise signal of the DUT without the disturbing inherent noise of the amplifier chain. The former is very small, and the latter very large. As mentioned earlier, the amplifier noise does not give an average value in the subsequent cross-correlation, because both channels are uncorrelated.

Let us first look at the voltage measurement with operational amplifier (Figure 18.2a). The DUT, here an ohmic resistor whose noise voltage is to be measured, is connected in parallel to the input of the two voltage amplifiers. Since their input resistance is practically infinite, the voltage source $\overline{e_n^2}$ does not generate any current in the DUT. The current source $\overline{i_n^2}$, however, drives a correlated current through the DUT whose voltage drop cannot be separated from the noise voltage to be measured.

When measuring current, the DUT is connected between the inputs of the two transimpedance amplifiers (Figure 18.2b). These have zero input resistance and amplify the short-circuit current of the DUT into an output voltage. In this system the voltage source $\overline{e_n^2}$ drives a correlated current through the DUT, while the

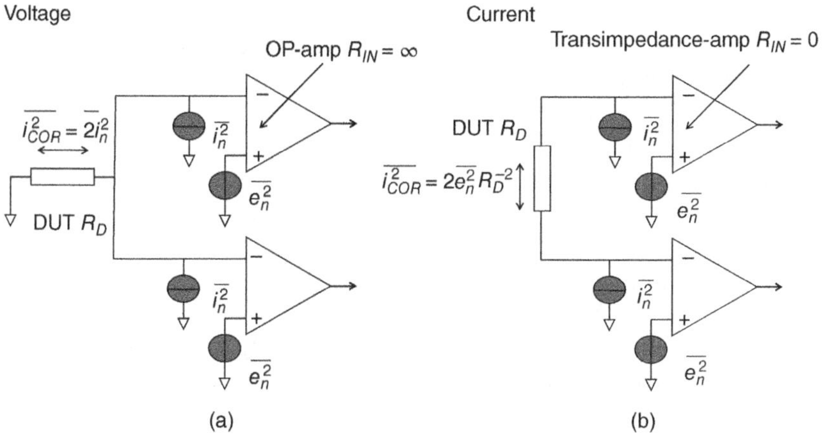

Figure 18.2 Correlation spectrum analyzer input stages for measuring voltage (a) and current (b).

current source operates in only one channel and does not generate a correlated current.

The selection of amplifiers for the two applications should be based on these aspects. For voltage measurement an operational amplifier with low noise current (FET technology) is favorable, for current measurement a circuit with low noise voltage (BJT technology).

We compare two OP amps for suitability in the CSA. The LT 1028 [3] is in BJT technology, and the LT 1113 [4] in FET technology. The data sheets provide the following information on the noise parameters (Table 18.1).

For comparison, the equivalent noise temperature of the setup is particularly suitable. If we use one OP-Amp in each of the usual single channel configuration (Chapter 12), we obtain a spectrum analyzer with the following noise temperature, which determines its detection limit

$$T_N = \frac{e_n^2 + \left(i_n R_{DUT}\right)^2}{4kR_{DUT}} \tag{18.1}$$

Table 18.1 Noise sources of operational amplifiers according to data sheet.

Type	Technology	Voltage density (nV/$\sqrt{\text{Hz}}$)	Current density (pA/$\sqrt{\text{Hz}}$)
LT 1028	BJT	0.9	1
LT 1113	FET	4.5	0.01

Source: Based on Linear Technology [4] and Linear Technology [3].

With the dual channel CSA, we have the limitation of sensitivity only by the correlated current that the previously quoted noise sources generate in the DUT. When measuring voltage, the current source acts directly.

$$\overline{|i_{COR}|^2} = 2\overline{|i_n|^2} \tag{18.2}$$

When measuring current, this source is ineffective because it is short-circuited. The voltage source generates the correlated current in the DUT

$$\overline{|i_{COR}|^2} = 2\overline{|e_n|^2}\left(\frac{1}{R_{DUT}^2} + \omega^2 C_{DUT}^2\right) \tag{18.3}$$

Since the LF noise ($1/f$ range) of transistors is usually measured by current measurement, a frequency dependence is taken into account here, which is given by the output capacitance C_{DUT} of the transistor. Not considered in (18.3) are stray capacitances of the measuring setup, e.g. cable capacitances, which can be considerable.

These two currents can be written as equivalent noise temperatures. For the voltage measurement

$$T_C = \frac{\overline{|e_n|^2}}{2k}\left(\frac{1}{R_{DUT}} + \omega^2 C_{DUT}^2 R_{DUT}\right) \tag{18.4}$$

and for current measurement

$$T_C = \frac{\overline{|i_n|^2} R_{DUT}}{2k} \tag{18.5}$$

These temperatures are shown in Figure 18.3. The dotted curves apply to the single channel system. Noise temperatures below 10 K can be achieved with the CSA for all possible resistances of the DUT if the appropriate components are used and the circuitry is designed accordingly for current or voltage measurement. For small values of R_{DUT} the voltage measurement according to Figure 18.2a is most suitable, and the current measurement according to Figure 18.2b is most sensitive for high values of R_{DUT}.

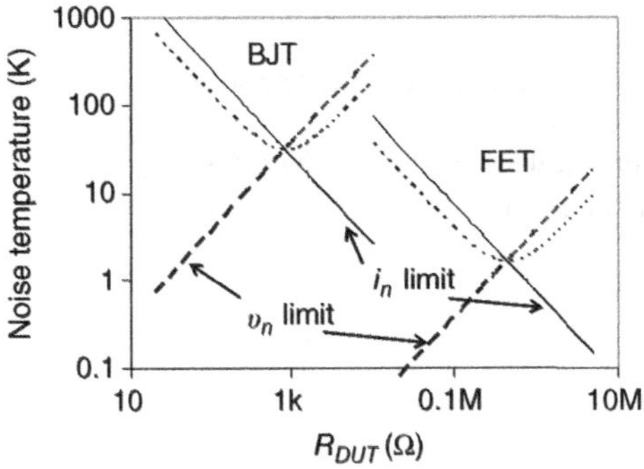

Figure 18.3 Noise temperatures achievable with the CSA. The limitation results from the interaction of v_n and i_n.

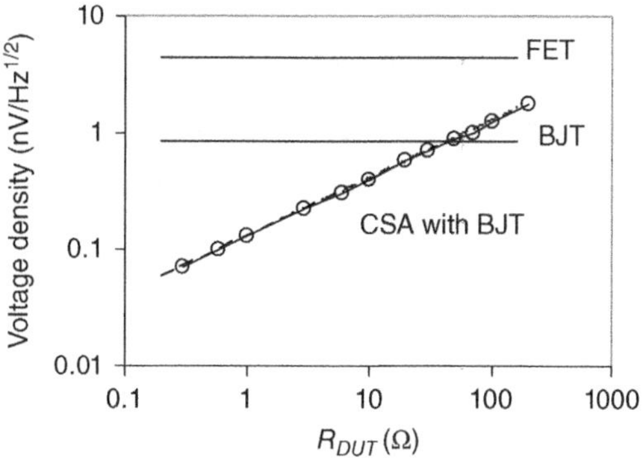

Figure 18.4 The CSA can measure the thermal noise of resistors down to $R < 1\,\Omega$.

Figure 18.4 shows an example measurement of the performance of the CSA [2]. Noise voltages of resistors $R_{DUT} < 1\,\Omega$ could be measured. However, a long integration time (about one hour) is required. The limits of the single-channel system are indicated by the horizontal straight lines.

The Low-Frequency Noise of Transistors

The low-frequency component of the noise of an active device plays an important role for microwave applications, too. Although initially outside the interesting frequency range, non-linear processes cause upward mixing. One then has, e.g. in the oscillator, the LF noise as sidebands beside the oscillation frequency. Also when a signal passes through an active two-port (amplifier, multiplier, etc.), the LF noise is superimposed to the useful signal. This is called residual phase noise. It is therefore important to measure the LF noise of transistors and to model it for CAD purposes, so that it can be used in design calculations. When developing new components or improving already established technologies, the measurement of LF noise is an important diagnostic tool.

Figure 18.5 shows typical examples of the spectral noise power density of the collector or drain current of transistors.

Let us consider the noise of a microwave transistor in the wide frequency range in Figure 18.6. For $f < 10\,\text{MHz}$ we see $1/f$ noise, above 10 GHz an increase, in between a wide minimum.

The $1/f$ range requires a different measurement technique than the microwave range with $f > 1\,\text{GHz}$. There are established techniques for both measuring ranges. Although the frequency range with minimum noise between them has been used for decades, component measurement technology for noise measurement is little established for several reasons. (i) The noise figures are very low and simple simulation models are usually sufficient. (ii) The microwave components, e.g. impedance tuners, are very large and unwieldy in the decimeter and meter wave

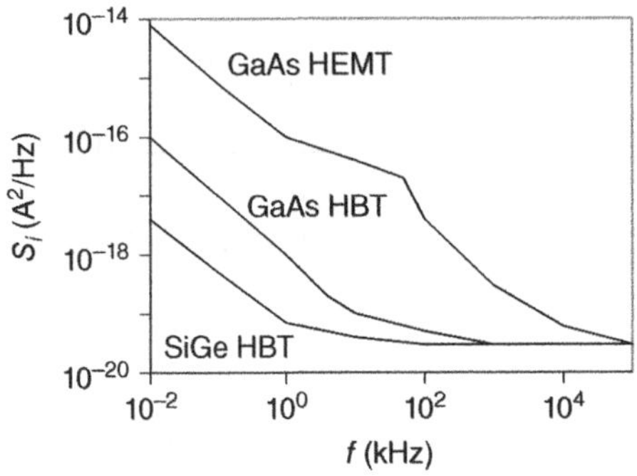

Figure 18.5 Spectral noise power density of the output current of transistor types in the LF range.

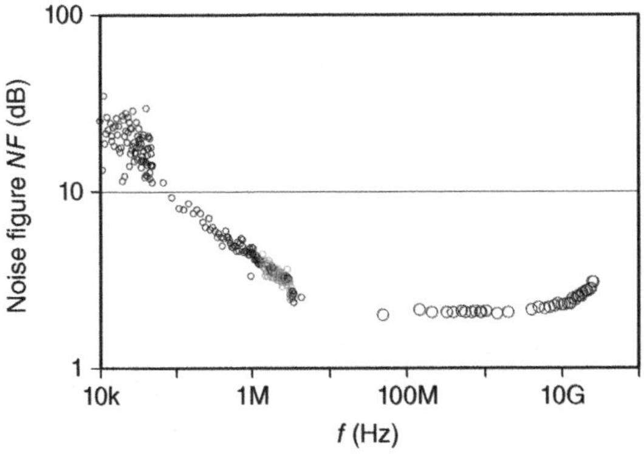

Figure 18.6 Noise figure of a microwave transistor in log frequency scale.

range. Here, improvisation is often used and an own measuring system adapted to the specific problem is built up.

Measurement Setup for LF Noise

The requirements of LF measurement of transistors are not so extreme that the CSA discussed earlier would be necessary. Nevertheless, when designing an LF noise measurement system, the choice of the measurement amplifier is important. It should meet the following criteria: (i) low inherent noise, (ii) 40–60 dB gain, and (iii) bandwidth up to 10 MHz. With a bandwidth of 10 MHz, the output is usually connected to a spectrum analyzer with 50 Ω input impedance. It is useful if the amplifier also has a 50 Ω output. Regarding the input impedance of the measuring amplifier, it is important to consider whether a voltage measurement or a current measurement is to be made. In the first case the resistance should be a few kiloohms, in the second case 0 Ω, e.g. for the operational amplifier [5, 6].

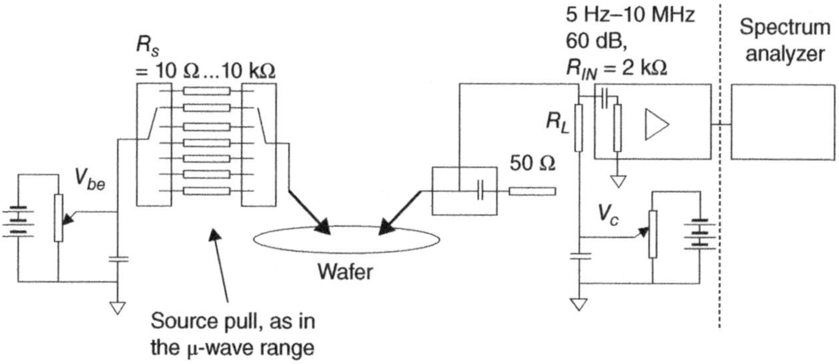

Figure 18.7 On-wafer system for LF noise modeling of a transistor. The input resistance can be varied. Bias is provided by batteries.

An example of an on-wafer system for measuring the low-frequency noise spectrum of a transistor is shown in Figure 18.7.

Interference and additional noise sources must be minimized as far as possible. With careful analysis of all sources of interference and selection of the correct grounding points, a Faraday cage is usually not necessary. The bias point of the transistor is adjusted with batteries. This is of course a low level of automation, as changes have to be made by manually turning potentiometers.

There is no recording of the values for the data files, as electronic measuring instruments should not be used for bias point measurement. However, if electronic sources, e.g. Keithley Source Meters, are used in the interest of high throughput, one must first examine very carefully what noise level these components contribute. Extensive filtering may be necessary, which in turn means that time constants are built in, which reduce the measuring speed.

The input can be wired with different resistors in the shown structure. This is useful because it can influence the effectiveness of the noise sources on the input side of the transistor. In the bipolar transistor, for example, the noise source of the base-emitter diode can be short-circuited by a small resistor (10 Ω) in the input circuit. On the other hand, it will be fully effective at open circuit operation (10 kΩ).

In this case a model of the Berlin company, FEMTO, was used as an amplifier. At the output of the transistor there is a Bias-Tee. The LF signal is conducted via the DC branch to the load resistor R_L. The microwave output is terminated with 50 Ω to prevent unwanted oscillations of the transistor.

This method is sufficient for LF noise measurements of most types of transistors, since the level is relatively high. If extremely low noise is expected, e.g. some Si-BJT or small series resistances of diodes, special correlation measurement methods should be used, e.g. the CSA discussed earlier.

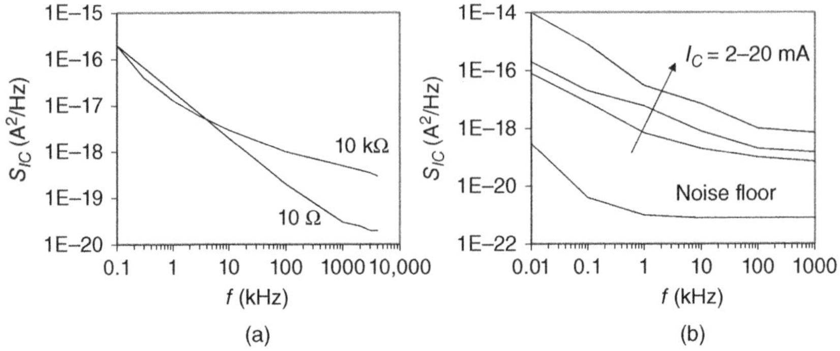

Figure 18.8 LF noise spectra of a GaAs HBT in common emitter circuit as a function of input resistance (a) and as a function of collector current (b). (Plotted are smoothed measurement curves.)

Examples of LF Noise Measurements on GaAs-HBT

The LF noise of transistors increases with the current, and it also depends on the input circuit. The result of a measurement must be to detect these dependencies of the spectrum, to understand them if possible, or at least to model them for the circuit design. Figure 18.8 shows the spectral noise power density of the collector current of an HBT measured with the setup Figure 18.7 [6].

If one wants to compare the LF noise characteristics of different technologies, e.g. for the suitability for oscillator design, one compares the levels of the spectral noise power density of the base current at a certain offset frequency. In Figure 18.9 bipolar transistors in common emitter circuit at $f = 100\,\text{Hz}$ versus the current

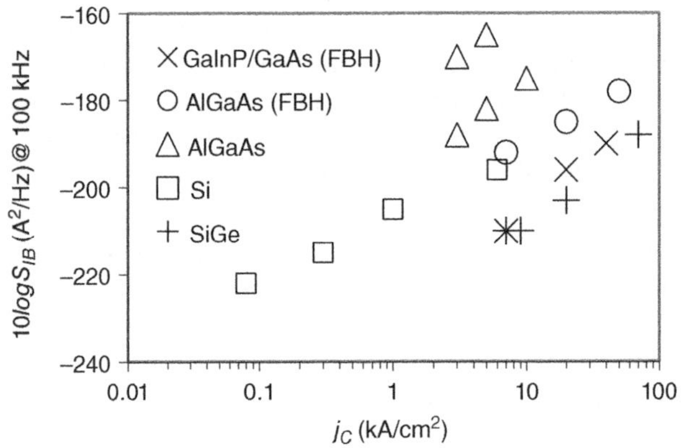

Figure 18.9 Comparing S_{IB} @ 100 Hz of different technologies versus current density.

density are compared. The values refer to the base, i.e. the input of the transistor. The values of the noise current density of the collector current are β^2 higher.

The field effect transistor (triangles) has a significantly higher LF noise. This is due to the influence of interface effects on the exposed surface of the channel.

Modeling of LF Noise

The LF noise characteristics of an HBT can be understood with the following noise equivalent circuit. For Si-BJT the base-emitter source i_B^2 is sufficient. In this manner the noise model is included in the usual design programs, i.e. without the additional collector source i_C^2 [7–10].

In Figure 18.10, only the shot noise sources are depicted. In specific cases the contributions of $1/f$ noise and gr-noise must be added. These can only be obtained by measurement. An example is given in Figure 18.11.

With this data, one can model the LF noise. The fitting to the formulas for S_{Ib} and S_{VR} used in CAD programs (Figure 18.11) usually works very well. The required temperatures and currents for this fitting are in general physically reasonable.

The Noise of the Microphone

The smallest level of an acoustic signal that can be detected by a microphone depends on the statistically distributed impact of the gas molecules in the surrounding air. However, this gas-kinetic effect is usually less than the electronic

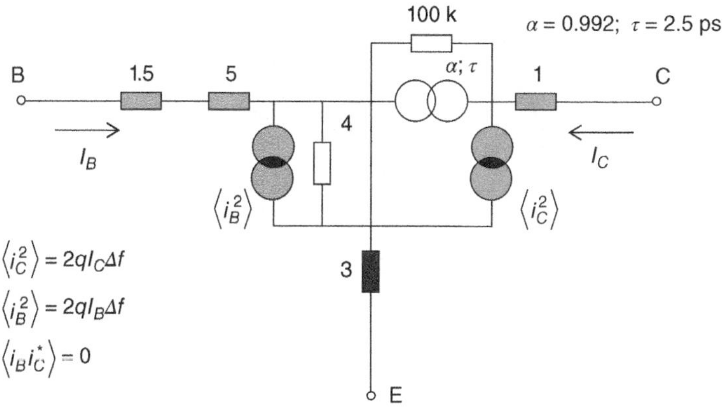

Figure 18.10 Equivalent circuit for the LF noise of an HBT. For Si-BJT the collector noise source is i_C^2 not required.

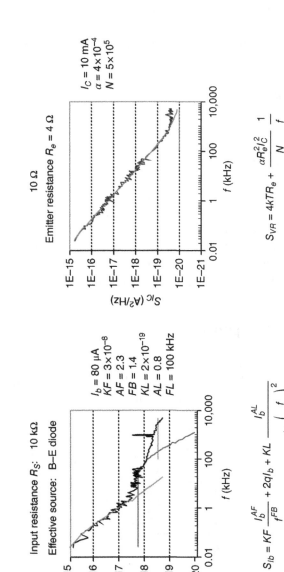

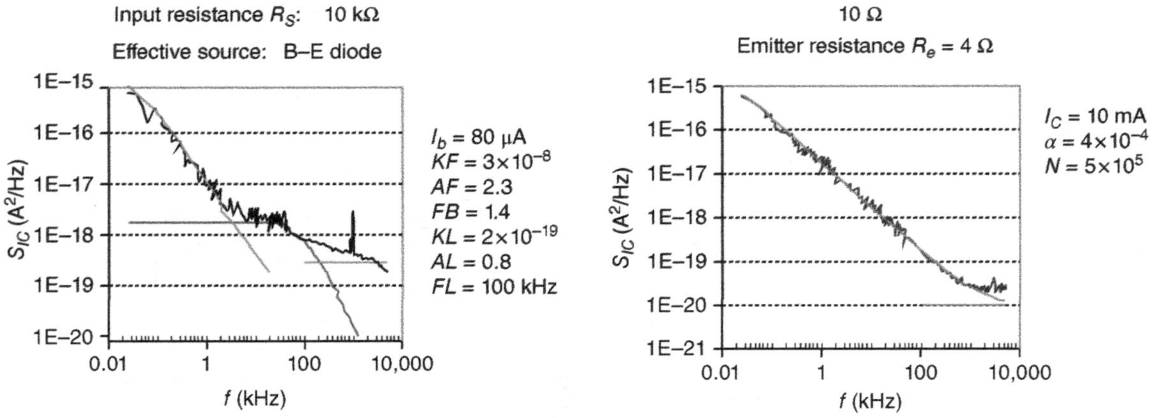

$$S_{Ib} = KF\,\frac{I_b^{AF}}{f^{FB}} + 2qI_b + KL\,\frac{I_b^{AL}}{1+\left(\dfrac{f}{FL}\right)^2}$$

$$S_{VR} = 4kTR_e + \frac{\alpha R_e^2 I_C^2}{N}\,\frac{1}{f}$$

Figure 18.11 Example of modeling measured curves of LF noise.

Figure 18.12 Movably mounted capacitor plate of the area A_C. Molecules of velocity v are reflected from it.

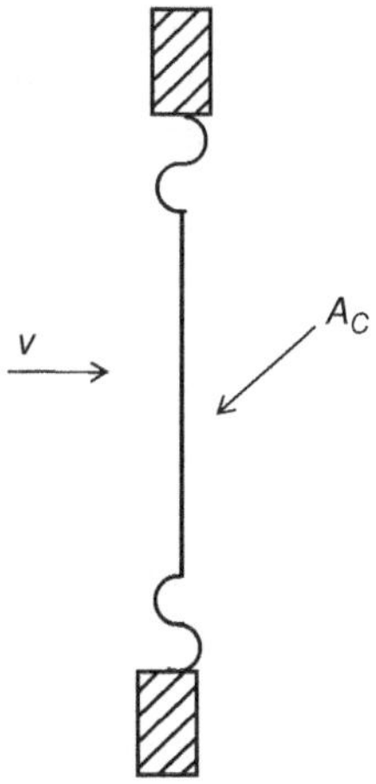

noise of the circuit elements. However, it represents the physical detection limit. The most sensitive design is the condenser microphone. The sensor is a membrane that can move in the direction of its surface normal and forms a plate of a capacitor (Figure 18.12) [11].

At each impact of a particle of mass M, the momentum $2Mv$ is transferred. In the time interval τ there collide N particles. Then the force is

$$F = N\frac{2Mv}{\tau} \tag{18.6}$$

This corresponds to the time derivative of the momentum. Due to the average action of force a pressure is applied to the membrane

$$\overline{F} = PA_C = \overline{N}\frac{2Mv}{\tau} \tag{18.7}$$

The average force $\overline{F}$ corresponds to the atmospheric air pressure P. There is a fluctuation superimposed to this force $F_\tau = F - \overline{F}$. It holds

$$\overline{F_\tau^2} = \overline{\left(F - \overline{F}\right)^2} = \left(\frac{2Mv}{\tau}\right)^2 \overline{(N - \overline{N})^2} = \frac{2MvPA_C}{\tau} \tag{18.8}$$

Accordingly, the following applies to pressure fluctuations

$$\overline{P_\tau^2} = \frac{\overline{F_\tau^2}}{A_C^2} = \frac{2MvP}{A_C\tau} \tag{18.9}$$

The Fourier analysis of the impulses that generate these pressure fluctuations leads us, by analogy with shot noise (Chapter 3), to the spectral density for low frequencies

$$S_P(f) = 2\tau\overline{P_\tau^2} \tag{18.10}$$

When calculating the pressure variations at a condenser microphone, we must take into account that the air molecules act on both sides of the membrane. One must integrate over the relevant frequency range B within the acoustic spectrum.

$$\overline{p^2} = \frac{8Mv P}{A_C} \int_{f_1}^{f_2} df \tag{18.11}$$

This is the square of variation of the pressure fluctuations

$$\overline{p} = \sqrt{\frac{8Mv P}{A_C} B} \tag{18.12}$$

As an example, we calculate this value for a condenser microphone of area $A_C = 1\,\text{cm}^2$ in air at room temperature in the acoustic frequency range. The values are: $M = 4.9 \times 10^{-23}$ g, $T = 300$ K, $v = 500$ m/s, $P = 10^5$ Pa. Thus we obtain the pressure fluctuation due to the gas-kinetic impact of the molecules on the membrane

$$B = 3\,\text{kHz}; \quad \overline{p} = 24\,\mu\text{Pa}$$

$$B = 20\,\text{kHz}; \quad \overline{p} = 63\,\mu\text{Pa}$$

The value decreases with increasing area A_C of the membrane. Very sensitive condenser microphones therefore have a larger area than usual. These large-diaphragm microphones have values below 10 dB(A) [12]. This value is based on the threshold of audibility of the human ear, here designated A, not to be confused with the area A_C. The eardrum also has a size of about $A_C = 1\,\text{cm}^2$. The bandwidth of the frequency range of highest sensitivity is about $B = 3$ kHz. The physiological threshold of audibility is about $p_A = 20\,\mu$Pa. It is set at 0 dB. This is the reference value to which the dB value refers. If we apply the full bandwidth of $B = 20$ kHz to the microphone, we obtain

$$10\,log\left(\frac{\overline{p}}{p_A}\right) \cong 5\,\text{dB(A)}$$

A microphone with a sensitivity better than 10 dB(A) therefore comes very close to the theoretical limit. Similarly, the human ear should not be much more sensitive, since it would otherwise constantly hear the thermal noise of the air.

The microphone is a sound converter. Pressure fluctuations are converted into electrical voltage changes. Accordingly, there is a transmission factor for each microphone type, which is quantitatively determined by the open loop transmission factor s_p with the unit $^{\text{mV}}/_{\text{Pa}}$. This increases with the area A_C. A value for condenser microphones with $^1/_4$ in. capsule is $s_p = 5\text{–}10$ mV/Pa. For moving-coil microphones about $s_p = 1\text{–}3$ mV/Pa [13].

The fluctuation due to the molecular collisions corresponds to a noise voltage, to which an equivalent noise resistance R_{eq} can be assigned on the input side. The alternating pressure $\overline{p}$ after (18.12) is converted into the noise voltage square $\overline{v_R^2}$:

$$\overline{v_R^2} = \overline{p}^2 s_P^2 \tag{18.13}$$

We can assign the equivalent noise resistance to this.

$$R_{eq} = \frac{2Mv P}{kTA_C} \times s_P^2 \tag{18.14}$$

With a realistic surface of the microphone $A_C = 10\ \text{cm}^2$ we get $R_{eq} \approx 1\ \Omega$. This is a very low noise contribution, which does not play any role for moving-coil microphones. In these, the diaphragm is mechanically connected to a small coil that is immersed in a magnetic field. The motion in the magnetic field induces the signal voltage in the coil. The ohmic resistance of this coil is about $R_{SP} = 20\ \Omega$ naturally also generates a noise voltage $\overline{v_{SP}}$. It determines the noise characteristic of this microphone, as its contribution clearly predominates. In the bandwidth of $B = 20\ \text{kHz}$ we have the parts from the molecular collisions $\overline{v_R} = 20\ \text{nV}$ and from the coil $\overline{v_{SP}} = 80\ \text{nV}$.

With the condenser microphone, we consider the low frequency circuit Figure 18.13. The condenser consisting of the diaphragm and counter-electrode is supplied with a DC voltage via a high resistance R. This ensures that its charge Q remains constant. If the capacitance changes as a result of the sound vibrations, an AC voltage $v(t)$ occurs. The relationship is $= {}^{Q}/_{C}$.

We consider the effect of the circuit in comparison to the gas kinetic fluctuations. The noise characteristics of the RC circuit we already derived in Chapter 3. There we had the integral (3.8)

$$\overline{v^2} = S_v \int_0^\infty \frac{df}{1 + (2\pi f CR)^2} \tag{18.15}$$

The frequency range here is very limited from $f_1 \approx 20$ Hz to $f_2 \approx 20$ kHz. Typical circuit elements are $C = 50\ \text{pF}$ and $R = 100\ \text{M}\Omega$. The value of the integral is

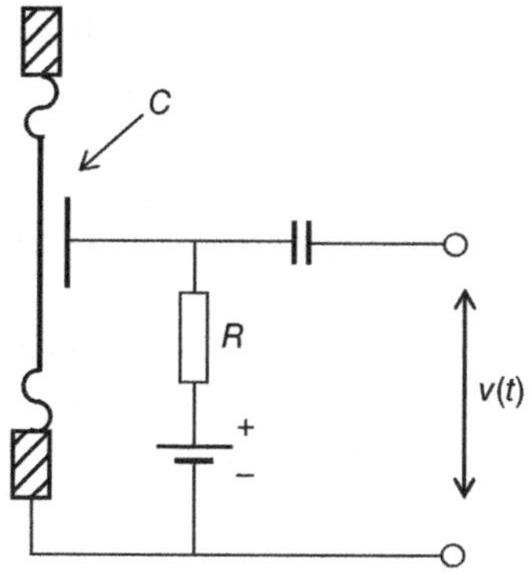

Figure 18.13 Low frequency circuit of the condenser microphone.

determined by

$$\overline{v^2} = \frac{S_v}{2\pi RC} \, arctan(2\pi fCR)|_{f_1}^{f_2} \tag{18.16}$$

The values for the integration limits are as follows: $arctan\left(2\pi f_2 CR\right) = {}^{\pi}/_{2}$ and $arctan\left(2\pi f_1 CR\right) = {}^{\pi}/_{4}$.

With $S_v = 4kTR$ we obtain the noise voltage of the low-frequency circuit of the condenser microphone without preamplifier

$$\overline{v^2} = \frac{kT}{2C} \tag{18.17}$$

In our example $\sqrt{\overline{v^2}} \approx 6.5\,\mu V$.

We compare this value with the noise voltage $\overline{v_G}$ which is generated by the fluctuations of the pressure (18.14). For $B = 20\,kHz$ the pressure is $\overline{p} = 63\,\mu Pa$. For the sensitivity of the condenser microphone, we use $s_P = 10\ {}^{mV}/_{Pa}$. With this we receive: $\overline{v_G} = 0.6\,\mu V$.

Again, the electronic noise is much higher than the influence of gas kinetics.

With condenser microphones, however, a different circuit technology can be helpful. In the high-frequency circuit, the capacity of the microphone together with a small coil forms the resonant circuit of an oscillator or a phase shifter. If the capacitance is changed, a frequency or phase shift occurs, which can be processed electronically directly at the microphone. This allows high sensitivities better than 10 dB(A) to be achieved and reaches the theoretical limit.

References

1 Ferrari, G. and Sampietro, M. (2002). Correlation spectrum analyzer for direct measurement of device current noise. *Rev. Sci. Instrum.* 73: 2717–2723.

2 Sampietro, M., Fasoli, L., and Ferrari, G. (1999). Spectrum analyzer with noise reduction by cross-correlation technique on two channels. *Rev. Sci. Instrum.* 70: 2520–2525.

3 Linear Technology (1992). LT1028 Ultra Low Noise Precision High Speed Op Amp. Datasheet 1992.

4 Linear Technology (1993). LT1113 Dual Low Noise Precision JFET Input OP Amp. Datasheet 1993.

5 Hansen, M. (May 2009). Achieving accurate on-wafer flicker noise measurement through 30 MHz. White Paper, Cascade Microtech.

6 Heymann, P., Rudolph, M., Doerner, R., and Lenk, F. (2001). Modeling of low-frequency noise in GaInP/GaAs hetero-bipolar transistors. *IEEE MTT-S Digest TH4C-4*. Phoenix: IEEE.

7 Kleinpenning, T.G.M. (1992). Location of low-frequency noise sources in submicrometer bipolar transistors. *IEEE Trans. Electron Devices* 39: 1501–1505.

8 Kleinpenning, G.M. and Holden, A.J. (1993). $1/f$-Noise in npn GaAs/AlGaAs heterojunction bipolar transistors: impact of intrinsic transistor and parasitic resistances. *IEEE Trans. Electron Devices* 40: 1148–1153.

9 Shin, J.H., Kim, J., Chung, Y. et al. (1998). Low frequency noise characterization of self-aligned AlGaAs/GaAs heterojunction bipolar transistors. *IEEE Trans. Microwave Theory Tech.* 46: 1604–1612.

10 Shin, J.H., Chung, Y., Suh, Y., and Kim, B. (1996). Extraction of low frequency noise model of self-aligned AlGaAs/GaAs heterojunction bipolar transistors. In: *IEEE MTT-S Digest TH1E-2*, vol. 3, 1309–1312.

11 van der Ziel, A. (1954). *Noise*. London: Chapman and Hall.

12 Neumann (2019). TLM103D Digitales Studiomikrofon. Datasheet 2019.

13 Bittel, H. and Storm, L. (1971). *Rauschen*. Springer Verlag.

19

Measurement Accuracy and Sources of Error

Accuracy of Measured Data

How many positions after decimal point make sense? One should not give more than one decimal place more than the accuracy of the result. With calculation programs the number can be entered, but this often has little to do with the accuracy of the result. Let us take a speed measurement with a stopwatch as an example. A vehicle drives through a distance of $x = 50$ m with constant speed about 50 km/h. The time $t = 3$ s is measured. This gives the speed $v = x/t = 16.6667$ m/s. In reality, the distance is only measured with an accuracy of about $\pm 1\%$, so it can be 49.5 or 50.5 m long. With high-precision time measurement, speeds of 16.5 or 16.83 m/s would result. The true value lies somewhere in between. The specification $v = 16.67$ m/s with two positions after decimal point is appropriate. One can also give as result $v = (16.67 \pm 0.17)$ m/s. There is betimes some carelessness in the use of positions after decimal point.

Error of Measurements

The true value of a physical quantity can never be determined exactly with one measurement. This deviation can be minimized by specific improvements of the measuring methods, but there is always an uncertainty. However, if the procedure can be developed to the point where the result is sufficient for the desired purpose, the problem is solved. There are systematic and random errors, which also occur together. Systematic deviations usually lead in one direction, i.e. the values are too high or too low. They are also present in repeated measurements and cannot be detected from the measured value itself. They can only be eliminated by systematic analysis of the measurement process. Causes can be, e.g. physical limits of the devices or components, manufacturing tolerances, retroaction of the measurement process on the object by reflection or own consumption, and

A Guide to Noise in Microwave Circuits: Devices, Circuits, and Measurement, First Edition.
Peter Heymann and Matthias Rudolph.

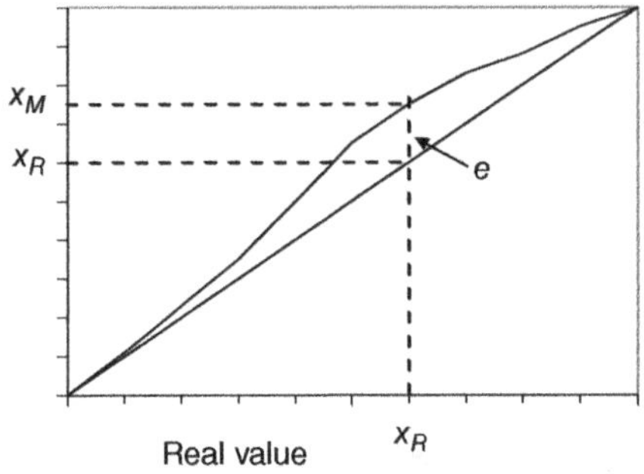

Figure 19.1 $e(x_R)$ gives the deviation of the real (top) from the ideal characteristic.

temperature influences. If these deviations are known, they can be eliminated by correction or calibration. The difference between the real and the ideal characteristic is designated in Figure 19.1 with $e(x_R) = x_M(x_R) - x_R(x_R)$. This deviation is generally a function of the magnitude of the measured value.

Random measurement deviations cause a scattering upwards and downwards. They cannot be predicted. With repeated measurement, there are always different values, but these lie within a certain range. A multitude of measurements give a good approximation to the true value via the arithmetic mean value. The best example is the noise measurement.

Inaccuracies of the Noise Measurement

Let us consider the noise measurement from the point of view of the law of error propagation according to Gauss [1, 2]. The quantity of interest NF_D, the noise figure of a DUT, cannot be measured directly. The setup is always a cascading of DUT and receiver. NF_D is only obtained by measuring other quantities. The error of NF_D results from the errors of these other quantities.

In general, the following procedure applies: The formula for the final result is partially derived with respect to the influencing variables. The squares of the derivatives are multiplied by the squares of the measuring error of these variables. These terms are added together and the square root is taken from this.

Our result is derived from the Friis formula.

$$F_D = F_M - \frac{F_{REC} - 1}{G_D} \tag{19.1}$$

with F_M: measured value; F_{REC}: noise factor of the receiver; G_D: gain of the DUT. The partial derivatives are as follows:

$$\frac{\partial F_D}{\partial F_M} = 1; \quad \frac{\partial F_D}{\partial F_{REC}} = -\frac{1}{G_D}; \quad \frac{\partial F_D}{\partial G_D} = \frac{F_{REC} - 1}{G_D^2} \tag{19.2}$$

Thus the error of the measurement results

$$\Delta F_D = \Delta F_M - \frac{1}{G_D}\Delta F_{REC} + \frac{F_{REC} - 1}{G_D^2}\Delta G_D \tag{19.3}$$

Since we want to have the deviation ΔF_D relative to the value F_D, we have to go to the relative values in dB. This is trivial for the absolute values F and G; for ΔF and ΔG the following consideration is useful.

Instead of the absolute measure of deviation ΔF, as shown in Figure 19.2, we want to arrive at the relative measure ΔNF in dB. We follow the Agilent Application Note 57_2 [1]. There we find

$$\Delta NF(\mathrm{dB}) = 4.34\frac{\Delta F}{F} \tag{19.4}$$

Since this transition is not immediately obvious, we will insert here a derivation based on the linear representation in Figure 19.2. As an example, repeated measurements of a housed amplifier, e.g. on different days and with repeated installation and removal, give an average value of $F = 7$ (lin) ($NF = 8.45$ dB) with a range of dispersion of $F = \pm 0.3$ (lin). The direct path to the dB scale is given by the difference of the maximum value and the minimum value, both in dB:

$$\Delta NF(\mathrm{dB}) = 10\left\{log\left(F + \frac{\Delta F}{2}\right) - log\left(F - \frac{\Delta F}{2}\right)\right\} \tag{19.5}$$

Due to the mathematical identity

$$log(x \pm y) = log(x) + log\left(1 \pm \frac{y}{x}\right) \tag{19.6}$$

We can we write

$$\Delta NF(\mathrm{dB}) = 10\left\{log\left(1 + \frac{\Delta F}{2F}\right) - log\left(1 - \frac{\Delta F}{2F}\right)\right\} \tag{19.7}$$

At this point we start a series expansion for the logarithm

$$ln(x) = \sum_{k=1}^{\infty}\frac{1}{k}\left(\frac{x-1}{x}\right)^k \tag{19.8}$$

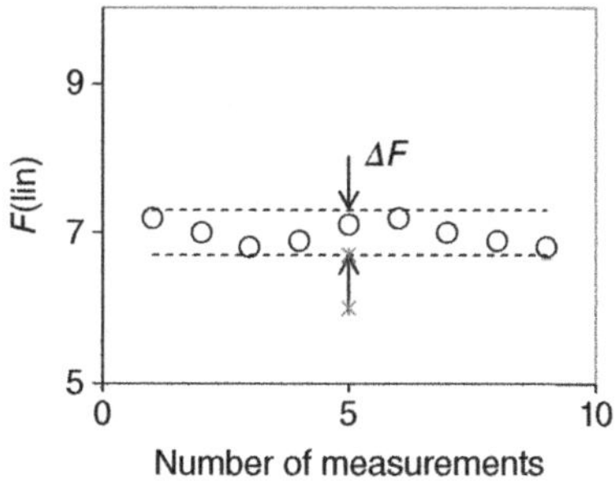

Figure 19.2 Example of a series of measurements of the noise factor F.

Of these we only use the first term. This is justified because it is an estimation.

$$log(x) = 0.434\frac{x-1}{x} \tag{19.9}$$

This gives

$$log\left(1+\frac{\Delta F}{2F}\right) = 0.434\left(\frac{\Delta F}{2F+\Delta F}\right) \tag{19.10}$$

Used in (19.7)

$$\Delta NF(\text{dB}) = 4.34\left\{\frac{\Delta F}{2F+\Delta F} + \frac{\Delta F}{2F-\Delta F}\right\} \tag{19.11}$$

After simplifying and neglecting a quadratic term ΔF^2, we arrive at the Eq. (19.4)

$$\Delta NF(\text{dB}) = 4.34\frac{\Delta F}{F}$$

After the transition to the dB scale, we have

$$\Delta NF_D = \frac{F_M}{F_D}\Delta NF_M - \frac{F_{REC}}{F_D G_D}\Delta NF_{REC} + \frac{F_{REC}-1}{F_D G_D}\Delta G_D(\text{dB}) \tag{19.12}$$

After the addition of the uncertainty of the ENR, which acts on the first two terms, a conclusion for the total uncertainty of the measurement is obtained via RSS.

$$\Delta NF_D(\text{dB}) = \sqrt{A^2+B^2+C^2+D^2} \tag{19.13}$$

with

$$A = \frac{F_M}{F_D}\Delta NF_M(\text{dB});\quad B = \frac{F_{REC}}{F_D G_D}\Delta NF_{REC}(\text{dB});\quad C = \frac{F_{REC}-1}{F_D G_D}\Delta G_D(\text{dB});$$

$$D = \left(\frac{F_M}{F_D} - \frac{F_{REC}}{F_D G_D}\right)\Delta ENR(\text{dB}) \tag{19.14}$$

What are the main errors in the individual terms?

$\Delta NF_M(\text{dB})$	Mismatch between noise source and DUT input during measurement.
$\Delta NF_{REC}(\text{dB})$	Mismatch between noise source and receiver during calibration.
$\Delta G_D(\text{dB})$	Mismatch between noise source and DUT input, and between DUT output and receiver
$\Delta ENR(\text{dB})$	Error in the ENR table, connector wear.
Rules of thumb	The noise figure of the measuring system must be as low as possible. Narrow band amplifiers are better than broadband amplifiers.

$$NF_{REC}(\text{dB}) < NF_D(\text{dB}) + G_D(\text{dB}) - 5\text{ dB} \tag{19.15}$$

Matching errors must be reduced as far as possible. Possible installation of insulator or attenuator.

One part of the problems with noise figure measurements comes from the measurement setup, and the other part comes from unwanted signals that are interpreted as additional noise. This can be, for example: (i) interference via the DUT, especially when a wafer is placed on a chuck, RF cables, and power cords (ii) the DUT oscillates (iii) the DUT is nonlinear (iv) the local oscillator introduces spurs or phase noise. When putting into operation a new measuring system, it is always recommended to monitor the relevant frequency range with a spectrum analyzer. In addition, a spectrum analyzer should be connected instead of the noise measuring receiver for monitoring purposes to ensure that the displayed noise level is correct. This must have the correct amplitude distribution without going into a limit, and it must be free of interference pulses. And it must display the expected levels: Low level without DUT and noise generator, medium level with active DUT and noise generator "off" and high level with active DUT and noise generator "on." Interference can arise from many sources, most of which are outside the actual frequency range. Some examples: Energy saving lamps based on gas discharge, mobile phones, computers in the laboratory, high-frequency plasma systems in neighboring technology laboratories, TV and radio stations, air traffic radar systems, etc. On the other hand, if the system is carefully analyzed for these possible interferences from outside, it is rarely necessary to use a Faraday cage, which usually causes great inconvenience.

Contact problems of the most general nature also play an important role. In on-wafer measurements they can be caused by worn or poorly adjusted probes or by poorly contactable materials of the pads. These are, e.g. hard vapor deposition layers instead of electroplating or aluminum metallization. The latter, however, is always in use in Si or Si/Ge technology. For this purpose there are special wafer probes, e.g. infinity probes [3]. For measurement setups in coaxial technology, special attention must be paid to perfect PC-7 or PC-3.5 connectors, which are always tightened with a torque wrench. When measuring single transistors in a test fixture, contact problems can easily occur within the fixture, which prevent a correct measurement. Contact problems can usually only be detected afterwards and after repeated measurements.

Uncertainty of the ENR Calibration

Despite careful calibration of the noise source, a certain error of the values in the ENR table remains [1]. This error increases slightly with frequency. It can be assumed that up to 10 GHz the $\Delta ENR \leq \pm 0.1$ dB. Above that it rises slightly. Figures 19.3 and 19.4 show a typical curve.

When the noise figure is calculated by a noise figure meter (e.g. N8975A or the older model HP8970B), the stored ENR table supplied with each noise source is

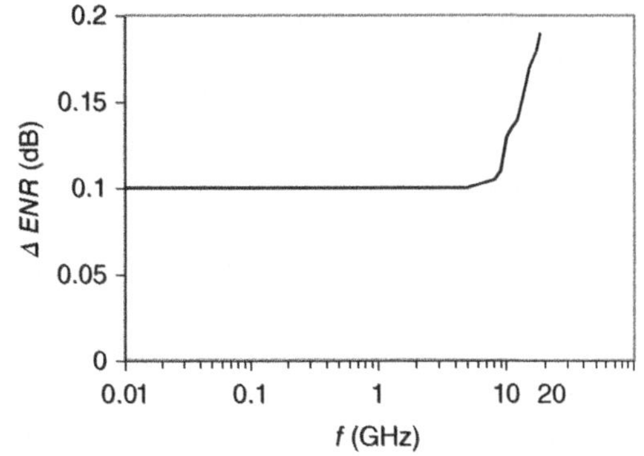

Figure 19.3 Uncertainty of the ENR specification for an 18 GHz noise generator.

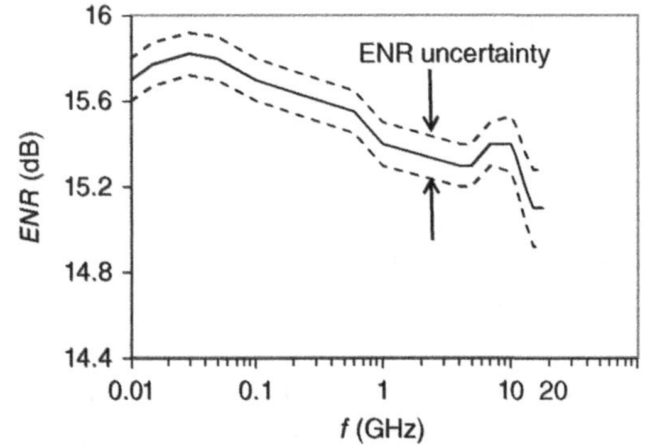

Figure 19.4 ENR curve of an 18 GHz noise generator with typical limits of variation.

Table 19.1 ENR Specification of a Keysight 346B noise source.

f (GHz)	0.01–1.5	1.5–3	3–7	7–18
ΔENR(dB)	0.2	0.19	0.2	0.23
VSWR	1.15	1.15	1.15	1.25

used. As shown in Figures 19.3 and 19.4, there are typical uncertainties whose RSS values are specified in the ENR table. The example in Table 19.1 is taken from the Keysight noise sources data sheet. These errors are also included in the total error.

Noise Source Mismatch

When using the Y-method, one source of error results from the mismatch of the noise source. Usually, sources with avalanche diodes are used, e.g. the Keysight 346 series [4]. Since the diode, the actual source of the noise power, is far away from the 50 Ω impedance, the matching is made by an attenuator. Of course, this is only imperfectly successful, especially the difference of the states "On" with T_H and "Off" with T_C remains. It has an effect on the Y-value. This results from two power measurements N_H and N_C at the output of the DUT. If the noise source is "On," we have the sum of the amplified noise of the DUT and the amplified contribution of the hot noise source

$$N_H = kT_0G_AB(F-1) + kG_ABT_H \tag{19.16}$$

Analogue with cold noise source

$$N_C = kT_0G_AB(F-1) + kG_ABT_C \tag{19.17}$$

The gain G_A of the DUT is not the same for both cases, as it depends on the reflection coefficient Γ_S of the source. It changes from Γ_{SH} in the "On" state to Γ_{SC} in the "Off" state. In general

$$G_A = |S_{21}|^2 \frac{1-|\Gamma_S|^2}{|1-\Gamma_S S_{11}|^2\left(1-|\Gamma_2|^2\right)} \tag{19.18}$$

We neglect the effect on Γ_2, the output reflection coefficient of the DUT, because the small changes at the input of the active DUT hardly affect the output.

Without taking into account the different G_A, the measured Y_1 is

$$Y_1 = \frac{N_H}{N_C} = \frac{T_E+T_H}{T_E+T_C} \tag{19.19}$$

As G_A is canceled from the fraction. If one takes into account the different G_A values, the following results are obtained

$$Y_2 = \frac{\left(1-|\Gamma_{SH}|^2\right)|1-\Gamma_{SC}S_{11}|^2}{\left(1-|\Gamma_{SC}|^2\right)|1-\Gamma_{SH}S_{11}|^2} \times \frac{T_E+T_H}{T_E+T_C} \tag{19.20}$$

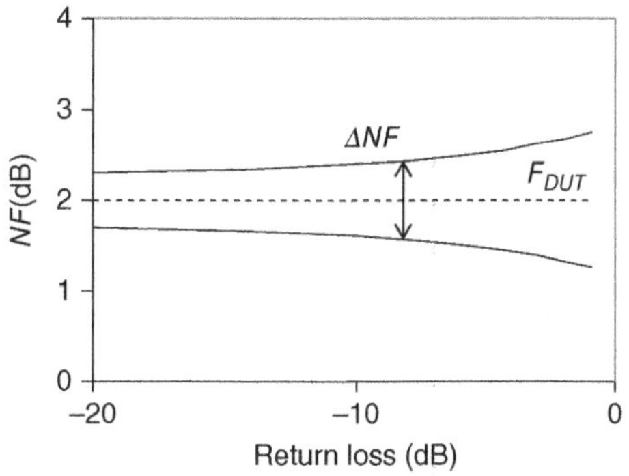

Figure 19.5 Noise measurement error due to mismatch of the noise source versus return loss of the DUT.

This error is by no means negligible [5]. Especially low-noise transistors, e.g. GaAs_FETs have a high input reflection coefficient. As Figure 19.5 shows, the error then increases and can take on values that are no longer tolerable. To minimize this effect, an attenuator or isolator can be connected behind the noise source. A noise source with reduced ENR already contains this additional attenuator. The option of choice could be to use the "Cold Source" method, where this error does not occur. However, this method requires a high level of technical equipment, because it uses an impedance tuner and determines the four noise parameters.

T_0 = 290 K Is not T_{OFF}

This is a rather small source of error, which only plays a role with extremely low-noise objects. For the sake of completeness we will deal with it here using the Agilent AN 57_2 [1]. The ENR table, which is provided by the manufacturer, results from a factory calibration. It has the following reference temperature T_0

$$ENR = \frac{T_H - T_0}{T_0} \tag{19.21}$$

ENR is not in dB here! However, calibration is not performed at T_0, but at the ambient temperature T_C. Agilent recommends using $T_C = 303\,\text{K}$ for most noise sources. This results in a slightly corrected value

$$ENR_{COR} = ENR + \frac{T_0 - T_C}{T_0} \tag{19.22}$$

When measuring according to the Y method, the noise generator switches between $T_{ON} = T_0 \times ENR_{COR} + T_C$ and the room temperature T_C. According to the Y-method, the noise temperature of the DUT is

$$T_D = \frac{T_{ON} - YT_C}{Y - 1} \tag{19.23}$$

We can now calculate the error caused by disregarding $T_0 \neq T_C$. The uncorrected value is

$$T_{DU} = \frac{T_H - YT_0}{Y - 1} \tag{19.24}$$

With $T_H = (ENR + 1)T_0$ (19.21). The corrected value is (19.22). Figure 19.6 shows that the difference between the results is small if the true ambient temperature (result T_D) is taken into account and if it is disregarded (result T_{DU}). However, if the DUT has a very low noise figure, the error can be around 0.1 dB.

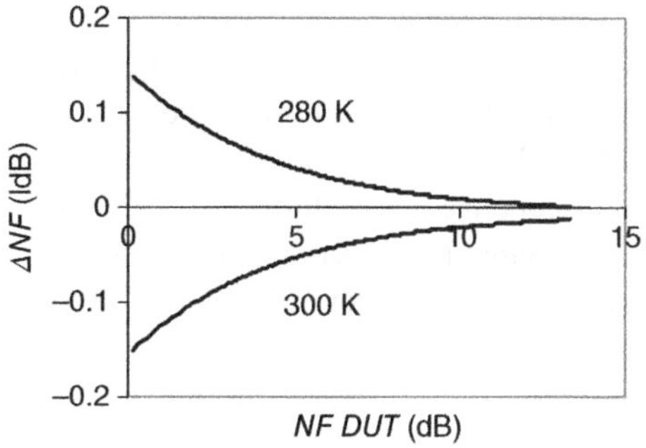

Figure 19.6 Error of a Y-measurement due to the deviation of T_C from $T_0 = 290\,\mathrm{K}$.

With modern noise figure measuring devices, the room temperature T_C can be entered so that this correction is automatically carried out.

Mismatch in the System

Microwave components are interconnected to form a system, and reflections and re-reflections occur between the individual components. This is particularly critical in a noise measurement system, because the power measurement involves extremely low levels and the power transmission is distorted due to mismatching at the junctions. This can only be partially compensated by vectorial network measurements and mathematical correction. The different matching situations between calibration and measurement, for example, are omnipresent. During calibration, the noise source is directly at the receiver, during measurement the object is at the receiver and the noise source at its input. These, even though often small, mismatches can cause considerable errors. We analyze exemplary the realistic situation of the measurement of a low-noise amplifier shown in Figure 19.7.

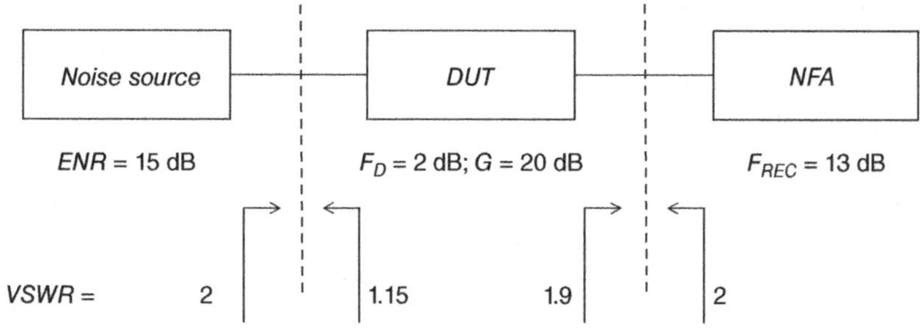

Figure 19.7 Scheme of the measurement of an amplifier at $f = 10\,\mathrm{GHz}$ with details of the interfaces.

Due to mismatch at the interface between network 1 and network 2 (e.g. Noise Source and DUT), only part of the power available at the output of network 1 is transferred to the input of network 2. Part of the reflected power is in turn reflected back to the output of network 1 because it is mismatched. This results in a sequence of re-reflections, which are a confusing source of error in a noise measurement. The remedy here is a best possible matching at this interface. Let us consider the power transmission from a source with reflection coefficient Γ_S to a load with reflection coefficient Γ_L.

$$P_L = \frac{|b_S|^2}{|1 - \Gamma_S \Gamma_L|^2} \tag{19.25}$$

P_L is the power absorbed by the load. $|b_S|^2$ is the power that the source delivers to a non-reflecting load. Since Γ_S and Γ_L are complex quantities, they must be measured with a VNA. If one does not have this possibility or wants to avoid the effort, one can use the scalar quantities VSWR, which are usually known for the components, to estimate the maximum error. With $\rho = |\Gamma|$ we have

$$\rho = \frac{VSWR - 1}{VSWR + 1} \tag{19.26}$$

The maximum error due to mismatch is therefore

$$M(\text{dB}) = \pm 20\, log\left(1 \pm \rho_S \rho_L\right) \tag{19.27}$$

This can be quickly taken from a nomograph [1] or calculated according to (19.27).

Table 19.2 lists all the variables we need for error analysis. This is done on the basis of the error propagation law.

If vector network measurements are used to avoid errors due to mismatch, only the uncertainty of the noise source and the measuring instrument remains. For this purpose, the complex reflection coefficients of the DUT (Γ_{IN}, Γ_{OUT}) of the receiver (Γ_{NFA}) are measured and Γ_{NS} is taken from the table. In this case the mismatch terms would be $\delta_{NS-DUT} = \delta_{NS-REC} = \delta_{DUT-REC} = 0$.

The result is then much more accurate: $NF_D = 2 \pm 0.12\,\text{dB}$.

The noise figure of the receiver also plays an important role, because it is included in the measurement result according to the Friis formula. The effect is particularly big if the DUT has only a low gain. An estimation of the measurement error ΔNF(dB) is shown in Figure 19.8. The low values on the left side of the diagram can be obtained using an NFA with a low-noise preamplifier. The error is small even with a moderate gain of the DUT. With receiver noise figures of $NF_{REC} > 20\,\text{dB}$ the error increases strongly, even if the DUT has a high gain. This is, for example, the case when using VNA or spectrum analyzer without preamplifier.

Table 19.2 Error terms from the example of the LNA measurement in Figure 19.7 at $f = 10\,\text{GHz}$.

Circuit parameters VSWR/dB	Noise source 1.15	DUT in:2 out:2	REC 1.9	NF_{DUT} 2 dB	NF_{REC} 13 dB	G_D 20 dB	NF_{12} (meas.) 2.5 dB
Formula (19.12)			$\Delta NF_D = \frac{F_{12}}{F_D}\delta NF_{12} - \frac{F_{REC}}{F_D G_D}\delta NF_{REC} + \frac{F_{REC}-1}{F_D G_D}\delta G_D$				
RSS uncertainty formula			$\delta NF_{12} = \sqrt{\delta^2_{NS\text{-}DUT} + \delta NF^2_{REC}}(\text{dB})$				
			$\delta NF_{REC} = \sqrt{\delta^2_{NS\text{-}REC} + \delta NF^2_{REC}}(\text{dB})$				
			$\delta G_D = \sqrt{\delta^2_{NS\text{-}DUT} + \delta^2_{NS\text{-}REC} + \delta^2_{DUT\text{-}REC} + \delta G^2_{REC}}(\text{dB})$				
Mismatch formula			$\delta_{Source-Load} = \pm 20\log(1 \pm \rho_S \rho_L)$				
Mismatch values (dB)	$\delta_{NS-DUT} = 0.2$		$\delta_{NS-REC} = 0.19$			$\delta_{DUT-REC} = 0.95$	
Instrument Uncertainty (dB)	$\delta NF_{REC} = 0.05$		$\delta G_{REC} = 0.15$			$\delta ENR = 0.1$	
RSS uncertaintyvalues	$\delta NF_{12} = 0.21$		$\delta NF_{REC} = 0.2$			$\delta G_D = 1$	
Formula (19.14) terms	$A = \frac{F_{12}}{F_D}\delta NF_{12}$	$B = \frac{F_{REC}}{F_D G_D}\delta NF_{REC}$		$C = \frac{F_{REC}-1}{F_D G_D}\delta G_D$		$D = \left\{\frac{F_{12}}{F_D} - \frac{F_{REC}}{F_D G_D}\right\}\delta ENR$	
Example values (dB)	$A = 0.24$	$B = 0.03$		$C = 0.12$		$D = 0.1$	
Result			$\Delta NF_D = \sqrt{A^2 + B^2 + C^2 + D^2} = 0.28\,\text{dB}$ $NF_D = 2 \pm 0.28\,\text{dB}$				

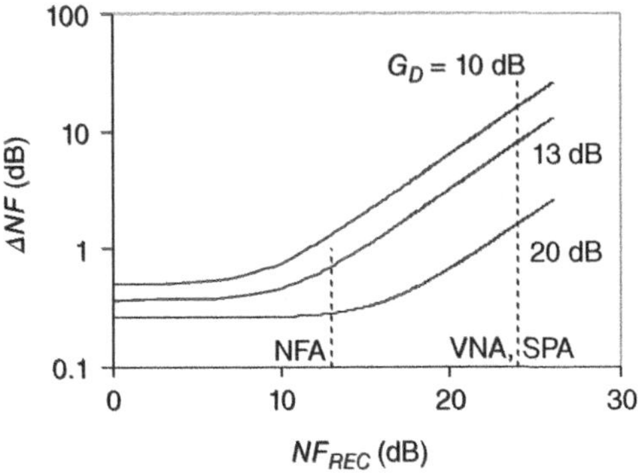

Figure 19.8 Error due to high receiver noise figure. Parameter is the gain of the DUT.

Linearity of the Receiver

Nonlinearities are not initially thought of in connection with noise signals. In a measuring system for transistors with impedance tuner, however, we have extremely different gain of the DUT, depending on the tuner setting. This leads to large differences in output power. The example in Figure 19.9 shows measurement results on a GaN–HFET with (0.25 × 100) μm² gate and a cut-off frequency $f_{MAX} \approx 100$ GHz.

The measurement was performed with an on-wafer system at $f = 9.5$ GHz [6]. A large number of tuner positions were used. The noise power displayed on the receiver has a range of about 20–40 dB, depending on the detuning of the source reflection factor Γ_S from the Γ_{OPT}. This is mainly due to the different gain. The circles apply to noise source "On," the crosses to the "Off" state. The dotted lines

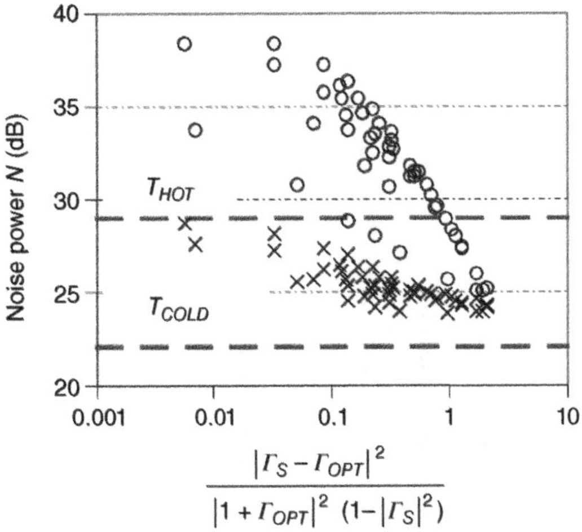

Figure 19.9 Output power for a transistor measurement with variation of Γ_S. Circles: Noise source "On," Crosses: "Off." Dotted lines: level during calibration.

show the two levels of calibration. In the level range $N = 20 \ldots 40\,\text{dB}$ a constant value for G_{REC} is assumed, although only the 7 dB range between T_H and T_C is scanned during calibration. In the NFA there is an internal system that ensures optimum linearity within the calibration range by means of switched attenuators or amplifiers. This is guaranteed when measuring amplifiers with an almost constant output level. However, in a typical transistor measurement, as in our example, we have a level range of approx. 15 dB within one measurement series. Theoretically G_{REC} can be adjusted to the level. However, this is time-consuming, changes the noise parameters of the receiver, and is therefore not very practical. If a spectrum analyzer or VNA is used as receiver, the situation is more critical. In these devices, the internal linearization effort is less than that of the NFA. The "Cold Source" method offers a way out. Here the high level of the noise generator does not occur with T_{HOT}. The level range at the receiver is only about 5 dB and is within the calibration range. This is a great advantage, especially when using SPA or VNA as test receiver.

References

1 Agilent Technologies (2014). Noise Figure Measurement Accuracy – The Y-Factor Method. Application Note 57-2.

2 Keysight Technologies (2017). Noise Figure Selection Guide, Minimizing the Uncertainties. Application Note.

3 FormFactor (2020). Cascade Infinity Probe-Coaxial. Datasheet.

4 Keysight Technologies (2016). 346 A/B/C Noise Sources. Operating and Service Manual.

5 Agilent Technologies (2004). Non Zero Noise Figure after Calibration. Application Note 1484.

6 Rudolph, M., Heymann, P., and Boss, H. (2010). Impact of receiver bandwidth and nonlinearity on noise measurement methods. *IEEE Microwave Mag.* 11: 110–121.

20

Phase Noise

Basics

Frequency stability is generally defined by two parameters: long-term stability and short-term stability. Long-term stability describes the frequency change over long periods of time. This change is called frequency drift. Short-term stability describes frequency changes within seconds and shorter. The boundary between the two terms depends on the intended use. Short-term stability refers to frequency changes, which do not appear as static deviations or drift, but as random, even periodic fluctuations around a mean value. Here there are two types of frequency fluctuations. Deterministic ones are modulation frequencies that appear as distinct components in the RF sideband spectrum. These frequencies are called "Spurious" and can usually be assigned to known causes, e.g. mains hum. The second form of appearance of frequency fluctuations is random in nature and is usually referred to as phase noise. The sources of phase noise of a frequency are thermal noise, shot effect, and flicker noise of active and passive components. Thus, in the frequency domain, a signal does not appear as a single monochromatic spectral line, but covers a spectral range above and below the nominal frequency with modulation sidebands caused by random phase variations (Figure 20.1) [1, 2].

Figure 20.2 shows the development of the phase of an oscillator signal in the time domain. The phasor rotates at the angular velocity $\omega = 2\pi f_S$. However, this is subject to small fluctuations, which are caused by the vectorial addition of the noise voltage v_N. In the diagram on the right, we see that the increase of the phase with time is not linear, but that rather a fluctuation is superimposed.

Depending on the application, fast or slow frequency fluctuations are disturbing. The spectrum of the phase noise is displayed over the offset frequency from the carrier. Slow changes determine the low-frequency range of the spectrum near the carrier, while fast changes determine the high-frequency range. In Doppler radar, for example, the resolution is determined by the noise near the carrier. Figure 20.3 shows the frequency shift of the transmitter frequency f_S as a function of the speed

A Guide to Noise in Microwave Circuits: Devices, Circuits, and Measurement, First Edition.
Peter Heymann and Matthias Rudolph.

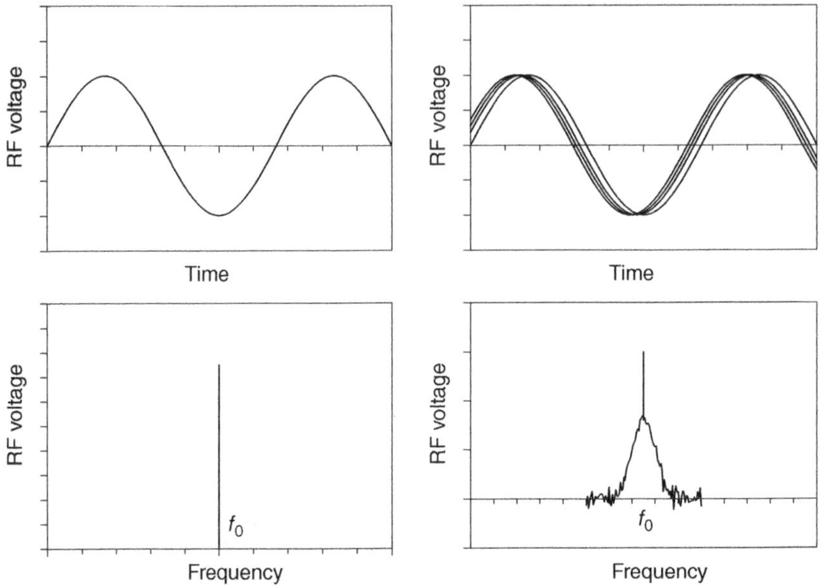

Figure 20.1 Ideal (left) and real signal (right) in time and frequency domain.

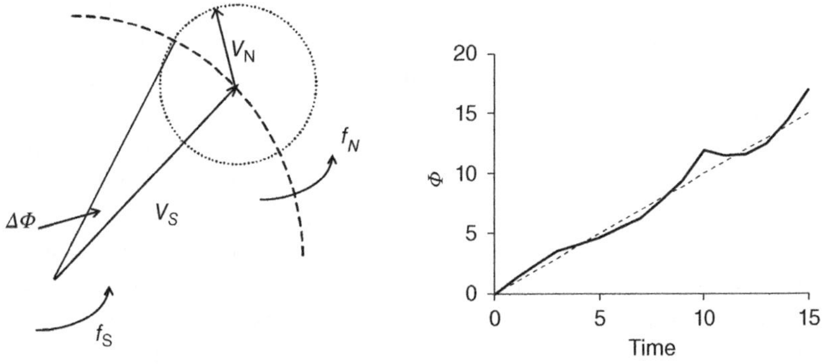

Figure 20.2 Left: Phasor of the signal voltage with superimposed noise voltage. Right: Time response of the phase.

of the object v according to (20.1). Since the phase noise is strongest in the immediate vicinity of the carrier, the signal from small velocities v can easily be covered.

$$\Delta f = 2f_S \frac{v}{c} \times cos(\alpha) \tag{20.1}$$

The angle between measuring beam and velocity vector is α.

There is a detection limit due to phase noise if a weak signal is to be measured near the carrier. It disappears in the sideband spectrum, as shown in Figure 20.4.

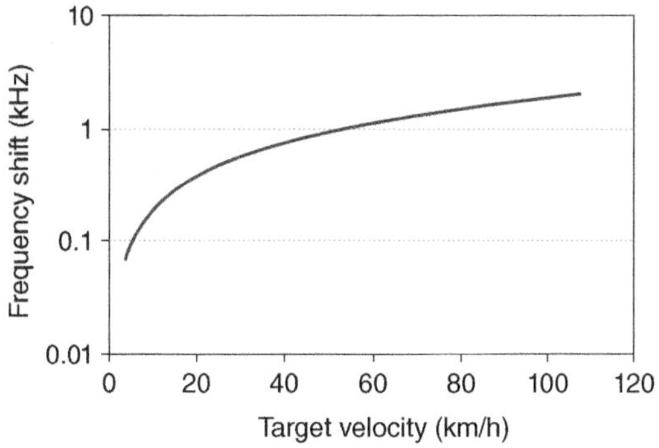

Figure 20.3 Frequency shift due to Doppler effect according to (20.1) at $f_S = 10\,\text{GHz}$.

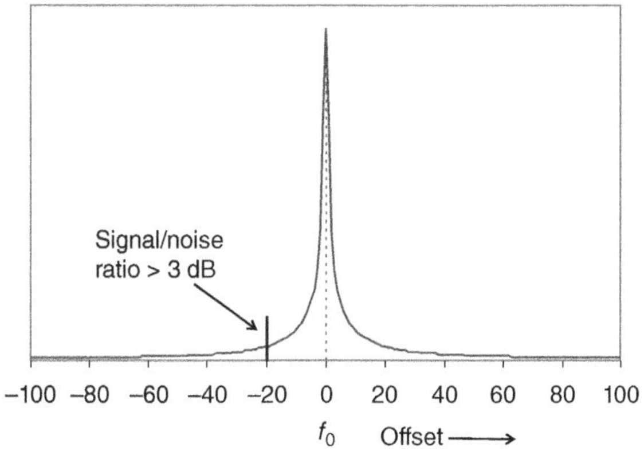

Figure 20.4 Detection problem of weak signals due to phase noise of LO.

This is also the case with heterodyne reception due to the phase noise of the local oscillator.

Even with UMTS, where noise problems are largely eliminated by spreading, attention must be paid to a minimum signal-to-noise ratio. Here it is important to maintain a precisely defined phase position, which is no longer given by the fluctuations. This fluctuation is sketched in Figure 20.5.

Reciprocal Mixing

There may be another effect with heterodyne reception. In addition to the disappearance of the wanted signal in the phase noise of the local oscillator, interference can occur if there is an interfering signal in the vicinity of the wanted signal. This is also mixed down into the baseband, due to the phase noise. This effect is called "reciprocal mixing." The explanation can be found in Figure 20.6. In the following diagram, only the wanted signal is present. There is no interference signal

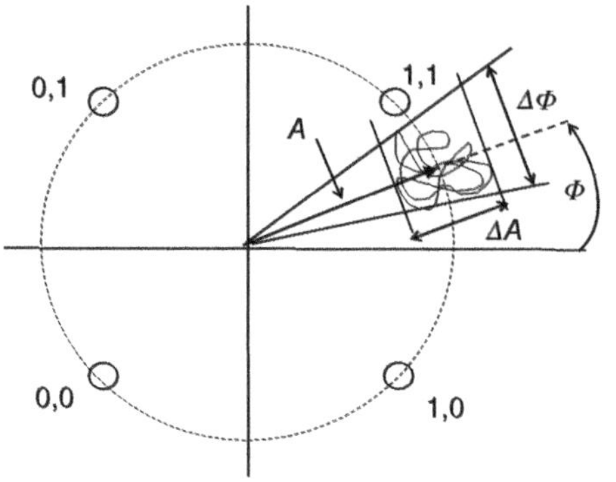

Figure 20.5 Phase modulation error due to phase variations.

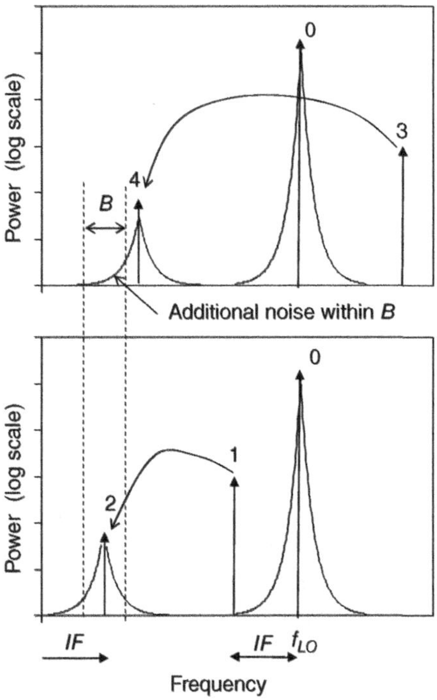

Figure 20.6 Reciprocal mixing. Below: No interference signal 3 present, 1 is mixed down to 2 in the baseband. Top: Noise signal 3 appears with phase noise in the baseband.

nearby. The desired signal 1 is located at the frequency spacing IF from the local oscillator 0. It enters the baseband by mixing and becomes the desired signal 2. The phase noise of the LO is found as sidebands of 2, provided they are within the reception bandwidth *B*. This is the ideal case of undisturbed heterodyne reception. In the upper diagram we see the influence of an interfering signal 3, which lies outside the reception band. The mixed product 4 is not within the passband *B*. Nevertheless, interference from reciprocal mixing can occur [3]. Since 4 also has

Table 20.1 Required phase noise levels for the demands of mobile phone standards.

f_m (MHz)	0.6	1.6	3
P_I (dBm)	−43	−33	−23
L(fm) (dBc/Hz)	−118	−128	−138

the noise sidebands of the LO, some of it can enter the reception band and thus interfere with the useful signal. From the permissible fault regulations for communication systems are rules for the local oscillator in the receiver derived. The permissible phase noise $L(f_m)$ is calculated using the following formula

$$L\left(f_m\right)\left({}^{\mathrm{dBc}}\!/_{\mathrm{Hz}}\right) = P_C(\mathrm{dBm}) - S(\mathrm{dB}) - P_I(\mathrm{dBm}) - 10\,log(B) \tag{20.2}$$

It means: P_C: level of the wanted signal; S: reduction of the interference signal; P_I: level of the interference signal; B: noise bandwidth.

An example for the mobile phone standard GSM. The signal-to-noise ratio S should be at least 9 dB. The bandwidth is $B = 200\,\mathrm{kHz}$, the signal level $P_C = -99\,\mathrm{dBm}$. The values in line 2 are assumed for the disturbance level. The requirements for the LO in line 3 result from (20.2) (Table 20.1).

These requirements are high. They are achieved at $f = 2\,\mathrm{GHz}$ by good BJT oscillators, but not by FET oscillators [1]. Because of the required frequency stability, synthesizers with a phase-locked loop must be used in any case.

Description of Phase Noise

It is a noise process, which means the instabilities are random in nature. They are therefore represented as spectral density distribution. As we know, the term "spectral density" describes the energy distribution as a continuous function expressed in units of the energy present in a given bandwidth. The phase noise of a carrier signal is interpreted as phase modulation with a small modulation index by a noise source. General observations as well as measurements are made in the frequency domain; nevertheless it is called phase noise. We consider the relationship between the fluctuating phase in Figure 20.2 and the fluctuating frequency in Figure 20.4. The phasor rotates with the angular velocity ω and the phase angle ϕ increases with time t. At the point τ it is $\phi = \omega\tau$. In general

$$\omega(t) = \frac{d\phi(t)}{dt} \tag{20.3}$$

Let us look at the fluctuations only

$$\Delta f(t) = \frac{1}{2\pi}\,\frac{d}{dt}(\Delta\phi(t)) \tag{20.4}$$

With (20.3) we move on to the frequency range

$$\Delta f(f) = \frac{\omega}{2\pi} \Delta\phi(f) \tag{20.5}$$

There are various mathematical terms for quantifying phase noise. All of them relate to the measurable quantity of the spectral power density [4]. It results from the square of the fluctuation quantity. For the phase, for example, in the unit rad^2/Hz.

$$S_\phi(f) = \frac{\Delta\phi(f)^2}{B} \tag{20.6}$$

$S_\phi(f)$: The spectral density of phase fluctuations.
$L(f)$: The single sideband phase noise as the ratio of the energy present in a sideband of $B = 1$ Hz bandwidth to the total signal power.
$S_{\Delta f}(f)$: The spectral density of frequency fluctuations.
$S_Y(f)$: The spectral density of fractional frequency fluctuations.

Spectral Power Density of Phase Fluctuations $S_\phi(f)$

This is the fundamental curve, but it is not directly visible on a spectrum analyzer. One needs a phase sensitive receiver with a mixer as phase detector. The signal of an oscillator, which is phase modulated by the noise, is demodulated in this way. The phase fluctuations are converted into voltage fluctuations.

$$\Delta V(f) = K_\phi \Delta\phi(f) \tag{20.7}$$

K_ϕ (V/rad) is the phase detector constant. The fluctuations of the phase of the oscillator signal are therefore converted into voltage fluctuations. In addition, the carrier disappears, so that the spectrum is now in the baseband. This signal is passed to an LF spectrum analyzer from which $S_\phi(f)$ can be read.

$$S_\phi(f) = \frac{\Delta\phi(f)^2}{B} = \frac{\Delta V^2(f)}{K_\phi^2 B} \quad \left(\frac{rad^2}{Hz}\right) \tag{20.8}$$

$S_\phi(f)$ is the measure of the phase stability of a signal. This type of description is appropriate to the analysis of effects of phase noise in systems with phase sensitive functions, e.g. in digital communication systems with phase modulation (UMTS) (Figure 20.7).

A transformation of the frequency dependencies takes place. For example, one finds the $^1/_f$ noise (Flicker FM) here with the slope $^1/_{f^3}$.

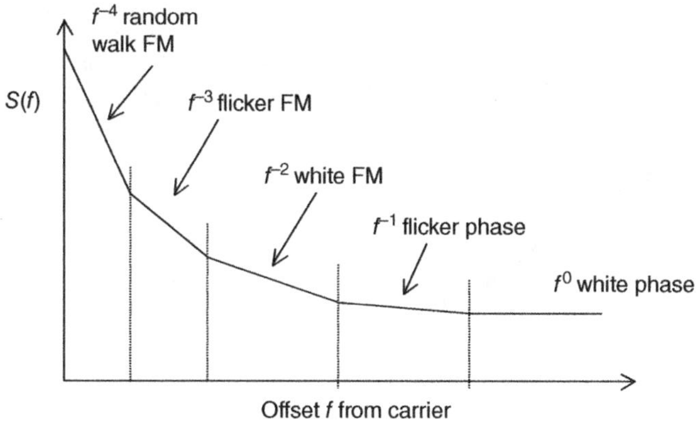

Figure 20.7 Spectral energy distribution of the phase noise of an oscillator. The contributions of the noise sources in the circuit are assigned.

The Single Sideband Phase Noise *L(f)*

$L(f)$ is an indirect measure of the noise energy. It corresponds to the trace on the spectrum analyzer. The relation to $S_\phi(f)$ is

$$L(f) = \frac{S_\phi(f)}{2P_C} \quad \left(\frac{dBc}{Hz}\right) \tag{20.9}$$

$L(f)$ is normalized to the output power P_C of the oscillator and therefore has the unit dBc/Hz. However, this only applies on condition that all energy is located in the first-order sidebands. The condition $\Delta\phi_{peak} \ll 1$ must be fulfilled. While the measurement of $S_\phi(f)$ is always correct, the specification of $L(f)$ is only correct at higher offset frequencies. The procedure is not correct if the instantaneous phase modulation exceeds the range of small phase angles. This can be the case at small offset frequencies (<10 kHz).

In Figure 20.8 the limit curve is drawn. Below the straight line (20.9) is valid, above not. It corresponds to an amplitude of the phase fluctuation $\Delta\phi_{peak} = 0.2$ rad and has a slope of −10 dB/dec.

Spectral Power Density of Frequency Fluctuations $S_{\Delta f}(f)$

It is derived from the phase noise $S_\phi(f)$ by the following consideration.

The frequency change in the time domain as derivative of the phase change (20.4) is

$$\Delta\omega(t) = \frac{d}{dt}\Delta\phi(t) \tag{20.10}$$

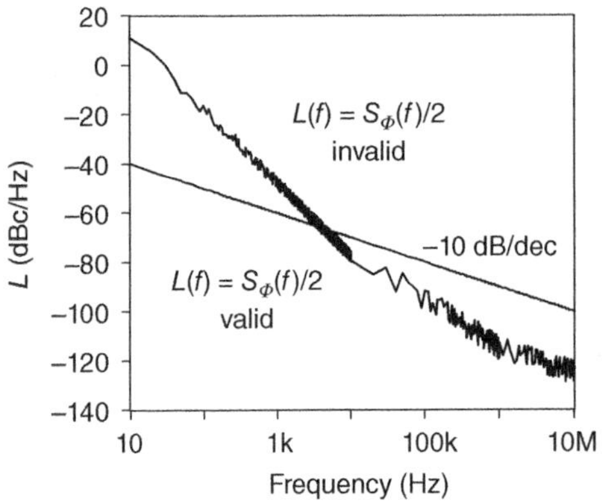

Figure 20.8 Phase noise spectrum $L(f)$ with the limit curve of the small angle range. Source: Based on Agilent Technologies [4].

Written in the frequency domain

$$\Delta f(f) = \frac{2\pi f}{2\pi} \Delta\phi(f) \tag{20.11}$$

The transition to power is made by squaring

$$\Delta f_{RMS}^2(f) = f^2 \Delta\phi_{RMS}^2(f) \tag{20.12}$$

Thus the spectral density of the frequency fluctuations is

$$S_{\Delta f}(f) = \frac{\Delta f_{RMS}^2(f)}{B} = S_\phi(f) \quad \left(\frac{Hz^2}{Hz}\right) \tag{20.13}$$

The spectral density of relative frequency fluctuations $S_Y(f)$ corresponds to the value of the previous variable normalized to the oscillator frequency f_S

$$S_Y(f) = \left(\frac{f}{f_S}\right)^2 S_\phi(f) \quad \left(\frac{1}{Hz}\right) \tag{20.14}$$

It allows the quality comparison of oscillators of different frequencies.

Excursus on Frequency and Phase Modulation

The instantaneous frequency of a real oscillator is

$$\omega_0 + \Delta\omega \tag{20.15}$$

The actual frequency ω_0 is superimposed by a fluctuation of the quantity $\Delta\omega$ with the frequency ω_f. Because of (20.4) the phasing results from the integration over time.

$$\phi = \int \left\{\omega_0 + \Delta\omega \cos\left\{\omega_f\, t\right\}\right\}\, dt = \omega_0 t + \frac{\Delta\omega}{\omega_f} sin\left(\omega_f t\right) \tag{20.16}$$

The time response of the oscillator voltage is thus

$$V_C(t) = V_P cos\left\{\omega_0 t + \frac{\Delta\omega}{\omega_f} sin\left(\omega_f t\right)\right\} \tag{20.17}$$

With addition theorem one can also write

$$V_C(t) = V_P\left\{cos\left(\omega_0 t\right) cos\left(\beta\, sin\left(\omega_f t\right)\right) - sin\left(\omega_0 t\right) sin\left(\beta\, sin\left(\omega_f t\right)\right)\right\} \tag{20.18}$$

The essential quantity is the modulation index

$$\beta = \frac{\Delta\omega}{\omega_f} \tag{20.19}$$

After (20.16) it stands for the amplitude of the phase fluctuation. In case of phase noise we cannot assume that this fluctuation is always very small, although it is the weak noise of a strong oscillator power. As long as $\beta \ll 1$, i.e. at high offset frequency ω_f, the approximations in (20.18) are

$$cos\left(\beta sin\left(\omega_f t\right)\right) = 1; \quad sin\left(\beta sin\left(\omega_f t\right)\right) = \beta sin\left(\omega_f t\right) \tag{20.20}$$

with

$$\beta sin\left(\omega_0 t\right) sin\left(\omega_f t\right) = \frac{\beta}{2}\left\{sin\left(\omega_0 + \omega_f\right) - sin\left(\omega_0 - \omega_f\right)\right\} \tag{20.21}$$

If we insert (20.20) and (20.21) in (20.18), we get

$$V_C(t) = V_P\left\{cos\left(\omega_0 t\right) + \frac{\beta}{2} cos\left(\omega_0 + \omega_f\right)t - \frac{\beta}{2} cos\left(\omega_0 - \omega_f\right)t\right\} \tag{20.22}$$

Phase disturbance occurs at the offset frequency ω_f. The limitation to one frequency is justified when considering the principle. The practical values are normalized to B = 1 Hz, too.

In (20.22) we see the formation of the sidebands which are offset by $\pm\omega_f$ against ω_0. The peak voltage of this sideband modulation is $V_P\beta/2$. Since $L(f)$ after (20.9) relates this value to the power of the oscillator, the result is

$$L(f) = \left(\frac{V_P\beta}{2}\right)^2 \frac{1}{V_P^2} \tag{20.23}$$

In the context of phase fluctuation (20.16), there is $\beta = \Delta\phi_{PEAK}$. With the RMS values $\Delta\phi_{PEAK} \approx \sqrt{2}\Delta\phi_{RMS}$ it results in

$$L(f) = \frac{1}{2}\Delta\phi_{RMS}^2 = \frac{1}{2}S_\phi(f) \tag{20.24}$$

As was made clear earlier, this only applies to higher offset frequencies ω_f. According to experience, for example, $\omega_f/2\pi > 10$ kHz. Below there is only the direct measurement of $S_\phi(f)$ with a phase detector correct. If the approximation

(20.20) is no longer valid, the general calculation of the modulation is carried out using Bessel functions. Instead of (20.22) the solution is then

$$V_C(t) = V_P \left\{J_0(\beta)cos\left(\omega_0 t\right) - J_1(\beta)sin\left(\omega_0 \pm \omega_f\right) t - \ldots\right\} \quad (20.25)$$

The small angle criterion for the validity of $L(f)$ is here

$$\frac{J_1(\beta)}{J_0(\beta)} = \frac{1}{2}\beta \quad (20.26)$$

The Allan Variance $\sigma_Y^2(\tau)$

For some applications, e.g. highly stable quartz oscillators, time standards, laser, or Doppler radar, it is useful to describe the frequency stability in the time domain. The Allan variance is a measure of the frequency stability in the range of a few seconds or more. In contrast, the short-term stability is described with the phase noise [5] (Figure 20.9).

Frequency measurements are carried out at intervals τ with a frequency counter and the difference between the measured values $y(i+1)$ and $y(i)$ is stored. The measurements are taken at the moments $t_0 + i\tau$ and $t_0 + (i+1)\tau$. The Allan variance calculated from this comprises all values measured in time T.

$$\sigma_y^2(\tau) = \frac{1}{2m}\sum_{i=0}^{m-1}\{y(i+1) - y(i)\}^2 \quad (20.27)$$

With $m = {}^T/_\tau - 1$ and the normalized frequency y_i measured at the moment $t_0 + i\tau$.

$$y(i) = \frac{f_i\left(t_0 + i\tau\right) - f_0}{f_0} \quad (20.28)$$

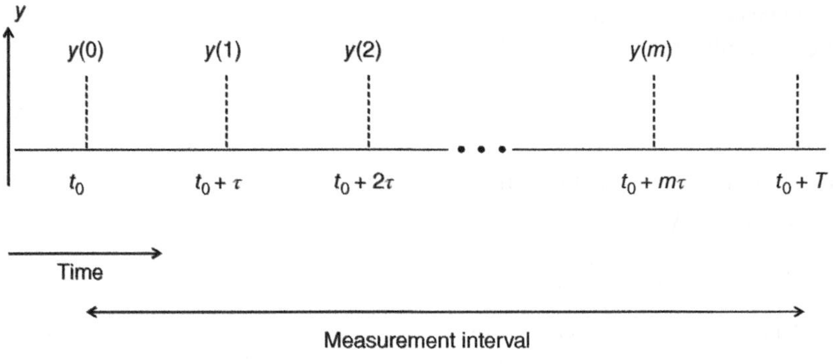

Figure 20.9 Time diagram of measurement of frequency values y at distance τ.

Here f_0 can be the start frequency or the frequency of a reference oscillator.

Allan variance and phase noise describe the same physical aspect: $\sigma_Y(\tau)$ in the time domain and $L(f)$ in the frequency range. Their conversion is complicated and practically not possible in a closed formula. The basic prerequisite for conversion is that the lowest frequency of the $L(f)$-curve corresponds to the highest value up to which the $\sigma_Y(\tau)$-curve is required. When measuring directly with a spectrum analyzer, one quickly reaches the limit of the small angle criterion discussed above. Better is the direct measurement of $S_\phi(f)$. Here is f the offset frequency from the carrier f_0.

First we need

$$S_y(f) = \frac{2f^2}{f_0^2} L(f) = \left(\frac{f}{f_0}\right)^2 S_\Phi(f) \tag{20.29}$$

Supposing that the total noise power above the highest offset frequency f_{MAX} is small, i.e.

$$\int_{f_{MAX}}^{\infty} S_\Phi(f)df \ll 1\ \text{rad}^2$$

holds the formula

$$\sigma_Y^2(\tau) = 2\int_0^{f_{MAX}} S_Y(f)\frac{sin^4(\pi\tau f)}{(\pi\tau f)^2}df \tag{20.30}$$

A single point $S_Y(f)$ can therefore not be converted; we need a range of the curve, which belongs appropriately to one of the ranges from Table 20.2.

Here is an approximation of $\sigma_Y(\tau)$ for the -20 dB/dec dependence of the phase noise spectrum, which is often present in a wide frequency range.

$$L(f) \propto \frac{1}{f^2};\quad L(f) = 10^{\frac{L(\text{dBc/Hz})}{10}}$$

$$\sigma_Y(\tau) = \frac{\sqrt{L(f)f^2}}{f_0}\tau^{-1/2} \tag{20.31}$$

Table 20.2 Frequency dependencies of the noise processes in a phase noise spectrum, as shown in Figure 20.7.

Noise process	$L(f)$	$S_Y(f)$	$\sigma_Y(\tau)$
Random walk FM	f^{-4}	f^{-2}	$\tau^{1/2}$
Flicker FM	f^{-3}	f^{-1}	τ^0
White FM	f^{-2}	f^0	$\tau^{-1/2}$
Flicker PN	f^{-1}	f^1	τ^{-1}
White PN	f^0	f^2	τ^{-1}

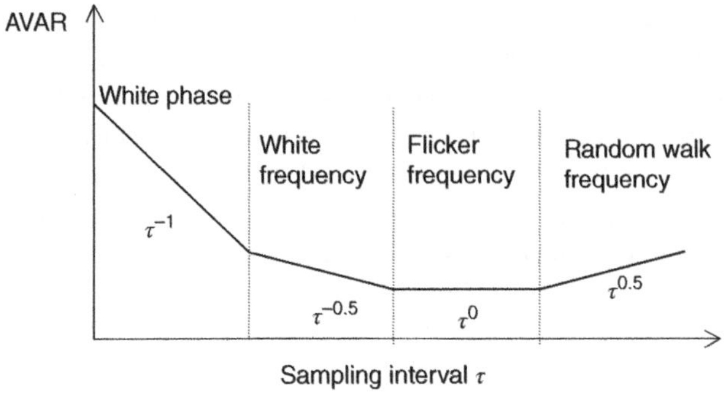

Figure 20.10 Allan variance of the instability of an oscillator.

Please note that the linear value for $L(f)$ must be used. The measured value is, e.g. for a very stable 10 MHz OCXO oscillator at 1 kHz offset $L = -155$dBc/Hz. Based on this fixed point, we get from (20.31) for $\tau = 1$ ms: σ (1 ms) = 5.6×10^{-11}.

The different components of the frequency spectrum of the phase noise can be found in the AVAR diagram. The high offset frequencies are now at low and vice versa. The following picture is obtained. The range "white frequency" corresponds, for example, to the white microwave noise from the calculation example. It transforms in the phase noise to f^{-2} and in the Allan variance to $\tau^{-1/2}$ (Figure 20.10).

This curve is obtained directly with a frequency counter with adjustable time interval between two measurements. "Random walk frequency" is, for example, the very slow frequency change of the carrier.

Residual FM

Residual FM (RFM) is a measure of the frequency stability of an oscillator, too. It specifies the noise within the bandwidth of a communication link. This is also related to $S_{\phi}(f)$. Residual FM is the RMS value of the total frequency fluctuations within a specific bandwidth between the frequencies f_L and f_U [6] (Figure 20.11).

$$RFM = \sqrt{\int_{f_L}^{f_U} f^2 S_{\Phi}(f) df} \tag{20.32}$$

Typical bandwidths of communication links are, e.g. 50 Hz–3 kHz; 20 Hz–15 kHz. Only short-term instabilities of frequencies within these bandwidths are covered. The spectral energy distribution within the band is lost with this specification.

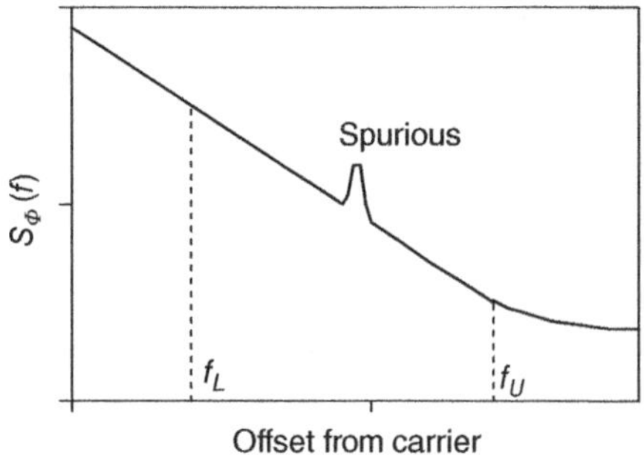

Figure 20.11 The frequency range between f_L and f_U of the phase noise spectrum contributes to the residual FM. "Spurious" signals within the relevant band can increase the value.

Multiplication and Division

Frequency multiplication and frequency division have a direct effect on the phase noise of the multiplied or divided frequency. The relation shown in (20.15) can be written in a simplified form with regard to frequency:

$$f(t) = cos(\omega t + \phi) \tag{20.33}$$

The argument contains the frequency term ωt and the phase term. Multiplication or division affects both terms $N \times \omega t + N \times \phi$. In the logarithmic scale of the single sideband phase noise $L(f)$ holds for $f_2 = N \times f_1$

$$L_2(f) = L_1(f) + 20\, log(N) \tag{20.34}$$

With each frequency doubling, the phase noise increases by 6 dB, with each division by 2 it decreases by 6 dB.

Amplitude Noise

Signal sources fluctuate not only in frequency but also in amplitude. In practice, however, this AM noise plays a minor role. Here are three reasons for this [7]:

1) Communication systems with high transmission capacity use phase angle modulation.
2) The AM noise of almost all oscillators is much lower than the phase noise, especially near the carrier. The reasons are amplitude saturation by nonlinearities and by means of automatic gain control (AGC).
3) A local oscillator of a receiver system transmits its phase and FM noise directly to the transposed frequency. In contrast, the effect of AM noise can be effectively suppressed by using balanced mixers.

According to the definition, $m(f)$ is defined as the noise power in an amplitude modulation sideband in relation to the total signal power with the numerical value

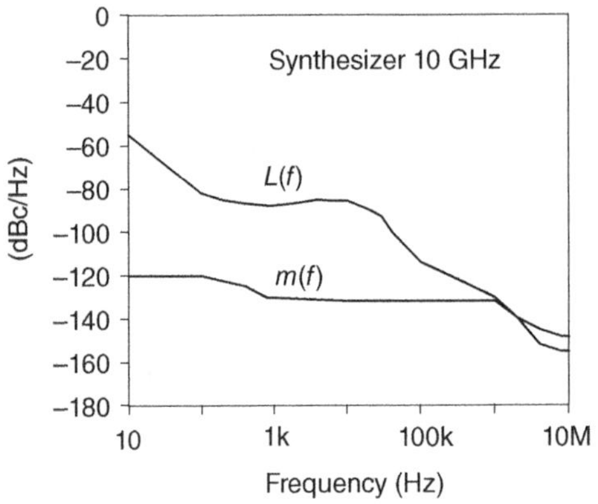

Figure 20.12 Phase and amplitude noise spectrum of a microwave synthesizer. Source: Based on Agilent Technologies [4].

in dBc/Hz. AM noise must, however, be taken into account in phase noise measurement, since it can influence the attainable sensitivity limit (Figure 20.12).

Phase Noise and Jitter

Digitally modulated signals are periodically regenerated along a transmission path. Signal errors are avoided by these regenerators. An essential task is the recovery of the clock. The phase noise of the clock signal is transferred as "jitter" to the time multiplex signal and can accumulate along a transmission path [8].

Phase noise and jitter therefore describe the same phenomenon (Figure 20.13).

The temporal fluctuations in the occurrence of the edges are called jitter. The same applies, of course, to sine waves and then affects the sequence of zero crossings. Jitter caused by noise processes is called "random jitter." This has a Gaussian distribution characterized by the standard deviation σ. Within a width of $\pm\sigma$ there are 68.26% of all events, within $\pm 3\sigma$ there are 99.73%.

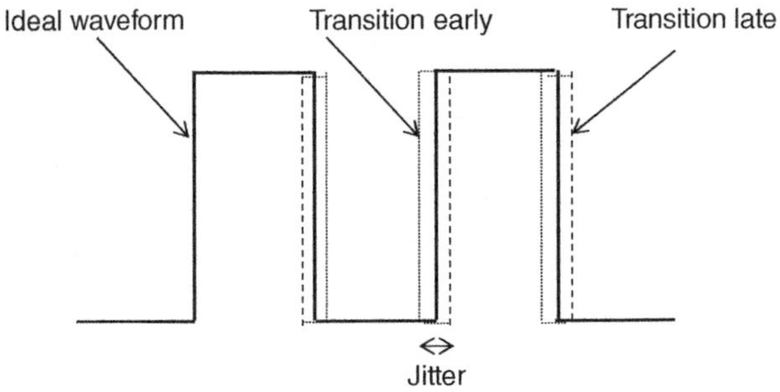

Figure 20.13 Square wave in the time domain. The edge distance is noisy.

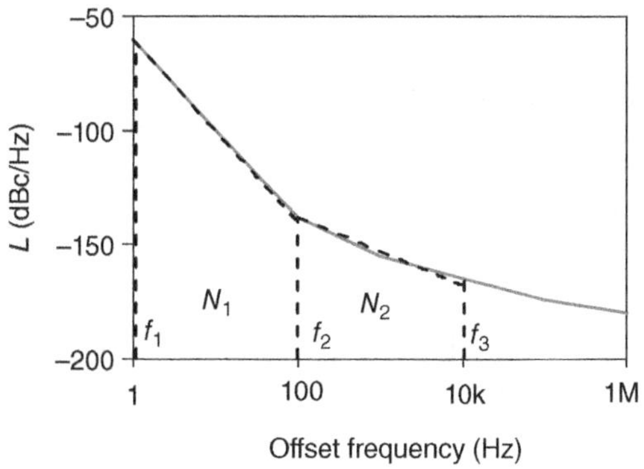

Figure 20.14 Phase noise spectrum *L*(f) of an oscillator divided into frequency ranges to calculate the jitter.

The specifications can refer to the "peak-to-peak jitter" or the "RMS jitter." The latter with $\sigma = 1$ is most suitable.

The quantitative relation of phase noise and jitter is given by the measured curve of the single sideband phase noise *L*(f) [9, 10]. An example is shown in Figure 20.14. It corresponds to a 10 MHz quartz oscillator.

The noise power *N* in a certain spectral range, e.g. from f_1 to f_3, is used for the calculation of the jitter. It is obtained by integration

$$N = \int_{f_1}^{f_3} L(f)df \tag{20.35}$$

From this the jitter of the phase is calculated according to (20.9)

$$J(Phase) = \sqrt{2 \times N} \quad \text{(rad)} \tag{20.36}$$

N is used here in linear representation, as shown in (20.39).

For the transition into the time domain, this value is divided by the frequency of the oscillator.

$$J(Time) = \frac{J(Phase)}{\omega_C} \quad \text{(s)} \tag{20.37}$$

The measured curve of the phase noise (e.g. Figure 20.14) is available in double-logarithmic representation. This has the advantage that in certain frequency ranges straight lines with a defined slope are characteristic. The corresponding physical reasons are noted in Figure 20.7. However, the areas N_1 and N_2 marked in Figure 20.14 do not correspond to the integral (20.35), so that graphic integration is not possible. For integration, however, we assume the double-logarithmic representation, since the sections can be well approximated by straight lines. This is the first step. For the straight lines the following applies

$$L_i(f) = a_i \left\{log(f) - log\left(f_i\right)\right\} + b_i = a_i log\left\{\frac{f}{f_i}\right\} + b_i \tag{20.38}$$

Table 20.3 Approximations of the curve $L(f)$ to the formula (20.38).

	Frequency range (Hz)	Slope a (dBc/dec)	Constant b (dBc)	Jitter (ps)
N_1	1–100	−40	−60	13
N_2	100–10 k	−16	−170	2×10^{-5}

With the values from Table 20.3, the two curve sections are well fitted. For the integration of (20.35) we use the linear representation.

$$L_i(f) = 10^{\frac{b_i}{10}} f_i^{-\frac{a_i}{10}} f^{\frac{a_i}{10}} \tag{20.39}$$

The integral is

$$N_i = \int_{f_i}^{f_{i+1}} L_i(f)df = 10^{\frac{b_i}{10}} f_i^{-\frac{a_i}{10}} \frac{1}{\frac{a_i}{10}+1} \left\{ f_{i+1}^{\frac{a_i}{10}+1} - f_i^{\frac{a_i}{10}+1} \right\} \tag{20.40}$$

In our case of the two frequency ranges $i = 1;2$. With (20.36) and (20.37) the contribution to jitter is as shown in Table 20.2. The low-frequency range provides practically the entire contribution of $J = 13$ ps, the second part is already negligible.

Usually, the total jitter of a clock signal is somewhat higher, since there are other influences besides noise. This "deterministic jitter" is mostly caused by interference of other signals by crosstalk from other transmission lines on the circuit board or by EMI.

References

1. Rohde, U.L., Poddar, A.K., and Böck, G. (2005). *The Design of Modern Microwave Oscillators for Wireless Applications*. Wiley Interscience.
2. Rubiola, E. (2011). *Phase Noise and Frequency Stability in Oscillators*. Cambridge University Press.
3. Grebenkemper, C.J. (1981). Local Oscillator Phase Noise and its Effect on Receiver Performance. Watkins-Johnson Comp., Vol. 8.
4. Agilent Technologies (2007). Phase Noise Characterization of Microwave Oscillators. Product Note 11729B.
5. Ramian, F. (2015). Time Domain Oscillator Stability Measurement, Allan Variance. Rohde & Schwarz Application Note 1EF69-E4.
6. Rohde, U.L. (1997). *Microwave and Wireless Synthesizers*. Wiley Interscience: Theory and Design.
7. Lance, A.L., Seal, W.D., and Labaar, F. (1984). Phase noise and AM-noise measurements in the frequency domain. *Infrared Millimeter Waves* 11: 239–289.

8 Abidi, A.A. (2011). Basic of Phase Noise and Jitter. IEEE SSCS Chapter, Toronto.

9 Maxim Integrated (2004). Clock Jitter and Phase Noise Conversion. Application Note 3359.

10 Kester, W. (2009). Converting Oscillator Phase Noise to Time Jitter. Analog Devices Tutorial MT-008.

21

Physics of the Oscillator

Oscillation Condition [1]

An oscillator generates an electrical oscillation for a specific application. The energy required for this is provided by the DC power supply. We are interested in frequencies in the RF and microwave range. Two physically different modes of operation can be distinguished. The one-port oscillator is based on the gain of a resonant circuit caused by a component with a negative characteristic. These are, e.g. Gunn or tunnel diodes or also space charge wave tubes. A two-port oscillator consists of an active element, e.g. transistor or operational amplifier and a frequency-determining element in the feedback loop. Noise and stray fields play an important role in this system. On the one hand, they prevent a purely monochromatic output signal due to phase noise and spurs; on the other hand they allow the circuit to oscillate by repeatedly amplifying the always present low noise level without the need for an additional input signal. The diagram of the two-port oscillator is shown in Figure 21.1.

The input voltage at the amplifier V_F is given by the part to be amplified and the feedback part.

$$V_F = \frac{V_O}{G(j\omega)} + V_O\beta(j\omega) \tag{21.1}$$

Solved for the output voltage

$$V_O = \frac{V_F G(j\omega)}{1 + G(j\omega)\beta(j\omega)} \tag{21.2}$$

The output voltage increases strongly at frequency ω_0, here to infinity, when the denominator disappears.

$$1 + G\left(j\omega_0\right)\beta\left(j\omega_0\right) = 0 \tag{21.3}$$

A Guide to Noise in Microwave Circuits: Devices, Circuits, and Measurement, First Edition.
Peter Heymann and Matthias Rudolph.

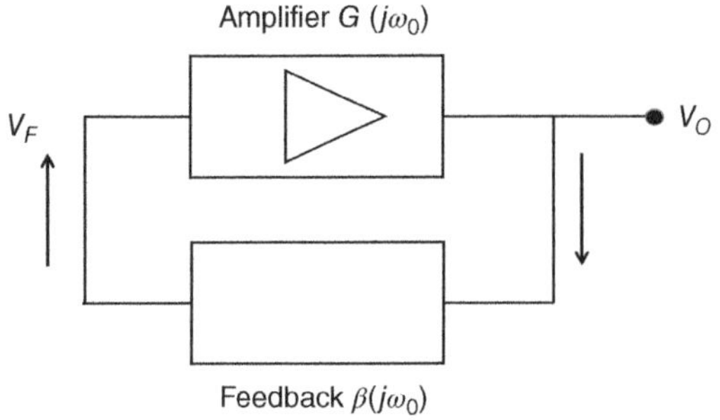

Figure 21.1 Scheme of an oscillator consisting of amplifier and feedback.

This is the simplest oscillation condition. The transfer function of the feedback loop has the value 1/180°.

$$\left|G\left(j\omega_0\right)\beta\left(j\omega_0\right)\right| = 1 \angle \left(G\left(j\omega_0\right)\beta\left(j\omega_0\right)\right) = 180° \tag{21.4}$$

The system is stable as long as the magnitude is less than 1, for values greater than 1 it will oscillate at the frequency $\omega_0/2\pi$. This simple, linear approach fails with more complicated systems. There the analysis is carried out using the stability theory of Nyquist. Here we are not going to analyze the operation of the oscillator itself. The main concern is the imperfection of the generated signal due to phase noise and other disturbances.

Simple Model of the Phase Disturbance [2]

The origin of the phase noise can be understood with a simple model. The basic oscillator circuits react differently to fluctuations in the current of the active element. We consider the three-point oscillator in inductive (Hartley) and capacitive (Colpitts) design (Figure 21.2).

The noise of the transistor current statistically feeds charges into the oscillating circuit. At a certain point in time, the additional charge q is fed into the LC circuit. This produces a voltage jump in the capacitor (Figure 21.3).

If this results in a statistical phase fluctuation of the quantity $\phi(t)$, phase noise is generated. However, phase noise only occurs when the zero crossing of the voltage fluctuates (jitter). This depends on the time of occurrence of the additional charge q.

If the disturbance occurs at maximum voltage, the zero crossing remains unchanged. In the context of Figure 21.4, there is $\phi = 0$, i.e. no phase noise is generated, only amplitude noise. Since oscillator circuits usually work in saturation, the effect on the amplitude is small. On the other hand, if the disturbance occurs at the zero crossing of the voltage ($\phi \neq 0$), a maximum phase change results. The conversion of the current noise of the transistor into phase noise is particularly

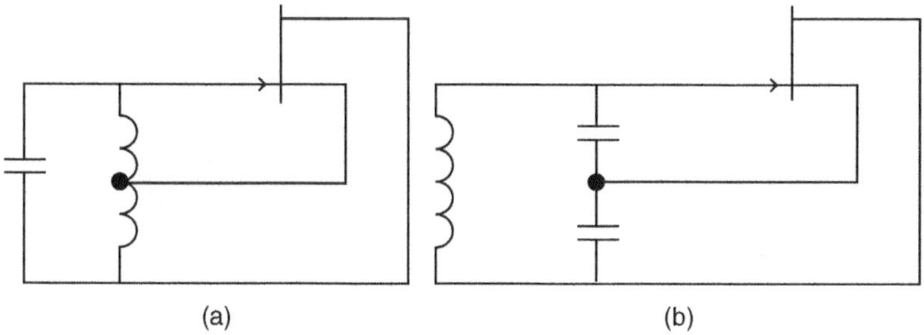

Figure 21.2 Three-point circuits of oscillators: Hartley (a) and Colpitts (b).

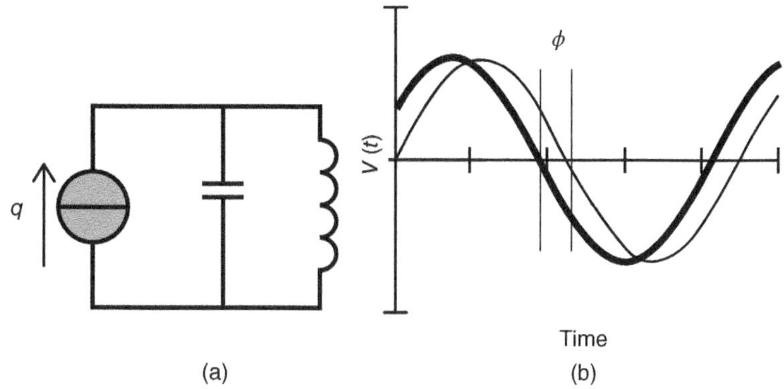

Figure 21.3 Resonant circuit with noise current source (a). Oscillator voltage with variable phase ϕ (b).

effective here (Figure 21.4b). This effect can be influenced by suitable circuit design.

If the current flows only at the voltage maximum and not during the zero crossing, the influence of this noise source is minimized. An example is shown in Figure 21.5. In a Clapp oscillator (variant of the Colpitts oscillator), the angle of current flow is very small, so that no current flows at the zero crossing of the voltage. This leads to good noise characteristics, independent of the transistor used.

Phase Slope, Resonator Quality, and Frequency Stability [3]

Equation (21.4) is the simplest oscillation condition: the transfer function of the feedback loop has the value 1/180°. The system is stable as long as the magnitude

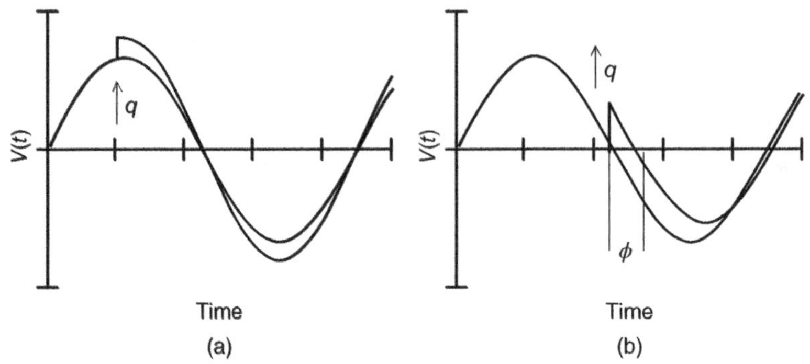

Figure 21.4 Feeding q at different phases of the oscillation. Maximum voltage (a) and zero crossing (b).

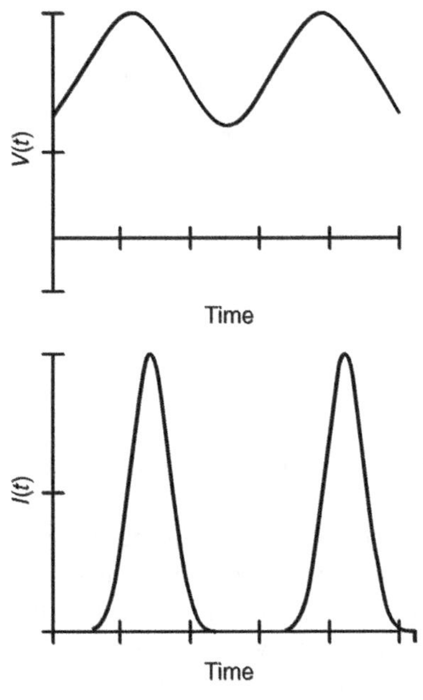

Figure 21.5 Voltage and current waveform in the Clapp oscillator. Current flows only at the voltage maximum.

is less than 1, for values greater than 1 it oscillates with $\omega_0/2\pi$. If a disturbance, e.g. noise, causes a small phase shift in the amplifier branch of the oscillator, this must be compensated in the feedback branch so that the oscillation condition (21.4) remains fulfilled at the frequency ω_0. If it is fulfilled at a neighboring frequency as a result of the phase disturbance, we have a frequency fluctuation and thus phase noise. A fluctuation of the phase angle $\Delta\varphi_A$ between input and output in the amplifier branch must be directly compensated in the feedback branch by $\Delta\varphi_\beta = -\Delta\varphi_A$.

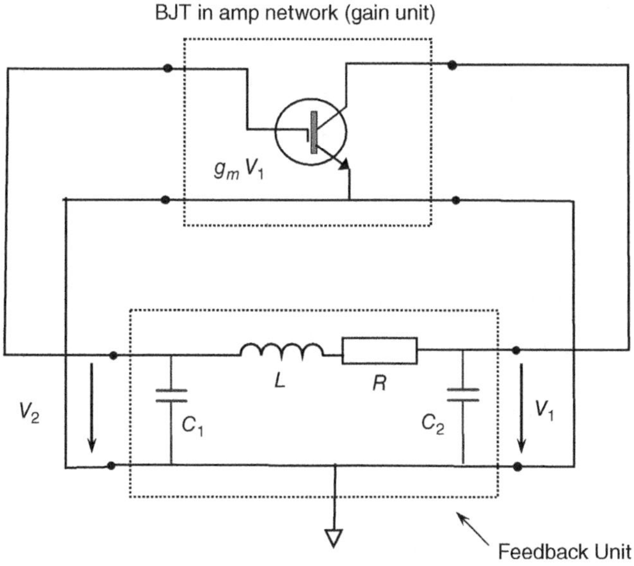

Figure 21.6 Capacitive three-point circuit with bipolar transistor and LC resonant circuit (Colpitts oscillator).

The frequency change $\Delta\omega$ required for this is the smaller the greater the phase slope of the resonant circuit is. It is defined by

$$s\varphi = \left|\frac{d\varphi_\beta}{d\omega}\right|_{\omega=\omega_0} \tag{21.5}$$

This results in

$$\Delta\omega = \frac{-\Delta\varphi_\beta}{s\varphi} \tag{21.6}$$

High phase slope of the resonator thus ensures good frequency stability. In Figure 21.6 we consider the Colpitts oscillator as an example. This capacitive three-point circuit is widely used from a few MHz up to the microwave range.

The feedback is an asymmetric two-port network. The following applies to its transmission factor, and the mutual impedance is

$$Z_{12} = \beta = \frac{-Z_1 Z_2}{Z_1 + Z_2 + Z_3} \tag{21.7}$$

With

$$Z_1 = \frac{1}{j\omega C_1}; \quad Z_2 = \frac{1}{j\omega C_2}; \quad Z_3 = R + j\omega L$$

we obtain

$$\beta = \frac{\omega^2 C_1 C_2 R + j\omega\left(C_1 + C_2 - \omega^2 C_1 C_2 L\right)}{\left(\omega^2 C_1 C_2 R\right)^2 + \omega^2\left(C_1 + C_2 - \omega^2 C_1 C_2 L\right)^2} \tag{21.8}$$

At the resonant frequency ω_0 the impedance is purely real, i.e. $Im(\beta)|_{\omega=\omega_0} = 0$. It follows

$$\omega_0 = \sqrt{\frac{C_1 + C_2}{C_1 C_2 L}} \tag{21.9}$$

The phase slope is obtained from the following formula.

$$\varphi_\beta(\omega) = tan\frac{Im(\beta)}{Re(\beta)} \approx \frac{Im(\beta)}{Re(\beta)} = \frac{C_1 + C_2}{\omega C_1 C_2 R} - \omega\frac{L}{R} \tag{21.10}$$

The magnitude of the derivative at frequency ω_0 is given by

$$\left|\frac{d\varphi_\beta}{d\omega}\right|_{\omega\approx\omega_0} = \frac{2L}{R} = \frac{2Q}{\omega_0} \tag{21.11}$$

A frequency change $d\omega$ results from a phase change at the input $d\varphi_\beta$

$$d\omega = \frac{\omega_0}{2Q} d\varphi_\beta \tag{21.12}$$

We have thus shown the relation between the resonator quality Q and the phase slope $s\varphi$. $\omega_0/2Q$ is half the 3 dB bandwidth of the resonator. From earlier, it is clear that Q has an extremely strong influence on the frequency stability and thus the phase noise of an oscillator. It is the loaded resonator quality in the real circuit.

The Formula of Leeson [4]

Leeson describes in an elementary way the relation between the phase noise spectrum $L(\omega_m)$ of a feedback oscillator and the noise spectrum of the amplifier stage, the characteristics of the resonator and the signal level at the output. The resonator is characterized by its phase slope (21.11). The frequency spectrum of the phase fluctuations $S_\phi(\omega_m)$ at the input of the resonator consists of the $1/f$ noise superimposed to the thermal noise of the amplifier stage (Figure 21.7a). The term $2kTF/P_S$ becomes understandable from the thermal limit of LF-noise. F must be at least as high as the noise factor of the amplifier. Usually, however, it must be regarded as a fitting factor. Furthermore, differences between RMS and peak value are not taken into account here, as it is a representation of the principle processes. When passing through the resonator, a phase change causes a frequency change, according to (21.12).

$$\Delta\omega = \frac{\omega_0}{2Q}\Delta\phi$$

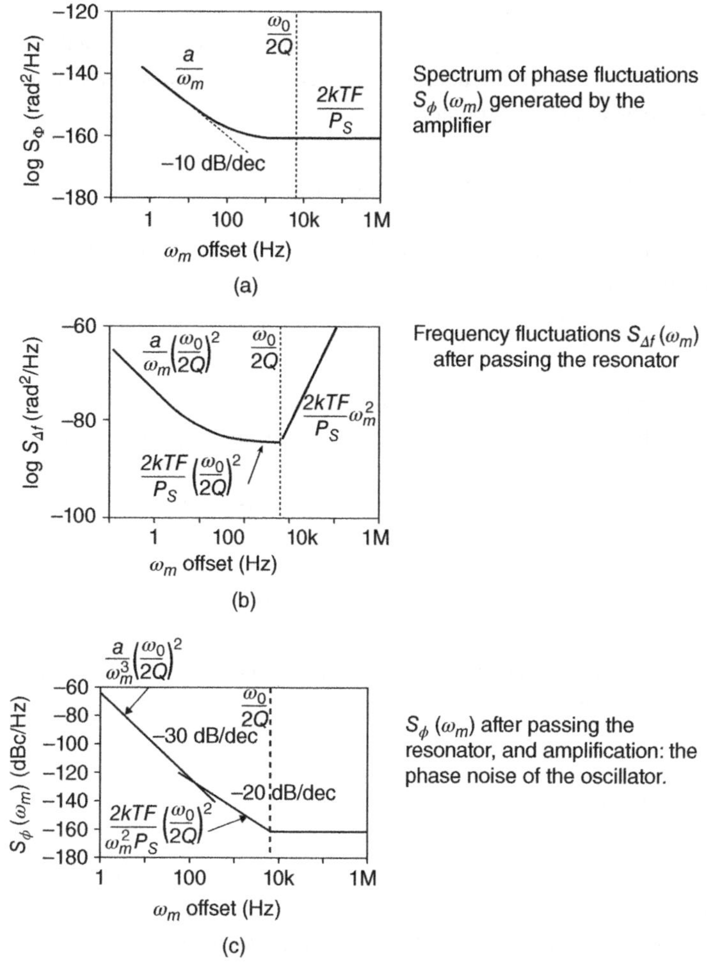

Figure 21.7 Change in the LF noise spectrum of the amplifying element (a) when passing through the resonant circuit (b) and resulting phase noise spectrum (c).

The following applies to the spectra:

Within the bandwidth of the resonator (near the carrier)

$$S_{\Delta f}\left(\omega_m\right) = \left(\frac{\omega_0}{2Q}\right)^2 S_\phi\left(\omega_m\right) \qquad \omega_m < \frac{\omega_0}{2Q}$$

Outside the transfer curve (away from the carrier)

$$S_{\Delta f}\left(\omega_m\right) = \omega_m^2 S_\phi\left(\omega_m\right) \qquad \omega_m > \frac{\omega_0}{2Q} \tag{21.13}$$

In Figure 21.7 in the center the diagram is shown.

Figure 21.7b is the typical curve of phase noise, which broadens the oscillator signal below and above ω_0 and which, as we have seen, is essentially determined by the transfer properties of the resonator.

$$S_{\Phi\,OUT}\left(\omega_m\right) = \left\{1 + \frac{1}{\omega_m^2}\left(\frac{\omega_0}{2Q}\right)^2\right\} S_{\phi\,IN}\left(\omega_m\right) \tag{21.14}$$

Leeson takes into account the $1/f$-noise and the noise factor F of the amplifier as input phase variations

$$S_{\phi\,IN}\left(\omega_m\right) = \frac{2kTF}{P_S}\left(1 + \frac{a}{\omega_m}\right) \tag{21.15}$$

Here, the signal power P_S is related to the input of the amplifier element, since it stands together with the fictitious noise factor F. In Leeson's analysis it is not the output power.

The measurement curve is the single sideband phase noise $L(\omega_m)$, i.e. 1/2 of this (Leeson's formula):

$$L\left(\omega_m\right) = \frac{1}{2}\left\{1 + \frac{1}{\omega_m^2}\left(\frac{\omega_0}{2Q}\right)^2 \frac{2kTF}{P_S}\left(1 + \frac{a}{\omega_m}\right)\right\} \tag{21.16}$$

These fundamental ideas have been critically discussed and further developed by a number of authors [5–7].

Components of Oscillators

We are particularly interested in the components of the oscillator that influence the phase noise. As one can see in Leeson's formula (21.16), these are the amplifier stage, which affect with the modified noise factor F and the resonant circuit with its (loaded) quality factor Q. No practical oscillator can operate without a tuning element for the frequency. This is usually a varactor diode, which also provides a noise contribution [8]. The amplifier stage is usually formed by a transistor. This transistor acts through its noise factor in the vicinity of the resonant frequency ω_0, i.e. in the RF and microwave range, and through its low-frequency noise in the $1/f$ range, which also gets there through upward mixing. Figure 21.8 shows typical values of the common technologies. It is essential for the design of the oscillator to minimize the influence of the $1/f$ noise. Although the Si-BJT is optimal here, it is not suitable for the upper microwave range. The GaAs-HEMT is ideal for highest frequencies, but has a strong LF noise. This is due to its planar structure [9].

The resonator usually consists of the actual resonant circuit and the varactor as tuning element. The use of discrete components, coils, and capacitors, is limited to

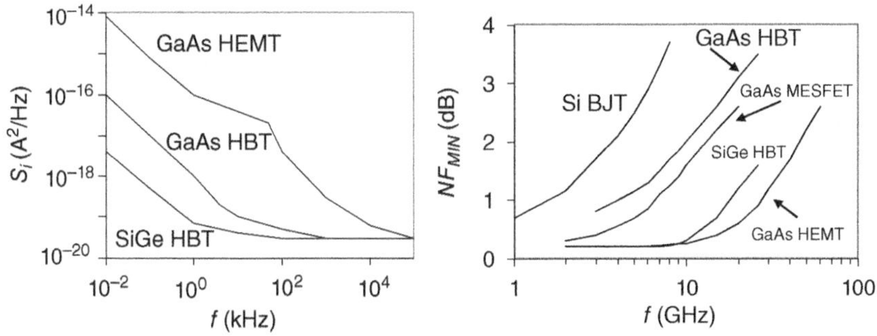

Figure 21.8 Noise characteristics of relevant transistor technologies.

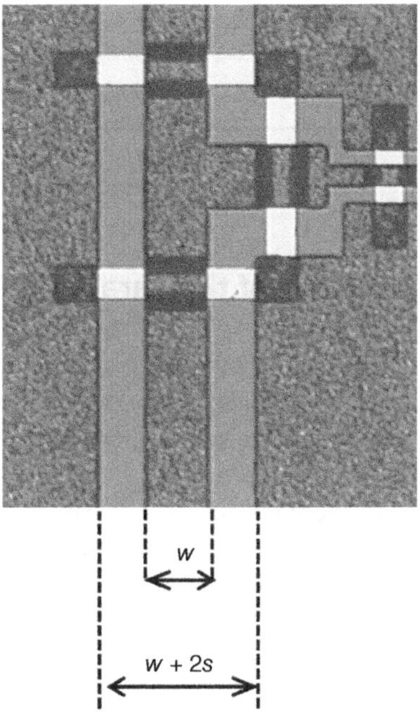

Figure 21.9 Transmission line resonator in coplanar technology. The bright rectangles are underpass elements which connect the grounds. They pass under the center conductors under air bridges. Source: Courtesy Ferdinand-Braun-Institut, Leibniz-Institut für Höchstfrequenztechnik.

the RF range ($f < 1$ GHz). In the microwave range resonant transmission lines are used. Figure 21.9 shows an example of an integrated transmission line resonator in coplanar waveguide (CPW) technology. The upper frequency limit is $w + 2s < \lambda/10$ determined by the geometry.

This technology achieves resonator quality values as shown in Figure 21.10.

The quality factors achievable with other techniques are shown in Table 21.1. They are approximate values, which apply depending on the technology and frequency range.

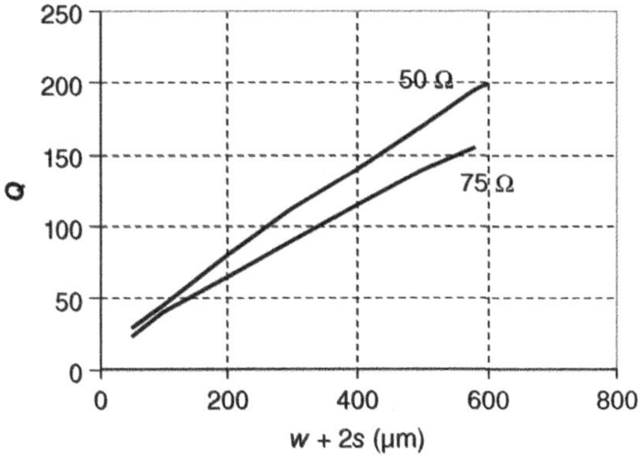

Figure 21.10 Quality values of line resonators in coplanar technology.

Table 21.1 Typical quality values of resonant circuits.

	LC	Microstrip	Varactor	YIG	Ceramic	Cavity	Quartz
Q-Factor	100	100	100	200	500	1000	10^5

Influence of the Varactor Diode

An important component of an oscillator circuit is a tuning element, necessary to adjust the frequency to a certain value. In the early days of high frequency technology, mechanical components such as variable capacitors or variable inductors (variometers) were used for this purpose. The varactor diode is a semiconductor component whose capacity can be changed up to a factor of 10 depending on the voltage. Figure 21.11 shows the equivalent circuit of the varactor with the parallel resistor R_p relevant for its noise contribution. With packaged diodes, which are commercially available, capacitance variations in the range 3 ... 300 pF are within reach. The tuning range depends on the doping profile. For the microwave range the varactors must be manufactured in MMIC technology. In the HEMT technology the gate-source capacitance is used, in the bipolar technology (BJT, HBT), the base-collector capacitance of the transistors, which are the basis of the circuit anyway.

Figure 21.12a shows the varactor diode BB833 in SMD technology for the RF range [10], Figure 21.12b in MMIC technology (FBH GaAs-HBT technology).

Figure 21.13 shows quality factors over frequency. With the HBT varactor one can achieve $Q = 20 \ldots 30$ at $f = 20\,\text{GHz}$.

The elements of the equivalent circuit (Figure 21.11) for a 4-finger HBT-varactor are shown in Table 21.2.

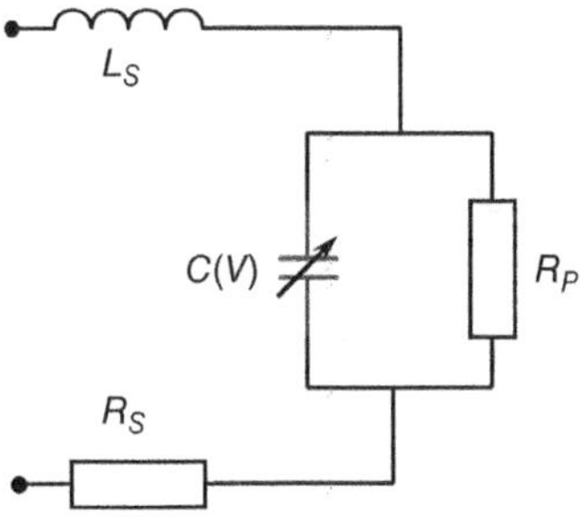

Figure 21.11 Equivalent circuit of a varactor diode.

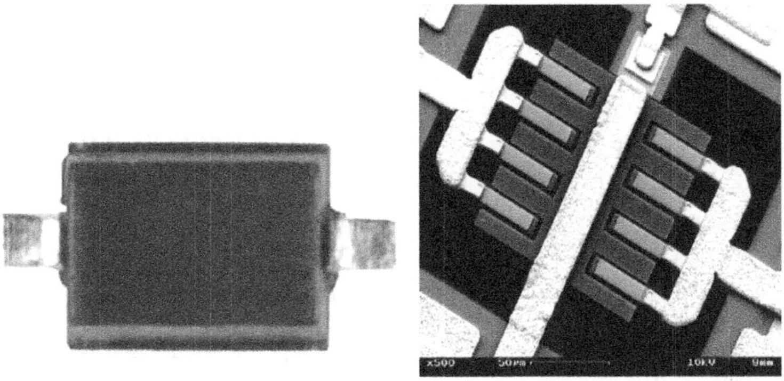

Figure 21.12 Varactor diode in SMD technology (a) and in MMIC technology (b). The integrated varactor is formed by the basic collector diode of an HBT. $C(V) = 220–400$ fF; $R_S = 1.4$; $R_P > 20$ kΩ. Source: Courtesy Ferdinand-Braun-Institut, Leibniz-Institut für Höchstfrequenztechnik.

Table 21.2 Elements of the equivalent circuit for a varactor in HBT MMIC technology.

V_{TUNE}	C (fF)	R_S (Ω)	R_P (Ω)	L (pH)	Q	f_{CO} (THz)
−1 … −10	350 … 210	1.4	>20k	53	18 … 28	0.5

The resonator quality is given by the formula

$$Q = \frac{\omega C R_P}{1 + \omega^2 C^2 R_P R_S} \tag{21.17}$$

Figure 21.14 shows two tunable resonant circuits for 20 GHz MMIC oscillators, (a) with lumped elements and (b) with transmission line resonator in CPW technology. Both are tunable with varactors. The LC resonator consists of the inductor with 1 Ω loss resistance and a MIM capacitance of 130 fF. The coupling is realized by a 20 fF interdigital capacitor. The quality values are $Q_U = 35$, $Q_L = 17$, and the

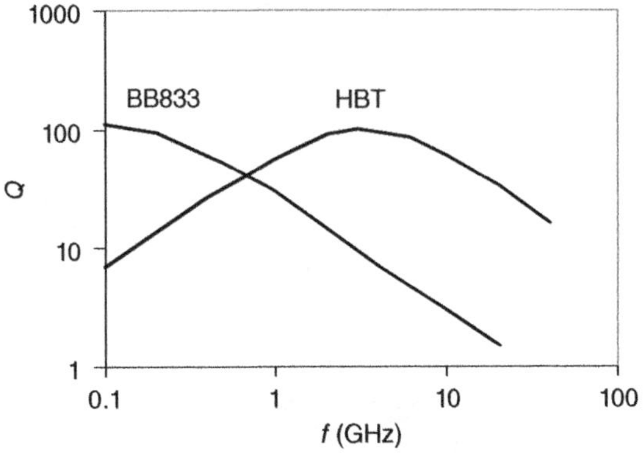

Figure 21.13 Frequency dependence of quality factors of varactor diodes.

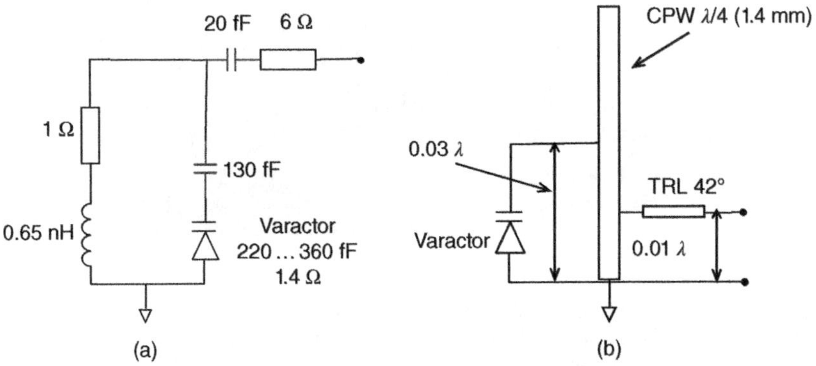

Figure 21.14 Resonant circuits tunable with varactor $f = 20$ GHz. (a) Lumped elements and (b) transmission line resonator.

tuning constant $K_{TUNE} = 240$ MHz/V. The CPW line resonator has the quality values $Q_U = 100$, $Q_L = 45$, and $K_{TUNE} = 130$ MHz/V. The tuning constant determines the effect of the varactor on the phase noise.

The contribution of the varactor diode to the phase noise comes mainly from the parallel resistor R_P.

$$v_n^2 = 4kTBR_P \tag{21.18}$$

The thermal noise voltage v_n of the resistance R_P modulates the actual capacitance $C(V)$ and thus ω_0.

The sensitivity depends on the tuning constant

$$K_{TUNE} = \frac{\Delta\omega_0}{V_{TUNE}} \tag{21.19}$$

The phase noise is calculated using the degree of modulation (Chapter 20):

$$\beta = \frac{\Delta\omega_0}{\omega_m} \qquad L = 20log\left(\frac{\beta}{2}\right) \tag{21.20}$$

Typical values are $K_{TUNE} = 0.05 \ldots 0.2$ GHz/V.

The Leeson formula was extended by Rohde to include the influence of the varactor. In Rohde's formulation it reads [7].

$$L(f_m) = 10log\left\{\left(1 + \frac{f_0^2}{(2f_mQ)^2}\right)\left(1 + \frac{f_C}{f_m}\right)\frac{FkT}{2P_S} + \frac{2kTR_PK_{TUNE}^2}{f_m^2}\right\} \tag{21.21}$$

The last term contains the parallel resistor R_P and the tuning constant K_{TUNE}.

An example of a 19 GHz MMIC oscillator in HBT technology is shown in Figure 21.15 [11]. On the right, the measured phase noise and the successfully minimized contribution of the varactor diode. The tuning range is $f = 0.5$ GHz with $V_{TUNE} = 1.5 \ldots 6$ V (source FBH).

The influence of the varactor on the phase noise is twofold. It reduces the resonator quality by its series resistance R_S and a conductivity occurring in the nonlinear range of the characteristic. It modulates the oscillation frequency by the thermal noise voltage of the parallel resistor R_P. Remedies are weak coupling of a large varactor, antiparallel circuit arrangement or short-out of the noise voltage.

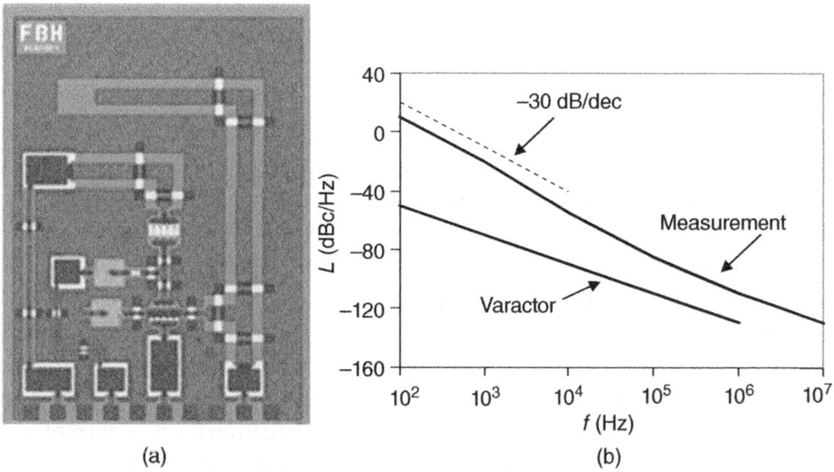

Figure 21.15 (a) 19 GHz MMIC oscillator in HBT technology. (b) Measured phase noise with part of the varactor. Source: Courtesy Ferdinand-Braun-Institut, Leibniz-Institut für Höchstfrequenztechnik.

Upward Mixing of LF Noise

The Leeson formula (21.21) is extended with regard to this effect [7]. We write

$$L(f_m) = 10\left\{\left[1+\left(\frac{f_0}{2Q_L f_m}\right)^2\right]\left[K_{UP}S_I(f_m)+\frac{2FkT}{P_S}\right]+\frac{2kTR_P K_{TUNE}^2}{f_m^2}\right\} \tag{21.22}$$

The upward mixing is included by the term $K_{UP}S_I(f_m)$. The coefficient K_{UP} is determined by the following effect. The condition (21.4) implied that for the oscillation frequency ω_0 the phase angle of the amplification must have a certain value ($\angle G(j\omega_0)\beta(j\omega_0) = 180\,°$). If the phase is not constant in time, the frequency is also not constant in time. With the previously shown interaction with the resonator quality (21.12), we obtain for the instantaneous oscillation frequency

$$\omega(t) = \omega_0 + \Delta\omega(t) = \omega_0\left(1+\frac{\Delta\phi(t)}{2Q}\right) \tag{21.23}$$

If we have a relatively low quality, which is always the case with MMIC oscillators, phase fluctuations in the amplifier element will be very effective. Let us consider the term of upward mixing in an elementary way in a transistor oscillator [12, 13]. The collector or drain current I_D has the maximum amplitude because the oscillator is a nonlinear circuit. The first term is of great influence, and the second term, which describes the AM-PM coupling, is usually of less influence.

$$K_{UP} = \left(\frac{d\phi}{dI_D} + K_{AM/PM}\frac{1}{G}\frac{dG}{dI_D}\right)^2 \tag{21.24}$$

The S-parameters of the transistor depend on the current. In this context there is the phase of the complex gain, i.e. $\angle S_{21}$ the most critical one. If this current dependency is large and I_D fluctuates strongly due to $1/f$ noise, K_{UP} has a high value. The consequence is a strong phase noise of the oscillator. The signal flow path is therefore the following: The transistor current I_D is modulated by the $1/f$ noise. This modulation is transmitted by the current dependence of the $\angle S_{21}$ via condition (21.4) to the oscillation frequency ω_0.

As already mentioned in Chapter 18, the modeling of the measured LF noise of the transistor is convenient carried out with the formalism used in simulation programs. We show it here once again as an example for a GaAs HBT. The LF-spectrum consists of generation recombination noise from the base current and a clear $1/f$ noise part of the emitter resistance. By means of the multi-impedance method for LF noise measurements described in Chapter 18 the sources can be separated. With Si-BJT one usually has only the term of the base-emitter diode. In Figure 21.16 the source resistance is $R_S = 10\,\text{k}\Omega$. In this case we have the noise

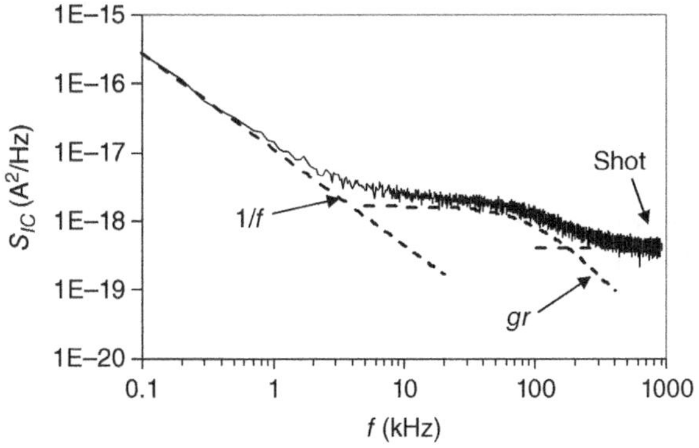

Figure 21.16 Contribution of the base–emitter diode to the LF noise current in the collector circuit.

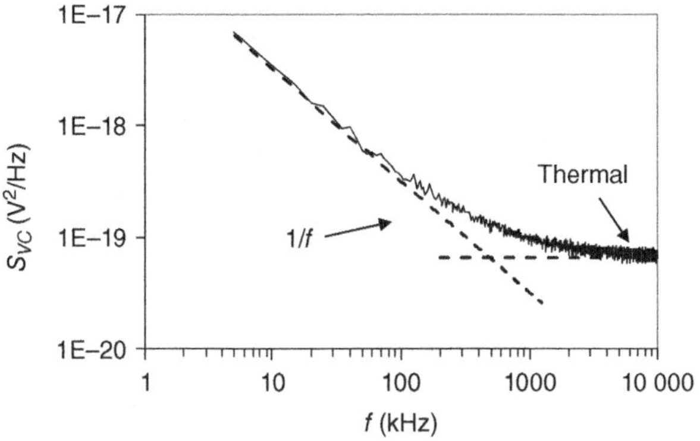

Figure 21.17 Spectral noise voltage density of the emitter resistance from measurement.

contribution of the base-emitter diode. In Figure 21.17 there is $R_S = 10\,\Omega$, we have the noise contribution of the emitter resistance $R_E = 4\,\Omega$.

The model data for (21.25) are $I_B = 80\,\mu\text{A}$; $I_C = 10\,\text{mA}$; $KF = 3\times10^{-8}$; $AF = 2.3$; $FB = 1.4$; $KL = 2\times10^{-19}$; $AL = 0.8$; $FL = 100\,\text{kHz}$.

The model data for the Hooge formula (21.26) are $I_C = 10\,\text{mA}$; $R_E = 4\,\Omega$; $\alpha = 10^{-5}$; $N = 5\times10^5$; $T = 300\,\text{K}$.

The formula for the simulation of the spectral noise current density for the base current with the gr-component is

$$S_{IB} = KF\frac{I_B^{AF}}{f^{FB}} + 2qI_B + KL\frac{I_B^{AL}}{1+\left(\frac{f}{FL}\right)^2} \tag{21.25}$$

For the noise voltage density of the emitter resistor the Hooge formula is used

$$S_{VR} = 4kTR_E + \frac{\alpha R_E^2 I_C^2}{N}\frac{1}{f} \tag{21.26}$$

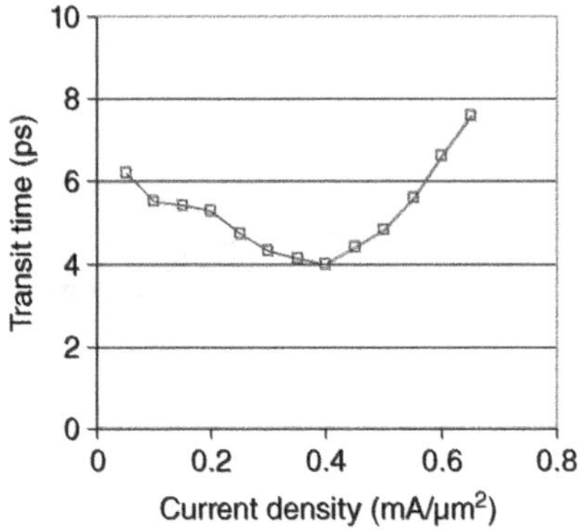

Figure 21.18 Emitter–collector transit time τ_{EC} of a GaAs HBT versus the collector current.

The coefficients must be taken from measurement curves, as shown in the example.

As mentioned earlier the frequency stability of the oscillator depends on the phase of S_{21}. That means, the complex gain as a function of the current is of particular interest. If this current has a noise component, which is always the case, this is transferred to the phase of the transmitted signal and produces phase noise. The physical transit time τ_{EC} refers to the particle velocity. The phase of the transmitted signal results from the complex current gain.

Let us consider this effect by an example using a GaAs-HBT microwave transistor as the basic element of an MMIC oscillator. The phase angle $\angle S_{21}$ is mainly determined by the emitter-collector transit time τ_{EC}.

$$\angle S_{21} \propto \frac{exp\left(-j\omega\tau_{EC}\right)}{1 + j\frac{\omega}{\omega_\alpha}} \tag{21.27}$$

This has a typical dependence on the transistor current I_C. Figure 21.18 shows a plot versus the collector current of a GaAs HBT.

The corresponding curve $\angle S_{21}(I_C)$ is shown in Figure 21.19. Their first derivative determines the coefficient of upward mixing via $^{d\Phi}/_{dI_C}$. At a current density of 0.4 mA/µm^2, the transit time is minimal and the derivative of the phase curve $^{d\angle S_{21}}/_{dI_C} = 0$. This should be the operating point for minimum phase noise.

That this effect does indeed occur is shown by the following example of a 19 GHz MMIC oscillator with a 2(3×30) µm^2 HBT in Figure 21.20. The phase noise level L (dBc/Hz) at $f_m = 100$ kHz offset shows a clear minimum at $I_C = 75$ mA. This corresponds to the maximum of the phase curve Figure 21.19. At this point the upward mixing of the LF-noise is ineffective or at least minimal. The plotted curve was calculated with the Leeson formula ($F = 4$; $Q = 18$; $G = 10$ dB; $P_{OUT} = 13$ dBm). The marked minimum value $L = -100$ dBc/Hz results without any LF-noise.

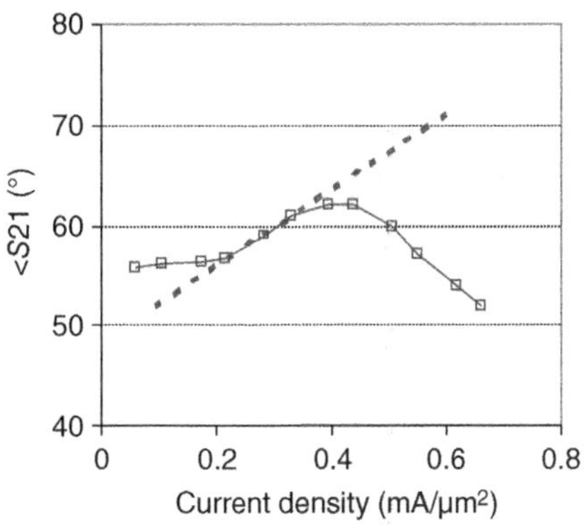

Figure 21.19 Phase angle of the gain ($\angle S_{21}$) of a GaAs HBT versus the collector current.

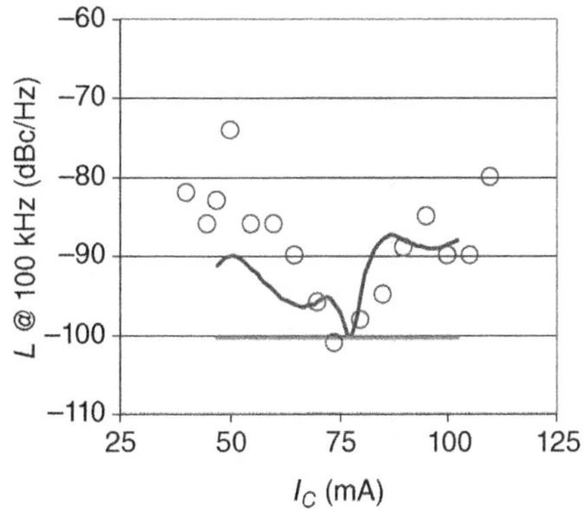

Figure 21.20 Phase noise of a 19 GHz MMIC oscillator versus the collector current. Minimum upward mixing at $I_C = 75$ mA.

A complete $L(f_m)$ curve of this oscillator is shown in Figure 21.21. At the optimum collector current ($I_C = 74$ mA), there is the lowest phase noise. This curve shows the small contribution of $1/f$ noise (f^{-3} curve) and the contribution of white microwave noise (f^{-2} curve). The $I_C = 50$ mA curve shows that this contribution is completely covered, because here the upward mixing is effective.

A detailed description of the computer simulation using the example of a microwave oscillator based on HBT can be found in [14].

The Influence of Microwave Noise on Phase Noise

Following the model of the influence of LF-noise on the phase noise of the oscillator, we examine the next term in the Leeson formula, microwave noise. We assume the best case that the microwave noise in the vicinity of the oscillation

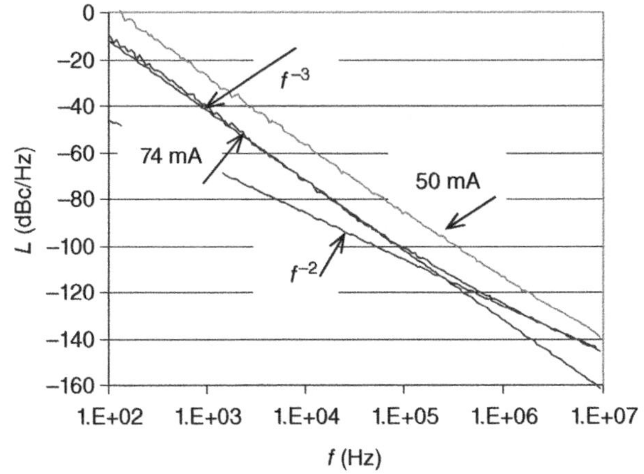

Figure 21.21 Measured phase noise of the 19 GHz oscillator at $I_C = 74$ mA (no upward mixing) and 50 mA (effective upward mixing).

frequency remains unchanged. With this assumption we can only find out the lower limit. More detailed considerations must take into account that in the case of the oscillating transistor we are dealing with cyclostationary LF-noise sources [15]. This means that the current passes through the entire output characteristic during oscillation. In the DC outer circuit, however, only an average direct current is measured. This means that the LF-noise level is periodically influenced and thus the oscillator always acts as a mixer, causing an upward mixing effect. Therefore one must assume that the microwave noise level is not undisturbed after all and simply corresponds to the noise of the transistor at the average operating current. Therefore the following consideration gives a lower limit. Of course, this also applies to the above considerations about the optimum current with the lowest upward mixing. However, oscillator calculation models of the common simulation tools sometimes do not give useful results, so that the estimation with simple models as a first orientation is quite justified.

We have already discussed noise models of transistors earlier. Here we have an application example. The noise of a HBT in a 19 GHz oscillator is calculated from the linear model. The effect of the correlation of the noise sources has to be considered. The dependency of the noise figure on the input circuitry, which is valid for every two-port, also applies to the transistor in the oscillator. As an example, we see in Figure 21.22 the scheme of a reflection oscillator, as it is often used in MMIC technology. The HBT sees the reflection coefficient $\Gamma_R(\omega)$ of the resonator at its input. At this reference plane we disconnect the circuit for the purpose of calculation. The capacity C against ground causes the transistor to become unstable.

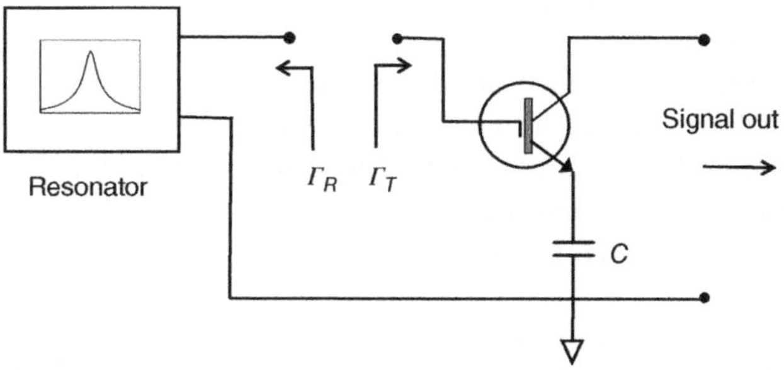

Figure 21.22 Scheme of a reflection oscillator. The frequency-determining resonator is located at the input of the transistor.

Figure 21.23 Noise figure of an HBT with mismatch at the input. Typical situation with the reflection oscillator.

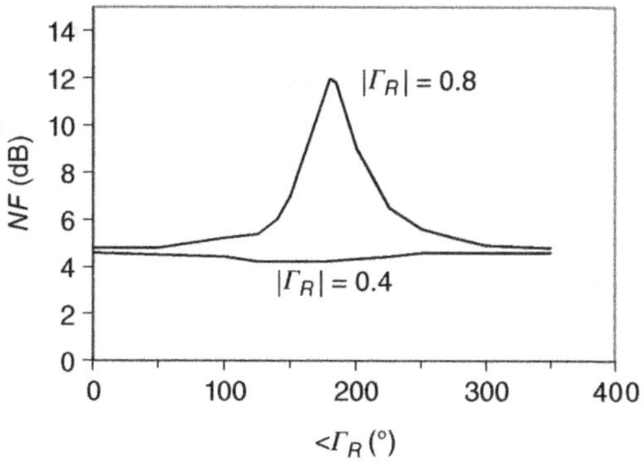

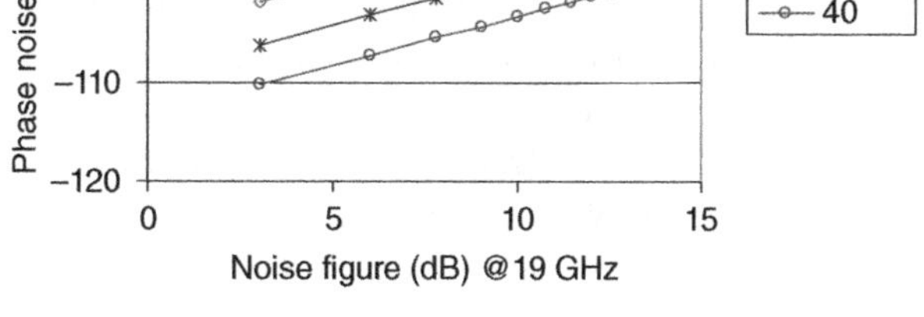

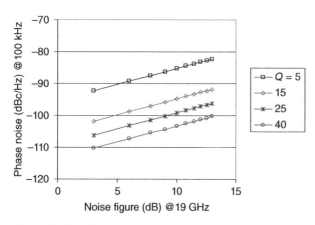

Figure 21.24 Phase noise after (21.28) above the noise figure of the transistor at the oscillation frequency. Parameter is the quality of the resonator.

The resonator at its input with its frequency-dependent $\Gamma_R(\omega)$ ensures that the oscillation condition is fulfilled.

As we can see from Figure 21.23, the value $\Gamma_R(\omega_0)$ has a strong influence on the microwave noise of the transistor. The transistor has a termination at its input, which can be far away from Γ_{OPT}, if no attention is paid to favorable design.

The pure microwave noise is obtained from the Leeson formula (21.16) by removing the $1/f$ term, i.e. $a = 0$. What remains is

$$L\left(\omega_m\right) = \frac{1}{2}\left\{1 + \frac{1}{\omega_m^2}\left(\frac{\omega_0}{2Q}\right)^2 \frac{2kTF}{P_S}\right\} \tag{21.28}$$

We see in Figure 21.24 that microwave noise can certainly have a determining influence.

The Leeson model is the linear treatment of a basically nonlinear system. Despite this simplification, there are significant results on the origin of the frequency response of the phase noise spectrum and the influences of circuit topology and components.

References

1 Wupper, H. (1996). *Elektronische Schaltungen 1*. Springer Verlag.

2 Lee, T.H. and Hajimiri, A. (2000). Oscillator phase noise: a tutorial. *IEEE J. Solid-State Circuits* 35: 326–336.

3 Zinke, O. and Brunswig, H. (1987). *Lehrbuch der Hochfrequenztechnik Band 2*. Springer Verlag.

4 Leeson, D.B. (1966). A simple model of feedback oscillator noise spectrum. *Proc. IEEE* 54: 329–330.

5 Nallatamby, J., Prigent, M., Camiade, M., and Obregon, J. (2003). Phase noise in oscillators – Leeson formula revisited. *IEEE Trans. Microwave Theory Tech.* 51: 1386–1394.

6 Nallatamby, J.C., Prigent, M., Camiade, M., and Obregon, J.J. (2003). Extension of the Leeson formula to phase noise calculation in transistor oscillators with complex tanks. *IEEE Trans. Microwave Theory Tech.* 51: 690–696.

7 Vendelin, G.D., Pavio, A.M., and Rohde, U.L. (2005). *Microwave Circuit Design Using Linear and Nonlinear Techniques*. Wiley Interscience.

8 Skyworks Solutions (August 2008). Varactor Diodes. Application Note.

9 Siweris, H.J. and Schiek, B. (1985). Analysis of noise up-conversion in microwave FET-oscillators. *IEEE Trans. Microwave Theory Tech.* 33: 233–242.

10 Infineon Technologies (2015). BB833 Silicon Tuning Diodes. Datasheet 2015.

11 Lenk, F., Schott, M., and Heinrich, W. (2001). Modeling and measurement of phase noise in GaAs HBT Ka-band oscillators. *Dig. Eur. Microwave Conf.* 1: 181–184.

12 Zhang, X. and Daryoush, A.S. (1994). Bias dependent noise up-conversion factor in HBT oscillators. *IEEE Microwave Guided Wave Lett.* 4: 423–425.

13 Zhang, X., Sturzebecher, D., and Daryoush, A.S. (1995). Comparison of the phase noise performance of HEMT and HBT based oscillators. In: *Proceedings of 1995 IEEE MTT-S International Microwave Symposium*, 697–700.

14 Rudolph, M., Lenk, F., Llopis, O., and Heinrich, W. (2006). On the simulation of low-frequency noise up-conversion in InGaP/GaAs HBTs. *IEEE Trans. Microwave Theory Tech.* 54: 2954–2960.

15 Phillips, J. and Kundert, K. (2019). *An Introduction to Cyclostationary Noise.* The Designers Guide Community.

22

Phase Noise Measurement

Basic Parameters

Basic variables of an oscillator are its oscillation frequency f_0 and the output level P_C. It is clear that both quantities show signs of fluctuation. Here we are interested in fluctuations around a mean value. In particular we are interested in the measurement of the fluctuations that occur at the offset frequency f_m. The fluctuations of the amplitude are mostly small, because the oscillator works in saturation. They can also be eliminated by suitable mixers. The frequency fluctuations, commonly called phase noise, have a spectral distribution versus the offset frequency f_m, which we discussed in Chapter 20. In Table 22.1 the measured quantities are summarized once again.

There are different methods of measuring phase noise, which are used according to the requirements. Such requirements are, e.g. frequency range of oscillator and offset, dynamic range and possible automation, on-wafer system, and budgetary limits.

Spectrum Analyzer

We first look at a direct measurement. The simplest method is to connect the signal from an oscillator under test to a spectrum analyzer. This is always recommended for initial screening measurements. The configuration is shown in Figure 22.1 [1].

The typical screen display, with f_0 as the center frequency, is shown in Figure 22.2. Due to the logarithmic scale, the phase noise is strongly enhanced compared with the signal level.

A spectrum analyzer is always required, even in more complex systems. If its frequency range covers the oscillator frequency f_0, the phase noise spectrum can be seen directly, as in Figure 22.2. Besides the actual measurement, this overview is very useful, because spurious or other interfering signals can be recognized

A Guide to Noise in Microwave Circuits: Devices, Circuits, and Measurement, First Edition.
Peter Heymann and Matthias Rudolph.

Table 22.1 Phase noise: quantities to be measured.

$L(f_m)$	Course of the curve is visible on the RF spectrum analyzer. Sum of phase and amplitude noise. The internal LO of the SPA must be better than the DUT. Suitable selection of the resolution bandwidth (RBW) is important. Offset frequency $f > 10 \times \text{RBW}$. Phase Noise Marker corrects quantitatively the display to $L(f_m)$
$S_\phi(f_m)$	Spectral density of phase fluctuations at the output of a phase discriminator (mixer in quadrature). Viewable on a LF spectrum analyzer
$S_{\Delta f}(f_m)$	Spectral density of frequency fluctuations at the output of a frequency discriminator (delay line). Viewable on a LF spectrum analyzer
$\sigma_Y(\tau)$	Allen variance. Periodic measurement of the frequency with a counter at time interval τ
RFM	Residual FM. The output of the frequency discriminator $S_{\Delta f}(f_m)$ is passed through a bandpass filter and the average value in this range Δf_m of offset frequencies is measured. Spurious and noise are not separated

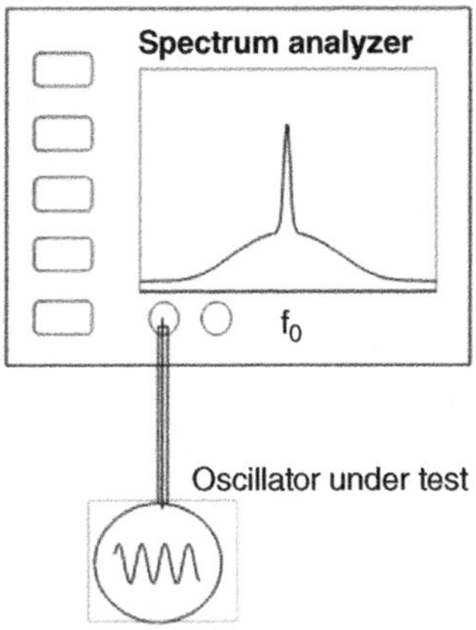

Figure 22.1 Direct connecting an oscillator to a suitable spectrum analyzer.

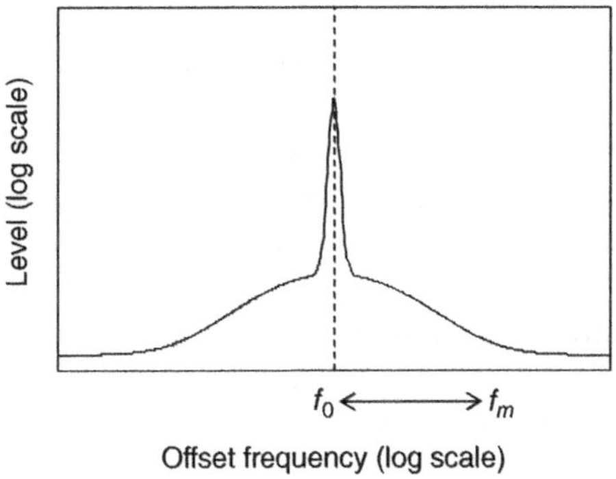

Figure 22.2 Screen display of a spectrum analyzer in log scale.

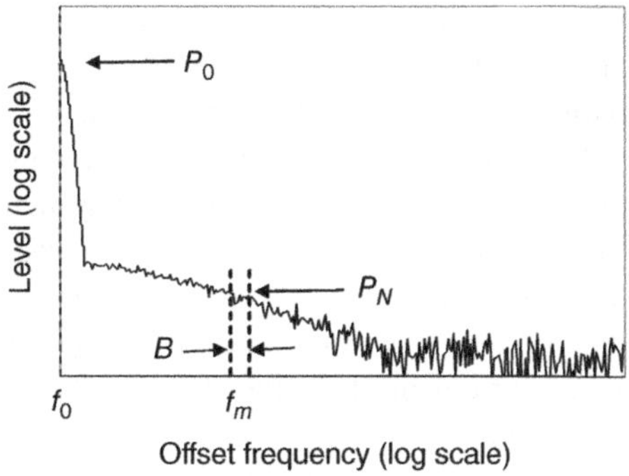

Figure 22.3 Display of the upper sideband in log scale. Oscillator frequency f_0 shifted to the origin of the scan.

immediately. In systems with phase discriminator the display of the LF-spectrum is always done on a spectrum analyzer, but with the frequency range of less than 10 … 100 MHz. The adjustment in Figure 22.3, in which only a sideband is displayed because f_0 is set at zero of the frequency scale, is more suitable for the measurement.

Figure 22.4 [3] shows the basic design of a spectrum analyzer using the superposition principle. This concept is universally applicable in an extremely broad frequency range. The input signal is transformed into the IF range in a mixer and subsequently amplified. It then passes through a filter of adjustable bandwidth. This is the resolution bandwidth (RBW). The mixer is driven by a swept local oscillator. This shifts the spectrum of the input signal across the filter and thus samples it. In most cases follow logarithmic amplification and detection. Averaging is carried out by a cascaded low-pass filter. As we have seen in Chapter 5, the ratio of the RBW and the cut-off frequency of the low-pass filter, i.e. the video bandwidth (VBW), determines the fluctuation of the displayed signal. This analogue principle is largely controlled by microprocessors in modern spectrum analyzers. In addition, the signal is digitized as far upstream as possible in the processing chain and further processed mathematically. The sweep of the local oscillator is realized by means of a phase-locked loop with variable divider

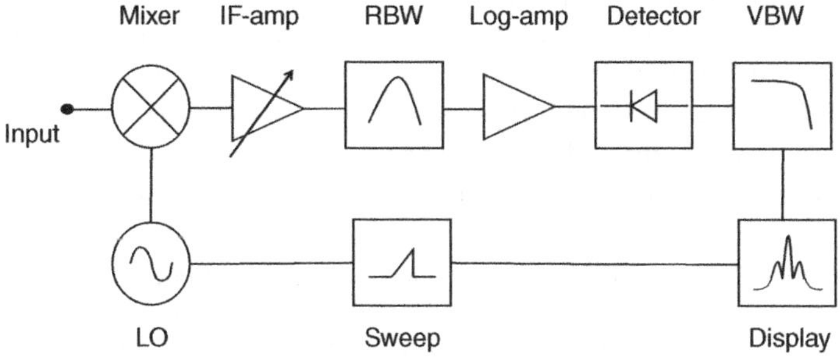

Figure 22.4 Scheme of a spectrum analyzer. Source: Rohde&Schwarz [3].

ratio. This allows a very high frequency accuracy and low phase noise of the analyzer. This is of great advantage, because the intrinsic phase noise competes with the phase noise of the test object.

To measure $L(f_m)$, the conditions given in Table 22.1 must be met. Since the sum of phase and amplitude noise is displayed, the amplitude noise must be negligible. This is usually the case. Both cannot be separated. The minimum offset frequency depends on the smallest technically reasonable RBW. With a small RBW the sweep may be very slow, so that a drift of the oscillator frequency f_0 distorts the measurement during this time. This can be the case when measuring free-running oscillators.

Another principle of spectrum analysis has become possible due to the progress of digital technology [4]. In the fast Fourier transformation (FFT) spectrum analyzer, the signal is measured in the time domain by sampling with an analogue-to-digital converter and stored temporarily. It is then converted into the frequency domain by means of FFT. The speed of digital signal processing limits the application to frequencies below 100 kHz. The use for phase noise measurements is therefore limited to special cases.

The unit of $L(f_m)$ is dBc/Hz. This is the noise level relative to the level of the carrier, referred to 1 Hz bandwidth. Since the measurement cannot be performed at $RBW = 1$ Hz, the conversion is performed internally, taking into account the noise bandwidth of the filter and the detector characteristics. The sample detector measures individual noise voltage values at a certain time interval and thus records the entire amplitude distribution, which is then stored as a sequence of numbers. In the case of trace averaging on the display, the noise level is estimated too low. In modern spectrum analyzers, the operator is relieved of these problems by internal data processing. Details of averaging are discussed below.

The spectrum analyzer must meet certain minimum requirements. Its own phase noise, that of the local oscillator, should be at least 10 dB below the DUT. During the measurement, the strong signal of the carrier and the extremely low noise signal are displayed simultaneously. This requires a high dynamic range of the logarithmic scale. A simple test shows the suitability of the measurement setup: First, the curve of the test object is saved. Then switch it off and compare the curve without signal at the same settings. From both pictures one can deduce the suitability of the device for this application, if there is a clear difference at the desired offset frequency. Basically one should pay attention to careful shielding and use high quality cables. The dangers of interference from power supplies, computers, and mobile phones are always present, as with any noise measurement.

During phase noise measurements with the spectrum analyzer, both the strong carrier signal and the usually very weak noise are displayed simultaneously. This high dynamic range can only be achieved in the logarithmic display mode. With weak signals, such as noise, averages are always formed from a large number of measurement points. Compared with averaging in the linear range, there are some special features to be considered with logarithmic scale, which we have already discussed in Chapter 14. Let us first look at a sampled signal according to Figure 22.5, which can already lead to some confusion.

The RF voltages 0.32 and 0.64 V (peak) follow each other in equally long time intervals. At 50 Ω they correspond to the power 0 and 6 dBm. If we calculate the linear average (two sections are sufficient), we obtain

$$M = \frac{0.32 + 0.64}{2} = 0.48\,\text{V} \triangleq 3.6\,\text{dBm}$$

If we average the dB values, we obtain

$$M_{\text{dB}} = \frac{0 + 6}{2} = 3.0\,\text{dBm}$$

A recommended procedure for averaging measured values that are available in dB is to convert them into linear values, calculate the average value, and then convert it again into dB. In our case we have another value

$$M = \frac{10^{\frac{0}{10}} + 10^{\frac{6}{10}}}{2}; \quad 10\,log(M) = 4\,\text{dBm}$$

This simple example shows that it is definitely important to look closely at the averaging process. While these differences do not occur with CW signals, the amplitude statistics also play a role with stochastic signals.

In phase noise measurements with the spectrum analyzer, the noise power is measured at the offset frequency f_m in dB within a defined bandwidth B. At first, we consider only the noise measurement without the carrier power P_C, to which the noise is normalized to obtain $L(f_m)$. At an 1 Ω-resistor there exists with the noise voltage v the known relation

$$N = 20\,log(v) \quad (\text{dB}) \tag{22.1}$$

The quantitative dB specification always requires a realistic reference value. The noise power $N = 1\,\mu\text{W}$ at the 50 Ω input of the spectrum analyzer corresponds to

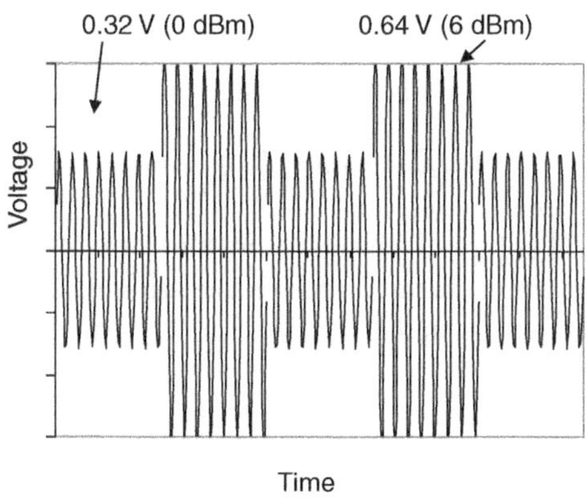

Figure 22.5 RF signal switched between the levels 0 and 6 dBm.

the voltage 7.07 mV.

$$N(\text{dBW}) = 20\,log\left(7.07 \times 10^{-3}\right) = -43\ \text{dBW}$$

In the more appropriate dBm specification, with reference to 1 mW (0.224 V at 50 Ω)

$$N(\text{dBm}) = 20\,log\left(\frac{7.07 \times 10^{-3}}{0.224}\right) = -30\ \text{dBm}$$

The level displayed by the spectrum analyzer at the offset frequency represents an average value over the noise power present in the RBW. The calculation of this average value leads to a problem of signal statistics. As we have seen in Chapter 2, the amplitude distribution for band-limited noise is no longer the Gaussian distribution of white noise, but the Rayleigh distribution. Measuring the amplitudes on a logarithmic scale complicates the interpretation of the mean value.

Modern spectrum analyzers have different types of detectors. The sample detector is the most suitable for noise measurements. It takes values from the envelope of the noise voltage at certain instants of time and thus records the entire distribution of amplitudes in a time range. We repeat here a part of the derivation already treated in Chapter 14. The root mean square of n measured voltage values v_i $(i = 1 \ldots n)$ is

$$\overline{v^2} = \frac{1}{n}\sum_{i=1}^{n} v_i^2 \tag{22.2}$$

Correspondingly for the power

$$\overline{N} = \frac{1}{n}\sum_{i=1}^{n} N_i \tag{22.3}$$

The measured voltages are fluctuations with the statistical distribution $W_v(v)$.

In general, the transition of the probability densities, here from voltage v to power N, is given by:

$$W_P(N) = W_v(v)\frac{dv}{dN} \tag{22.4}$$

The power is normalized to the mean square value of v.

$$N - \overline{N} = 10\,log\left(\frac{v^2}{\overline{v^2}}\right) \quad \text{(dB)} \tag{22.5}$$

For further consideration, the shape of the probability density distribution W is essential. In our case, W is a logarithmic Rayleigh distribution, since the measured values are in the dB scale. The Rayleigh distribution of the voltages is

$$W_v(v) = \frac{2v}{\overline{v^2}}\,exp\left(-\frac{v^2}{\overline{v^2}}\right) \tag{22.6}$$

Since the measured power N is in dB, the following conversion is necessary. We remain at the 1 Ω-reference, since the resistance is not important for the consideration. Instead of the decadic logarithm (*log*) we move on to the natural (*ln*).

Equation (22.1) then reads

$$N = 20 \times 0.434\, ln(v) \tag{22.7}$$

The transition from the distribution (22.6) to the logarithmic Rayleigh distribution of the power is achieved by transformation (22.4). For this we need $^{dv}/_{dN}$. From (22.7) results

$$v = exp(aN) \quad a = 0.115$$

And the derivation

$$\frac{dv}{dN} = a \times exp(aN) \tag{22.8}$$

The logarithmic distribution is

$$W(N) = \frac{2a}{\overline{v^2}} exp(2aN)\, exp\left\{-\frac{exp(2aN)}{\overline{v^2}}\right\} \tag{22.9}$$

Instead of the root mean square of the voltage (22.2) we introduce the mean value of the power corresponding to (22.7)

$$N_0 = 4.34 \times ln\left(\overline{v^2}\right); \quad \overline{v^2} = exp\left(2aN_0\right)$$

With this we can write (22.9):

$$W(N) = 2a\, exp\left\{2a\left(N - N_0\right) - exp\left[2a\left(N - N_0\right)\right]\right\} \tag{22.10}$$

As already shown in Chapter 14, the averaging of the logarithmic measured values N is different from the linear average N_0 according to (22.2). By averaging the logarithmic distribution, the band-limited noise is underestimated by 2.5 dB compared to linear averaging.

The spectrum analyzer method is easy to use. It has the advantage of direct measurement at the signal frequency and an overview over a wide frequency range. All conversion and normalization functions are provided internally. In addition, it is a very widely used, low-cost instrument, which is available in almost every laboratory.

A disadvantage is the possibly too high own phase noise of the local oscillator and the limited dynamics. The gain of the logarithmic amplifier is limited to approx. 100 dB, so that the dynamic display may be insufficient. In addition, the inherent noise is limited by $NF \approx 20$ dB relatively high.

However, these disadvantages can be overcome, but also at higher cost. An advanced measuring instrument based on a spectrum analyzer with signal processing using cross-correlation is the R&S FSWP [2]. This provides excellent sensitivity (−172 dBc/Hz @ f_0 = 1 GHz, at f_m = 10 kHz offset). Furthermore, extremely low-noise DC sources are available for VCO measurements. The quality of the control voltage for the VCO is of great importance as it has a direct effect on the phase noise. Batteries are largely noise-free but are difficult to tune and not suitable for automation. Tunable DC power supplies usually have too much noise superimposed on the DC voltage. Low-pass filters are a remedy, but reduce the speed of the tuning due to their high time constant. Therefore, a voltage source specially designed for VCO tuning is very important. It is also important to avoid ground loops between the DUT and the power supply unit, as they can cause mains hum.

Phase Detector Method

This very sensitive and in a wide frequency range applicable method is based on the use of a double balanced mixer as phase detector. This mixer is particularly suitable due to its DC-coupling at the IF-port. Two signals of the same frequency f_0, which are held in quadrature (90° phase difference) are fed to one input each. In the ideal case, the frequency 0 Hz with amplitude 0 V appears at the output. The sum frequency $2f_0$, which also appears, is removed by the low-pass filter. Since the frequencies of both sources show phase noise, this appears as a low-frequency signal and is displayed on a baseband spectrum analyzer. The scheme of the system is shown in Figure 22.6. The carrier with its high amplitude is eliminated [5].

Of great importance is the quadrature monitor, which must be used to ensure that DUT and reference signal are in quadrature. This is a practical difficulty which cannot be solved with the principle circuit Figure 22.6. In a real circuit both sources must be coupled in a phase-locked loop.

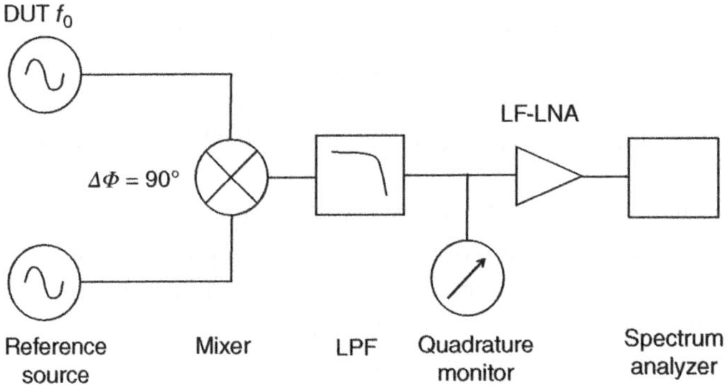

Figure 22.6 Principle of a phase detector setup.

The Sensitivity of the Phase Detector

The signal with the phase noise to be measured and the much stronger local oscillator are fed to the two inputs of the mixer. Both have the same frequency and a phase shift ϕ

$$P_1 = Asin(\omega t); \quad P_2 = Bsin(\omega t - \phi) \tag{22.11}$$

In the mixer they are multiplied with each other.

$$P_1P_2 = \frac{AB}{2}\{cos\,(\omega t - \omega t + \phi) - cos\,(2\omega t - \phi)\} \tag{22.12}$$

This produces a signal with twice the frequency and a DC component $V(\phi)$, which depends on the phase shift.

$$V(\phi) = V_{MAX}cos\,(\phi) \tag{22.13}$$

V_{MAX} results from the amplitude of the signals and the conversion loss of the mixer.

In quadrature, i.e. $\phi = 90°$, the output voltage is 0 V. Only the phase noise spectrum remains, since the quadrature is not fulfilled for this. The 2ω signal is separated by the low pass. The sensitivity K_ϕ is determined by the slope of the *cos*-function at zero crossing at $\phi = 90°$ (Figure 22.7). This phase detector constant is

$$K_\phi = \frac{\Delta V(\phi)}{\Delta \phi} = -V_{MAX}sin(\phi) \tag{22.14}$$

Since $sin(90°) = 1$, and the sign does not matter, K_ϕ can be determined by measuring V_{MAX} and thus has determined the phase detector constant. According to Figure 22.7 a phase difference of $\phi = 180°$ is adjusted and V_{MAX} and thus K_ϕ is obtained. This is achieved by slightly detuning the two input signals. The slope of the sine curve at the zero crossing, which is decisive for K_ϕ, is equal to the peak

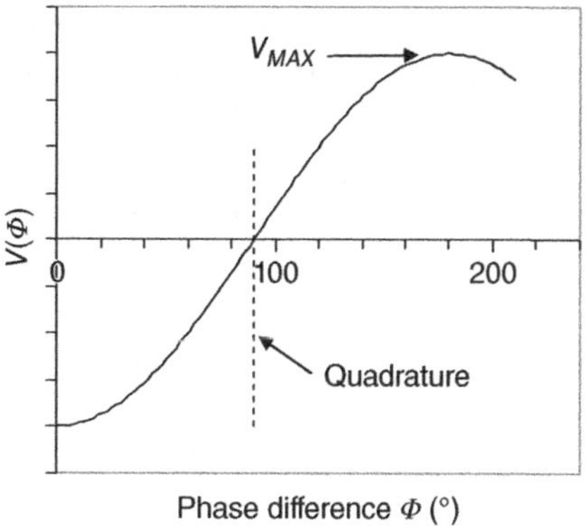

Figure 22.7 Determining the phase detector constant from V_{MAX} with slight detuning from the quadrature.

value of this curve, which is obtained according to Figure 22.7. An example for a value would be $V_{MAX} = 0.5$ V the phase sensitivity $K_\phi = 0.5$ V/rad.

The relation of the phase fluctuation $\Delta\phi(t)$ to the fluctuation of the measured output voltage V_{OUT}(t) is

$$\Delta V_{OUT}(t) = K_\phi \Delta\phi(t) \tag{22.15}$$

The output signal of the phase detector in the system Figure 22.6 consists of the pure phase noise of the DUT. Any amplitude noise is eliminated by the design of the mixer.

There are minimum requirements for proper operation. The DUT signal should be > – 5 dBm and the LO signal of the reference source should be > + 5 dBm.

The low pass filters out the baseband signal and eliminates a possible LO transfer and the 2ω component in the output spectrum. However, it must be ensured that these frequencies do not return to the mixer by reflection and disturb its function.

The baseband amplifier increases the sensitivity of the spectrum analyzer to the required level. When selecting this device, special attention must be paid to low $1/f$ noise and sufficient dynamic range. Designs based on Si-BJT are well suited.

For the display, an analogue spectrum analyzer with a frequency range of approx. 100 MHz or a digital FFT analyzer can be used. The latter works very fast with a small RBW and allows measurements with an offset of less than 1 Hz. However, the frequency range is limited to 100 kHz, which is particularly suitable for the audio range. The analogue spectrum analyzer is used for overview measurements and large offset frequencies.

A simple oscilloscope with DC coupling is suitable as a quadrature monitor. It has the advantage that one can see the phase noise in the time domain, which is useful for monitoring purposes. A voltmeter would also be suitable. However, analogue pointer instruments are rarely found in a laboratory and digital multimeters are of limited use. They can interfere with the sensitive measurement setup. Furthermore, the continuous numbers on the display are often confusing. The slight detuning of the two sources against each other results in a beat note with the difference frequency. Its peak value is the V_{MAX} we are looking for. If this beat note is measured with an RF voltmeter or a spectrum analyzer, e.g. in an automatic system, the effective value V_{RMS} is measured. The peak value is $V_{MAX} = \sqrt{2}\, V_{RMS}$.

When evaluating the phase noise spectrum

$$\Delta\phi_{RMS}(f_m) = \frac{1}{K_\phi}\Delta V_{RMS}(f_m) = \frac{1}{V_{MAX}}\Delta V_{RMS}(f_m) \tag{22.16}$$

The actual measured variable is the spectral noise power density of the phase fluctuations.

$$S_\phi(f_m) = \frac{(\Delta\phi_{RMS}(f_m))^2}{B} \tag{22.17}$$

As well as the single sideband phase noise

$$L\left(f_m\right) = \frac{1}{2} S_\phi\left(f_m\right) = \frac{1}{2} \frac{\Delta V_{RMS}^2\left(f_m\right)}{V_{MAX}^2} \tag{22.18}$$

In this conversion, the restriction of the small angle criterion must be taken into account (Chapter 20). The reference generator is of particular importance. One wants to measure the phase noise of the DUT and therefore needs a second oscillator, which of course also has a phase noise spectrum. The sum of these two levels appears at the output of the phase detector. The spectrum analyzer measures the sum of the RMS values. A reference generator whose phase noise is 10 dB lower than that of the DUT at all offset frequencies can be used without any problems. If both have approximately the same phase noise, the display can be corrected by 3 dB. To calibrate the system, it is useful to vary the power of a generator in a defined way to adjust the amplitude of the beat. This can usually be done by an internal attenuator.

Example Calibration and Measurement

Hardware	
DUT	10 GHz YIG oscillator; $P_S = -5$ dBm
Phase detector	Double balanced mixer 2–18 GHz; $IF = 0$–750 MHz, conversion loss = 6.5 dB; $VSWR$ RF- IF port 1.5
LNA	$G = 20$ dB
Calibration	
K_ϕ	Peak value of beat: $P_S = -5\text{ dBm} \triangleq V_S = 0.126$ V. −6.5 dB of which at IF output: $V_O = V_S - 6.5\text{ dB} = 59$ mV. Peak value: $V_{OPK} = \sqrt{2}V_O = 84$ mV. $K_\phi = V_{OPK} = \frac{84\text{ mV}}{\text{rad}}$
Measurement	Spectrum analyzer $f_m = 100$ kHz; $RBW = 10$ kHz; $Z = 50\,\Omega$ Displayed value: $P_M = -47.5$ dBm
Corrections	Log. Averaging: 2.5 dB; LNA: 20 dB, normalization to $N_{BW} = 1$ Hz: $10\,log(1.2\times 10^4) = 40.8$ dB, $P_{MC} = P_M + 2.5 - 20 - 40.8 = -105.8$ dBm/Hz
Evaluation	Conversion to voltage/Hz: $P_{MC} = 2.6\times 10^{-14}$ W/Hz, $V_{MC} = \sqrt{P_{MC}Z} = 1.1\ {}^{\mu V}/_{Hz}$. Phase fluctuations: $\Delta\phi = {}^{V_{MC}}/_{K_\phi} = 1.4\times 10^{-5}\ {}^{\text{rad}}/_{\text{Hz}}$ Single sideband phase noise: $L = 10\,log\left({}^{\Delta\phi^2}/_2\right) = -100\ {}^{dBc}/_{Hz}$

Keeping the Quadrature by a PLL

A crucial point for practice has not yet been considered. One of the two sources must be electronically tunable and both must be coupled together in a phase-locked loop [6]. In the simple configuration shown in Figure 22.6, quadrature for the duration of a measurement of the entire spectrum, even with very stable oscillators, can only be maintained in exceptional cases. The circuit Figure 22.6 must be completed by a phase-locked loop. The principle is shown in Figure 22.8. If the DUT is a VCO, the PLL can also be connected to it. The effort of circuitry and theory of corrections increases considerably, because the PLL eliminates the phase noise within its bandwidth.

At first it seems to be contradictory to insert a PLL into a system for measuring phase noise. In oscillator circuits this is used to stabilize the frequency and thus also to suppress the phase noise. The effect of the PLL in a 10 GHz synthesizer is shown as an example in Figure 22.9. Near the carrier there is a significant reduction in phase noise. In a system as shown in Figure 22.8, it ensures that the quadrature is maintained at the phase detector. It is therefore necessary to precisely analyze the effect of the PLL on the measurement.

The system with PLL (Figure 22.8) is characterized by several frequency ranges. The widest is the

Peak Tune Range (PTR): The VCO that is stabilized by the PLL has a tuning constant K_{VCO} (Hz/V). Together with the voltage V_{TUNE} provided by the PLL, the maximum tuning range is $PTR = K_{VCO}V_{TUNE}$. The PTR must be related to the frequency range and drift of the DUT. The maximum is a few MHz. The frequency range required for the PTR can be estimated from Figure 22.10. The higher the phase noise of the DUT, the higher the PTR must be. One first

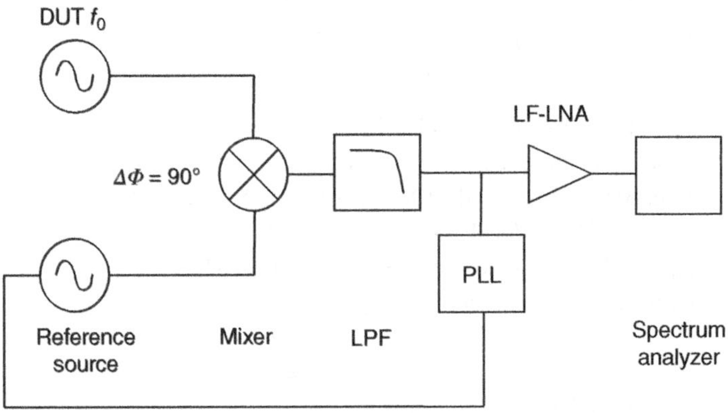

Figure 22.8 Setup with phase-locked loop to ensure quadrature.

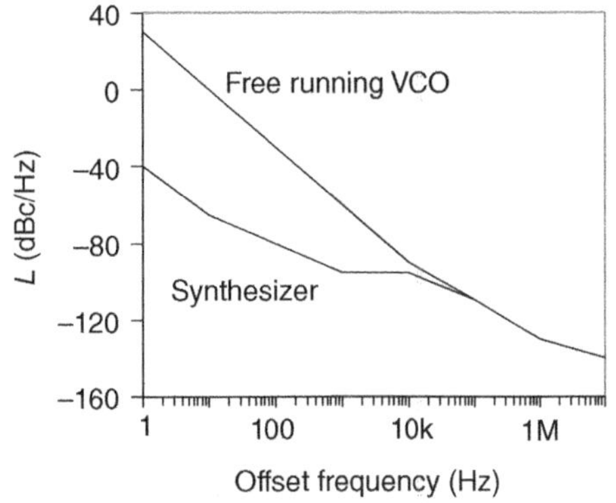

Figure 22.9 Reduction of phase noise near the carrier using a PLL.

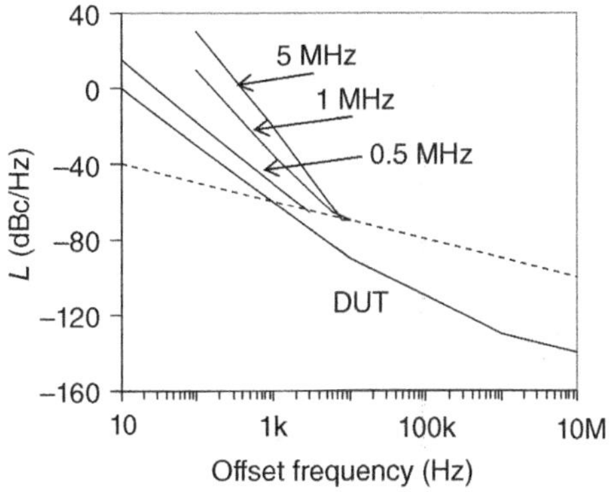

Figure 22.10 Required PTR (0.5–5 MHz) to ensure that the PLL is locked. Higher phase noise requires a higher PTR. Source: Based on Agilent Technologies [1].

needs an estimated curve of the expected phase noise spectrum, which actually should be measured. The procedure is an iteration. A screening measurement with a spectrum analyzer is helpful here. The $L(f)$-value at which this curve intersects the small angle straight line is taken from the diagram Figure 22.10. In our example we have $L(f) = -60$ dBc/Hz @ 1 kHz. We increase this value by 10 dB and arrive at the curve $PTR = 0.5$ MHz. This is an estimate for dimensioning the synchronization range of the system.

An empirical formula for estimating the required PTR can be derived from Figure 22.10:

$$PTR = 1.5 f^{1.7} + 3 \ \text{(Hz)} \tag{22.19}$$

where f is the frequency at which the curve of the DUT intersects the small-angle limit. This value increases with higher phase noise and therefore a higher PTR is required.

Capture range (CR): This is the frequency range in which the PLL locks in and can phase-lock the two sources. This is significantly lower than the PTR: $CR \leq 0.1$

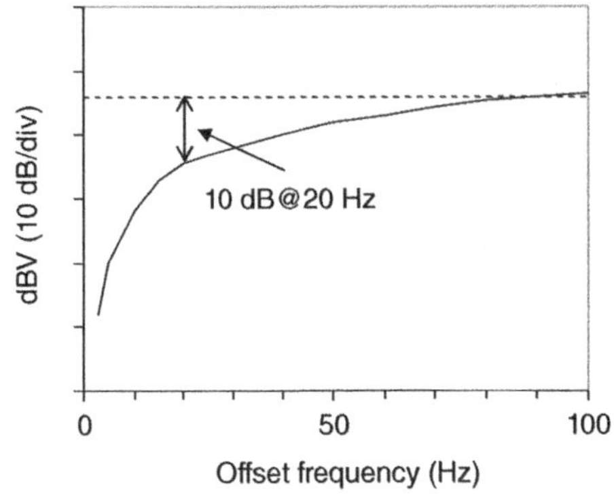

Figure 22.11 PLL transfer function measured with noise generator. Here a reduction by 10 dB at 20 Hz.

PTR. In a measurement of drifting oscillators it can become a problem that the PLL does not lock.

Drift tracking range (DTR): Once the PLL is locked, it can maintain this state even beyond the capture range. As a guide value $DTR \cong 0.2\ PTR$

Loop bandwidth (LBW): The bandwidth of the control loop is also an important quantity. Within this frequency range, not only quadrature is forced, but the phase noise of the DUT is also effectively suppressed. Beyond the cut-off frequency there is no interference and we have the undisturbed phase noise. This can be seen in Figure 22.9. The LBW of a measuring system is usually much lower (≈100 Hz) than in a synthesizer (≈10 kHz). The transfer function of the PLL can be determined by feeding a noise signal into the phase detector. The response of the PLL appears on the spectrum analyzer. It typically is a plot shown in Figure 22.11. One can read off the correction values for $f_m < 100$ Hz, e.g. 10 dB at 20 Hz. Beyond $f_m > 100$ Hz no correction is required.

The injection of white noise into the PLL for the purpose of measuring the transfer function encounters the practical difficulty that a noise generator that delivers white noise in the range below 100 Hz is not part of the standard equipment of an RF laboratory. Semiconductor components show the well-known $1/f$ spectrum, vacuum diodes have strong flicker noise. Some older devices have built in suitable sources, e.g. the HP 3582A LF spectrum analyzer. One component that delivers white noise in the range 0.01 Hz–3 kHz is the MAXIM 4238 amplifier [7], which is optimized for minimal drift.

If precise measurements below 100 Hz offset are not intended, this correction is not necessary.

Delay Line as Frequency Discriminator

The effort for the phase detector method is not insignificant. A reference phase is required, which is provided by the reference oscillator. This is not necessary

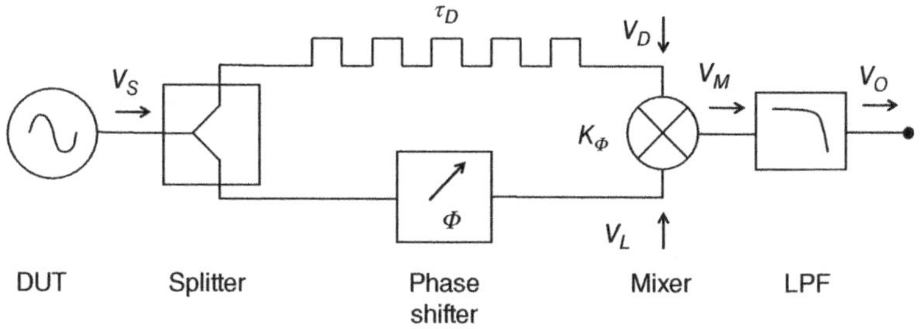

Figure 22.12 Scheme of the FM discriminator technique with delay line.

when using the frequency discriminator method. Instead of an additional generator, a reference frequency is required. It can be realized by a passive element, e.g. a resonance curve or the cutoff-frequency of a filter. In addition, free-running oscillators with resonators of low quality, as frequently found in MMIC technology, often cannot be locked to a PLL. When using a frequency discriminator, no reference oscillator is required, since the signal of the DUT is to some extend compared with itself. Oscillators with frequency drift and higher phase noise can also be measured by this method (Figure 22.12) [8].

For comparison on the phase detector, there are the measurement signal at time t and the delayed measurement signal at time $-\tau_D$. We analyze the signal $V_S(t)$ modulated with phase noise when passing through the circuit with delay line and assume (20.17).

$$V_S(t) = V_P cos\left\{\omega_o t + \frac{\Delta\omega}{\omega_m} cos\left(\omega_m t\right)\right\} \tag{22.20}$$

The splitter causes a reduction of at least 3 dB. We do not consider further attenuation, e.g. in the delay line. We then have the signals at the mixer input:

$$V_L(t) = v\, cos\left\{\omega_o t + \frac{\Delta\omega}{\omega_m} cos\left(\omega_m t\right)\right\} \tag{22.21}$$

$$V_D(t) = v\, cos\left\{\omega_0\left(t-\tau_D\right) + \frac{\Delta\omega}{\omega_m} cos\left(\omega_m\left[t-\tau_D\right]\right)\right\} \tag{22.22}$$

Due to multiplication in the mixer, difference and sum frequencies are generated, as well as harmonics. The output amplitude is determined by the phase detector constant of the mixer. The difference frequencies pass the low-pass filter.

$$V_O(t) = K_\phi cos\left\{\omega_0\left[t-\tau_D\right] + \frac{\Delta\omega}{\omega_m} cos\left(\omega_m\left[t-\tau_D\right]\right) - \omega_0 t - \frac{\Delta\omega}{\omega_m} cos\left(\omega_m t\right)\right\} \tag{22.23}$$

Using the addition theorem:

$$V_O(t) = K_\phi cos\left\{-\omega_0\tau_D + 2\frac{\Delta\omega}{\omega_m} sin\left(\omega_m\tau_D\right) sin\left\{\omega_m\left[t-\frac{\tau_D}{2}\right]\right\}\right\} \tag{22.24}$$

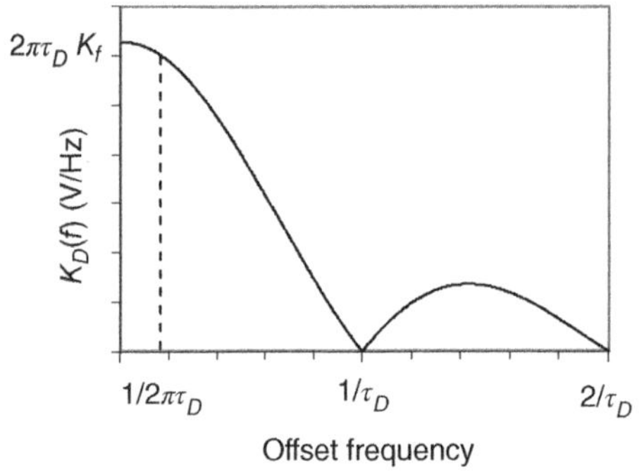

Figure 22.13 Frequency discriminator constant over the inverse line length. In the range $f < 1/2\pi\tau_D$ no correction is necessary.

If both signals are hold in quadrature at the mixer, then is $\omega_0\tau_D = {}^{\pi}/_{2}$. Inserted in Eq. (22.24) and application of an addition theorem results in

$$V_O(t) = K_\phi sin\left\{\frac{2\Delta\omega}{\omega_m} sin\left\{\omega_m \frac{\tau_D}{2}\right\} sin\left(\omega_m \left[t - \frac{\tau_D}{2}\right]\right)\right\} \tag{22.25}$$

The small angle criterion ${}^{\Delta\omega}/_{\omega_m} < 0.2$ rad leads to $sin\left({}^{\Delta\omega}/_{\omega_m}\right) = {}^{\Delta\omega}/_{\omega_m}$. We omit the time-dependent term in (22.25). Then we obtain the amplitude of the output voltage. i.e. the sensitivity of the system to phase fluctuations.

$$\Delta V = K_\phi 2\pi\tau_D \Delta f \frac{sin\left(\pi f_m \tau_D\right)}{\pi f_m \tau_D} \tag{22.26}$$

This formula contains the typical delay line ${}^{\sin x}/_{x}$ -term. For offset frequencies $f_m < {}^{0.5}/_{2\pi\tau_D}$ the method is applicable without correction, because then

$$\frac{sin\left(\pi f_m \tau_D\right)}{\pi f_m \tau_D} \cong 1 \tag{22.27}$$

One can then define a frequency discriminator constant K_D:

$$K_D = K_\phi 2\pi\tau_D = \frac{\Delta V}{\Delta f} \quad ({}^{\mathrm{V}}/_{\mathrm{Hz}}) \tag{22.28}$$

For offset frequencies that are not too large and fulfill condition (22.27), the phase fluctuation is proportional to the frequency fluctuation of the output signal. For each range of offset frequencies, the length of the delay line must therefore be chosen appropriately (Figure 22.13).

The Sensitivity of the Delay-Line Method

In (20.17) we had described the phase fluctuation for a singular frequency ω_f. The term there is ${}^{\Delta\omega}/_{\omega_f} sin\left(\omega_f t\right)$. If we look at the whole of the fluctuations, we can write for the signal of the source [8]:

$$V_C(t) = V_P cos\left(\omega_0 t + \phi(t)\right) \tag{22.29}$$

The small angle criterion ($^{\Delta\omega}/_{\omega_f} = \beta = |\phi(t)| \ll 1$) and conversion leads to the representation with the two sidebands, whose voltage is $^{V_P\beta}/_2$. Accordingly, we can write for the noise spectrum

$$V_N(\omega) = \frac{1}{2} V_P \phi(\omega) \tag{22.30}$$

To estimate the sensitivity of the delay-line method, we compare the decrease in signal voltage as it passes through the system with the inherent noise of the system. The calculation is not done in the dB scale but linearly related to the voltage. For example, an attenuation of $\alpha = 3$ dB stands for a voltage reduction of 0.708 ($L = 10^{\frac{-\alpha(\text{dB})}{20}}$).

According to Figure 22.12 the signal passes the splitter. Both components are reduced in power by 3 dB: $L_1 = 0.708$. The reference branch passes through the phase shifter and ideally experiences no further attenuation.

$$V_{REF}(t) = L_1 V_P cos\left(\omega_0 t + \phi(t) + \psi\right) \tag{22.31}$$

The signal of the delay line is reduced by the attenuation L_2 and shifted in phase by τ_D in relation to the reference signal.

$$V_{DEL}(t) = L_1 L_2 V_P cos\left(\omega_0 \left[t - \tau_D\right] t + \phi\left[t - \tau_D\right]\right) \tag{22.32}$$

Both are multiplied in the mixer. The mixing losses are entered with L_3. The filter eliminates high-frequency components and the amplifier increases the baseband signal by G and at the same time reduces the (relatively high) noise figure of the spectrum analyzer. The signal is fed to the spectrum analyzer

$$V_{SPA}(t) = L_1 L_2 L_3 G V_p cos\left(\omega_0 \tau_D + \phi(t) - \phi\left[t - \tau_D\right] + \psi\right) \tag{22.33}$$

The phase shifter is used to adjust quadrature, i.e. $\psi + \omega_0 \tau_D = {}^{\pi}/_2$. With this, and with transition to sine, (22.33) move on to

$$V_{SPA}(t) = L_1 L_2 L_3 G V_P sin\left(\phi(t) - \phi\left[t - \tau_D\right]\right) \tag{22.34}$$

The products in the amplitude term are summarized in the mixer losses L_3. Because of the small angle approximation, we can write

$$V_{SPA}(t) = L_1 L_2 L_3 G V_P \left\{\phi(t) - \phi\left[t - \tau_D\right]\right\} \tag{22.35}$$

The measurement is performed in the frequency domain, therefore we move on to

$$V_{SPA}(\omega) = L_1 L_2 L_3 G V_P \omega \tau_D \phi(\omega) \tag{22.36}$$

After (22.30) we can write for the noise spectrum:

$$V_N(\omega) = \frac{1}{2\,L_1 L_3 L_3 G} \frac{V_{SPA}(\omega)}{\omega\tau_D} \tag{22.37}$$

As in any system, the sensitivity is limited by its inherent noise. The signal $V_N(\omega)$ (22.37) must exceed this inherent noise level. If we add the amplifier to the spectrum analyzer, we have the rms-value

$$V_O = \sqrt{2}L_1 L_2 L_3 V_P \omega\tau_D \phi(\omega) \tag{22.38}$$

The mixer and the spectrum analyzer with pre-amplifier contribute to the inherent noise of the system. With a Schottky diode mixer, we must account for the $1/f$-noise in the frequency range below $f_C \approx 20$ kHz. Beyond this, there is thermal noise of its output resistance $R_M = 50\ \Omega$ with $\overline{v_{RM}^2} = 4kTBR_M$

$$\overline{v_M^2} = \frac{1.7 \times 10^{-14}}{f} + \overline{v_{RM}^2} \tag{22.39}$$

The spectrum analyzer with pre-amplifier has the noise factor F. There is the uncorrelated part $F\overline{v_{RM}^2}$ additional in (22.39).

$$\sqrt{2}L_1 L_2 L_3 V_P \omega\tau_D \phi(\omega) > \sqrt{F\,\overline{v_{RM}^2} + \overline{v_M^2}} \tag{22.40}$$

The following formula holds for the minimum detectable voltage V_{OMIN} in dBc/Hz

$$V_{OMIN}(\omega) = 20\,log\left\{\frac{1}{2}\frac{\sqrt{F\,\overline{v_{RM}^2} + \overline{v_M^2}}}{\sqrt{2}L_1 L_2 L_3 V_P \omega\tau_D}\right\} \tag{22.41}$$

The sensitivity can be improved by a small F, by a high power of the DUT as well as by an extension of the delay-line, i.e. a higher τ_D. However, lengthening the line has the negative effect of increasing the attenuation. I.e. τ_D becomes larger, but L_2 becomes smaller. There is obviously an optimum. This correlation is occasionally found in the literature as a limit curve of the delay line method, but mostly without details of the derivation. For example: Is $1/f$ noise accounted for or not?

Let's consider as an example a cable RG223 as delay line. It has the data $\tau = 5$ ns/m and $\alpha = 0.5$ dB/m @ 1 GHz. The quantities L_2 and τ_D in (22.41) depend on the length of the cable. The sensitivity is maximum, if the product $L_2 \times \tau_D$ is maximum. In Figure 22.14 the product $C = L_1 L_2 L_3 \tau_D$ is plotted in dependence of the cable length l. The losses in the splitter are assumed to be 3 dB ($L_1 = 0.71$) and in the mixer 6 dB ($L_3 = 0.5$). The following applies to the line $L_2 = 10^{-\alpha\, l/20}$.

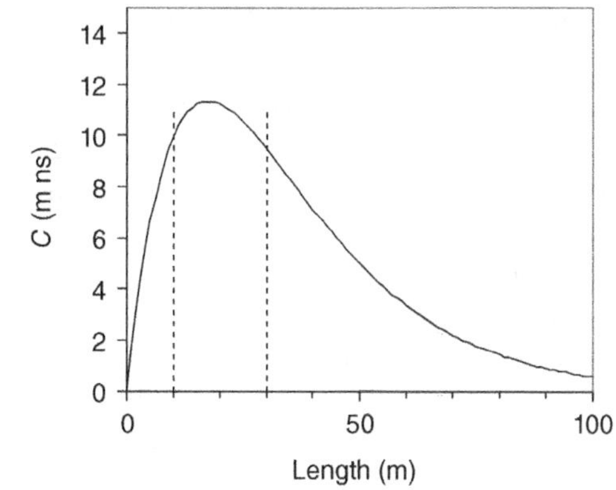

Figure 22.14 Sensitivity of a system with delay line versus the length of the line. Optimum is 10 … 30 m.

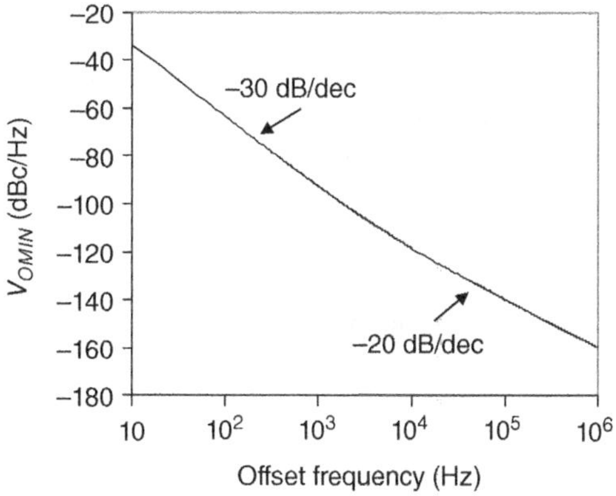

Figure 22.15 Detection limit of the phase noise of a $P = 10$ dBm oscillator ($f = 10$ GHz) with a 20 m long delay line to (22.41).

One can see that an optimum is at $l = 10 \ldots 30$ m. A longer cable has too high attenuation, and a shorter cable would only be sensitive enough at high offset frequencies. At the optimum length $l = 20$ m we have a delay time $\tau_D = 100$ ns. The detection limit (22.41) for an oscillator with $P = 10$ mW is shown in Figure 22.15. The part of the curve with −30 dBc/dec below 10 kHz is determined by the $1/f$ noise (22.39). Above this the white noise of the losses and the receiver restricts the sensitivity. The noise figure of the spectrum analyzer with preamplifier is assumed to be $NF = 6$ dB. A free-running VCO can be measured well with such an configuration.

When selecting the components of the delay line setup, the length of the delay line determines the maximum offset frequency. Accordingly (22.27) we have $2\pi f_m \tau_D < 0.5$. In our example with $l = 20$ m ($\tau_D = 100$ ns) one can measure up to $f_m = 1$ MHz without correction.

Configuration and Calibration

For measuring a real device the principle scheme Figure 22.12 must be completed. A low-noise RF amplifier for boosting the signal amplitude $V_C(t)$ is useful if it is too low so that the mixer does not operate in the optimum range. This applies to both the LO and the signal input. Typical are +10 dBm for the LO. The signal should not be less than 8 dB below the compression point of the mixer. If this is +5 dBm, the output of the delay line should be −3 dBm, i.e. the effective value $V_D > 0.16$ V. In addition, a quadrature monitor is required and a baseband spectrum analyzer, preferably with a low-noise preamplifier. An oscilloscope or voltmeter is suitable as a quadrature monitor. We then proceed to the setup shown in Figure 22.16.

After (22.26), the voltage spectrum in the baseband measured by the spectrum analyzer is converted into the frequency spectrum of the phase noise of the oscillator by $\Delta V = K_\Phi 2\pi\tau_D \Delta f$. There are two ways of calibration. One can determine K_Φ and τ_D individually. On the other hand it is possible to calibrate the sensitivity as a whole by feeding a well-defined signal. For the single determination we need an additional tunable source. With this source one can generate a beat signal and determine K_Φ according to (22.14) from its maximum value. The tunability of the source also allows the determination of τ_D in the real circuit. The frequency change between two successive quadrature states is $\Delta f_C = {}^1/_{2\tau_D}$.

The simplest method is the direct measurement of K_D. In this way, the two variables K_Φ and τ_D are recorded together.

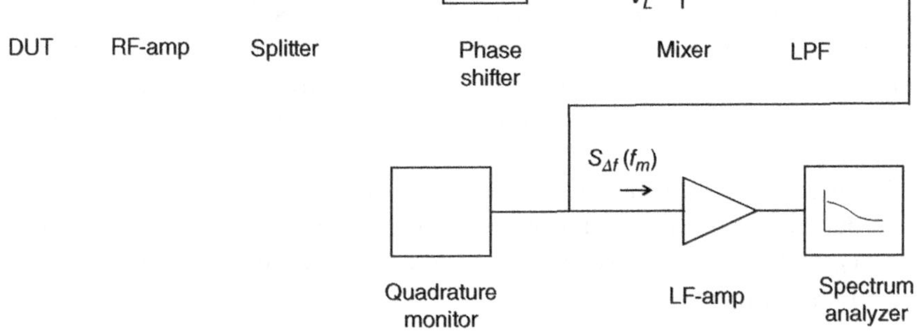

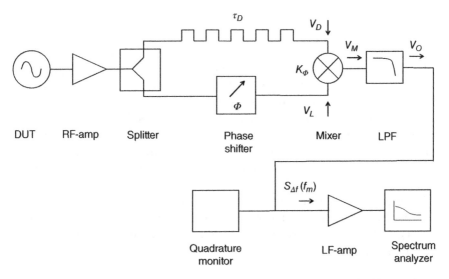

Figure 22.16 Setup of the delay-line method with all components required in practice.

Needed is a source with defined FM modulation. The modulation level $\beta = {}^{\Delta f_C}/_{f_m}$ determines the amplitude of the sideband that appears in the baseband and is measured as V_{RMS}. However, it must be $\beta < 0.2$ rad so that no higher sidebands occur. For a weakly frequency modulated signal, the ratio of the voltage V_{SB} in the sideband at the offset frequency f_m to the voltage V_C of the carrier at the frequency f_C is given by

$$\frac{V_{SB}}{V_C} = \frac{1}{2}\beta = \frac{\Delta f_C}{2f_m} \tag{22.42}$$

K_D is defined as the conversion coefficient of voltage fluctuation ΔV at the output to frequency fluctuation Δf at the input:

$$K_D = \frac{\Delta V}{\Delta f_C} \tag{22.43}$$

If we take Δf_C from (22.42) and set ΔV to the value V_{RMS} read on the spectrum analyzer, we obtain

$$K_D = \frac{\sqrt{2}V_{RMS}}{2f_m {}^{V_{SB}}/_{V_C}} \tag{22.44}$$

which in turn becomes

$$K_D = \frac{\sqrt{2}V_{RMS}}{f_m {}^{\beta}/_{2}} \tag{22.45}$$

The parameters f_m and β can be set on the generator (Figure 22.17).

During the measurement, the baseband voltage spectrum $S_V(f_m)$ must be converted into the spectral density of the frequency fluctuations of the DUT $S_{\Delta f}(f_m)$ or into the single-sideband phase noise $L(f_m)$. With $S_V(f_m) = {}^{\Delta V^2_{RMS}(f_m)}/_{B}$ as

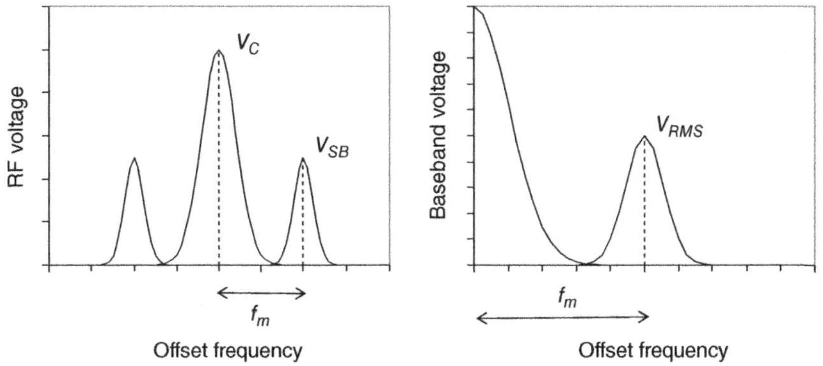

Figure 22.17 Calibrating the system with a frequency-modulated source according to (22.45).

measure it results in

$$S_{\Delta f}\left(f_m\right) = \frac{\Delta V_{RMS}^2\left(f_m\right)}{K_D^2 B} \tag{22.46}$$

And for $L(f_m)$:

$$L\left(f_m\right) = 10\, log \frac{S_V\left(f_m\right)}{2 f_m^2 K_D^2} \tag{22.47}$$

This equation shows why with increasing offset frequency f_m the phase noise curve decreases with −20 dB/dec. A longer delay line increases the sensitivity near the carrier, but on the other hand reduces the power at the mixer. The maximum offset frequency is also reduced. The Eq. (22.47) corresponds to the value directly readable from the measurement curve

$$L\left(f_m\right) = 10\, log\left(\frac{P_N\left(f_m\right)}{P_C}\right) \quad \left(\frac{\mathrm{dBc}}{\mathrm{Hz}}\right) \tag{22.48}$$

Resonator as Frequency Discriminator

Instead of the delay line a resonator with high quality can be used. In the microwave range these are line or cavity resonators [9]. An equivalent circuit using the quality Q has the impedance

$$Z = R\left(1 + jQ\frac{2\Delta f}{f_0}\right) \tag{22.49}$$

The phase angle is

$$\phi = arctan\left(\frac{2Q}{f_0}\Delta f\right) \tag{22.50}$$

Its sensitivity to frequency change is derived from the derivative

$$\frac{d}{dx} arctan(Ax) = \frac{A}{1 + (Ax)^2} \tag{22.51}$$

In our case

$$\frac{d\phi}{df} = \frac{2Q}{f_0 + \frac{(2Q\Delta f)^2}{f_0}} \tag{22.52}$$

Near the resonance frequency, i.e. at a small offset frequency Δf in (22.52), the term $(2Q\Delta f)^2/f_0$ is very small and the resonator acts as a linear frequency discriminator. It converts a frequency change Δf proportionally into a phase change.

$$\Delta\phi = \frac{2Q\Delta f}{f_0} \tag{22.53}$$

The higher the quality of the resonator, the more effective this is. With the restriction that at higher Δf the denominator of (22.52) can become effective. In quadrature, the signal at the spectrum analyzer is

$$V_O(t) = 2QK_D G\frac{f_m}{f_0} \tag{22.54}$$

We measure the spectral power density

$$S_{\Delta f}\left(f_m\right) = \left(\frac{V_O(f)f_0}{2K_D QG}\right)^2 \tag{22.55}$$

The spectral power density of the phase fluctuations results in $S_\phi\left(f_m\right) = {S_{\Delta f}(f_m)}/{f_m^2}$ and thus the single sideband phase noise to

$$L\left(f_m\right) = \frac{1}{2}\left(\frac{f_0}{f_m}\right)^2 \frac{S_{\Delta f}\left(f_m\right)}{\left(2K_D QG\right)^2} \tag{22.56}$$

The advantage of the delay line, that one can measure relatively strongly drifting oscillators, is lost here. Frequency of the DUT and resonance frequency of the resonator must agree, so that (22.53) is valid. The calibration can be done by a small change of the input frequency. This causes a voltage change at the output. The quotient is the sensitivity constant of the system (Figure 22.18).

Detection Limit

As everywhere, thermal noise determines the detection limit of the measurement. The measured variable $L(f_m)$ with the unit dBc/Hz describes the quotient of noise power at the offset frequency f_m and the carrier power. In logarithmic scale we have

$$L\left(f_m\right) = N_P(\text{dBm/Hz}) - P_S(\text{dBm}) \tag{22.57}$$

The thermal noise power in $B = 1$ Hz bandwidth at room temperature is known to be

$$N_0 = 10\,log\left(kTB \times 10^3\right) = -174\,\text{dBm/Hz}.$$

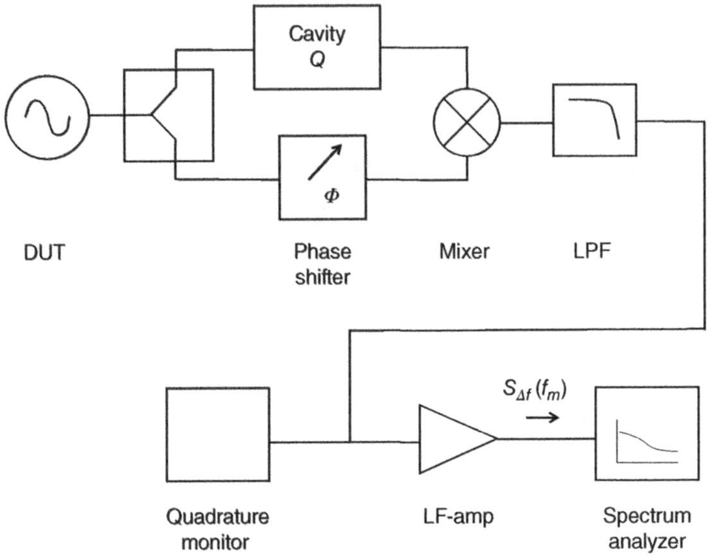

Figure 22.18 Setup with resonator as frequency discriminator.

Table 22.2 The effect of thermal noise.

Signal power P_S (dBm)	+10	0	−10	−20
Detection limit L (dBc/Hz)	−187	−177	−167	−157

This noise power generates phase and amplitude noise in equal parts. The minimum phase noise power is therefore

$$N_P = N_0 - 3\text{ dB} = -177\text{ dBm/Hz}.$$

The best sensitivity is therefore at high carrier power P_S. If this is not given, we can use an additional amplifier. This, however, adds a portion of residual phase noise. Table 22.2 gives some examples for the thermal limit of the phase noise measurement.

Comparison of Measurement Systems

Table 22.3 and Figure 22.19 give an overview.

In Figure 22.19 the curves of the inherent noise 1 (spectrum analyzer), 2 (100 ns delay line), and 3 (phase detector with high quality reference source) are plotted versus the offset frequency f_m. They mark the limit sensitivity of the methods in the

Table 22.3 Phase noise measurement systems.

	Spectrum analyzer	Phase detector	Delay-line	Resonator
Expense	Low	High	Medium	Medium
Measured variable	$L(f_m)$	$S_\Phi(f_m)$	$S_{\Delta f}(f_m)$	$S_{\Delta f}(f_m)$
Sensitivity	−−	+++	++	++
Drift	−	−−−	++	−−
Reference source	No	Yes	No	No
AM contribution	Yes	No	No	No

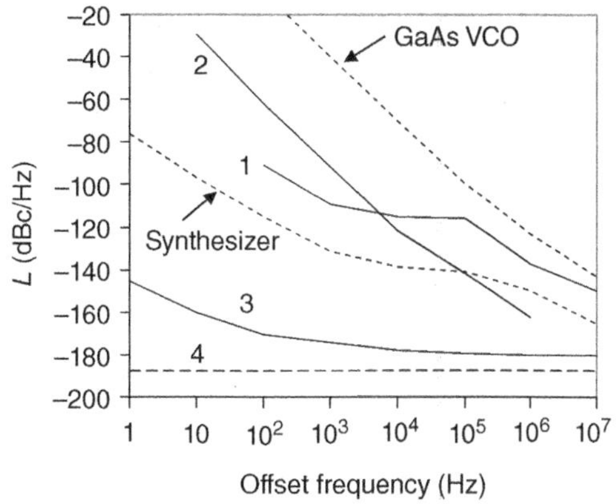

Figure 22.19 Detection limit of the methods [1]: 1 Spectrum analyzer; 2 Delay line; 3 Reference generator. Dotted line: Two typical sources. Synthesizer R&S SMA @1 GHz $P_S = 10$ dBm. GaAs-VCO [10] Figure 9.13, 2 GHz free running MMIC, $P_S = 2$ dBm. Source: Based on Refs [1, 10].

range around $f_0 \approx 1$ GHz. If the phase noise to be measured is below this curve, this method cannot be used. The detection limit due to thermal noise is shown by the dashed straight line 4 according to Table 22.2. As an example, two typical oscillators are shown in dashed lines: The free-running GaAs-FET VCO in MMIC technique can be measured with all methods with regard to noise. However, in case of strong drift the delay line is recommended. The modern synthesizer has such low phase noise that it can only be measured with the phase detector method using a very good reference source.

Cross-Correlation Technique

The most sensitive method with phase detector and reference source reaches its limits when the reference source is noisier than the measured object or when the level P_S is very low overall. This can be the case, for example, with downmixed millimeter-wave oscillators. As already discussed in Chapter 18 for the LF-range, the two-channel method with subsequent cross-correlation offers a way out here. Let us first consider the noise contributions in the single-channel method [11, 12].

$$S_\phi(f) = S_{\phi,DUT}(f) + S_{\phi,REF}(f) + \frac{v_N^2(f)}{K_D^2 B} + \{S_{A,DUT}(f) + S_{A,REF}(f)\}\,\beta^2 \quad (22.58)$$

It means: $S_{\phi,DUT}(f)$ phase noise of the DUT; $S_{\phi,REF}(f)$ phase noise of the reference source; $v_N^2(f)$ noise contribution of mixer, amplifier, and spectrum analyzer; $S_{A,DUT}(f)$, $S_{A,REF}(f)$ amplitude noise of DUT and reference source; β suppression of amplitude noise in the mixer.

The last two terms stand for the system's own noise. This can only be influenced by optimizing the components.

A two-channel cross correlation system is shown in Figure 22.20. The signal from the DUT is divided into two equal parts. Both parts pass through identically designed phase detector systems, each with a high quality reference source. These are the channels Chapters 1 and 2. The noise terms in the two channels correspond to (22.58).

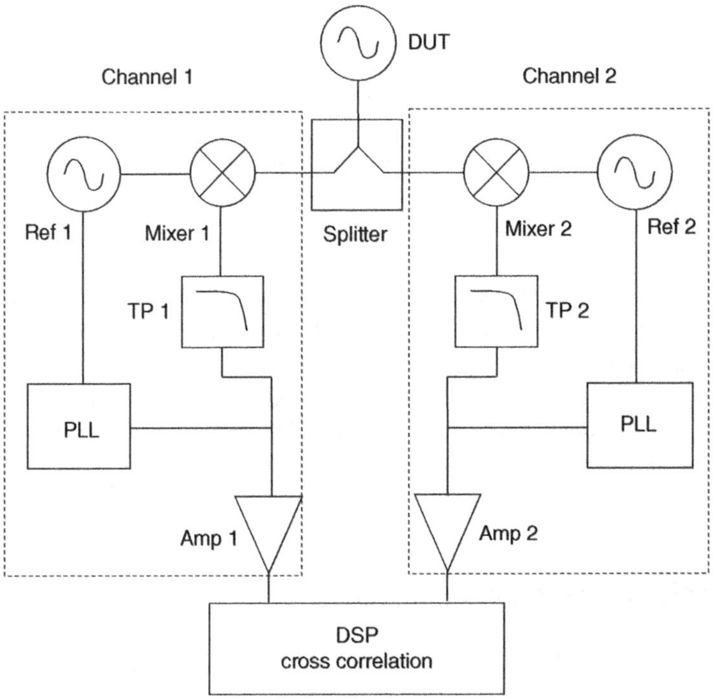

Figure 22.20 Scheme of a cross correlation system for measuring phase noise.

These two signals are compared in the DSP Cross Correlator. They contain the portion of the DUT that is identical in both channels and the uncorrelated noise contributions of the components of the two channels, which can be much larger. The DUT portion is then hidden in the noise of the system. The cross-correlation in the time domain is performed mathematically by convolution of the two signals. The value of the resulting weighting function indicates the extent to which the value of one function in relation to the other determines this value. The uncorrelated noise makes no contribution and only the signal of the DUT becomes visible. This effect becomes stronger the more such operations M are performed. From (22.58), neglecting the terms of the amplitude noise

$$S_\phi(f) = S_{\phi,DUT}(f) + \frac{1}{\sqrt{M}}\left\{S_{\phi,REF1}(f) + S_{\phi,REF2}(f) + \frac{v_N^2(f)}{K_D^2\,B}\right\} \tag{22.59}$$

The contributions of the reference sources and the noise of the mixers and amplifiers are reduced by the factor $\sqrt{M}$ because they are uncorrelated. Quantitatively, the improvement in dB goes with 5 $log(M)$. An increase of the measurement time by a factor of 10 results in an improvement of 5 dB. This also shows the disadvantage of the method: The reduction of the detection limit requires more time.

Amplitude Noise

A noisy sinusoidal signal contains amplitude and phase noise. Amplitude noise is caused by the fluctuations of the peak voltage, and phase noise is caused by the fluctuations of the zero crossings. Amplitude noise is usually considered to be of little importance, as it is small compared to phase noise and causes little interference in communication channels. This has the following reasons: Oscillators usually work in saturation, therefore amplitude fluctuations are small. Double balanced mixers are insensitive to this and suppress it by about 20 dB. The effect of AM-PM conversion in amplifiers can be disturbing. In addition, the measurement of phase noise with a spectrum analyzer can be incorrect, as it does not separate the two components [13].

The setup for measuring the amplitude noise is comparatively simple. We need a detector with which the AM noise is converted into voltage fluctuations. This can be a low barrier Schottky diode, which is suitable for the frequency range f_0 of the carrier. Figure 22.21 shows a setup with amplifier and spectrum analyzer.

The capacitance is inserted to block the DC component at the diode output. The lower frequency limit depends on its capacitance. At least 1000 μF are recommended. The voltage at the detector output is

$$v_D(t) = K_A \frac{\Delta v(t)}{V_0} \tag{22.60}$$

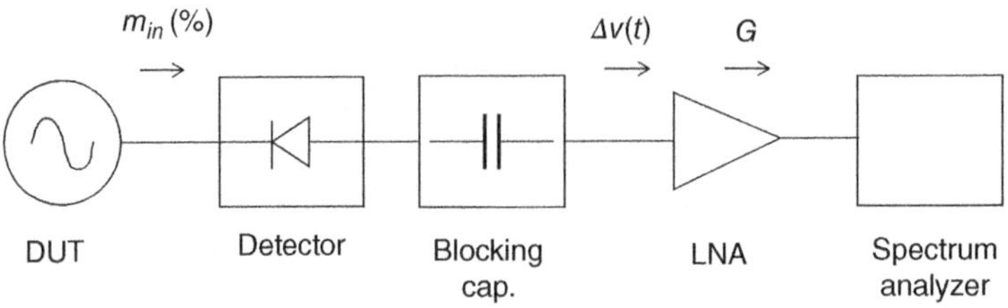

Figure 22.21 Scheme of an amplitude noise measurement setup.

At the spectrum analyzer one measures, the spectral power density

$$S_{\Delta f}(f) = K_A^2 G^2 S_A(f) \tag{22.61}$$

To obtain the desired spectrum of the amplitude noise $S_A(f)$, the calibration factor K_A must be determined. This is best done with a signal with known AM-sidebands which has the same level as the signal to be measured. The amplitude noise of the DUT corresponds to an amplitude modulation of the carrier. The detector constant K_A indicates the extent to which this amplitude modulation by noise is converted into a voltage spectrum. If a calibration signal with defined amplitude modulation is fed to the input, the voltage at the spectrum analyzer at the modulation frequency is given by the detector constant K_A and the gain of the LNA. We have an upper and a lower sideband, as shown schematically in Figure 22.22.

The relative level of the sidebands results from the theory of amplitude modulation. The modulation depth m gives the ratio of the modulation voltage v_m to the

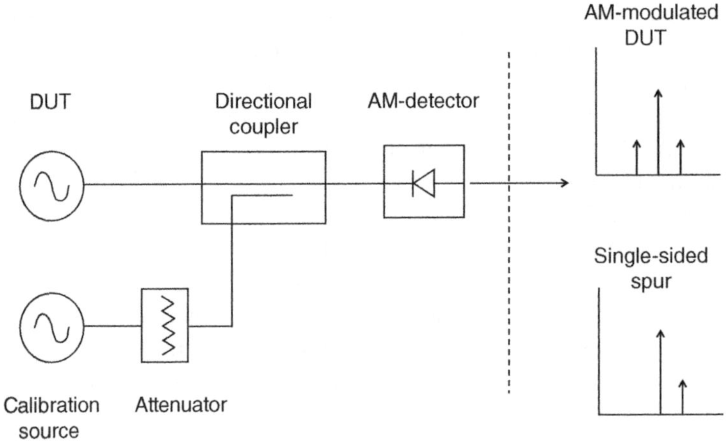

Figure 22.22 Scheme of calibration for measuring amplitude noise.

voltage of the carrier V_0

$$m = \frac{v_m}{V_0} \tag{22.62}$$

The total power of the amplitude-modulated signal is

$$P = P_0 + 2P_m \tag{22.63}$$

With the carrier power $P_0 = V_0^2/R$ and the sideband power $P_m = v_m^2/4R$. The ratio expressed in terms of the modulation factor is

$$\frac{P_m}{P} = \frac{m^2}{4 + 2m^2} \tag{22.64}$$

In logarithmic measure dBc

$$\frac{P_m}{P}\,(\text{dBc}) = 10\,log\left(\frac{m^2}{4 + 2m^2}\right) \tag{22.65}$$

Thus, a modulation factor of 1% ($m = 0.01$) results in a sideband power of −46 dBc. When calibrating we have m_{IN} (%) at the source and measure the voltage fluctuation at the modulation frequency $\Delta v(t)$ before the amplifier. The detector constant results to

$$K_A = \frac{\Delta v(t)}{m_{IN}}\left(\frac{V}{\%AM}\right) \tag{22.66}$$

With the square of this value and the gain G, the measured spectral power density $S_{\Delta f}(f)$ can be corrected to obtain the desired amplitude noise $S_A(f)$. If the DUT does not have the possibility of amplitude modulation, an additional source can be used. Its signal is fed via a directional coupler in addition to the signal from the DUT. This method is called Single Sided Spur. The second signal is shifted by about 100 kHz to 1 MHz against the DUT signal and is lower due to the losses in directional coupler and attenuator. A sideband appears on the display.

Figure 22.23 shows the phase and amplitude noise of a synthesizer [14]. At higher offset frequencies, both components can be comparable. See also Figure 20.12.

Of course, the detection limit is also given here by the inherent noise of the components. The detector, amplifier, and spectrum analyzer contribute to this. At low offset frequency it is especially the flicker noise. It is summarized in the frequency-dependent noise voltage $v_n(f)$.

$$S_A(f) = S_{A,DUT}(f) + \frac{v_n^2(f)}{K_A^2\,B} \tag{22.67}$$

The spectral power density $S_A(f)$ measured on the spectrum analyzer contains this component, which cannot be separated.

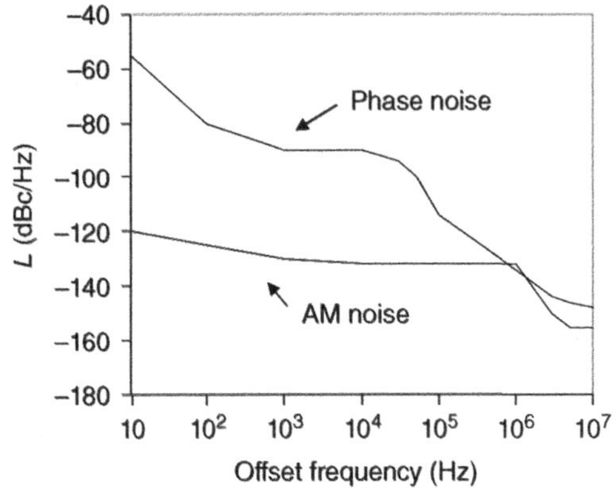

Figure 22.23 Typical levels of phase and amplitude noise of a synthesizer. Source: Based on Hewlett Packard [14].

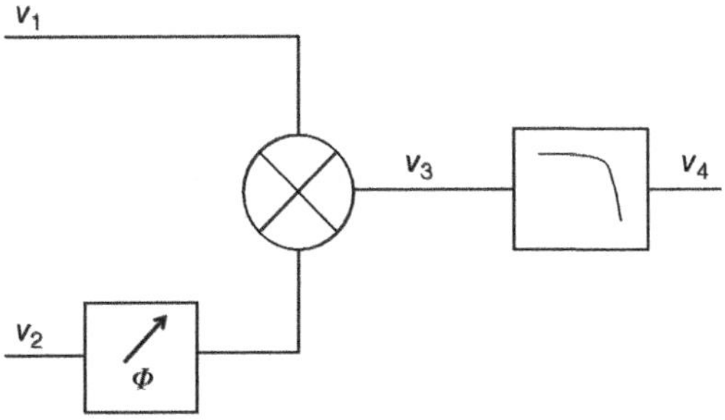

Figure 22.24 Section of the measuring setup with phase detector.

The amplitude noise can also be measured with the setup based on the mixer as a phase detector. This means that phase and amplitude noise are measured with the same equipment. The scheme is shown again in Figure 22.24.

Based on Eqs. (22.11)–(22.13), we add a term of fluctuation to the signal voltage amplitude $\varepsilon(t)$ a: $A(t) = V + \varepsilon(t)$

$$v_1(t) = A_1(t)cos\left[\omega_0 t + \Phi_1 + \Phi_1(t)\right]$$
$$v_2(t) = A_2(t)cos\left[\omega_0 t + \Phi_2 + \Phi_2(t)\right] \tag{22.68}$$

In the mixer both are multiplied with each other. This leads to

$$v_3(t) = \frac{A_1(t)A_2(t)}{2}\left\{cos\left[2\omega_0 t + \Phi_1(t) + \Phi_2(t)\right] + cos\left[\Phi_1 - \Phi_2\right]\right\} \tag{22.69}$$

Behind the low pass we have

$$v_4(t) = \frac{A_1(t)A_2(t)}{2}cos[\Delta\Phi + \Delta\Phi(t)] \tag{22.70}$$

With $\Delta\Phi = \Phi_1 - \Phi_2$ and $\Delta\Phi(t) = \Phi_1(t) - \Phi_2(t)$.

Depending on the phase setting between the two inputs of the mixer, the phase noise is measured: $\Delta\Phi = \pi/2$ (quadrature) or the amplitude noise at $\Delta\Phi = 0$.

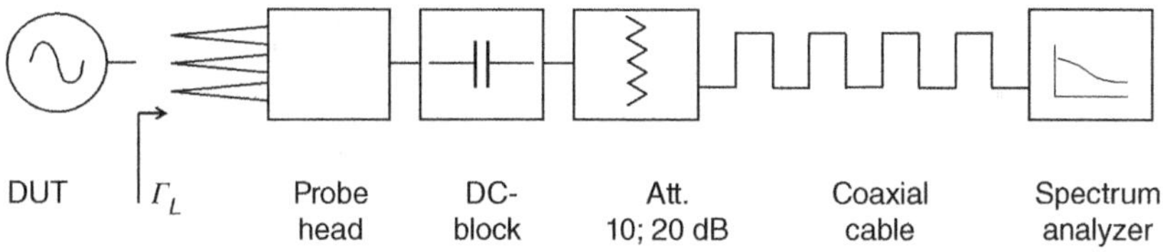

Figure 22.25 Schematic of an on-wafer measurement of an MMIC oscillator with spectrum analyzer.

Problems with On-Wafer Measurement

In early stage of development of integrated microwave oscillators in MMIC technology, it is obvious to carry out its characterization on wafer. Oscillation frequency and phase noise can be measured with a high-quality spectrum analyzer. Figure 22.25 shows a suitable setup.

A flexible cable is used to bridge the unavoidable distance between wafer prober and spectrum analyzer. Under certain circumstances this can lead to considerable errors in phase noise measurement. At the end of the cable there is no ideal termination, because the input of the spectrum analyzer is slightly mismatched. The cable acts as a resonator with relatively high quality. With longer cables, the resonance frequencies are very close together. If the DUT is the pure oscillator without buffer amplifier or other decoupling, this external resonator influences the oscillation condition and leads to an increased phase slope. This results in a higher quality, which in turn reduces the phase noise. Figure 22.25 shows as an example the reflection coefficient Γ_L, measured via a Thru into the measuring probe, which the DUT "sees." This fine structure can only be seen when measuring with an extremely small step size of the network analyzer.

Figure 22.26 Reflection factor at the input of the measuring tip, caused by the cable in Figure 22.25.

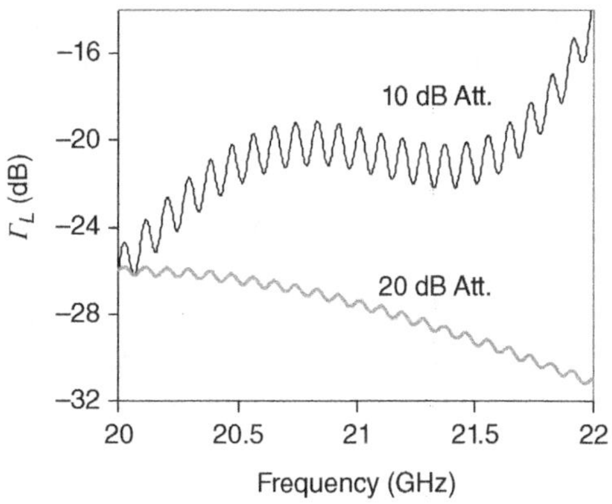

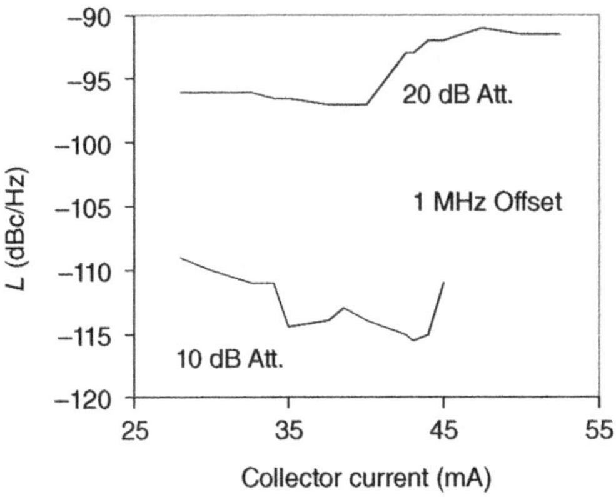

Figure 22.27 Effect of the transmission line on phase noise at 1 MHz offset. Decoupling of 10 dB is not sufficient and leads to errors up to 20 dB.

It is necessary to insert an attenuator between measuring probe and cable. Figure 22.26 shows that only 20 dB lead to a sufficient attenuation of the disturbing resonances. Example of a measurement of a 20 GHz HBT oscillator in MMIC technology without buffer amplifier at $f_m = 1$ MHz offset is shown in Figure 22.27.

Even with the 10 dB attenuator, the line resonator is still effective and feigns a low phase noise, which is sometimes 20 dB better than the real value. This can only be measured at about 20 dB attenuation.

Residual Phase Noise

In a high frequency system, the signal generated by the oscillator usually passes through a number of components. These all have noise characteristics and reduce the quality of the original signal. This effect is quantitatively described by the residual phase noise, also called two-port phase noise. These components in a system are, e.g. amplifiers, frequency dividers, frequency multipliers, filters, and mixers or also SAW delay lines. It is also possible that amplifiers are driven into the non-linear range, in which case the small-signal noise figure NF is no longer valid. Flicker and shot noise are added. Insufficient screening of the DC supply voltage can also contribute to residual phase noise. In general, there are additive and multiplicative contributions. Additive contributions of two-port noise near the carrier superimpose linearly. Multiplicative contributions result from the non-linear properties of the two-port, e.g. from AM-PM coupling. The combination of the measurement of oscillator phase noise of the source and two-port phase noise of the components provides the information about the dominant noise sources of the whole system [15, 16]. A measurement setup is depicted in Figures 22.28 and 22.29.

If the DUT is frequency-converting, i.e. the output frequency is different from the input frequency, two identical DUTs must be available. This is, for example,

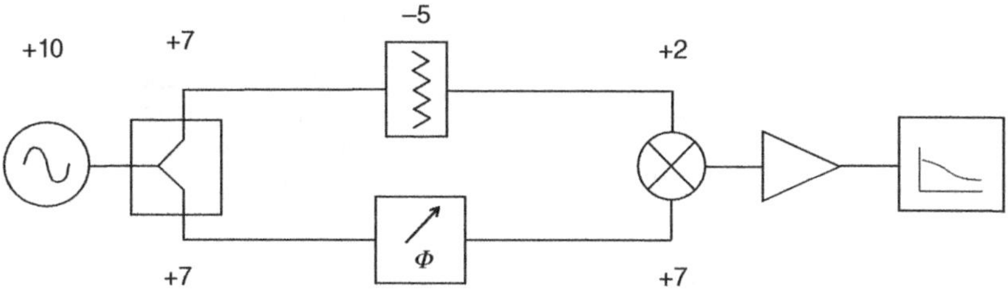

Figure 22.28 Calibrating the system Figure 22.29 with typical levels (dBm). Source: Based on Breitbarth and Koebel [15].

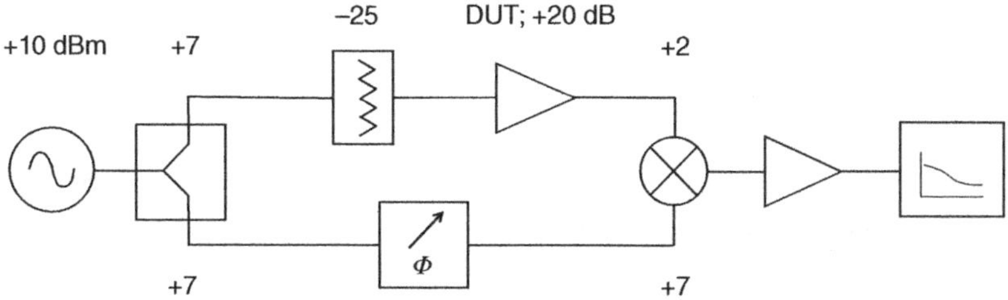

Figure 22.29 Measurement of the residual phase noise of a 20 dB amplifier at the same levels as with calibration.

the case with frequency dividers, multipliers, mixers, or PLL synthesizers. The configuration is shown in Figure 22.30.

The residual phase noise measurement is based on the following consideration. The signal of the oscillator is divided into both branches. One acts as local oscillator at the mixer, the other as signal. Both branches must have the same electrical length. Then the noise of the source is correlated at the two inputs and cancels itself out. Only the uncorrelated noise of the DUT remains. Therefore it is important to take care that the phase noise of the source is not decorrelated by different transit times. In the case of two DUTs, (Figure 22.30), the two noise levels add up in the display because they are uncorrelated. The measured value is then 3 dB

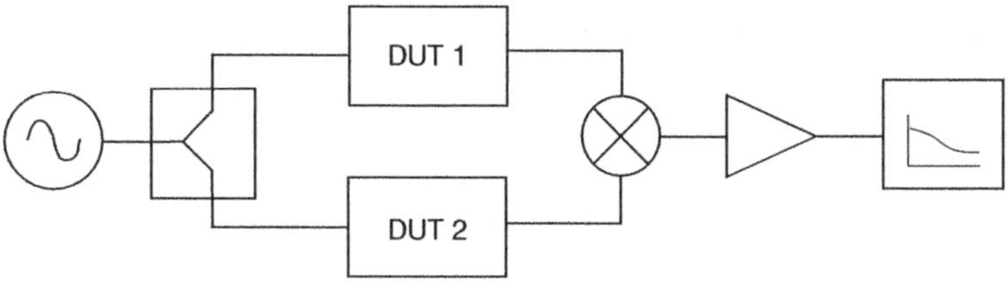

Figure 22.30 Setup for measuring frequency-converting objects. Two identical models are required.

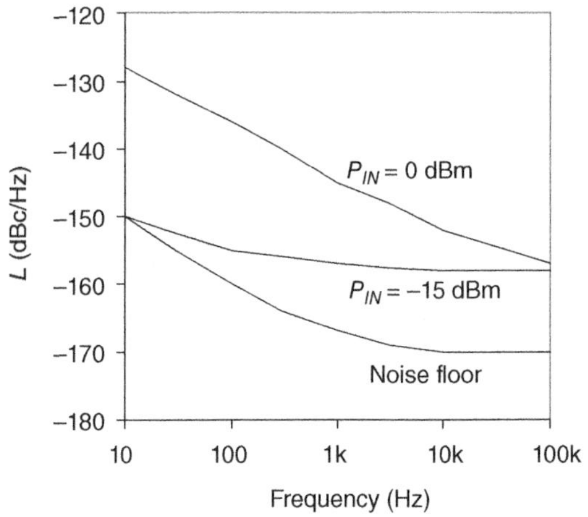

Figure 22.31 Residual phase noise of an amplifier. Plot −15 dBm: Linear operation. Plot 0 dBm: Nonlinear operation. Source: Based on Breitbarth and Koebel [15].

too high. The calibration of the system is done in the same way as the calibration of the phase detector described earlier.

An example of measuring an amplifier with a noise figure $NF = 4\,\text{dB}$ is shown in Figure 22.31. A $f_0 = 100\,\text{MHz}$ signal is fed in at two different levels. With $P_1 = -15\,\text{dBm}$ it is in small-signal linear range. With $P_2 = 0\,\text{dBm}$ it is in 1 dB compression.

In linear operation, the phase noise level beyond the flicker limit corresponds approximately to the noise figure $NF = 4\,\text{dB}$. In non-linear operation a much higher flicker noise with a significantly increased cut-off frequency is observed. This example shows the usefulness of these measurements for evaluating the noise contributions of individual components in their effect in the entire system, especially in the immediate vicinity of the carrier.

References

1 Agilent Technologies and Gheen, K. (2012). Phase Noise Measurement Methods and Techniques.

2 Rohde&Schwarz (2017). FSWP Phase Noise Analyzer and VCO Tester. Product Brochure.

3 Rauscher, C. (2007). *Grundlagen der Spektrumanalyse*. München: Rohde & Schwarz.

4 Keysight Technologies (2017). 35670A Dynamic Signal Analyzer. Datasheet.

5 Agilent Technologies (2007). Phase Noise Characterization of Microwave Oscillators. Phase Detector Methods. Product Note 11729B-1.

6 Kester, W. 2009. Fundamentals of Phase Locked Loops. Analog Devices Tutorial MT-086.

7 Maxim Integrated (2010). White Noise Generator has no 1/f-Component. Application Note 4527.

8 Schiebold, C. (1983). Theory and design of the delay line discriminator for phase noise measurements. *Microwave J.* 26: 103–112.

9 Rizzi, A. (1988). *Microwave Engineering*. Prentice Hall International.

10 Rohde, U.L., Poddar, A.K., and Böck, G. (2005). *The Design of Modern Microwave Oscillators for Wireless Applications*. Wiley Interscience.

11 Breitbarth, J. (2011). Cross Correlation in Phase Noise Analysis. *Microwave J.* 54: 78–85.

12 Rubiola, E. and Giordano, V. (2000). Correlation-based phase noise measurements. *Rev. Sci. Instr.* 71: 3085–3091.

13 Faulkner, T.R. Temple, R.E. (1990). Residual phase noise and am noise measurements and techniques. RF & Microwave Measurement Symposium. Hewlett Packard.

14 Hewlett Packard (1998). Practical considerations for modern RF & microwave phase noise measurements. Seminar.

15 Breitbarth, J. and Koebel, J. (2008). Additive (residual) phase noise measurement of amplifiers, frequency dividers and frequency multipliers. *Microwave J.* 51: 66–82.

16 Brandon, D. and Cavey, J.2008 The Residual Phase Noise Measurement. Analog Devices Application Note AN-0982.

Appendix

Noise Signals and Deterministic Signals

All measurable data of a physical object are either deterministic or stochastic (random). We are concerned here with noise signals and these are random. We will occasionally use deterministic signals to explain basic relationships in a mathematical simple way. Noise signals in technical systems have two additional important properties. They are stationary and ergodic [1, 2].

Here is a brief explanation of the terms.

Deterministic signals	The variation in time is described by a mathematical relation (example: sinusoidal voltage)
Stochastic signals	No mathematical description, and therefore prediction. Only probability statements are possible about the variation in time (noise voltage)
Stationary	Characteristic values of the signal are independent of the time (resistance at constant temperature)
Ergodic	For a stationary signal, the time average is the same as the ensemble average(characteristic values of the noise voltage measured N times at a resistor are the same as measured once at N resistors)

In the practice of electronics and RF-technology, all random stationary processes can be assumed to be ergodic. Characteristic values and characteristic functions are used for description and quantitative processing.

Characteristic values	Mean values (average)
Characteristic functions	Probability density function
	Correlation function
	Power spectral density

A Guide to Noise in Microwave Circuits: Devices, Circuits, and Measurement, First Edition.
Peter Heymann and Matthias Rudolph.

The characteristic functions provide information about the dependence of the signals on a parameter. They can refer to a single signal, and then one speaks of the autocorrelation function (ACF) or the auto-spectral density. If two different signals are considered, one speaks of the cross-correlation function (CCF) or the cross-spectral density. More than two signals are also possible, but are not dealt with here.

Random Signals

Let us first consider an electrical signal $y(t)$ obtained from the observation of a time-varying physical process. It consists of a mean value, which is determined by the process, and a fluctuation component, which is usually caused by the measurement technique. The fluctuation can also result from the process itself. Figure A.1a shows a non-stationary process [3].

In Figure A.1b, mean value and fluctuation component are separated. Technically, this can be done by filters. The mean value appears at the output of a low-pass filter, the noise at the high-pass. In this general process, which is non-stationary, one must bear in mind that we are dealing with short-time averages, in which only a short time T is averaged. In this time T, one obtains a value

$$\mu\left(t_1\right) = \frac{1}{T}\int_{t_1}^{t_1+T} y(t)\,dt \tag{A.1}$$

The sequence for all t_1 then yields the curve of the short-term mean values. This also results in a criterion for the stationarity of a process: the short-term averages must be constant, i.e. independent of t_1.

Despite the time dependence of the short-term averages, the fluctuation component can be stationary, as can be seen in the example. Non-stationary fluctuation

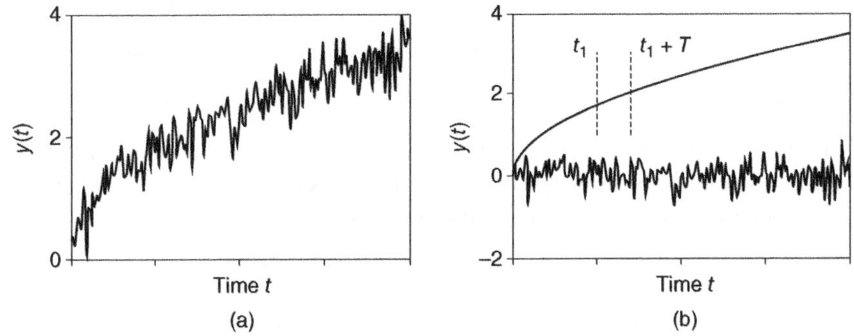

Figure A.1 (a) Measurement of a time-varying process. (b) Mean value and fluctuation are separated from each other.

processes are also possible, e.g. an exponential increase or decrease, but we are not concerned with these here. As mentioned earlier, we will deal with stationary and ergodic processes.

Characteristic Values

The simplest and most frequently used quantities of random signals are average μ, variance σ^2, and standard deviation σ. In the following, we restrict ourselves to the time dependence of a stationary random signal $y(t)$, since we assume that time average and ensemble average are the same. If we have recorded N measurement points $y_i(t)$ at constant time intervals (Figure A.2), the mean value is

$$\mu_y = \frac{1}{N}\sum_{i=1}^{N} y_i\,(t) \tag{A.2}$$

and the variance

$$\sigma_y^2 = \frac{1}{N}\sum_{i=1}^{N} \left[y_i\,(t) - \mu_y\right]^2 \tag{A.3}$$

Strictly speaking, of course, (A.2) and (A.3) are only approximations. If we measure only a few points, the values thus obtained will fluctuate around the true average. Only when N is very large does (A.2) give the correct result. Mathematically correct is the limit

$$\mu_y = \lim_{N\to\infty}\frac{1}{N}\sum_{i=1}^{N} y_i\,(t) \tag{A.4}$$

In practice, however, apart from exceptions, one will always have a sufficient number of measurement points.

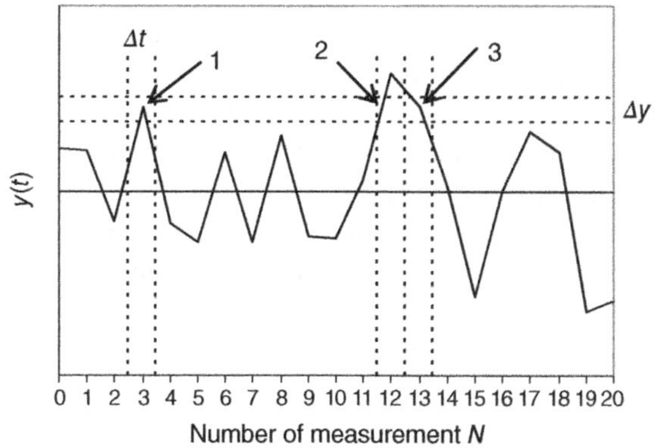

Figure A.2 A sequence of N measurement points of a noise voltage scatters around a mean value.

A more general definition results with the probability density function $p(y)$ (see the following text). The characteristic values are moments of the probability density.

$$\mu_y^{(\nu)} = \int_{-\infty}^{\infty} y^\nu p(y)\,dy \tag{A.5}$$

$\nu = 1$ is the linear mean value (A.1) and $\nu = 2$ is the quadratic mean value.

$$\sigma_y^2 = \mu_y^{(2)} - \left[\mu_y^{(1)}\right]^2 \tag{A.6}$$

The root mean square is a measure of the scatter of the values and thus of the amplitude of the variation. The variance σ^2 is equal to the root mean square if the linear average is zero or has been previously subtracted. If $y(t)$ is an electrical voltage, then σ^2 is the noise power at an $1\,\Omega$ resistor and σ is the rms value of this noise voltage.

Random signals, however, contain more information than is given by the mean values. The signals are described more precisely by the characteristic functions. Characteristic functions are time averages depending on a single variable. This can be, for example, the value of the function itself (probability density function), a time delay (correlation function), or the frequency (power spectral density).

The Probability Density Function

For each signal $y(t)$, the values are distributed in a characteristic way between the minimum value and the maximum value. The function value is a parameter of the probability density function (y). It indicates the probability that a signal value $y(t)$ lies in an interval $y < y(t) < y + \Delta y$ at any given time t. Let us consider a series of measurement points of a noise source measured with an AD converter with a sampling rate Δt. According to the resolution of the ADC, it measures a function value within the interval Δy. This is shown in Figure A.2 [4].

The total number of measurements is N. In the selected y-interval of width Δy we see a number of "hits." In our example there are 3. We denote this number of "hits" by ΔN. According to the definition of probability, this is the number of favorable cases. The number of possible cases is N. Of course, we still have to consider the width of the interval Δy. As the interval increases, the probability of encountering a value in this Δy becomes greater and greater and the resolution decreases. The probability density distribution of the function values is thus

$$p(y) = \lim_{\Delta y \to 0} \frac{\Delta N}{N \Delta y} \tag{A.7}$$

For deterministic signals, the probability density can be calculated analytically, because its time dependence is mathematically given. With stochastic signals, the future values are not known. Therefore, a histogram analysis must be performed with the measured signals. On the other hand, we know that noise signals of thermal origin or from the shot effect have a Gaussian distribution of the voltages.

Example Sine Function

We calculate the probability density of a sine signal as an example. A period is shown in Figure A.3. During a period of length T, there are instants of time 1 and 2 where the function value $y(t)$ during the time interval Δt is in the range Δy. According to definition (A.7)

$$p(y) = 2\frac{\Delta t}{T\Delta x} \tag{A.8}$$

${}^{\Delta t}/_{\Delta x}$ we obtain from the derivative of the sine function

$$y(t) = y_0 \sin(2\pi ft); \quad \frac{dy}{dt} = 2\pi f y_0 \cos(2\pi ft) \tag{A.9}$$

The reciprocal is

$$\frac{dt}{dy} = \frac{T}{2\pi y_0 \cos(2\pi ft)} \tag{A.10}$$

With $\cos(x) = \sqrt{1 - \sin^2(x)}$ we get

$$\frac{dt}{dy} = \frac{T}{2\pi y_0 \sqrt{1 - \left(\frac{y(t)}{y_0}\right)^2}} \tag{A.11}$$

Since we consider one period, the time dependence does not matter and we obtain for the distribution of the function values (A.7) of the sine function:

$$p(y) = \frac{1}{\pi y_0 \sqrt{1 - \left(\frac{y}{y_0}\right)^2}} \tag{A.12}$$

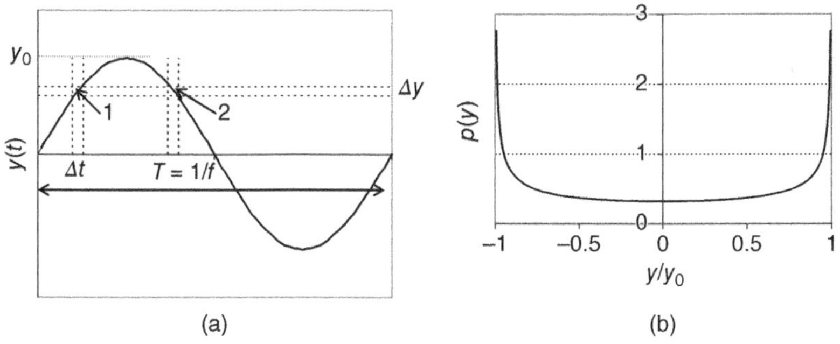

Figure A.3 (a) Period of a sine signal. (b) Probability density $p(y)$ of the sine function according to (A.12).

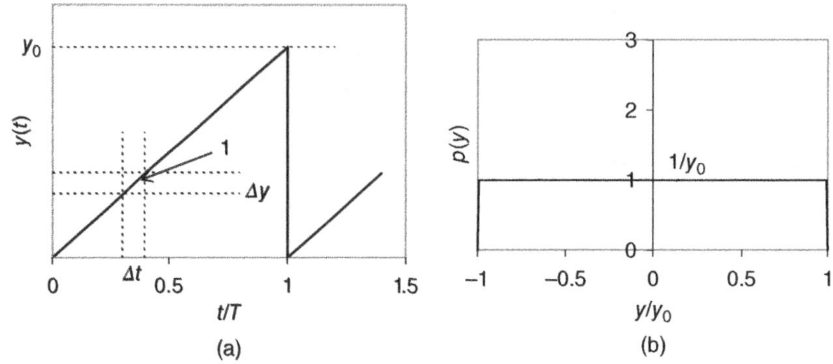

Figure A.4 (a) Period of a sawtooth curve (A.13). (b) Probability density $p(y)$ of the sawtooth curve according to (A.15).

This probability density distribution of the sine function is shown in Figure A.3b.

Example Sawtooth Voltage

The sawtooth is also a deterministic signal. An example with the period T and the peak value y_0 is shown in Figure A.4a. During a period, a certain function value occurs only once. The time variation is given by

$$y(t) = y_0 \frac{t}{T}; \quad \frac{dy(t)}{dt} = \frac{y_0}{T} \tag{A.13}$$

The reciprocal of the time derivative is constant.

$$\frac{dt}{dy} = \frac{T}{y_0} \tag{A.14}$$

Thus the probability density distribution also becomes constant, i.e. all function values of the sawtooth function occur with the same occurrence $^1/_{y_0}$.

$$p(y) = \frac{1}{T}\frac{T}{y_0} = \frac{1}{y_0} \tag{A.15}$$

The plot is shown in Figure A.4b.

Example White Noise

A noise voltage, e.g. from an ohmic resistor, is caused by the thermal fluctuation of electrons. An elementary process is the free flight of an electron between two collisions. The number of electrons involved is very high and the contribution of a single elementary process is very small. The generation of the noise voltage is therefore the ideal case of a stochastic process. According to the central

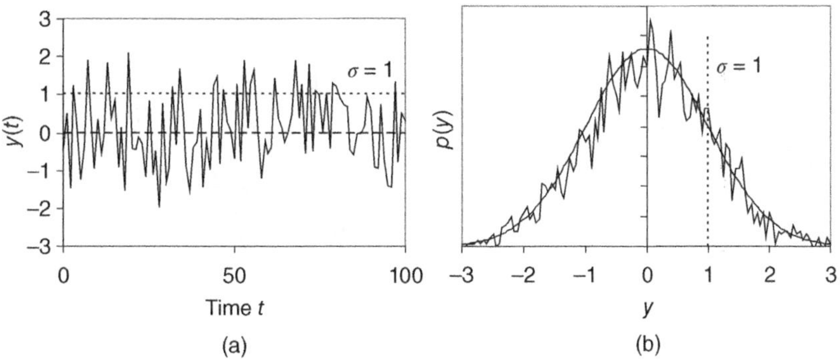

Figure A.5 (a) White noise in the time domain ($\mu_y = 0, \sigma = 1$). (b) Probability density distribution with plotted Gaussian curve.

limit theorem of statistics, the probability density has exactly a Gaussian distribution when a very large number of elementary processes are involved. A precondition is that each individual elementary process makes only a small contribution. Figure A.5a shows a sequence of numbers with a Gaussian distribution, which was generated with a mathematics program. It corresponds to a noise voltage with the time average (direct current power) $\mu_y = 0$. The effective value (standard deviation) is $\sigma = 1$, the alternating current power (variance) is also $\sigma^2 = 1$. A histogram analysis of this number sequence results in the Gaussian distribution shown in Figure A.5b. For comparison, an analytically calculated bell shaped curve.

The probability density function $p(y)$ is thus completely determined by the linear average (here = 0) and by the standard deviation.

Example Sinusoidal Signal with Noise

A signal with superimposed noise is the normal case in a communication system or measuring device. Characteristic probability densities occur, which lie between Figure A.3 (pure sine) and Figure A.5 (pure noise). We first consider the superposition of a sinusoidal and a broadband noise voltage. With a low noise signal, the probability density is still clearly characterized by the curve (A.12). The histogram in Figure A.6b is a characteristic for this signal. With stronger noise as in Figure A.7, the probability density quickly loses its significance. While the signal is still clearly identifiable in the time domain (Figure A.7a), the probability density (Figure A.7b) is only noisy. This method it is not suitable for finding a periodicity.

Example Narrowband Noise

A similar situation occurs when analyzing narrowband noise. This is the case, for example, when white noise has passed through a band-limiting filter. When

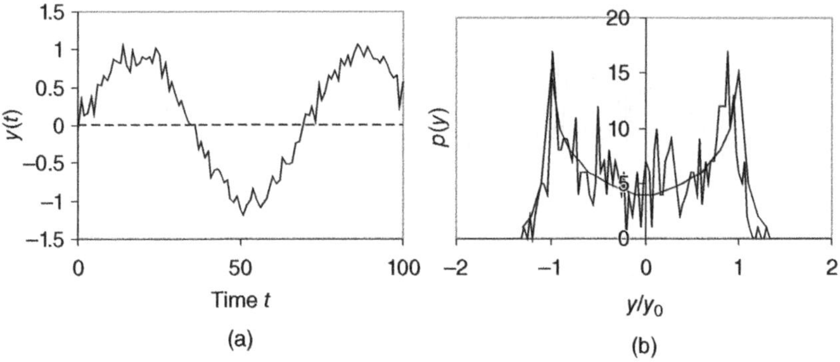

Figure A.6 (a) Weakly noisy sine in the time domain. (b) Corresponding probability density.

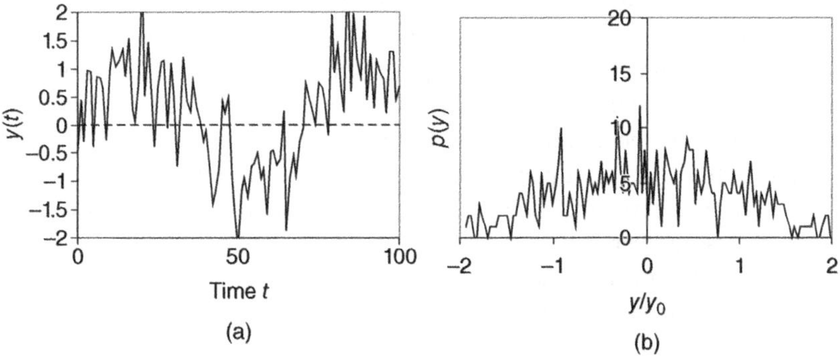

Figure A.7 (a) Highly noisy sine in the time domain. (b) Corresponding probability density.

passing through a filter, the Gaussian distribution of the scalar voltage values does not change. However, a blurred sinusoidal oscillation with the center frequency f_0 of the filter appears from the white noise, because only noise frequencies near f_0 can pass. This is the case, for example, in a spectrum analyzer. A phasor rotating with f_0 now arises, whose real and imaginary parts are noise voltages with Gaussian distribution. Here we write for a noise signal after a narrow band filter

$$y(t) = v_I(t) \cos\left(\frac{2\pi}{T}t\right) + v_Q(t) \sin\left(\frac{2\pi}{T}t\right) \tag{A.16}$$

$v_I(t)$, $v_Q(t)$ are the fluctuating parts of the real and imaginary parts (I/Q components) of the phasor rotating with the angular frequency $2\pi f_0 = {}^{2\pi}/_{T}$. Both are uncorrelated and have a Gaussian distribution of the amplitudes. By detection and data processing in the spectrum analyzer, the magnitude of this phasor is measured. A law of statistics states: If the components of a two-dimensional random

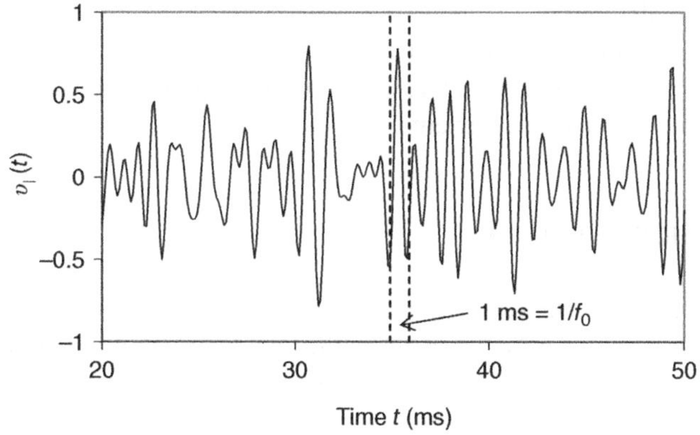

Figure A.8 Original white noise after passing a bandpass filter with center frequency f_0.

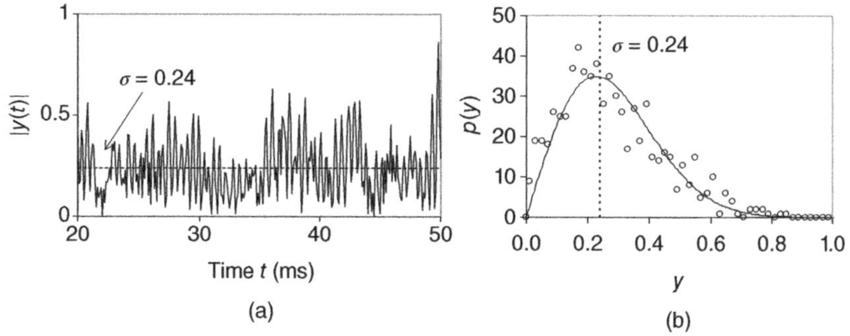

Figure A.9 (a) Magnitude of narrowband noise in the time domain. (b) Corresponding probability density with Rayleigh distribution.

vector are Gaussian distributed and statistically independent, then the magnitude is Rayleigh distributed. Figure A.8 shows our calculation example for $f_0 = 1$ kHz and a noise signal with $\sigma = 0.24$. The I-component of the phasor (A.16) is shown in Figure A.8. Its mean value is zero. One can find out the oscillation with frequency $^1/_T$, whose envelope fluctuates irregularly. The occurrence of the periodicity with T is understood from the analysis of the ACF. In Figure A.9a, the variation in time of the magnitude of (A.16) is plotted. The probability density (Figure A.9b) was calculated with a histogram function. It corresponds to a Rayleigh distribution. A spectrum analyzer averages over this distribution. Therefore, this relation must be taken into account in the quantitative evaluation (see Chapter 13).

When analyzing the probability density of deterministic and stochastic signals, we saw that this quantity is only conditionally suitable to discover and separate signals from each other. Purely deterministic and purely stochastic signals can be discovered very well, deterministic signals of different frequencies not. White noise and low-pass noise both have Gaussian distributions. A special feature is the magnitude of narrowband noise. A sinusoidal oscillation with superimposed noise can also only be discovered to a limited extent.

The Autocorrelation Function

The ACF is better suited to distinguish signal types than the probability density function. It characterizes the internal relationship of the signal in the time domain. Parameter of the ACF is a time shift. It is an extended averaging. This consists of forming not only the product of the function values $y(t)$ with itself, but also the function values $y(t+\tau)$ that are offset in time. If this time offset increases from $\tau = 0$, one obtains the ACF $\rho(\tau)$ dependent on τ. With N discrete measured values as in Figure A.2, one uses [5]

$$\rho_k = \frac{1}{N-k} \sum_{n=1}^{N-k} y_n y_{n+k} \tag{A.17}$$

Here the index k stands for the time shift.

In the case of a continuous-time function $y(t)$, the definition of the ACF applies

$$\rho(\tau) = \overline{y(t)\,y(t+\tau)} = \lim_{\Delta t \to \infty} \frac{1}{\Delta t} \int_0^{\Delta t} y(t)\,y(t+\tau)\,dt \tag{A.18}$$

For k or $=0$, we obtain the root mean square, i.e. the variance. Let us first look at simple examples again.

Example Sine

Here we draw the sine function with amplitude 1 also for negative time values. This makes it easier to see the properties of the ACF (Figure A.10).

The ACF is a real, even function $\rho(-\tau) = \rho(\tau)$. The maximum value of the ACF is at $= 0$. It is equal to the variance of the signal $\rho(0) = \sigma^2$.

The ACF of a periodic signal has the same time-period T as the signal itself.

The phase information is lost during the transition to the ACF.

Example Sawtooth

To calculate the ACF over several periods according to (A.17), we start from the Fourier series of the sawtooth function in the time domain with the peak value 1.

$$y(t) = \frac{1}{2} - \frac{1}{\pi} \sum_{n=1}^{12} \frac{1}{n} sin\left(n \frac{2\pi}{T} t\right) \tag{A.19}$$

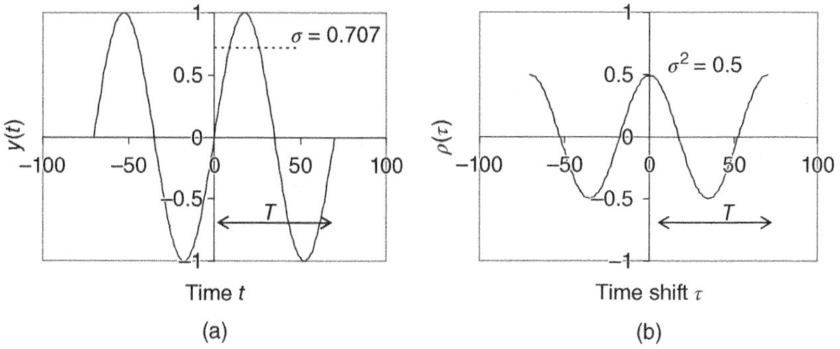

Figure A.10 (a) Sine function in the time domain. (b) Corresponding ACF.

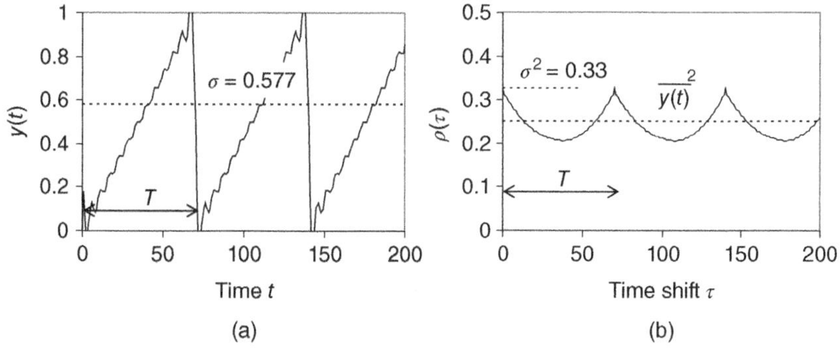

Figure A.11 (a) Sawtooth calculated according to (A.19) with 12 harmonics. (b) Corresponding ACF according to (A.17).

The summation over 12 harmonics leads to a good approximation of the sawtooth plot in Figure A.11. The ACF is calculated with (A.17).

We see here another property of the ACF. If the mean value of the time function is $\overline{y(t)} \neq 0$, the values of the ACF increase by $\overline{y(t)}^2$. If the voltage is DC with superimposed AC voltage, the ACF is increased by the DC power. In our example, $\overline{y(t)} = 0.5; \overline{y(t)}^2 = 0.25$. For the sawtooth voltage, the rms value is $= y_{max}/\sqrt{3}$, in our case $\sigma = 0.577$. For the variance, i.e. the maximum value of the ACF, we have $\sigma^2 = 0.33$.

Example Noisy Sine

The two examples of sine and sawtooth are deterministic functions and do not show the advantages of the ACF, as their time dependence can be calculated. No mathematical prediction is possible for noisy signals. One has to analyze a

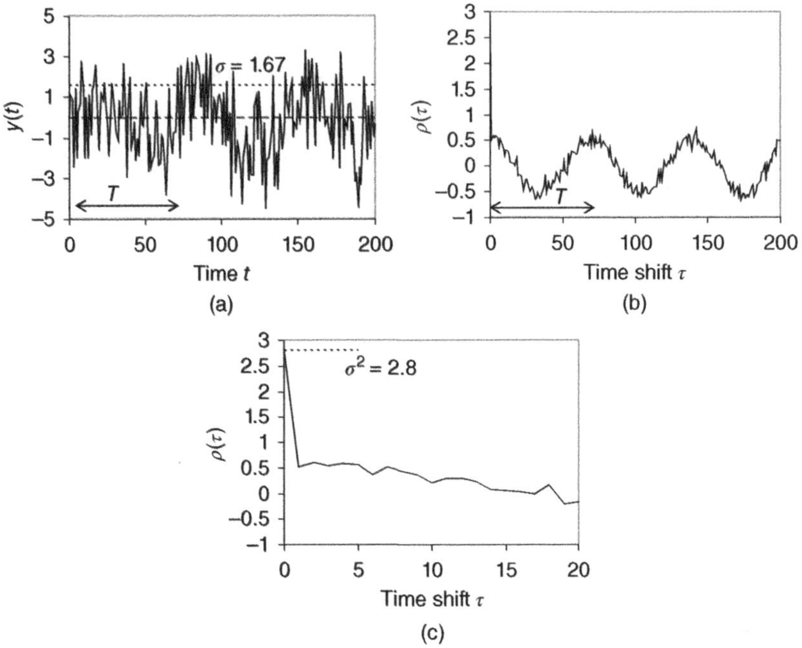

Figure A.12 (a) Noisy sinusoidal signal in the time domain. (b) ACF of the same. (c) ACF at small time offset, at $\tau = 0$ we have the variance σ^2.

measurement, i.e. the already existing signals. With a sinusoidal signal superimposed by a weak noise component, one can perform a histogram analysis of the probability density function $p(y)$. At best (Figure A.6) one recovers the sine in the weak noise. However, this can be seen much better in the time domain, e.g. with an oscilloscope. With stronger noise components (Figure A.7), nothing can be seen in the probability density function. This is where the advantages of the ACF become apparent. In Figure A.12, we look at a very noisy signal whose periodic component can only be seen with difficulty in the time domain.

The ACF clearly shows the periodic structure of the very noisy signal (Figure A.12b). In the detail at small offset (Figure A.12c), we see the value of the variance at =0.

Example White Noise

In the case of white noise, there is no correlation of the function values at different points in time. Here, the immediate decay of the ACF to zero when $\tau > 0$ shows that white noise is present within the scope of observation. Figure A.13 shows an

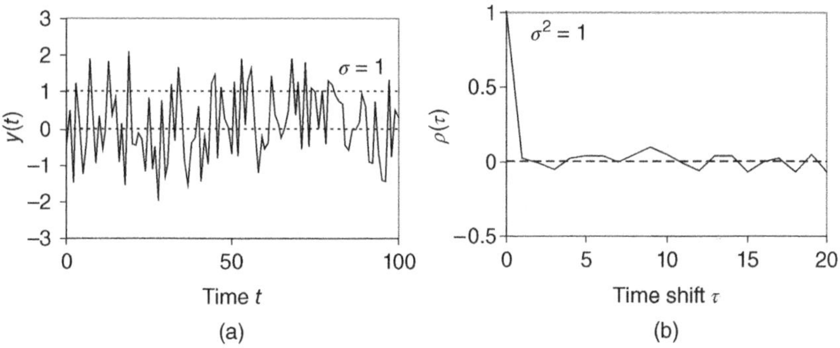

Figure A.13 (a) White noise in the time domain. (b) The ACF is $\rho(\tau) = 0$ for all $\tau > 0$.

example of the ACF calculated with (A.17) from a normally distributed sequence of numbers.

Example Low-Pass Noise

Low-pass noise occurs when white noise passes a low-pass filter. Beyond the cut-off frequency, the noise level disappears. With a lossless passive low-pass filter, the level of the white noise at the input within the passband is kept. The probability density function, i.e. the Gaussian distribution of the amplitudes, remains unchanged. So far, we have generated the example noise signals from a sequence of random numbers to which we have assigned a time axis. This method is not suitable for band-limited signals, because a frequency dependence cannot be simulated easily. Here, a sum of harmonic oscillations with statistically distributed frequencies and phases is more suitable [6]. The frequency dependence of the amplitude at the output is calculated according to the transfer function of the network.

Let us first consider the synthesis of a noise signal from the sum of harmonic signals with statistical frequency distribution within the transfer bandwidth of the network $H(f)$. As an example, we use 15 frequencies f_n, which can be generated by equally distributed random numbers. In the case of a low-pass filter, they lie between $0 < f < f_g$. Similarly, the phase angles φ can be generated by random numbers. By choosing the frequency limits and the corresponding transfer function $H(f)$, one can simulate band-limited noise.

$$y(t) = \sum_{n=1}^{15} H(f) \cos\left(2\pi f_n t + \varphi_n\right) \tag{A.20}$$

At the input of the band-limiting network there should be a white noise that is as well simulated as possible. For this we analyze (A.20) with $H(f) = 1$. The 15 frequencies are statistically distributed in the range $(200 < f_n < 1200)$ Hz. The phases

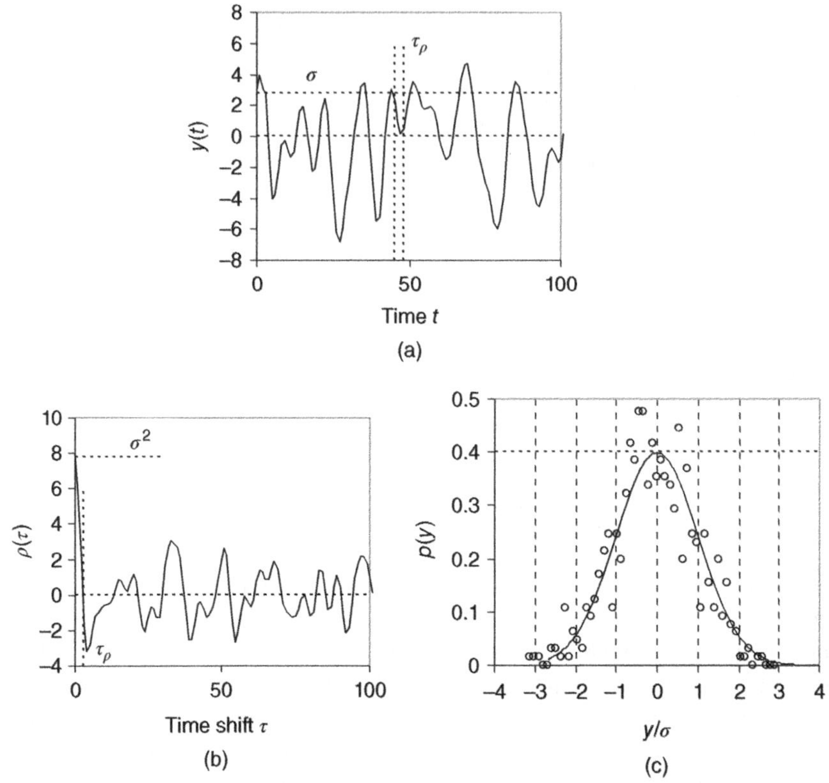

Figure A.14 Example of a noise signal generated with (A.20). (a) Time domain. (b) ACF. (c) Gaussian distribution of amplitudes.

φ_n also have a random distribution. The frequency range can be varied depending on the application.

Criteria for the good simulation of a noise signal with (A.20) in the time domain are the fast sign changes ($\tau_\rho \to 0$). That means a fast decaying ACF and the Gaussian distribution of the amplitudes. The time dependence and the characteristic functions ACF and probability density distribution in Figure A.14 show that the simulation of the noise signal is sufficient. It can be used for the analysis of the frequency dependence. For this purpose, the corresponding transfer function is introduced and the frequency range is adapted to the passband. Figure A.14a shows a small section of $y(t)$. The standard deviation σ (effective value) and the very small correlation time τ_ρ can be seen as typical values. This results from the ACF calculated with (A.17). The ACF has as maximum value the variance σ^2 (AC power) at $\tau = 0$. A histogram confirms the Gaussian distribution of the amplitudes.

Figure A.15 Transfer function of a low-pass filter according to (A.21).

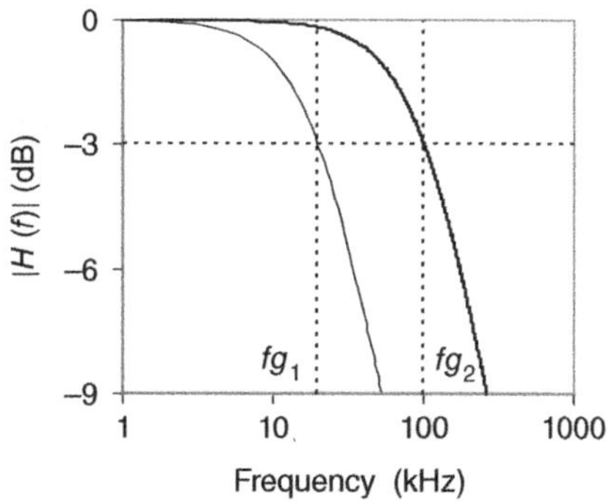

We pass a corresponding signal with an appropriate frequency distribution through a low-pass filter and look at the characteristic functions at the output. The transfer function of the amplitudes of the low pass is given by

$$|H(f)| = \frac{1}{\sqrt{1 + \frac{f^2}{f_g^2}}} \tag{A.21}$$

In Figure A.15 the function is shown for two cutoff frequencies $f_{g1} = 20$ kHz and $f_{g2} = 100$ kHz. The settling time of the ideal lowpass filter is $=1/2f_g$. Figure A.16 shows the time response of the noise voltage at the output. In the signal limited to $f_{g1} = 20$ kHz, the zero crossings can only occur rarely ($\tau_1 = 25$ μs) compared with the signal limited to $f_{g2} = 100$ kHz ($\tau_2 = 5$ μs) or in a white noise ($\tau = 0$) due to the longer settling time.

The normalized ACF for both signals is shown in Figure A.16. The theoretical curve is

$$\frac{\rho(\tau)}{\sigma^2} = exp\left(-\frac{\tau}{\tau_\rho}\right) \tag{A.22}$$

It is not surprising that the simulation with our generated noise voltage (A.20) does not exactly reproduce this curve. However, the typical drop of the ACF with the time constant of the low-pass filter is clearly visible.

Example Bandpass Noise

If we pass the white noise through a bandpass filter, the lower frequency limit is no longer zero, as it is with the low-pass filter. Earlier, we had already mentioned this example and shown that there is a Rayleigh distribution of the magnitude of fluctuation. We now simulate the bandpass noise through an LC resonant circuit with the resonance frequency $f_0 = 1$ kHz and the 3 dB bandwidth $B = 200$ Hz. The

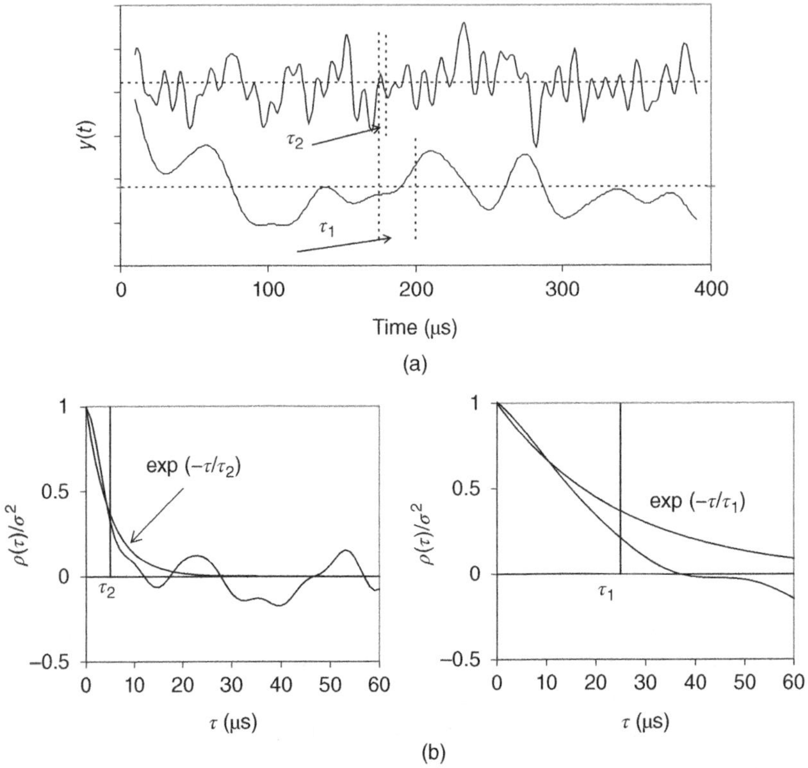

Figure A.16 White noise behind a low-pass filter. (a) Time response for the time constants $\tau_1 > \tau_2$. (b) Normalized ACF. The higher the cut-off frequency, the faster the decay.

transfer function of the amplitudes is

$$|H(f)| = \frac{1}{\sqrt{1 + \left(2\frac{|f - f_0|}{B}\right)^2}} \tag{A.23}$$

The plot is shown in Figure A.17. The corresponding time dependence of the band-limited noise is shown in Figure A.18a. The corresponding decay of the ACF in Figure A.18b. It has a different character than the low-pass noise in Figure A.16. The irregular fluctuations now affect the envelope of the resonance frequency of the filter $f_0 = 1$ kHz. The ACF is periodic with the frequency f_0. It decays according to the bandwidth with the time constant $\tau_B = {}^1/_B$.

This example leads us to the relation of time domain and frequency domain, which is generally given by the Fourier transformation.

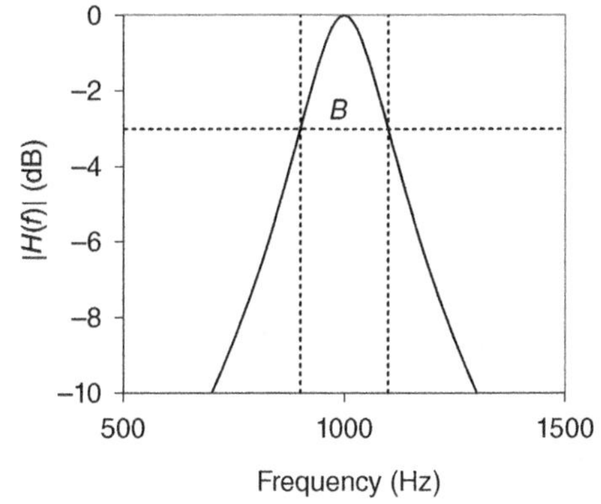

Figure A.17 Transfer curve of the bandpass filter according to (A.23).

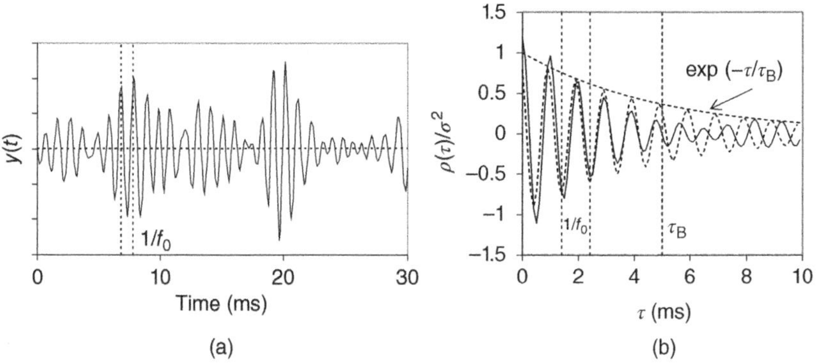

Figure A.18 White noise after passing through the bandpass with center frequency $f_0 = 1$ kHz. (a) Time domain. (b) The normalized ACF decays according to the bandwidth.

The characteristic function "power spectral density" has the frequency as a parameter. Let us first summarize some basic concepts of the Fourier transform. The Fourier transformation establishes the connection of a signal in the time domain with the representation in the frequency domain. In measurement technology, this corresponds to visualization with the oscilloscope for the time domain and with the spectrum analyzer for the frequency domain.

Fourier Series

Periodic time functions are the simplest deterministic signals. They can be represented by Fourier series. In addition to the fundamental frequency $\omega_1 = {}^{2\pi}/_{T}$ (first harmonic), they contain the integer multiples of this fundamental frequency $k \times \omega_1$ (harmonics, higher harmonics). While the frequency of the harmonics is fixed, amplitude and phase have different values depending on the waveform. In the frequency range, it is a line spectrum. The distance between two spectral lines is given by [7]

$$\Delta f = {}^{1}/_{T} = f_1 = {}^{\omega_1}/_{2\pi} \tag{A.24}$$

There are different notations in use.

Sine–Cosine Spectrum

We have a periodic function $y(t)$ in the time domain with period T and $\omega_1 = {}^{1}/_{2\pi T}$.

$$\begin{aligned} y(t) = {} & P_1 \sin(\omega_1 t) + P_2 \sin(2\omega_1 t) + P_3 \sin(3\omega_1 t) + \cdots \\ & + Q_0 + Q_1 \cos(\omega_1 t) + Q_2 \cos(2\omega_1 t) + Q_3 \cos(3\omega_1 t) + \cdots \end{aligned} \tag{A.25}$$

The coefficients, i.e. the amplitudes of the harmonics, are calculated by

$$\begin{aligned} P_k &= \frac{2}{T}\int_0^T y(t)\sin(k\omega_1 t)\,dt \\ Q_0 &= \frac{1}{T}\int_0^T y(t)\,dt \\ Q_k &= \frac{2}{T}\int_0^T y(t)\cos(k\omega_1 t)\,dt \end{aligned} \tag{A.26}$$

with $k = 1, 2, 3, \ldots$.

Amplitude–Phase Spectrum

The superposition of sine and cosine functions can also be understood as phase-shifted sine functions. The following notation is equivalent to (A.25).

$$y(t) = Q_0 + Y_1 \sin(\omega_1 t + \phi_1) + Y_2 \sin(2\omega_1 t + \phi_2) + Y_3 \sin(3\omega_1 t + \phi_3) + \cdots \tag{A.27}$$

The amplitudes and phase angles result from

$$Y_k = \sqrt{P_k^2 + Q_k^2}; \quad \tan\phi_k = \frac{Q_k}{P_k} \tag{A.28}$$

Complex Fourier Series

An elegant notation results from Euler's formula

$$\cos(z) = \frac{1}{2}\exp(jz + \exp(-jz)) \tag{A.29}$$

With this, both series can be transformed into a representation in polar coordinates. But the frequency axis must now go down to $-\infty$, because the sum (A.29) only gives a real cosine with a negative and a positive frequency.

$$y(t) = \sum_{-\infty}^{+\infty} C_k exp\left(jk\omega_1 t\right) \tag{A.30}$$

The coefficients C_k are now complex amplitudes which also contain the phase information

$$C_k = \frac{1}{T}\int_0^T y(t)\, exp\left(-jk\omega_1 t\right) dt; \quad C_0 = Q_0 \tag{A.31}$$

The following applies for $k > 0$

$$C_k = \frac{1}{2}\left(Q_k - jP_k\right) \tag{A.32}$$

For our purposes of describing noise, the Fourier series are not suitable, since noise as a stochastic signal has no defined periodicity. The Fourier series are only the preliminary stage for the following introduction of the Fourier integral. In the frequency range, noise does not have a discrete line spectrum. Rather, it has a continuous spectrum, as a look at the noise floor of a spectrum analyzer shows.

The Fourier Integral

The Fourier sum becomes the Fourier integral when the period goes $\rightarrow\infty$. Then the distance between the spectral lines (A.24) becomes infinitesimally small. They now form a continuous spectrum. Mathematically, this is the transition from the sum to the integral. This transition to the Fourier integral with its amplitude spectrum can be well understood with the following example.

We assume a sequence of rectangular pulses in the time domain (Figure A.19). It could be, e.g. a square wave voltage whose duty cycle can be changed. In our case, the pulse width $\tau = 0.5$ seconds should be constant and the period T should

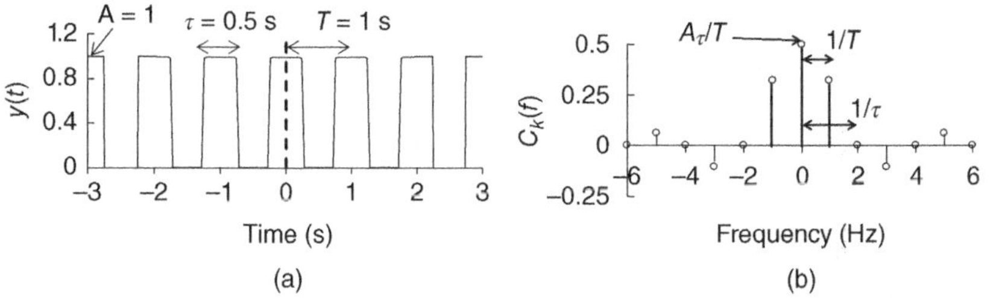

Figure A.19 (a) Sampled signal in the time domain with period $T = 1$ second. (b) Fourier coefficients (A.32) on the frequency axis.

be increased until only a single pulse remains. This pulse sequence with the peak value A is described by the following Fourier series (A.30).

$$y(t) = C_0 + \sum_{k=-\infty}^{\infty} C_k \cos\left(\frac{k2\pi}{T}t\right) \tag{A.33}$$

With the coefficients

$$C_0 = \frac{\tau}{T}A; \quad C_k = \frac{A}{k\pi}\sin\left(\frac{k\pi\tau}{T}\right) \tag{A.34}$$

We have here the mathematically oriented description $-\infty < k < +\infty$ with the frequency range $-\infty < f < +\infty$. In engineering, of course, there are only positive frequencies.

For a typical square wave voltage with duty cycle 50% $\tau/T = 0.5$, we have the time dependence in Figure A.19a and the spectral lines in Figure A.19b. The frequency spacing of the spectral lines corresponds to the period $\Delta f = 1/T$. The zeros in the frequency spectrum result from the pulse width at the frequencies $f_k = k/\tau$. If we now double the period duration to $T = 2$ seconds (Figure A.20), the pulses move further apart and the spectral lines move together. The frequency spacing is now $\Delta f = 1/T = 0.5$ Hz. The zeros of the spectrum remain at the frequencies 2, 4, 6, … Hz given by the pulse width, since the pulse width has not changed.

If we continue, the frequency spacing becomes smaller and smaller until, in the limiting case of a single pulse, i.e. $T \to \infty$, a continuous spectrum arises with an infinite number of lines which can no longer be detected individually. Here, in Figure A.21 is an example for $T = 10$ seconds (duty cycle 5%).

Besides the convergence of the spectral lines to a continuum, one also observes the decrease of the amplitudes C_k. Figure A.21 shows that with $T \to \infty$, i.e. with the transition to a single pulse, the spectral lines become smaller and smaller and finally disappear completely. The spectrum of aperiodic signals, regardless of their form, thus consists of a continuum of lines with zero amplitude. Of course, there is nothing to be done with this. The Fourier series is therefore not applicable in the limiting case of aperiodic signals. Figure A.22 shows the decrease of the amplitudes with increasing T according to (A.34).

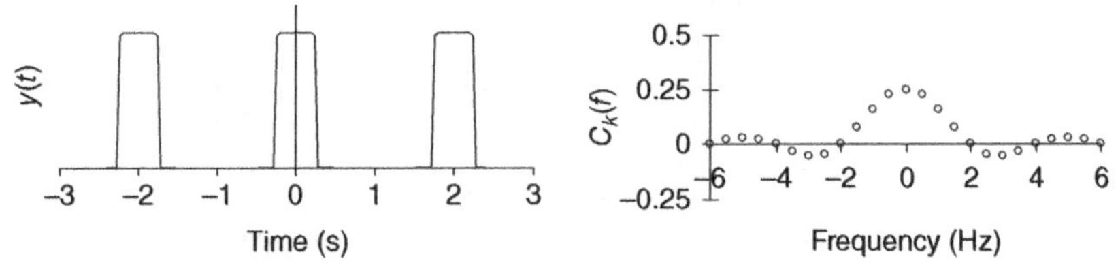

Figure A.20 Time dependence and Fourier coefficients as in Figure A.19, but for twice the period $T = 2$ seconds.

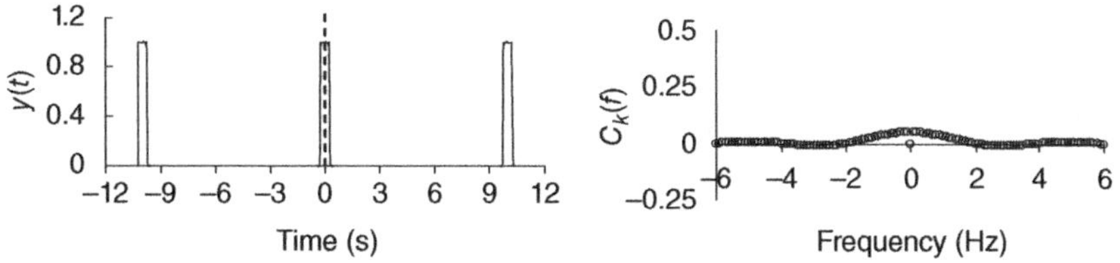

Figure A.21 Time dependence and Fourier coefficients as in Figure A.19, but for the period $T = 10$ seconds. The C_k change into a continuous spectrum.

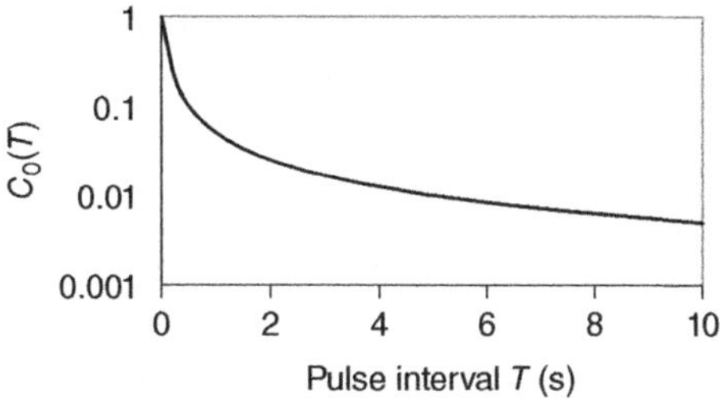

Figure A.22 Decrease in the amplitude of the Fourier coefficients with increasing period.

This can also be seen in the expression for the Fourier coefficients (A.31) in the limiting case $\omega_1 \to 0$:

$$\lim_{\omega_1 \to 0} C_k = \lim_{\omega_1 \to 0} \frac{1}{T} \int_{-\frac{T}{2}}^{\frac{T}{2}} y(t)\, exp\left(-jk\omega_1 t\right) dt = 0 \tag{A.35}$$

The single pulse is a signal with finite energy. With $T \to \infty$ the area of amplitude × time remains finite. The numerator of (A.35) therefore remains finite, while the denominator (T) approaches infinity. The amplitudes of the spectral lines therefore disappear.

The Fourier approach, which introduces the Fourier transform $F(\omega)$ for energy signals, provides a remedy.

$$F(\omega) = 2\pi \lim_{T \to \infty} \frac{C_k}{\omega_1} = \lim_{T \to \infty} \left(TC_k\right) \tag{A.36}$$

The amplitudes thus remain finite. If we insert this into (A.35), we get the Fourier integral.

$$F(\omega) = \int_{-\infty}^{+\infty} y(t)\, exp(-j\omega t)\, dt \tag{A.37}$$

The Fourier transform $F(\omega)$ consists of the continuous amplitude spectrum $A(\omega)$ and the phase spectrum $\varphi(\omega)$, which in general is also continuous.

$$F(\omega) = A(\omega)\, exp\,(j\varphi(\omega)) \tag{A.38}$$

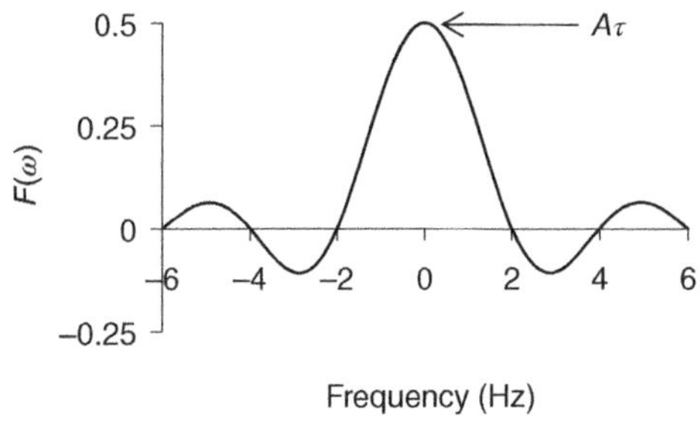

Figure A.23 Amplitude spectrum for an impulse train $T = 10$ seconds.

with the values for magnitude and phase

$$A(\omega) = \sqrt{(ReF)^2 + (ImF)^2}; \quad \varphi(\omega) = arctan\frac{ImF}{ReF} \tag{A.39}$$

A precondition for this transformation is that the integral $\int_{-\infty}^{+\infty} |y(t)|\, dt$ converges. The time signal $y(t)$ must be "absolutely integrable." Such signals are called energy signals.

$F(\omega)$ is therefore generally complex. In our example, however, it is not, since the time function (A.33) is even. The magnitude $|F(\omega)|$ now no longer describes individual, separate spectral lines but a density spectrum. The density of the amplitudes in a certain frequency range, e.g. 1 Hz. For an electrical voltage, the spectral voltage density $F(\omega)$ has the unit

$$[F(\omega)] = \mathrm{V} \times \mathrm{s} = \frac{\mathrm{V}}{\mathrm{Hz}} \tag{A.40}$$

For our example earlier with $T = 10$ seconds, which is already very similar to the single pulse, the amplitude spectrum $F(\omega)$ results as shown in Figure A.23.

Unlike the sum of monochromatic frequencies (each has bandwidth zero) in the Fourier series, in the integral we have the Fourier transform $F(\omega)$. Its magnitude $|F(\omega)|$ gives the density of amplitudes within a certain bandwidth. In technology, the specifications usually refer to 1 Hz.

Energy and Power Signals

At this point, a short excursion to energy and power signals. Everyone is familiar with the electrical power $P = {}^{v^2}/_{R}$ with the unit Watt (W). Normalized to $=1\,\Omega$, v^2 represents the power. From a physical point of view, the following generally applies: Energy (work) equals power multiplied with that fraction of time, in which that power is active. Electrical energy (work) is therefore electrical power multiplied by the time t over which this power is consumed. $W = P \times t$ with the unit watt second (W s) or Joule (J) (1 W s = 1 J). In everyday life also kWh, the energy consumption that is periodically paid to the electricity provider. An energy signal $y_W(t)$ only exists in a finite time interval $t_1 < t < t_2$, thus its energy is finite.

$$W = \int_{t_1}^{t_2} y_W^2(t)\,dt = \int_{-\infty}^{+\infty} y_W^2(t)\,dt < \infty \tag{A.41}$$

If this energy is averaged over the entire period from $-\infty$ to $+\infty$, the power is zero:

$$P = \lim_{T\to\infty} \frac{1}{T} \int_{-\frac{T}{2}}^{+\frac{T}{2}} y_W^2(t)\,dt = 0 \tag{A.42}$$

Energy signals are pulses, as in our example earlier, also slowly decaying signals, and also series of measurements which of course can only ever contain a finite number of measured values. In measurement technology, one therefore practically always has to deal with finite time ranges in which signals are recorded. Even if these exist for an infinite time, such as the noise of an ohmic resistor.

A power signal $y_P(t)$ exists for an infinitely long time or is infinitely large in a small time interval. It has an infinitely high energy, but the power remains finite.

$$W = \int_{-\infty}^{+\infty} y_P^2(t)\,dt = \infty \tag{A.43}$$

$$P = \lim_{T\to\infty} \int_{-\frac{T}{2}}^{\frac{T}{2}} y_P^2(t)\,dt < \infty \tag{A.44}$$

Power signals are all DC signals, periodic signals, e.g. the always available voltage of the mains and noise. But also δ-pulses with infinitely high power.

The Fourier integral therefore exists for power signals. This is important for theoretical analysis; the problem does not exist in data processing of measured noise.

Example Transient Time Function

Let us take as an example a transient time function that rises from the value $y(0) = 0$ at $t = 0$ and then decays exponentially.

$$y(t) = t\,exp(-\alpha t) \tag{A.45}$$

For two values of the decay constant $\alpha = 0.5$ and $1.5\ s^{-1}$, it is plotted in Figure A.24.

For better comparison, both curves are normalized to the maximum value 1. The prerequisite for the existence of the Fourier integral is $Q_0 = 0$ (A.26). In contrast to the periodic sequence of rectangular pulses (A.33), we only have a single pulse with finite energy.

The integral over the time function (A.45) is of the type

$$\int_0^T x\,exp(ax)\,dx = \frac{ax-1}{a^2} exp(ax)\Big|_0^T \tag{A.46}$$

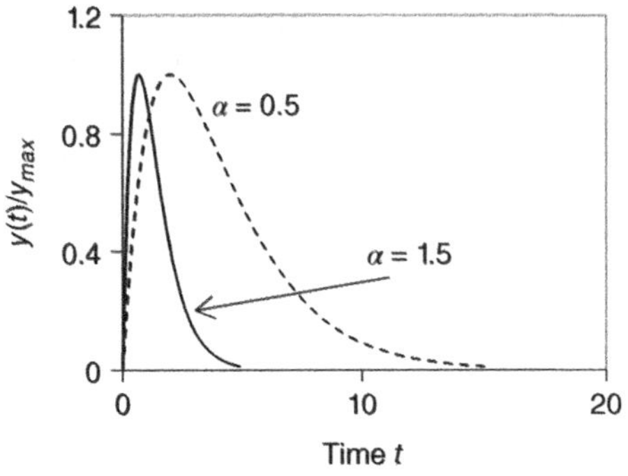

Figure A.24 Transient time function according to (A.45) as an example for the application of the Fourier integral.

We leave out the normalization to $y_{max} = 1$ here; it does not play a role in the limiting value. With $x = t$ and $a = -\alpha$ we obtain after integration

$$\int_0^T t\,exp\,(-\alpha t)\,dt = \left\{\frac{exp\,(-\alpha T)}{\alpha^2}\,[-\alpha T - 1]\right\} - \left\{\frac{1}{\alpha^2}\,[0-1]\right\} \tag{A.47}$$

thus, we obtain in the limit $T \rightarrow \infty$

$$Q_0 = \lim_{T\rightarrow\infty}\left\{\frac{1}{\alpha^2 T}\left[1 - exp\,(-\alpha T)\,(1+\alpha T)\right]\right\} = 0 \tag{A.48}$$

The constraint of the Fourier integral is therefore fulfilled. We integrate here only over positive times, i.e. only where $y(t) \neq 0$.

The Fourier transform is

$$F\,(\omega) = \int_{-\infty}^{\infty} y\,(t)\,exp\,(-j\omega t)\,dt \tag{A.49}$$

With the normalized time function we obtain

$$F\,(\omega) = \frac{1}{c}\int_0^{\infty} t\,exp\,(-\alpha t)\,exp\,(-j\omega t)\,dt = \frac{1}{c}\int_0^{\infty} t\,exp\left(-\left[\alpha + j\omega\right]t\right)\,dt \tag{A.50}$$

$c = y_{max}$. Again, we must leave out the lower limit $-\infty$ because our time function (A.45) does not hold for $t < 0$. The integral is of the same type as previously.

We now set $a = -(\alpha + j\omega)$ and $x = t$.

$$F\,(\omega) = -\frac{1}{c}\,\frac{(\alpha + j\omega)\,t + 1}{(\alpha + j\omega)^2}\,exp\,(-(\alpha + j\omega)\,t)\Bigg|_0^{\infty} \tag{A.51}$$

For the numerator, with $t \rightarrow \infty$ real and imaginary parts quickly go to zero (Figure A.25). What remains is the denominator with

$$F\,(\omega) = \frac{1}{c}\,\frac{1}{(\alpha + j\omega)^2} \tag{A.52}$$

The spectrum of the steep pulse ($\alpha = 1.5$) extends over a wider frequency range than the spectrum of the flat pulse. A Dirac pulse would have an infinitely

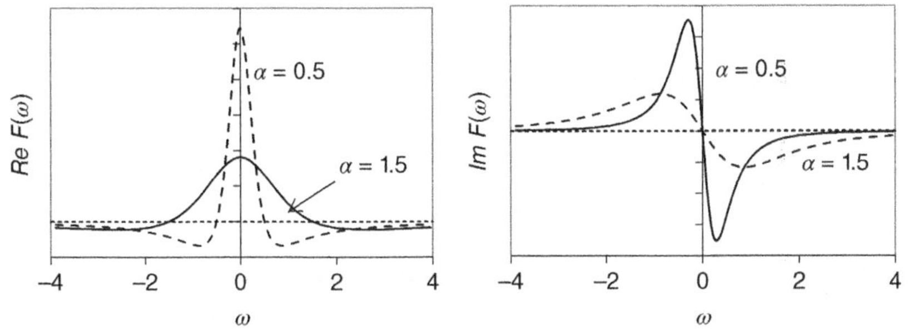

Figure A.25 Real and imaginary part of the Fourier transform of the time function from Figure A.24.

wide spectrum. We also have an imaginary part of $F(\omega)$ here, since the time function (A.45) is not even. In the example earlier with the pulse sequence (A.33), $y(t) = y(-t)$, i.e. an even function where the Fourier transform is real.

The Parseval Equation

The energy of a signal in the time domain and in the frequency domain is the same. This is actually obvious, since these are two ways of looking at the same process. Let us write for the energy W of a transient time signal $y(t)$

$$W = \int_{-\infty}^{\infty} y^2(t)\,dt \tag{A.53}$$

For example, if $y(t)$ is an electric voltage across a resistor of $R = 1\ \Omega$, then $P(t) = {}^{y^2(t)}/_{R}$ is the instantaneous power. Thus, there is a power under the integral.

The integration over time gives the energy W of the signal. One can see it also at the unit.

$$[P(t) \times t] = \frac{\text{V}^2}{\Omega}\text{s} = \frac{\text{V}^2\,\text{A}\,\text{s}}{\text{V}} = \text{W}\,\text{s} = \text{J} \tag{A.54}$$

Relating the energy to the time unit T, we have the average power P in this time interval.

$$P = \frac{1}{T}\int_{-T/2}^{T/2} y^2(t)\,dt \tag{A.55}$$

Now the unit is

$$[P] = \frac{1}{\text{s}}\frac{\text{V}^2}{\Omega}\text{s} = \frac{\text{V}^2\,\text{A}}{\text{V}} = \text{Watt} \tag{A.56}$$

The transient signal $y(t)$ has a Fourier transform.

$$F(\omega) = \int_{-\infty}^{\infty} y(t)\, exp\,(-j\omega t)\, dt \tag{A.57}$$

and backward

$$y(t) = \frac{1}{2\pi} \int_{-\infty}^{\infty} F(\omega)\, exp\,(j\omega t)\, d\omega \tag{A.58}$$

We now want to analyze the relation of the energy in the time domain (A.53) with the energy in the frequency domain. For this purpose we write (A.53)

$$W = \int_{-\infty}^{\infty} y(t)\, y(t)\, dt \tag{A.59}$$

and replace one $y(t)$ by (A.58)

$$W = \frac{1}{2\pi} \int_{-\infty}^{\infty} y(t) \left\{ \int_{-\infty}^{\infty} F(\omega)\, exp\,(j\omega t)\, d\omega \right\} dt \tag{A.60}$$

Since $F(\omega)$ does not depend on t, we can reverse the order of integration.

$$W = \frac{1}{2\pi} \int_{-\infty}^{\infty} F(\omega) \left\{ \int_{-\infty}^{\infty} y(t)\, exp\,(j\omega t)\, dt \right\} d\omega \tag{A.61}$$

The term in curly brackets is

$$F(-\omega) = F^*(\omega) \tag{A.62}$$

so we have

$$W = \frac{1}{2\pi} \int_{-\infty}^{\infty} F(\omega)\, F^*(\omega)\, d\omega = \frac{1}{2\pi} \int_{-\infty}^{\infty} |F(\omega)|^2 d\omega \tag{A.63}$$

this brings us to Parseval's theorem

$$\int_{-\infty}^{\infty} y^2(t)\, dt = \frac{1}{2\pi} \int_{-\infty}^{\infty} |F(\omega)|^2 d\omega \tag{A.64}$$

The energy of a signal in the time domain (left) is equal to the energy of the signal in the frequency domain (right). M.A. Parseval (1755–1835) did not prove this theorem mathematically because he found its statement trivial.

Example Voltage Pulse

Let's take again our transient time function (A.45) as an example. We now consider electrical power at a resistor $R = 1\ \Omega$. If the time function is to describe a voltage waveform, we have to introduce a normalization factor for dimensional reasons

$$v(t) = \frac{v_0}{t_0} t\, exp\,(-\alpha t) \tag{A.65}$$

This is necessary only for the fact that $v(t)$ has the unit "Volt." Under the integral there is the instantaneous power

$$\int_{-\infty}^{\infty} v^2(t)\,dt = \frac{v_0^2}{t_0^2}\int_0^{\infty} t^2\,exp(-2\alpha t)\,dt \tag{A.66}$$

We also set the lower limit of integration to zero here, since our formula for $v(t)$ would give an incorrect contribution for $t < 0$. In reality, however, it vanishes.

The integral is of the type

$$\int x^2\,exp(ax)\,dx = exp(ax)\left\{\frac{x^2}{a} - \frac{2x}{a^2} + \frac{2}{a^3}\right\} \tag{A.67}$$

With $x = t$ and $a = -2\alpha$.

$$\frac{v_0^2}{t_0^2}\int_0^{\infty} t^2\,exp(-2\alpha t)\,dt = \frac{v_0^2}{t_0^2}exp(-2\alpha t)\left\{-\frac{t^2}{2\alpha} - \frac{2t}{4\alpha^2} - \frac{2}{8\alpha^3}\right\}\Bigg|_0^{\infty} \tag{A.68}$$

the upper term with $t = \infty$ vanishes due to $exp(-\infty) = 0$. From the lower term with $t = 0$ remains

$$\frac{v_0^2}{t_0^2}\frac{1}{4\alpha^3} \tag{A.69}$$

A dimensional consideration should reveal that this is an energy because we have integrated an instantaneous power over a period of time.

$$\left[\frac{v_0^2}{t_0^2}\frac{1}{4\alpha^3}\right] = \frac{\mathrm{V}^2}{\mathrm{s}^2}\mathrm{s}^3 \tag{A.70}$$

this at the 1 Ω resistor

$$\frac{\mathrm{V}^2}{\Omega}\mathrm{s} = \frac{\mathrm{V}^2\,\mathrm{A}}{\mathrm{V}}\mathrm{s} = \mathrm{W\,s} = \mathrm{J} \tag{A.71}$$

thus the unit of energy.

Let us now consider the other side of the Parseval equation, the energy in the frequency domain.

We had derived above for the Fourier transform (A.52):

$$F(\omega) = \frac{1}{c}\frac{1}{(\alpha + j\omega)^2} \tag{A.72}$$

we set for the illustration of the dimension $^1/_c = {^{v_0}/_{t_0}}$.

We arrive at $|F(\omega)|$ by using the scientific notation in the denominator of (A.72).

$$\alpha + j\omega = \rho\,exp(j\phi) \tag{A.73}$$

(A.72) now reads

$$F(\omega) = \frac{v_0}{t_0}\frac{1}{\rho^2}exp(-j2\phi) \tag{A.74}$$

and thus for the magnitude

$$|F(\omega)| = \frac{v_0}{t_0}\frac{1}{\rho^2} \tag{A.75}$$

with $\rho^2 = \alpha^2 + \omega^2$ it becomes

$$|F(\omega)|^2 = \frac{v_0^2}{t_0^2}\frac{1}{\left(\alpha^2+\omega^2\right)^2} \tag{A.76}$$

The integral is of the type

$$\int \frac{dx}{\left(a^2+x^2\right)^2} = \frac{x}{2a^2\left(a^2+x^2\right)} + \frac{1}{2a^3} arctan\left(\frac{x}{a}\right) \tag{A.77}$$

With $a = \alpha$ and $x = \omega$ we have

$$\int_{-\infty}^{\infty} |F(\omega)|^2 d\omega = \left.\frac{\omega}{2\alpha^2\left(\alpha^2+\omega^2\right)} + \frac{1}{2\alpha^3} arctan\left(\frac{\omega}{\alpha}\right)\right|_{-\infty}^{\infty} \tag{A.78}$$

The first term goes rapidly toward zero as ω increases. Arctan goes toward $\pm\pi/2$. From the second term remain

$$\left\{\frac{1}{2\alpha^3}\frac{\pi}{2}\right\} - \left\{-\frac{1}{2\alpha^3}\frac{\pi}{2}\right\} \tag{A.79}$$

The result is

$$\frac{1}{2\pi}\frac{v_0^2}{t_0^2}\frac{1}{\alpha^3}\frac{\pi}{2} = \frac{v_0^2}{t_0^2}\frac{1}{4\alpha^3} \tag{A.80}$$

thus exactly like the result (A.69) in the time domain. It is to be noted that always also over the negative frequencies of the spectrum must be integrated.

Fourier Transform and Power Spectral Density

In measurement technology, we always get time-limited signals. For example, we have measured a time function $y(t)$ during the period T. Outside this time period, $y(t) = 0$. A Fourier integral exists for this time function. According to the Parseval's theorem (A.64) is

$$\int_{-\infty}^{\infty} y^2(t)\,dt = \int_{-\frac{T}{2}}^{\frac{T}{2}} y^2(t)\,dt = \frac{1}{2\pi}\int_{-\infty}^{\infty} |F(\omega)|^2 d\omega \tag{A.81}$$

In the limiting case $T \to \infty$ we can write

$$\overline{y^2(t)} = \lim_{T\to\infty}\frac{1}{T}\int_{-\frac{T}{2}}^{\frac{T}{2}} y^2(t)\,dt = \frac{1}{2\pi}\int_0^{\infty} G(\omega)\,d\omega \tag{A.82}$$

This is valid because $|F(\omega)|^2$ is an even function, i.e. $|F(\omega)|^2 = |F(-\omega)|^2$.

This is the definition of the one-sided power spectral density $G(\omega)$ for positive (technically meaningful) frequencies. The lower integration limit is zero.

$$G(\omega) = \lim_{T\to\infty} \frac{2}{T} |F(\omega)|^2 \tag{A.83}$$

Keeping the original integration limits in the Parseval equation, we arrive at the definition of the two-sided power spectral density $S(\omega)$.

$$\lim_{T\to\infty} \frac{1}{T}\frac{1}{2\pi} \left\{ \int_{-\infty}^{\infty} |F(\omega)|^2 d\omega \right\} = \frac{1}{2\pi} \int_{-\infty}^{\infty} S(\omega)\, d\omega \tag{A.84}$$

With the definition of the two-sided power spectral density

$$S(\omega) = \lim_{T\to\infty} \frac{1}{T} |F(\omega)|^2 \tag{A.85}$$

This definition is mathematically oriented. However, $G(\omega)$ is practically relevant since only positive frequencies exist in engineering.

The relation is trivial:

$$G(\omega) = 2 \times S(\omega) \tag{A.86}$$

Let us additionally consider the units in our example

$$[\alpha] = \mathrm{s}^{-1}; [\omega] = \mathrm{s}^{-1}; [F(\omega)] = \mathrm{V\,s}$$

$$[G(\omega)] = \frac{\mathrm{V}^2\,\mathrm{s}^2}{\mathrm{s}} = \frac{\mathrm{V}^2}{\mathrm{Hz}}$$

It is a spectral density of the voltage square. At an 1 Ω resistor it is also the power spectral density in W/Hz. Integrating over the entire spectrum, we have the power present in the signal in the time domain. This again is the statement of the Parseval theorem.

Example Rectangular Pulse

We want to analyze the relation between the time domain and the frequency domain via the Fourier transformation once again using a rectangular pulse. It leads to the physically important sinc-function. As we have seen, a wide pulse has a narrow frequency spectrum, a narrow pulse has a wide frequency spectrum. Figure A.22 shows the examples a with long pulse duration T, b with average pulse duration $0.1T$, and c with $0.01T$ as the limiting case of the Dirac pulse. The pulse amplitude of a is $^1/_T$ so that the area is $^1/_T \times 2\,^T/_2 = 1$. For b and c, the pulse amplitude increases according to the contraction, so that the area remains constant. The log measure in Figure A.26 enables the large amplitude differences to be represented.

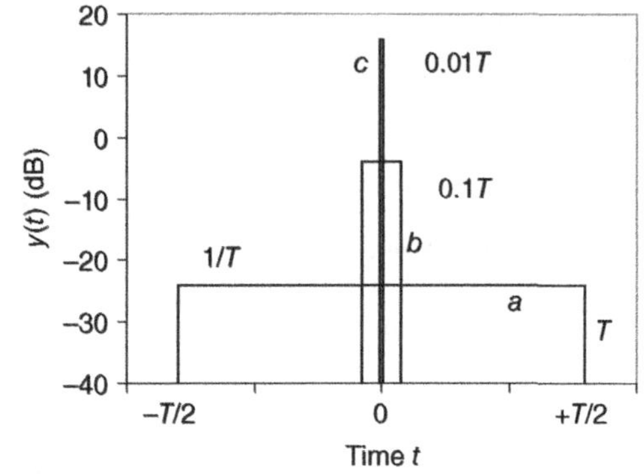

Figure A.26 Individual pulses of different durations in dB measure. *c* approximates the Dirac pulse.

The time function is

$$y(t) = \frac{1}{T} \quad \text{if} -\frac{T}{2} < t < \frac{T}{2}$$
$$y(t) = 0 \quad \text{outside} \tag{A.87}$$

If this is a voltage pulse, the Fourier transform has the unit V s. We do not consider other normalizations.

The frequency spectrum is

$$F(\omega) = \frac{1}{T}\int_{-\frac{T}{2}}^{+\frac{T}{2}} exp\,(-j\omega t)\,dt \tag{A.88}$$

Solution of the integral:

$$F(\omega) = \frac{-1}{j\omega T} exp\,(-j\omega T)\Big|_{-\frac{T}{2}}^{+\frac{T}{2}} = \frac{1}{j\omega T}\{exp\,(j\omega T) - exp\,(-j\omega T)\} \tag{A.89}$$

with Euler's formula

$$sin\,(x) = \frac{1}{2j}\{exp\,(jx) - exp\,(-jx)\} \tag{A.90}$$

we obtain the sinc-function

$$F(\omega) = \frac{sin\left(\frac{\omega T}{2}\right)}{\frac{\omega T}{2}} \tag{A.91}$$

The limit case $T \to 0$ to the Dirac pulse requires the application of the l'Hospital rule

$$F(\omega) = \lim_{T\to 0}\frac{sin\left(\frac{\omega T}{2}\right)}{\frac{\omega T}{2}} = \frac{\omega}{2}\frac{cos\left(\frac{\omega T}{2}\right)}{\frac{\omega}{2}} = 1 \tag{A.92}$$

The spectral distribution of the frequencies comprised in the square pulse is shown in Figure A.27. The Fourier transform of the broad pulse *a* is already known.

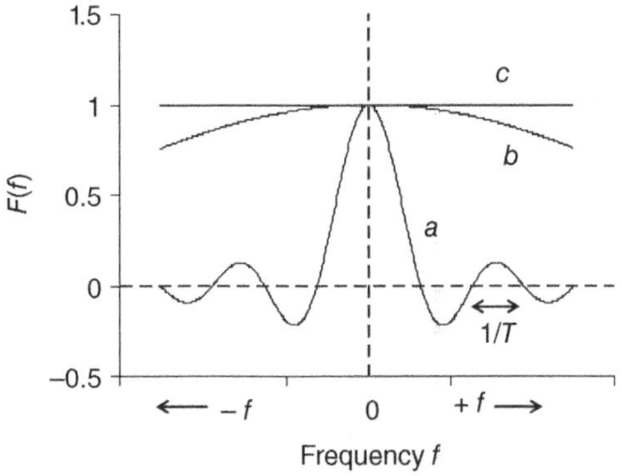

Figure A.27 Spectral distribution of the frequencies contained in the square pulse. The Dirac pulse *c* has a white spectrum.

Besides the DC component at $f = 0$, it contains only a few high-frequency components that decay rapidly with increasing frequency. The zeros follow at a distance of $1/T$. The narrower the pulse becomes (b), the more high-frequency components are contained. In the limiting case of the Dirac pulse (c) we have a white spectrum.

If one wants to obtain the two-sided power spectral density from (A.85), one must note that the time T there must not be confused with the pulse width T in this example. In (A.85), T means a long time over which the time function $y(t)$ must be measured to obtain the most accurate result.

Time-Limited Noise Signal

The Fourier transform only exists for energy signals. Our two examples concern time-limited impulses, i.e. they are energy signals. However, we are interested in noise signals, which, at least theoretically, exist without time limits. In the real case, of course, we are always dealing with finite signal sequences. A device is switched off or a time-resolved measurement delivers a limited set of voltage values. Therefore, one cuts out a finite time section from the infinitely long power signal and thus obtains an energy signal to which the formalism of the Fourier transformation can be applied. Figure A.28 shows an example of the square of a noise voltage in the time domain. We use $y^2(t)$, since the power is of interest in noise observations. The amplitudes $y(t)$ are stochastic and can only be recorded by means of the statistical characteristic functions. Since the signal is stationary (its characteristic values are constant in time), we can take out the time period T. It has the same properties as any other time section.

In this section, we have an energy signal to which the Fourier transformation can be applied. One can also place these sections one after the other and thereby artificially generate a periodicity T. This turns the non-periodic noise signal into a periodic signal. The Fourier analysis results in a line spectrum with the frequency

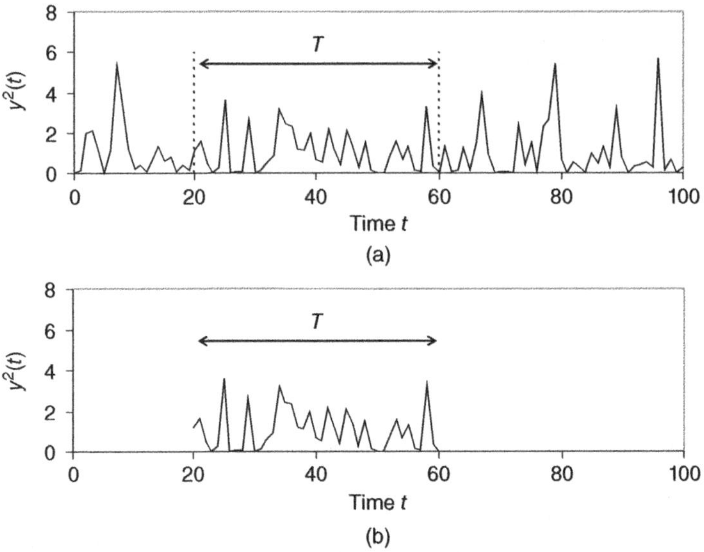

Figure A.28 (a) Unlimited fluctuating time signal $y^2(t)$. (b) Time segment of length T.

spacing $\Delta f = {}^1/_T$. Fundamental wave and harmonics have defined amplitudes. By increasing the section T, the frequency spacing can be reduced more and more. In the limiting case $T \to \infty$, a continuous spectrum results. Individual spectral lines with a defined amplitude no longer exist. As we have seen, the power spectral density takes their place.

Example of a Time-Limited Wave Train

Let us first consider a simple example of a deterministic signal. A wave train limited to the time T with the frequency $f_0 = {}^{\omega_0}/_{2\pi}$ and the peak value 1 (Figure A.29):

The time function is

$$y(t) = \cos^2(\omega_0 t) \quad -\frac{T}{2} < t < +\frac{T}{2} \tag{A.93}$$

$$y(t) = 0 \quad \text{outside}$$

Let us first consider the existence condition of the Fourier integral for our example function. We write here again

$$Q_0 = \lim_{T\to\infty} \frac{1}{T} \int_{-\frac{T}{2}}^{\frac{T}{2}} y(t)\,dt = 0 \tag{A.94}$$

and first show that this condition is not fulfilled for the unlimited time function (A.93). It is only fulfilled if we take out the time segment $-{}^T/_2 < t < +{}^T/_2$.

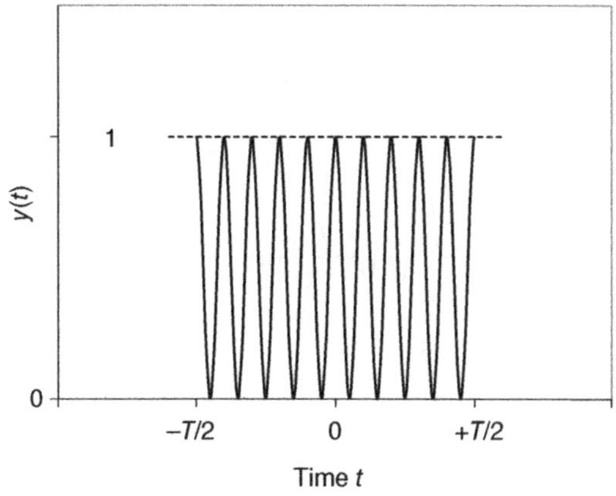

Figure A.29 Time-limited signal (A.93).

The integral is of the type

$$\int \cos(ax)\,dx = \frac{1}{2}x + \frac{1}{4a}\sin(2ax) \tag{A.95}$$

(A.94) is then

$$Q_0 = \lim_{T\to\infty}\frac{1}{T}\left\{\frac{t}{2} + \frac{1}{4\omega_0}\sin\left(2\omega_0 t\right)\Big|_{-\frac{T}{2}}^{\frac{T}{2}}\right\} = \lim_{T\to\infty}\frac{1}{T}\left\{\frac{T}{2} + \frac{1}{2\omega_0}\sin\left(\omega_0 T\right)\right\} \tag{A.96}$$

so we arrive at the limit (A.94)

$$Q_0 = \lim_{T\to\infty}\left\{\frac{1}{2\omega_0 T}\sin\left(\omega_0 T\right) + \frac{1}{2}\right\} = \frac{1}{2} \neq 0 \tag{A.97}$$

With the finite limit value ½.

If we limit the signal to the length T, we again integrate from $-{}^{T}/_{2}$ to $+{}^{T}/_{2}$, since there is no contribution outside. However, we calculate the mean value (A.94) over the entire time axis t, while T is a constant value.

$$Q_0 = \lim_{t\to\infty}\frac{1}{2t}\left\{\frac{1}{\omega_0}\sin\left(\omega_0 T\right) + T\right\} = 0 \tag{A.98}$$

Now the limit is zero and the Fourier integral exists.

The Fourier integral is

$$F(\omega) = \frac{1}{2\pi}\int_{-\frac{T}{2}}^{\frac{T}{2}} \cos^2\left(\omega_0 t\right) \exp(-j\omega t)\,dt \tag{A.99}$$

with the transformations

$$\cos^2 x = \frac{1}{2}\left[1 + \cos(2x)\right]; \quad \cos x = \frac{1}{2}\left[\exp(jx) + \exp(-jx)\right] \tag{A.100}$$

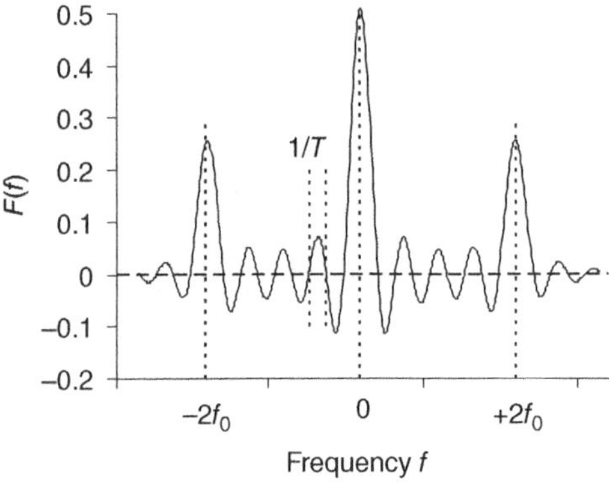

Figure A.30 Fourier spectrum of the signal limited in time to the length T from Figure A.29.

We obtain

$$F(\omega) = \frac{1}{2\pi} \int_{-\frac{T}{2}}^{+\frac{T}{2}} \left\{ \frac{1}{2} + \frac{1}{4} \left[exp\left(j2\omega_0 t\right) + exp\left(-j2\omega_0 t\right) \right] \right\} exp(-j\omega t)\, dt$$

$$= \frac{1}{4\pi} \int_{-\frac{T}{2}}^{\frac{T}{2}} exp(-j\omega t)\, dt + \frac{1}{8\pi} \int_{-\frac{T}{2}}^{\frac{T}{2}} exp\left(-j2\left[\frac{\omega}{2} + \omega_0\right] t\right) dt$$

$$+ \frac{1}{8\pi} \int_{-\frac{T}{2}}^{\frac{T}{2}} exp\left(-j2\left[\frac{\omega}{2} - \omega_0\right] t\right) dt \tag{A.101}$$

three integrals of the same type as above for the rectangular pulse result

$$\int exp(-jax)\, dx = \frac{-1}{ja} exp(-jax) \tag{A.102}$$

In the integration limits $-{}^{T}/_{2}$ to $+\,{}^{T}/_{2}$ we again obtain the modified sinc-function.

$$F(\omega) = \frac{1}{2\pi\omega} sin\left(\frac{\omega T}{2}\right) + \frac{1}{8\pi\left(\frac{\omega}{2} + \omega_0\right)} sin\left(\left[\frac{\omega}{2} + \omega_0\right] T\right)$$

$$+ \frac{1}{8\pi\left(\frac{\omega}{2} - \omega_0\right)} sin\left(\left[\frac{\omega}{2} - \omega_0\right] T\right) \tag{A.103}$$

The Fourier transform $F(f)$ is shown versus the frequency $f = {}^{\omega}/_{2\pi}$ in Figure A.30. In the physically relevant range $f > 0$ we have enhanced spectral components at $f = 0$ (DC term) and at twice the fundamental frequency $f = 2 \times f_0$.

The Wiener–Khinchin Theorem

For the noise signal, $y(t)$ is not determined. A time function only exists if a time-resolved measurement has been made with a transient recorder. Since this prerequisite is unrealistic in most cases, the ACF $\rho(\tau)$ takes the place of the time function [5].

This is also not a function of the real time t, but a function of the time shift τ. The ACF is related to the power spectral density via the Wiener Khinchin theorem. It results from the stochastic signal $y(t)$ as the generalized average, as we have already seen in (A.17). We write here once again

$$\rho(\tau) = \lim_{T\to\infty} \frac{1}{T} \int_{-\frac{T}{2}}^{\frac{T}{2}} y(t)\, y(t+\tau)\, dt \tag{A.104}$$

The signal existing in the period $-{}^{T}\!/_{2} < t < {}^{T}\!/_{2}$ has the Fourier transform

$$F(\omega) = \int_{-\infty}^{\infty} y(t)\, exp(-j\omega t)\, dt \tag{A.105}$$

and reverse

$$y(t) = \frac{1}{2\pi} \int_{-\infty}^{\infty} F(\omega)\, exp(j\omega t)\, d\omega \tag{A.106}$$

for the function shifted in time τ

$$y(t+\tau) = \frac{1}{2\pi} \int_{-\infty}^{\infty} F(\omega)\, exp(j\omega[t+\tau])\, d\omega \tag{A.107}$$

We insert this the ACF (A.104)

$$\rho(\tau) = \lim_{T\to\infty} \frac{1}{T} \int_{-\frac{T}{2}}^{\frac{T}{2}} y(t) \frac{1}{2\pi} \int_{-\infty}^{\infty} F(\omega)\, exp(j\omega[t+\tau])\, d\omega dt \tag{A.108}$$

since $F(\omega)$ does not depend on t, we can also write

$$\rho(\tau) = \lim_{T\to\infty} \frac{1}{2\pi T} \int_{-\infty}^{\infty} F(\omega) \left[\int_{-\frac{T}{2}}^{\frac{T}{2}} y(t)\, exp(j\omega t)\, dt \right] exp(j\omega\tau)\, d\omega \tag{A.109}$$

The integral term in the square bracket is $F^*(\omega)$. Because of $F(\omega)F^*(\omega) = |F(\omega)|^2$ we have

$$\rho(\tau) = \lim_{T\to\infty} \frac{1}{2\pi T} \int_{-\infty}^{\infty} |F(\omega)|^2 exp(j\omega\tau)\, d\omega \tag{A.110}$$

with the above definition of the two-sided power spectral density (A.85) we are at the Wiener-Khinchin theorem, in which mostly over f instead of ω is integrated.

$$\rho(\tau) = \int_{-\infty}^{\infty} S(f)\, exp(j2\pi f\tau)\, df \tag{A.111}$$

and reverse

$$S(f) = \int_{-\infty}^{\infty} \rho(\tau)\, exp(-j2\pi f\tau)\, d\tau \tag{A.112}$$

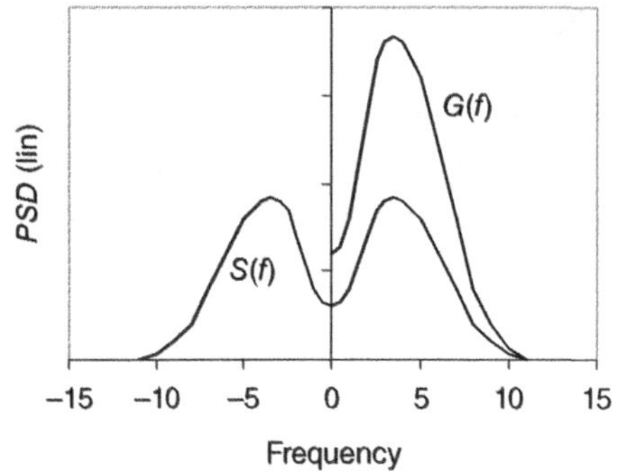

Figure A.31 Typical curve of the two-sided $S(f)$ and the one-sided $G(f)$ power spectral density.

The integral of the spectral power density $S(f)$ over the entire frequency range is equal to the total noise power. If one sets in (A.111) = 0, one obtains

$$\rho(0) = \int_{-\infty}^{\infty} S(f)\, df = \sigma^2 \tag{A.113}$$

Since the ACF is a real, even function ($\rho(\tau) = \rho(-\tau)$), $S(f)$ is also an even function ($S(f) = S(-f)$) (Figure A.31). It is therefore sufficient to use the real form

$$\rho(\tau) = 2\int_{0}^{\infty} S(f) \cos(2\pi f\tau)\, df \tag{A.114}$$

$$S(f) = 2\int_{0}^{\infty} \rho(\tau) \cos(2\pi f\tau)\, d\tau \tag{A.115}$$

The one-sided power spectral density $G(f)$ is the experimental equivalent to the mathematically oriented two-sided power spectral density $S(f)$. The Wiener–Khinchin theorem reads for the practically meaningful positive frequencies

$$\rho(\tau) = \int_{0}^{\infty} G(f) \cos(2\pi f\tau)\, df \tag{A.116}$$

$$G(f) = 4\int_{0}^{\infty} \rho(\tau) \cos(2\pi f\tau)\, d\tau \tag{A.117}$$

The designation chosen here: $S(f)$ for the two-sided (Signal and system theory) and $G(f)$ for the one-sided (Measuring technique) power spectral density is not under an uniform use. One also finds the reverse in the literature. One has to clarify this for the respective case.

Cross Correlation

The CCF $\rho_{12}(\tau)$ is a measure for the similarity of two signals $y_1(t)$ and $y_2(t)$. It describes to what extent a signal value $y_1(t)$ at any time t influences the signal value $y_2(t+\tau)$ at a later time $t+\tau$. Its definition corresponds to the ACF [2].

$$\rho_{12}(\tau) = \lim_{T\to\infty} \frac{1}{T} \int_{-\frac{T}{2}}^{\frac{T}{2}} y_1(t) y_2(t+\tau)\, dt \tag{A.118}$$

It is a real function for which holds:

1) Reversal of the order of the signals corresponds to a change of sign of the time shift.

$$\rho_{12}(-\tau) = \rho_{21}(\tau) \tag{A.119}$$

2) The peak values are given by

$$|\rho_{12}|^2 \leq \rho_1(0)\, \rho_2(0) \tag{A.120}$$

3) $\rho_{12}(\tau)$ is also denoted covariance in statistics.

In the case of complete correlation, the equal sign applies.

If one normalizes the CCF, one gets a value between 0 (uncorrelated) and ±1 (fully correlated).

$$-1 \leq \frac{\rho_{12}(\tau)}{\sqrt{\rho_1(0)\, \rho_2(0)}} \leq +1 \tag{A.121}$$

The CCF can be measured for several important engineering applications.

1) Measurement of a time delay. The passage of a signal through any transmission system causes a time delay. Measuring the CCF between input and output will result in a maximum in the time delay τ corresponding to the transit time.
2) Identification of a transmission path. If an input signal can be transmitted to an output via different paths, this also means different transit times. For example, the transmission path of mechanical vibrations can be identified and specifically influenced.
3) Detection and reconstruction of a signal in noise. For periodic signals, the ACF is usually sufficient. With the CCF, a stochastic signal can also be extracted from the noise.

Example of Two Sine Functions

This is the simplest example of complete correlation: Two phase-shifted sinusoidal functions with the same frequency. These can be, for example, current and voltage in a complex load

$$\begin{aligned} y_1(t) &= V \sin(\omega t) \\ y_2(t) &= I \sin(\omega t + \phi) \end{aligned} \tag{A.122}$$

The CCF reads

$$\rho_{12}(\tau) = \lim_{T\to\infty} \frac{VI}{T} \int_{-\frac{T}{2}}^{\frac{T}{2}} sin\,(\omega\,[t+\tau])\,sin\,(\omega t+\varphi)\,dt \tag{A.123}$$

The integral is of type

$$\int sin\,(ax+b)\,sin\,(ax+d)\,dx = \frac{x}{2} cos\,(b-d) - \frac{1}{4a} sin\,(2ax+b+d) \tag{A.124}$$

with $x = t$, $a = \omega$, $b = \omega\tau$, $d = \varphi$ we have

$$\rho_{12}(\tau) = \lim_{T\to\infty} \frac{VI}{2T} \left\{ t\,cos\,(\omega\tau-\varphi) - \frac{1}{2\omega} sin\,(2\omega t+\omega\tau+\varphi)\Big|_{-\frac{T}{2}}^{\frac{T}{2}} \right\} \tag{A.125}$$

Inserting the results we have

$$\rho_{12}(\tau) = \lim_{T\to\infty} VI \left\{ \frac{1}{2} cos\,(\omega\tau-\phi) - \frac{1}{4\omega T} [sin\,(\omega\,(T+\tau)+\phi) + sin\,(-\omega\,(T+\tau)+\phi)] \right\} \tag{A.126}$$

with $T \to \infty$ the second term disappears, so that we get for the CCF of current I and voltage V

$$\rho_{12}(\tau) = \frac{VI}{2} cos\,(\omega\tau-\varphi) \tag{A.127}$$

For $\tau = 0$ the ACF gives the variance, the CCF gives the formula known in electrical engineering for the active power in a complex load.

$$P = \frac{VI}{2} cos\,\varphi \tag{A.128}$$

Example of Two White Noise Signals

As a simple example of no correlation: Two white noise signals $y_1(t)$ and $y_2(t)$ (Figure A.32).

The ACF $\rho_{11}(\tau) = \rho_{22}(\tau)$ of the white noise (see Figure A.13) has the maximum value σ^2 at $\tau = 0$ and is zero otherwise. The CCF is zero for all τ, as expected.

Example of Two Bandpass Noise Signals

As a further example, we consider the cross-correlation of two originally white noise signals that have passed a bandpass filter of center frequency $f_0 = 100$ kHz (Figure A.33).

At the output, we identify the frequency f_0 in the time functions $y_1(t)$ and $y_2(t)$ (Figure A.34), while the envelope contains the statistical fluctuations of the noise

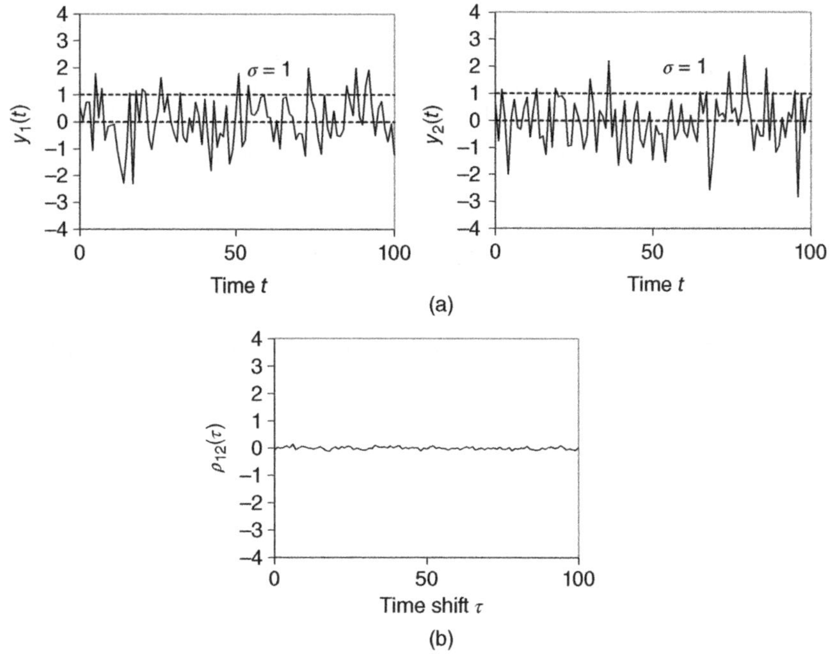

Figure A.32 The two white noise signals $y_1(t)$ and $y_2(t)$ (a) show no cross-correlation (b) $\rho_{12}(\tau) = 0$.

Figure A.33 Frequency spectrum of the white noise after passing the bandpass filter corresponds to the transfer characteristic.

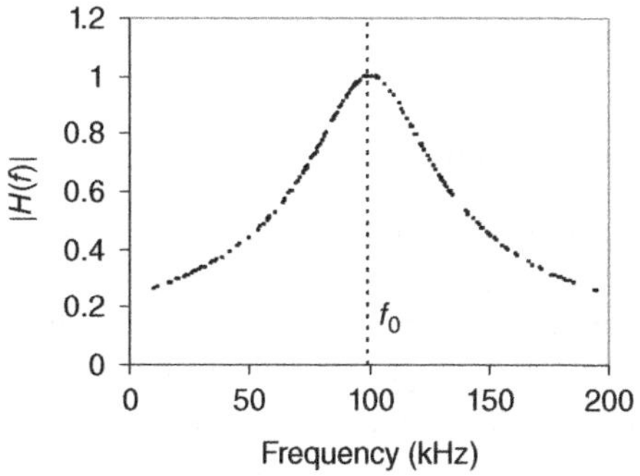

within the passband. The low and high frequencies are now missing in the fluctuation, because no faster changes can be transmitted than those, corresponding to the time constant of the filter. Let the effective value be $\sigma = 6$.

In the ACF we again see the fundamental frequency f_0 and the maximum value $\rho_{11}(0) = \sigma^2 = 36$. In the CCF we see the fundamental frequency f_0, too (Figure A.35).

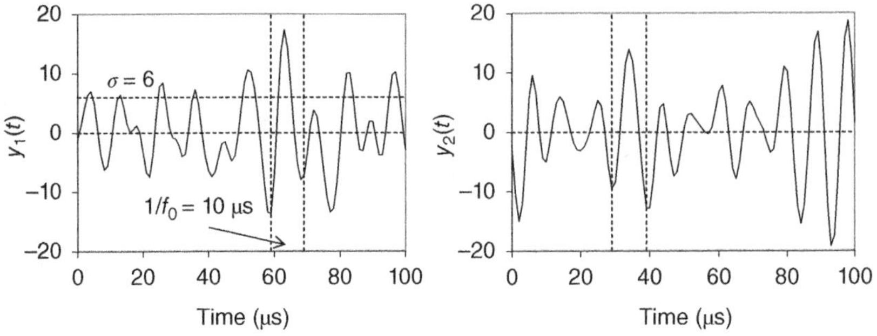

Figure A.34 Time dependence of the two noise signals $y_1(t)$ and $y_2(t)$ behind the filter Figure A.33.

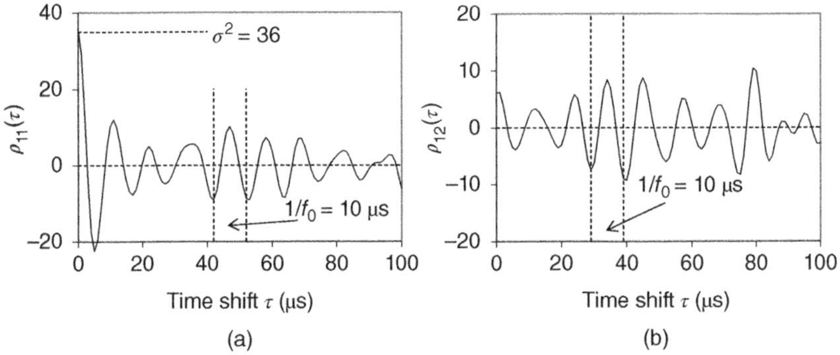

Figure A.35 (a) ACF of the signal from Figure A.34. (b) CCF of the two signals behind the filter.

Example White Noise and Bandpass Noise

As another example in Figure A.36, we consider the cross correlation of white noise $y_2(t)$ with a bandpassed noise $y_1(t)$. This corresponds to the identification of a weak signal in a noise floor. Both signals have the same variance $\sigma = 6$.

The time dependence is well reflected in the periodicity of the CCF versus τ (Figure A.37). With the addition of the signals $y_1(t) + y_2(t)$, as is the case in the superposition without correlation analysis, a time structure can only be identified with difficulty (Figure A.38).

Cross-Correlation After Splitting into Two Branches

Let us consider another weak signal $S(t)$ to be measured, e.g. an electric voltage. The level is very low and must be highly amplified. With amplification,

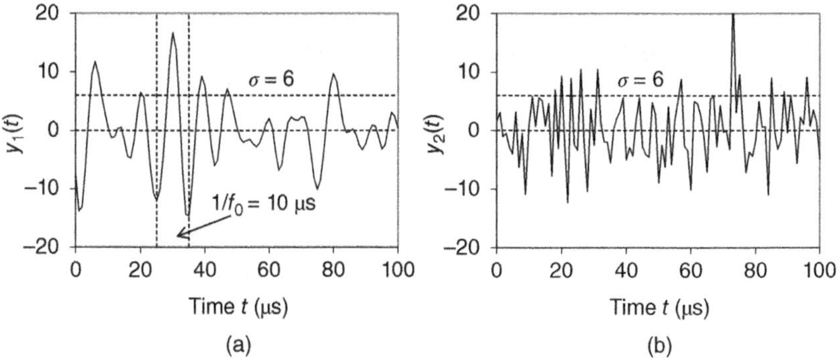

Figure A.36 (a) Time dependence of the noise signal behind the filter. (b) White noise without band limitation.

Figure A.37 CCF of the two signals from Figure A.36. The center frequency f_0 can be clearly identified.

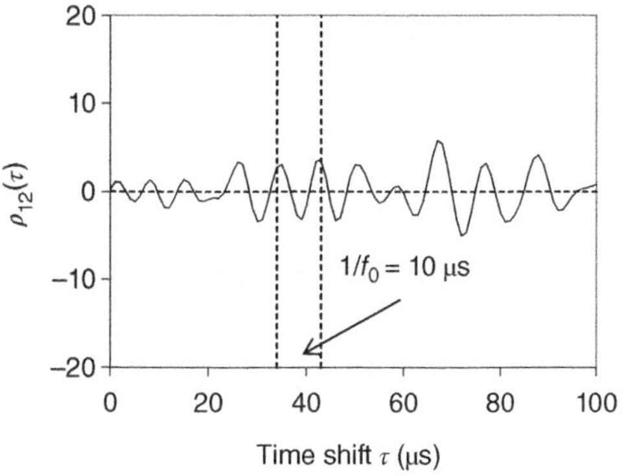

Figure A.38 Addition of the two signals $y_1(t) + y_2(t)$. The frequency f_0 cannot be seen.

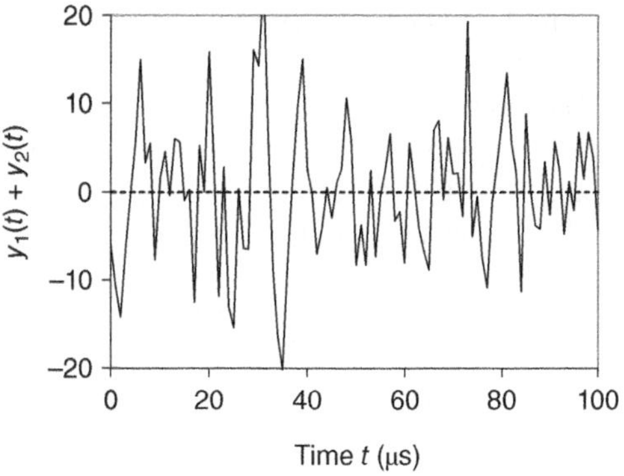

the inherent noise of the amplifier $N(t)$ adds up. In the worst case, the signal disappears in the noise and is no longer detectable. The signal can be split and sent through two amplifiers connected in parallel. If the CCF of the two very noisy signals $y_1(t)$ and $y_2(t)$ is formed at the output, the searched-for signal appears and the noise is averaged out. The noise components $N_1(t)$ and $N_2(t)$ of the two branches are uncorrelated, the signal components are correlated [8].

$$y_1(t) = S(t) + N_1(t)$$
$$y_2(t) = S(t) + N_2(t) \tag{A.129}$$

The CCF is defined as

$$\rho_{12}(\tau) = \overline{y_1(t)y_2(t+\tau)} = \overline{[S(t) + N_1(t)]\,[S(t+\tau) + N_2(t+\tau)]} \tag{A.130}$$

expanded

$$\rho_{12}(\tau) = \overline{S(t)S(t+\tau)} + \overline{S(t)N_2(t+\tau)} + \overline{S(t+\tau)N_1(t)} + \overline{N_1(t)N_2(t+\tau)} \tag{A.131}$$

The significant contribution is provided by the first term, since the two noise contributions N_1 and N_2 are uncorrelated with each other and with the signal $S(t)$ and their mean values are zero.

In the CCF, therefore, the searched-for signal appears without noise.

$$\rho_{12}(\tau) = \overline{S(t)S(t+\tau)} \tag{A.132}$$

We calculate an example with two different levels of white noise. A sinusoidal signal $sin(\omega t + \varphi)$ is split. Both parts pass through similar amplifiers and are then correlated with each other. In the first case the amplifiers have low inherent noise ($N_1, N_2, \sigma = 0.5$), in the second case they have strong inherent noise ($N_1, N_2, \sigma = 2$).

$$y_1(t) = \sin(\omega t + \phi) + N_1(t)$$
$$y_2(t) = \sin(\omega t + \phi) + N_2(t) \tag{A.133}$$

Figure A.39 shows an example of $y_1(t)$ for weak noise (a) and $y_2(t)$ for strong noise (b).

The desired sine signal is no longer visible after passing through the amplifiers with strong noise (Figure A.39b). However, the CCF clearly reproduces the signal again. The CCF of the output signals of the two branches is shown in the normalized form, which we denote here by $\rho n_{12}(\tau)$ in Figure A.40.

$$\rho n_{12}(\tau) = \frac{\rho_{12}(\tau)}{\sqrt{\rho_{11}(0)\,\rho_{22}(0)}} \tag{A.134}$$

This is very useful because it graphically represents the degree of correlation, as we have already seen above (A.121).

$$-1 \le \rho n_{12}(\tau) \le +1$$

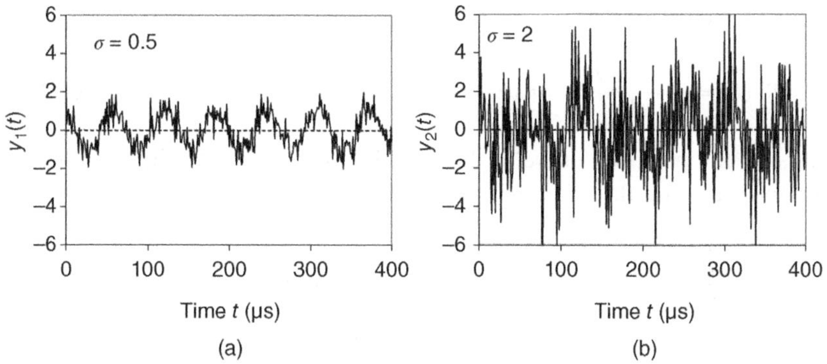

Figure A.39 Sinusoidal signal after passing through noisy amplifiers. (a) Low noise figure of the amplifier. (b) High noise figure.

Figure A.40 Normalized CCF of the signals from Figure A.39 after passing two parallel amplifiers. With high $\sigma = 2$, respectively low noise figure $\sigma = 0.5$.

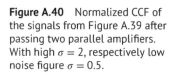

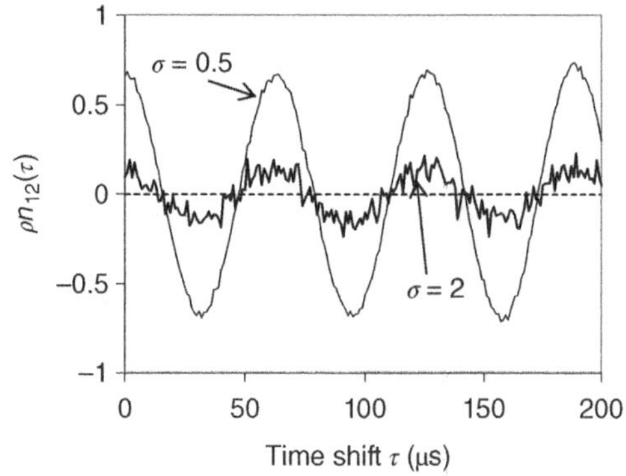

The maximum value ± 1 means complete correlation. The value zero means no correlation. Even in the very noisy signal, the input sine can still be clearly identified.

Power Spectral Density Real and Complex

If the time dependence is an even function, i.e. if $y(t) = y(-t)$, the spectral function is real. This is true for our examples (A.30) and (A.85). If this is not true, as in example (A.45), then the spectral function is complex. Let us look again at the sinc-function (A.87). If the rectangular transmission curve is symmetrical to the ordinate axis, and we integrate from $-{}^{T}/_{2}$ to $+{}^{T}/_{2}$, we get

$$F(\omega) = \frac{-1}{j\omega T} exp(-j\omega T)\Bigg|_{-\frac{T}{2}}^{+\frac{T}{2}} = \frac{1}{j\omega T}\{exp(j\omega T) - exp(-j\omega T)\} \tag{A.135}$$

The imaginary part is zero.

$$Re(F) = \frac{sin\left(\frac{\omega T}{2}\right)}{\frac{\omega T}{2}}; \quad Im(F) = 0 \tag{A.136}$$

If we shift the passband curve to the right into the range $t \geq 0$, it is no longer axisymmetrical and thus odd. For $t < 0$ it disappears. We now integrate from 0 to T.

$$F(\omega) = \frac{-1}{j\omega T} exp(-j\omega T)\Big|_0^T = \frac{1}{j\omega T}\{1 - exp(-j\omega T)\} \tag{A.137}$$

With $exp(-j\omega T) = cos(\omega T) - j\,sin(\omega T)$ we obtain a non-zero imaginary part. The spectral function is therefore complex.

$$Re(F) = \frac{sin(\omega T)}{\omega T}; \quad Im(F) = \frac{-1}{\omega T}(1 - cos(\omega T)) \tag{A.138}$$

The ACF also has these properties of an even function. Therefore, the power spectral density of the noise is real as its Fourier transform according to the Wiener–Khinchin theorem.

The Cross-Spectral Density

To the ACF $\rho_{11}(\tau)$ of the signal $y_1(t)$ belongs the auto-spectral density $S_{11}(f)$. To the CCF $\rho_{12}(\tau)$ of the two signals $y_1(t)$ and $y_2(t)$, one can accordingly define the cross-spectral density $S_{12}(f)$ [9].

$$S_{12}(f) = \int_{-\infty}^{\infty} \rho_{12}(\tau)\, exp(-j2\pi f\tau)\, d\tau \tag{A.139}$$

This is the Wiener–Khinchin relation for the cross spectral density.

The ACF are even functions

$$\begin{aligned} \rho_1(-\tau) &= \rho_1(\tau) \\ \rho_2(-\tau) &= \rho_2(\tau) \end{aligned} \tag{A.140}$$

In contrast, the following applies to the CCF

$$\rho_{12}(-\tau) = \rho_{21}(\tau) \tag{A.141}$$

Because of this symmetry of ACF and CCF, it is also valid for the two-sided spectral functions:

$$\begin{aligned} S_{11}(-f) &= S_{11}(f) \\ S_{12}(-f) &= S_{21}(f) = S_{12}^*(f) \end{aligned} \tag{A.142}$$

The CCF, on the other hand, is asymmetric, therefore the cross spectral density is complex.

Complex Representation of the Cross-Spectral Density

In (A.139) we had defined S_{12}. For G_{12} we write

$$G_{12}(f) = 2\int_{-\infty}^{\infty} \rho_{12}(\tau)\, exp(-j2\pi f\tau)\, d\tau = C_{12}(f) - jQ_{12}(f) \tag{A.143}$$

and vice versa

$$\rho_{12}(\tau) = \int_{0}^{\infty} \left[C_{12}(f)\cos(2\pi f\tau) + Q_{12}sin(2\pi f\tau)\right] df \tag{A.144}$$

C_{12} is a real and even function of f, Q_{12} is a real and odd function of f.
Furthermore holds

$$C_{12}(f) = \frac{1}{2}\left[G_{12}(f) + G_{21}(f)\right] \tag{A.145}$$

$$Q_{12}(f) = j\frac{1}{2}\left[G_{12}(f) - G_{21}(f)\right] \tag{A.146}$$

We can also write the complex cross-spectral density G_{12} in magnitude and phase:

$$G_{12}(f) = \left|G_{12}(f)\right| exp\left(-j\theta_{12}(f)\right) \quad 0 \leq f < \infty \tag{A.147}$$

it holds for the magnitude

$$\left|G_{12}(f)\right| = \sqrt{C_{12}^2 + Q_{12}^2} \tag{A.148}$$

And for the phase angle

$$\theta_{12}(f) = arctan\frac{Q_{12}}{C_{12}} \tag{A.149}$$

The spectral description of two stationary fluctuation quantities $y_1(t)$ and $y_2(t)$ can be done by the two autospectral densities S_{11} and S_{22} and the cross-spectral density S_{12} or by the four quantities S_{11}, S_{22}, C_{12}, and Q_{12}. The latter have the advantage that only positive frequencies are required, since the symmetry conditions (A.142) and (A.145) apply.

Transmission of Noise by Networks

The complex cross-spectral density discussed in Section "Power Spectral Density Real and Complex" plays an important role in the superposition of noise powers. We will consider this here using transmission through linear networks. The analysis of the transmission characteristics of a network can be done in the frequency domain as well as in the time domain. Both approaches are equivalent and are interrelated by the Fourier transformation. Depending on the problem, one or the other method is appropriate. In the time domain (Figure A.41a), the

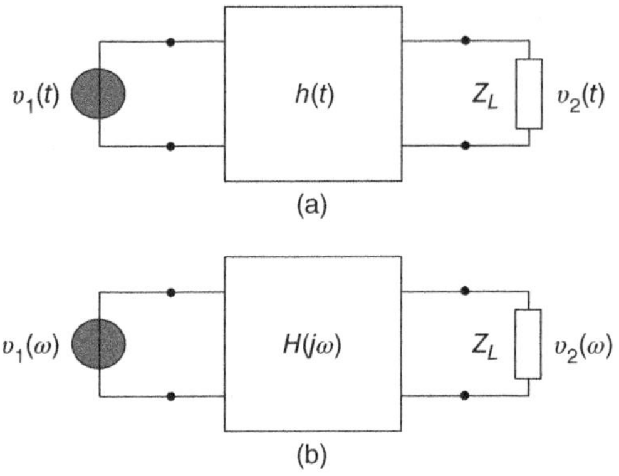

Figure A.41 Transmission of an input voltage v_1 by a linear network. (a) Time domain. (b) Frequency domain.

signal transmission through a linear two-port is described by the weight function $h(t)$. It is also called impulse response. In the frequency range (Figure A.41b) we have the complex transfer function $(j\omega)$. Complex because both amplitude and phase of the transmitted signal are influenced [9, 10].

In the time domain, the transfer from input 1 to output 2 is given by a convolution integral.

$$v_2(t) = \int_{-\infty}^{\infty} h(t')\, v_1(t - t')\, dt' \tag{A.150}$$

where a time axis t' shifts against the time axis t and all contributions of the input voltage for any t' deliver a contribution to the output voltage weighted by $h(t)$.

If the input voltage is sinusoidal with defined phase:

$$v_1(t) = V_1 exp(j2\pi ft); \quad V_1 = |v_1|\, exp(j\phi) \tag{A.151}$$

one obtains from (A.150)

$$\begin{aligned} v_2(t) &= \int_{-\infty}^{\infty} h(t)\, V_1 exp\left(j2\pi f[t - t']\right) dt \\ &= V_1 exp(j2\pi ft) \int_{-\infty}^{\infty} h(t')\, exp(-j2\pi ft')\, dt' \end{aligned} \tag{A.152}$$

The last integral is the Fourier transform of the weight function $h(t')$. In the case of an amplifier this is the complex voltage gain $H(f)$.

In the frequency domain we have for the complex amplitudes simply

$$v_2 = H(f)\, v_1 \tag{A.153}$$

Fourier forward and inverse transforms are

$$\begin{aligned} h(t) &= \int_{-\infty}^{\infty} H(f)\, exp(j2\pi ft)\, df \\ H(f) &= \int_{-\infty}^{\infty} h(t)\, exp(-j2\pi ft)\, dt \end{aligned} \tag{A.154}$$

The weight function $h(t)$ is a real function. The following therefore applies to the Fourier transform

$$H(-f) = H^*(f) \tag{A.155}$$

When analyzing noise processes, the ACF takes the place of the sinusoidal signal (A.151). As discussed earlier, the ACF is the product of a function at one time t with a value of the same function at another time $t+\tau$. We have at the input the ACF

$$\rho_1(\tau) = \lim_{T\to\infty} \frac{1}{T} \int_{-\frac{T}{2}}^{\frac{T}{2}} v_1(t) v_1(t+\tau)\, dt \tag{A.156}$$

When transferring by the two-port, a second convolution integral is now added, since we have another time axis t''.

$$\rho_2(\tau) = \int_{-\infty}^{\infty} \int_{-\infty}^{\infty} h(t')\, h(t'')\, \rho_1(\tau + t' - t'')\, dt'dt'' \tag{A.157}$$

Since $h(t)$ can only exist for times $t \geq 0$, one can also set the lower integration limit to 0 in (A.157).

If one wants to measure the ACF of a noise source, one usually uses amplifiers to obtain a sufficient level. From (A.157) one can derive the requirements for an amplifier. The factor A in (A.158) should be as constant as possible.

$$\rho_2(\tau) = A\rho_1(\tau) \tag{A.158}$$

This is the case if $h(t)$ decreases to zero in a very short time t_0 without the ACF $\rho_1(\tau)$ having changed noticeably in this time. It is then sufficient to integrate (A.157) only in the time interval $0 \leq t < t_0$ where $\rho_1(\tau + t' - t'') \to \rho_1(\tau)$ holds. In this case we obtain from (A.157)

$$\rho_2(\tau) = \left\{ \int_0^{t_0} h(t)\, dt \right\}^2 \rho_1(\tau) \tag{A.159}$$

With the above restrictions, the integral is the Fourier transform of the frequency response $H(f)$ of the amplifier. Strictly speaking, the gain at frequency =0. So we have for the factor A from (A.158)

$$A = H^2(f = 0) \tag{A.160}$$

For the measurement of the ACF, an amplifier with the broadest possible bandwidth and a lower frequency limit of 0 Hz is required. This can be quite a problem in practice.

When transmitting via a lossless two-port without internal noise sources, the input and output signals will be completely correlated. If noise is added along the

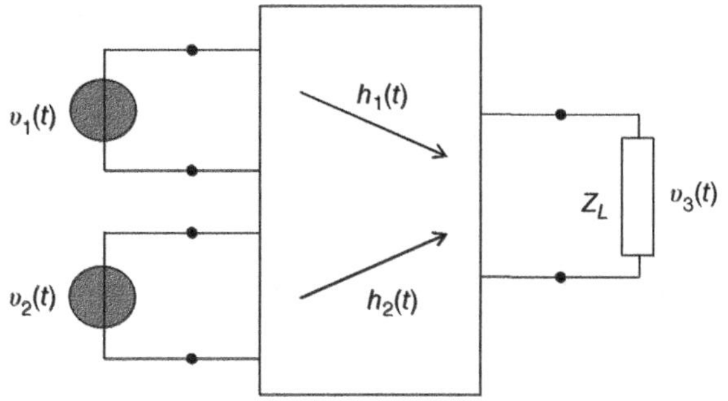

Figure A.42 Superposition of two noise voltages in a network. Source: Schiek et al. [10] / John Wiley & Sons.

transmission path, the degree of correlation at the output will decrease. One measure is the cross-correlation of the signals in the input and output reference planes. Analogous to (A.150) we write

$$v_1(t)\,v_2(t+\tau) = v_1(t)\int_{-\infty}^{\infty} h(t')\,v_1(t+\tau-t')\,dt' \tag{A.161}$$

We make no distinction between time average and ensemble average because of assumed ergodicity. The average leads to the cross-correlation

$$\rho_{1,2}(\tau) = \int_{-\infty}^{\infty} h(t')\,\rho_1(\tau-t')\,dt' \tag{A.162}$$

In noise analysis and modeling of circuits, the superposition of noise voltages in the output plane is of crucial importance. The contributions generally come from different sources, as shown in Figure A.42. They may also originate from one source and have reached the output along different paths. When superimposing them to the total noise power, one must know the degree of correlation of the individual contributions.

In our simple example we have a linear network with two inputs and one output. The two input voltages $v_1(t)$ and $v_2(t)$ superimpose in different ways to form the output voltage $v_3(t)$. The weight functions h_1 and h_2 apply to the two paths.

$$v_3(t) = \int_{-\infty}^{\infty} h_1(t')\,v_1(t-t')\,dt' + \int_{-\infty}^{\infty} h_2(t')\,v_2(t-t')\,dt' \tag{A.163}$$

We get at the ACF at the output analogous to (A.157) with the product

$$\begin{aligned} v_3(t)\,v_3(t+\tau) = \int_{-\infty}^{\infty}\int_{-\infty}^{\infty} &\{h_1(t')\,v_1(t-t') + h_2(t')\,v_2(t-t')\} \\ &\times \{h_1(t'')\,v_1(t+\tau-t'') + h_2(t'')\,v_2(t+\tau-t'')\}\,dt'dt'' \end{aligned} \tag{A.164}$$

Expanding and averaging leads to the following expressions for the ACF at the output, in which we find auto and cross correlation of the input signals.

$$\rho_3(\tau) = \int_{-\infty}^{\infty}\int_{-\infty}^{\infty} h_1(t')\,h_1(t'')\,\rho_1(\tau+t'-t'')\,dt'dt''\ a$$
$$+\int_{-\infty}^{\infty}\int_{-\infty}^{\infty} h_2(t')\,h_2(t'')\,\rho_2(\tau+t'-t'')\,dt'dt''\ b$$
$$+\int_{-\infty}^{\infty}\int_{-\infty}^{\infty} h_1(t')\,h_2(t'')\,\rho_{1,2}(\tau+t'-t'')\,dt'dt''\ c$$
$$+\int_{-\infty}^{\infty}\int_{-\infty}^{\infty} h_2(t')\,h_1(t'')\,\rho_{2,1}(\tau+t'-t'')\,dt'dt''\ d \quad \text{(A.165)}$$

The terms a and b correspond to the transformation of the ACF from the input to the output as already derived in (A.157). Terms c and d contain the CCFs $\rho_{1,2}(\tau)$ and $\rho_{2,1}(\tau)$ of the two input signals, which must therefore also be known.

In most practical cases, one works with the noise spectra rather than with the correlation functions. The transformation properties of linear networks presented so far in the time domain can be transferred to the frequency domain by Fourier transformation. The relation is given by the already known Wiener–Khinchin theorem.

$$S(f) = \int_{-\infty}^{\infty} \rho(\tau)\,exp(-j2\pi f\tau)\,d\tau \quad \text{(A.166)}$$

We obtain the power spectrum at the output of the network by applying the Fourier transform (A.166) to the ACF $\rho_2(\tau)$ according to (A.157).

$$S_2(f) = \int\int_{-\infty}^{\infty}\int h(t')\,h(t'')\,\rho_1(\tau+t'-t'')\,exp(-j2\pi f\tau)\,d\tau dt' dt'' \quad \text{(A.167)}$$

This expression becomes much clearer if we expand in the exponential function with t' and $-t''$.

$$S_2(f) = \int\int_{-\infty}^{\infty}\int h(t')\,h(t'')\,\rho_1(\tau+t'-t'')\,exp\left(-j2\pi f\left[\tau+t'-t''\right]\right)$$
$$\times\, exp\left(j2\pi f t'\right)\,exp\left(-j2\pi f t''\right)\,d\tau dt' dt'' \quad \text{(A.168)}$$

We can write (A.168) so that we find out three terms.

$$\int_{-\infty}^{\infty} \rho_1(\tau+t'-t'')\,exp\left(-j2\pi f\left[\tau+t'-t''\right]\right)d\tau = S_1(f) \quad \text{(A.169)}$$

$$\int_{-\infty}^{\infty} h(t')\,exp\left(j2\pi f t'\right)dt' = H^*(f) \quad \text{(A.170)}$$

$$\int_{-\infty}^{\infty} h(t'')\,exp\left(-j2\pi f t''\right)dt'' = H(f) \quad \text{(A.171)}$$

We obtain the simple relation of the power spectra at the input and output of a linear two-port

$$S_2(f) = |H(f)|^2 S_1(f) \tag{A.172}$$

When superimposing two input signals in a load resistor, we have the correlation function in (A.165). After transformation into the frequency domain, we get the power spectrum of this superposition

$$S_3(f) = |H_1|^2 S_1(f) + |H_2|^2 S_2(f) + H_1^* H_2 S_{1,2}(f) + H_2^* H_1 S_{2,1}(f) \tag{A.173}$$

Since $S_{1,2} = S_{2,1}^*$ holds, we can write (A.173):

$$S_3(f) = |H_1|^2 S_1(f) + |H_2|^2 S_2(f) + 2Re\left(H_1^* H_2 S_{1,2}(f)\right) \tag{A.174}$$

In addition to the two transformed power spectra, the complex cross spectra also occur. This is a very important fact for the noise analysis of circuits or active components. There we usually have to deal with band-limited noise. If we look at the voltages, the auto-spectra $S_1(f)$ and $S_2(f)$ correspond to the voltage squares $|v_1|^2$ and $|v_2|^2$ and the cross-spectra $S_{1,2}(f)$ and $S_{2,1}(f)$ correspond to the mixed products $v_1^* v_2$ and $v_1 v_2^*$. Calculations with noise spectra can therefore be carried out with the methods of complex analysis used in the mathematics of AC-technique. We know from Chapter 2 that the magnitude of the narrowband noise has a Rayleigh distribution. The IQ-components of this vector still have Gaussian distribution. We can therefore apply for the complex amplitude of the noise voltage:

$$v(f) = V(f)\, exp(j\varphi(f)) \tag{A.175}$$

where the magnitude of $V(f)$ is, of course, a fluctuation variable whose expected value results from the integral over the Rayleigh distribution.

$$\langle V(f)\rangle = \int_0^\infty V(f) f_R(v)\, dv;\ f_R(v) = \frac{v}{\sigma^2} exp\left(-\frac{v^2}{2\sigma^2}\right) \tag{A.176}$$

The phases are statistically distributed and are not of interest in this context. The IQ-components of the voltage phasor are

$$v(f) = v_X(f) + jv_Y(f) \tag{A.177}$$

Their integral of $v^2(f)$ over the Gaussian distribution gives the square of fluctuation σ^2 and hence the noise power N.

$$\langle v_X^2\rangle = \langle v_Y^2\rangle = \int_0^\infty v^2(f) f_G(v)\, dv = \sigma^2 \tag{A.178}$$

The relationship with the spectral density S_V at a resistor R is given by

$$N = \frac{1}{R}\langle v^2(f)\rangle = \frac{1}{R}\langle v(f) v^*(f)\rangle = \frac{1}{R} S_V B \tag{A.179}$$

Here S_V is the spectral density of the voltage square with the unit $^{V^2}/_{Hz}$ and B is the bandwidth in Hz.

References

1 van der Ziel, A. (1954). *Noise*. London: Chapman and Hall.
2 Bendat, J.S. and Piersol, A.G. (1971). *Random Data: Analysis and Measurement*. Wiley Interscience.
3 Mildenberger, O. (2020). *Grundlagen der Informationstechnik*. Vieweg&Sohn.
4 Meinke, H.H. and Gundlach, F.W. (1992). *Taschenbuch der Hochfrequenztechnik*. Springer Verlag.
5 Müller, R. (1990). *Rauschen*. Springer Verlag.
6 Rudolph, D. *Modulation und Rauschen*, Series of Lectures. TFH Berlin.
7 Rudolph, D. (2002). Die Fouriertransformation und ihre Anwendung. *Wissen Heute*, H.4–H.11.
8 Rubiola, E. and Giordano, V. (2000). Correlation-based phase noise measurements. *Rev. Sci. Instr.* 71: 3085–3091.
9 Bittel, H. and Storm, L. (1971). *Rauschen*. Springer Verlag.
10 Schiek, B., Rolfes, I., and Siweris, H.-J. (2006). *Noise in High-Frequency Circuits and Oscillators*. Wiley Interscience.

Glossary of Symbols

Some symbols are used with different meanings. The respective assignment results clearly from the context in the text.

4PSK	Quadrature phase shift keying
A	Area, also linear absorption
$\boldsymbol{A}$	Chain matrix
$\boldsymbol{ABCD}$	Chain matrix
ADC	Analog to digital converter
B	Bandwidth, also susceptance
BER	Bit error rate
C	Capacity
c	Speed of light
c_{12}	Cross correlation coefficient
$\boldsymbol{C}^{\mathrm{A}}$	Noise correlation matrix in chain form
$E(\nu)$	Expected value of a distribution
E, W	Energy
EIRP	Equivalent isotropically radiated power
e_n, v_n	Noise voltage normalized to $B = 1$ Hz
ENR	Excess noise ratio
f	Frequency
F	Noise factor (lin), also force
$F(\omega)$	Fourier transform of a function $y(t)$
G	Conductance, also gain
$G(f)$	One sided power spectral density

A Guide to Noise in Microwave Circuits: Devices, Circuits, and Measurement, First Edition.
Peter Heymann and Matthias Rudolph.

g_m	Transconductance of the FET
G_N	Noise conductance (uncorrelated part)
$\boldsymbol{H}$	Hybrid matrix
h	Planck's constant
H	Transfer function
I	Direct current
i	Alternating current also index
I, Q	Real part and imaginary part of a complex signal
i_n	Noise current normalized to $B = 1$ Hz
J	Jitter, also current density
k	Boltzmann constant
L	Inductance, also gate length, also attenuation linear
$L(f)$	Single sideband phase noise (dBc/Hz)
LNB	Low noise block
m	Electron mass
M	Noise measure
$m(f)$	Amplitude noise (dBc/Hz)
N	Noise power, also particle number, also parameter in FET model
n	Particle number
NF	Noise figure (dB)
P	Electrical power, also probability
$p(y)$	Probability density distribution of the values y
P, R, C	Factors in the FET noise model
q	Elementary charge
Q	Resonator quality factor
R	Ohmic resistance, also resistance
$R(t)$	Envelope of a fluctuating voltage
r_E	Resistance of the base-emitter diode
R_{EQ}	Equivalent noise resistance
R_N	Noise resistance of a two-port
$S(f)$	Two sided power spectral density
$S_{i,j}$	S-Parameters
S_{IN}, S_{OUT}	Signal powers
SINR	Signal plus interference plus noise ratio
SNR	Signal to noise ratio
T	Temperature, also time period

$\boldsymbol{T}$	Transformation matrix
t	Time
$\boldsymbol{T}^{\mathrm{H}}$	Hermitian of the matrix $\boldsymbol{T}$
T_N	Noise temperature
u_x	x-Component of the velocity
V	Electrical voltage, also volume
v	Alternating voltage, also velocity
$W(v)$	Statistical distribution of voltages
X	Reactance
Y	Admittance, also ratio of noise powers
$\boldsymbol{Y}$	Admittance matrix
Y_{COR}	Correlation conductance
Y_{OPT}	Generator conductance for noise matching
Z	Impedance
$\boldsymbol{Z}$	Impedance matrix
Z_0	Characteristic impedance of a transmission line
Φ	Diffusion voltage
Ω_A	Solid angle
α	Attenuation, also current amplification
β	Current amplification
$\beta(j\omega_0)$	Complex feedback
λ	Wavelength, also noise coefficient
μ	Mobility
$\rho_{12}(\tau)$	Cross correlation function
$\rho(\tau)$	Correlation function
σ	Standard deviation, also conductivity
τ	Time displacement
ω	Angular frequency
$s\varphi$	Phase slope
Ψ, ϕ, φ	Phase angles

Index

A Guide to Noise in Microwave Circuits: Devices, Circuits, and Measurement, First Edition.
Peter Heymann and Matthias Rudolph.

Printed and bound by CPI Group (UK) Ltd, Croydon, CR0 4YY

16/04/2025

14658579-0003